THE CHRYSOKAMINO METALLURGY WORKSHOP AND ITS TERRITORY

HESPERIA SUPPLEMENTS

6* J. H. Oliver, *The Sacred Gerusia* (1941)
7* G. R. Davidson and D. B. Thompson, *Small Objects from the Pnyx:* I (1943)
8* *Commemorative Studies in Honor of Theodore Leslie Shear* (1949)
9* J. V. A. Fine, *Horoi: Studies in Mortgage, Real Security, and Land Tenure in Ancient Athens* (1951)
10* L. Talcott, B. Philippaki, G. R. Edwards, and V. R. Grace, *Small Objects from the Pnyx:* II (1956)
11* J. R. McCredie, *Fortified Military Camps in Attica* (1966)
12* D. J. Geagan, *The Athenian Constitution after Sulla* (1967)
13 J. H. Oliver, *Marcus Aurelius: Aspects of Civic and Cultural Policy in the East* (1970)
14 J. S. Traill, *The Political Organization of Attica* (1975)
15 S. V. Tracy, *The Lettering of an Athenian Mason* (1975)
16 M. K. Langdon, *A Sanctuary of Zeus on Mount Hymettos* (1976)
17 T. L. Shear Jr., *Kallias of Sphettos and the Revolt of Athens in 268 B.C.* (1978)
18* L. V. Watrous, *Lasithi: A History of Settlement on a Highland Plain in Crete* (1982)
19 *Studies in Attic Epigraphy, History, and Topography Presented to Eugene Vanderpool* (1982)
20 *Studies in Athenian Architecture, Sculpture, and Topography Presented to Homer A. Thompson* (1982)
21 J. E. Coleman, *Excavations at Pylos in Elis* (1986)
22 E. J. Walters, *Attic Grave Reliefs That Represent Women in the Dress of Isis* (1988)
23 C. Grandjouan, *Hellenistic Relief Molds from the Athenian Agora* (1989)
24 J. S. Soles, *The Prepalatial Cemeteries at Mochlos and Gournia and the House Tombs of Bronze Age Crete* (1992)
25 S. I. Rotroff and J. H. Oakley, *Debris from a Public Dining Place in the Athenian Agora* (1992)
26 I. S. Mark, *The Sanctuary of Athena Nike in Athens: Architectural Stages and Chronology* (1993)
27 N. A. Winter, ed., *Proceedings of the International Conference on Greek Architectural Terracottas of the Classical and Hellenistic Periods, December 12–15, 1991* (1994)
28 D. A. Amyx and P. Lawrence, *Studies in Archaic Corinthian Vase Painting* (1996)
29 R. S. Stroud, *The Athenian Grain-Tax Law of 374/3 B.C.* (1998)
30 J. W. Shaw, A. Van de Moortel, P. M. Day, and V. Kilikoglou, *A LM IA Ceramic Kiln in South-Central Crete: Function and Pottery Production* (2001)
31 J. K. Papadopoulos, *Ceramicus Redivivus: The Early Iron Age Potters' Field in the Area of the Classical Athenian Agora* (2003)
32 J. Wiseman and K. Zachos, eds., *Landscape Archaeology in Southern Epirus, Greece* I (2003)
33 A. P. Chapin, ed., *ΧΑΡΙΣ: Essays in Honor of Sara A. Immerwahr* (2004)
34 F. Zarinebaf, J. Bennet, and J. L. Davis, *A Historical and Economic Geography of Ottoman Greece: The Southwestern Morea in the 18th Century* (2005)
35 Gloria S. Merker, *The Greek Tile Works at Corinth: The Site and the Finds* (2006)

Out of print

Hesperia Supplement 36

THE CHRYSOKAMINO METALLURGY WORKSHOP AND ITS TERRITORY

Philip P. Betancourt

WITH CONTRIBUTIONS BY

Eleni A. Armpis · Yannis Bassiakos · Curt Beck · Ruth F. Beeston · Mihalis Catapotis · Brigit Crowell · Doniert Evely · William R. Farrand · Susan C. Ferrence · Cheryl R. Floyd · Noel Gale · William B. Hafford · Donald C. Haggis · Glynis Jones · Byron Koukaras · James D. Muhly · George H. Myer · Eleni Nodarou · Lada Onyshkevych · Joe Palatinus · Natalia Poulou-Papadimitriou · Robert S. Powell · David S. Reese · Ann Schofield · Elizabeth B. Shank · Evi Sikla · Zofia Stos · Edith C. Stout · Charles P. Swann · Christine M. Thompson · Tanya Yangaki

The American School of Classical Studies at Athens
2006

Copyright © 2006
The American School of
Classical Studies at Athens

All rights reserved.

To order, contact:
(in North America)
The David Brown Book Company
www.davidbrownbookco.com
Tel. 800-791-9354

(outside North America)
Oxbow Books
www.oxbowbooks.com
Tel. +44 (0) 1865-241-249

Out-of-print *Hesperia* supplements
may be purchased from:
Royal Swets & Zeitlinger
Swets Backsets Service
P.O. Box 810
2160 SZ Lisse
The Netherlands
E-mail: backsets@nl.swets.com

Cover illustration: The metallurgy site during excavation

Library of Congress Cataloging-in-Publication Data

Betancourt, Philip P., 1936–
 The Chrysokamino Metallurgy Workshop and its territory / Philip P. Betancourt.
  p. cm. — (Hesperia supplement ; 36)
 Includes bibliographical references and index.
 ISBN-13: 978-0-87661-536-2 (alk. paper)
 ISBN-10: 0-87661-536-1 (alk. paper)
 1. Chrysokamino Site (Greece) 2. Metallurgy—Greece—Chrysokamino Site. 3. Minoans—Greece—Chrysokamino Site. 4. Excavations (Archaeology)—Greece—Kavousi Region. 5. Kavousi Region (Greece)—Antiquities. I. Title. II. Series: Hesperia (Princeton, N.J.). Supplement ; 36.
DF221.C8B55 2006
939´.1801—dc22   2006047906

CONTENTS

List of Illustrations — ix
List of Tables — xvii
Preface and Acknowledgments — xix

PART I: THE CHRYSOKAMINO TERRITORY

Chapter 1
INTRODUCTION
by Philip P. Betancourt — 3

Chapter 2
THE NATURAL ENVIRONMENT
by Philip P. Betancourt and William R. Farrand — 19

PART II: THE METALLURGY WORKSHOP

Chapter 3
THE EXCAVATION OF THE METALLURGY WORKSHOP
by Philip P. Betancourt, James D. Muhly, Eleni A. Armpis, Robert S. Powell, Elizabeth B. Shank, Evi Sikla, and Tanya Yangaki — 47

Chapter 4
THE APSIDAL STRUCTURE
by Philip P. Betancourt — 55

Chapter 5
THE POTTERY
by Philip P. Betancourt — 67

Chapter 6
THE STONE TOOLS
by Doniert Evely — 99

Chapter 7
THE FURNACE CHIMNEY FRAGMENTS
by Philip P. Betancourt — 109

Chapter 8
THE POT BELLOWS
    by Philip P. Betancourt and James D. Muhly   125

Chapter 9
MISCELLANEOUS CERAMIC ARTIFACTS
    by Susan C. Ferrence and Byron Koukaras   133

Chapter 10
OTHER METALLURGICAL MATERIALS
    by Philip P. Betancourt   137

Chapter 11
FAUNAL REMAINS
    by David S. Reese   149

Chapter 12
EVIDENCE FOR THE USE OF THRESHING REMAINS AT THE EARLY MINOAN METALLURGICAL WORKSHOP
    by Glynis Jones and Ann Schofield   153

Chapter 13
CHRYSOKAMINO IN THE HISTORY OF EARLY METALLURGY
    by James D. Muhly   155

Chapter 14
DISCUSSION OF THE WORKSHOP AND RECONSTRUCTION OF THE SMELTING PRACTICES
    by Philip P. Betancourt   179

PART III: THE SURFACE SURVEY

Chapter 15
INTRODUCTION TO THE SURFACE SURVEY
    by Philip P. Betancourt   193

Chapter 16
TOPOGRAPHY OF THE CHRYSOKAMINO REGION
    by Lada Onyshkevych and William B. Hafford   197

Chapter 17
A SUMMARY OF THE HABITATION SITE AT CHRYSOKAMINO-CHOMATAS
    by Cheryl R. Floyd   205

Chapter 18
EDITH HALL'S EXCAVATIONS IN THE THERIOSPELIO CAVE
    by Philip P. Betancourt and Cheryl R. Floyd   215

Chapter 19
CHRYSOKAMINO IN CONTEXT: A REGIONAL ARCHAEOLOGICAL SURVEY
    by Donald C. Haggis   221

Chapter 20
THE GEOGRAPHIC BOUNDARIES OF THE
CHRYSOKAMINO FARMSTEAD TERRITORY
    by Philip P. Betancourt   233

Chapter 21
LAND USE ON THE CHRYSOKAMINO FARMSTEAD
    by Philip P. Betancourt   241

Chapter 22
SURVEY CONCLUSIONS
    by Philip P. Betancourt   257

APPENDIXES

Appendix A
PETROGRAPHY AND X-RAY DIFFRACTION
ANALYSES OF SLAGS AND FURNACE CHIMNEYS
    by George H. Myer and Philip P. Betancourt   281

Appendix B
SEM/EDAX ANALYSIS
    by Yannis Bassiakos   293

Appendix C
LEAD ISOTOPE AND CHEMICAL ANALYSES OF
SLAGS FROM CHRYSOKAMINO
    by Zofia Stos and Noel Gale   299

Appendix D
ARSENIC CONTENT OF COPPER PRILLS:
A STUDY APPLYING PIXE
    by Susan C. Ferrence and Charles P. Swann   321

Appendix E
SLAG ANALYSIS BY WAVELENGTH DISPERSIVE
SPECTROMETRY
    by Christine M. Thompson   325

Appendix F
RECONSTRUCTION OF THE COPPER SMELTING
PROCESS AT CHRYSOKAMINO BASED ON THE
ANALYSIS OF ORE AND SLAG SAMPLES
    by Yannis Bassiakos and Mihalis Catapotis   329

Appendix G
REGISTER OF ANTHROPOGENIC FEATURES
    by Philip P. Betancourt, Lada Onyshkevych, and
    William B. Hafford   355

Appendix H
THE MINOAN POTTERY FROM THE SURVEY
    by Philip P. Betancourt   377

Appendix I
EVIDENCE FOR BEEKEEPING
    by Susan C. Ferrence and Elizabeth B. Shank   391

Appendix J
THE BYZANTINE TO OTTOMAN POTTERY FROM
THE SURVEY
    by Natalia Poulou-Papadimitriou   393

Appendix K
THE EXCAVATION OF CAVE AF 9 AND
TERRACE AF 22b
    by Brigit Crowell and Philip P. Betancourt   399

Appendix L
SOILS AND SEDIMENTS FROM NATURAL DEPOSITS
AT CHRYSOKAMINO
    by Eleni Nodarou   403

Appendix M
ORGANIC RESIDUE ANALYSIS OF POTTERY SHERDS
FROM CHRYSOKAMINO
    by Ruth F. Beeston, Joe Palatinus, Curt Beck, and
    Edith C. Stout   413

Appendix N
PETROGRAPHIC ANALYSIS OF TWO FINAL
NEOLITHIC SHERDS FROM THE CHRYSOKAMINO
METALLURGY LOCATION
    by Eleni Nodarou   429

References   433
Index   457

ILLUSTRATIONS

1.1.	Map of eastern Crete	4
1.2.	Three-dimensional topographic plan of the Chrysokamino metallurgy and habitation sites and the surrounding territory	5
1.3.	Aerial photograph showing Chrysokamino, the village of Kavousi, and the surrounding region	6
1.4.	Map of the Chrysokamino territory	7
1.5.	Toponyms of the area near Chrysokamino	9
1.6.	Olive trees growing in Chordakia, looking northeast from the habitation site	10
1.7.	The site of the metallurgical activity at Chrysokamino	11
1.8.	Chylopittes, a travertine formation at the coast near the Theriospelio cave	11
1.9.	The Kambos as seen from Kavousi, with the coastal hills behind it	12
1.10.	Lakkos Ambeliou, in the Kambos, looking south from the habitation site	13
1.11.	The burial cave called Theriospelio	13
1.12.	A fragment of industrial pottery published by Angelo Mosso	15
2.1.	Detailed bedrock map of the Chrysokamino area	23
2.2.	Bedrock geology of eastern Crete	24
2.3.	Schematic section through the nappe pile of eastern Crete	25
2.4.	Plattenkalk limestone with calcite veins	25
2.5.	Metacarbonate formation near the metallurgy site	26
2.6.	Stratigraphic column through the Phyllite nappe somewhat east of the Chrysokamino-Kavousi area	28

2.7.	Cliffs north of the metallurgy site, looking toward the island of Pseira	30
2.8.	Caves and cliffs south of Chrysokamino	31
2.9.	Red clay and the entrance to the ravine leading to Agriomandra	31
2.10.	Agriomandra, a small harbor at the foot of the ravine that connects the coast with the Kambos	31
2.11.	Soft, powdery phyllite exposed in a trench at the metallurgy workshop	32
2.12.	Kink fold in phyllite	33
2.13.	Paleoravine uphill from the metallurgy site	35
2.14.	Detail of the paleoravine showing the contact between the reddish soil in the ravine and the underlying pale-colored phyllite	36
2.15.	Colluvium consisting of reworked phyllite and other rocks	36
2.16.	Terrace system **AF 22**, located near the habitation site	40
3.1.	The surface of the metallurgy site before excavation	48
3.2.	The metallurgy site during excavation	48
3.3.	Sieving at the metallurgy workshop	49
3.4.	James Muhly cleaning the upper part of the apsidal building	50
3.5.	Plan of the grid-squares excavated at the metallurgy site	51
4.1.	Aerial view of the apsidal building	56
4.2.	Plan of the lower floor	56
4.3.	The apsidal building	57
4.4.	Level 4 in grid-square N 19, looking north, showing the lower floor at the eastern side of the apsidal building	57
4.5.	Apsidal building, lower floor, showing the location of the doorway	58
4.6.	Plan of the apsidal building's middle floor and cross section of the building	59
4.7.	Plan of the apsidal building's lower floor	60
5.1.	Location of Neolithic dark-burnished sherds within the slag pile	69
5.2.	Sherds of conical cup **75**	72
5.3.	Pottery from the slag pile (**1–23**)	74
5.4.	Pottery from the slag pile (**24–43**)	80

5.5.	Pottery from the slag pile (**44–59**)	82
5.6.	Pottery from the slag pile (**60–72**) and the apsidal structure (**73–82**)	85
5.7.	Pottery from the apsidal structure (**83–90**)	90
5.8.	Pottery from the apsidal structure (**91–101**)	92
5.9.	Pottery from the apsidal structure (**102–121**)	95
6.1.	Stone tools (**122–127**)	102
6.2.	Stone tools (**128–133**), obsidian tools (**142–144**), and cutting in bedrock	104
6.3.	Cutting in bedrock at the metallurgy workshop	105
6.4.	Stone tools (**122–124** and **127–129**) from the metallurgy site	106
7.1.	Front and back of chimney fragment **180**	110
7.2.	Chimney fragment no. X 213, preserving the impression of an olive leaf	110
7.3.	Alternative reconstructions of the industrial ceramics fragments	111
7.4.	Profile drawings of chimney fragments (**145–162**)	114
7.5.	Profile drawings of chimney fragments (**163–178**)	118
7.6.	Profile drawings of chimney fragments (**179–188**)	122
8.1.	Drawing of workmen using bellows, from the Tomb of Rekh-mi-re	126
8.2.	Reconstruction of the pot bellows in use	127
8.3.	Bellows fragment **189**	128
8.4.	Profile drawings of pot bellows (**189–194**)	129
8.5.	Profile drawings of pot bellows (**195–203**)	131
9.1.	Miscellaneous artifacts (**204–209**)	135
10.1.	Lumps of slag from unit M 22-3	138
10.2.	Slag resting on bedrock, looking south	139
14.1.	East Cretan White-on-Dark Ware from Chrysokamino and nearby sites	181
14.2.	Reconstruction of a bowl furnace from Chrysokamino	184
16.1.	Map of the greater region around Chrysokamino	198
16.2.	Map of anthropogenic features in the Chrysokamino farmstead	199
16.3.	Topographic map showing terraced areas	200

16.4.	Map of grid points for the Chrysokamino farmstead	201
16.5.	Map of the Chrysokamino farmstead, with indications of all points measured electronically	202
16.6.	Topographic map of the territory of the Chrysokamino farmstead	203
17.1.	Western portion of the plan of the habitation site	206
17.2.	State plan of the habitation site showing the LM III architecture	207
17.3.	The LM I bronze dagger from Room A (X 830)	208
17.4.	Examples of EM ceramics: X 1321, EM I pyxis lid fragment; X 1179, EM IIB Vasiliki Ware closed vessel base	208
17.5.	Examples of MM ceramics: X 541 and X 1448, fragments from carinated cups; X 1497, base of a tumbler; X 932, body sherd of a large, closed vessel	208
17.6.	Examples of LM IB ceramics: X 1480, base of a bell cup; X 808, cup rim; X 836, neck of a jug	208
17.7.	Stone bowl (X 818) from LM I stratum	209
17.8.	LM IIIA hearth above a LM I wall (Wall C), but below a LM IIIA wall (Wall 9)	210
17.9.	Example of LM IIIA wall construction at the habitation site	210
17.10.	East entrance into the LM IIIA complex	210
17.11.	Examples of LM IIIA ceramics	212
17.12.	Sealstone (X 208) from Room 2	212
17.13.	Examples of ground stone tools from the habitation site	213
17.14.	The ritual deposit in Room 11	213
18.1.	Profile drawings of sherds from the Theriospelio cave	219
19.1.	Map of the Kavousi survey area in the late Prepalatial period	222
19.2.	Map of the Kavousi survey area in the Protopalatial period	222
19.3.	Map of the Kavousi area in the late Prepalatial period	223
19.4.	The phyllite terraces of Chordakia and the Kavousi plain (the Kambos), seen from the west	224
19.5.	An olive crushing basin *(trapetus)* at Kephalolimnos	224
19.6.	A millstone *(orbis)* at Kephalolimnos	224
19.7.	Theriospelio cave from the west	225

19.8.	Map of the Kavousi survey area in the Neopalatial period	230
19.9.	Map of the north Kambos and Tholos Bay in the Neopalatial period	231
20.1.	Ravine and Agriomandra harbor	237
20.2.	Boundaries for the farmstead in LM III	239
21.1.	Land of Type 1, used for horticulture	243
21.2.	Land of Type 2, used for pasture	249
21.3.	Land of Type 3, used for dry farming	251
21.4.	Steep or rocky areas unsuitable for farming	253
22.1.	Map of Clusters 1–3, with closely related anthropogenic features	275
22.2.	Pressures influencing greater or lesser use of marginal land for agriculture	277
A.1.	Slag sample CHR 97	282
A.2.	Slag sample CHR 87	283
A.3.	Slag sample CHR 99 showing pyroxene crystals, magnetite, and cuprite	283
A.4.	Slag sample CHR 99 showing pyroxene crystals and cuprite	283
A.5.	Furnace chimney sample CHR 97	291
C.1.	The Anorthite-Wollastonite-FeO ternary phase diagram	309
C.2.	The Anorthite-SiO_2-FeO ternary phase diagram	309
C.3.	Lead isotope compositions of ores from Crete compared with those for slags from Chrysokamino	312
C.4.	Lead isotope compositions of slags from Chrysokamino and ores from Lavrion and the Cyclades	315
C.5.	Lead isotope analyses of some Prepalatial and Protopalatial copper-based alloy artifacts	317
D.1.	PIXE analyses of copper prills arranged by increasing arsenic content	322
D.2.	PIXE measurements showing the variation within individual prills	323
E.1.	Backscattered electron image of CHR 55	326
E.2.	White light photomicrograph of CHR 55	326
E.3.	Backscattered electron image showing core of CHR 48	326
E.4.	Photomicrograph of CHR 50	327
E.5.	Photomicrograph of CHR 8	328

E.6.	Backscattered electron image of CHR 74	328
F.1.	Copper ore (ORE 1) from Chrysokamino showing residual sulfides	333
F.2.	Iron ore from Chrysokamino showing rhomboid crystals of calcite	333
F.3.	The position of ores and slag from Chrysokamino in the ternary system $CaO\text{-}FeO\text{-}SiO_2$	335
F.4.	The position of slags from Chrysokamino and other EBA smelting sites from the southern Aegean in the ternary system $CaO\text{-}FeO\text{-}SiO_2\ (+\ Al_2O_3)$	335
F.5.	Sample A04: magnetite "bands" forming three slag "layers"	340
F.6.	Sample B18: Copper prills showing massive precipitation of a second arsenic-rich phase (Cu_3As)	342
F.7.	The position of matte inclusions present in slags from Chrysokamino in the ternary system Cu-Fe-S	342
F.8.	Typical microstructures found in slags from Chrysokamino	345
F.9.	Schematic representation of the proposed copper smelting process at Chrysokamino	352
G.1.	The metallurgical workshop and its nearby territory	356
G.2.	Location of anthropogenic features **AF 1**, **AF 2**, and **AF 3**	358
G.3.	Well (**AF 2**)	358
G.4.	Segment of a wall, possibly a boundary marker (**AF 3**)	359
G.5.	Destroyed threshing floor (**AF 4**)	359
G.6.	Collapsed field house (**AF 5**)	360
G.7.	One of the terrace walls (**AF 6**)	361
G.8.	Collapsed field house (**AF 7**), looking west	362
G.9.	Collapsed field house (**AF 7**), looking south	362
G.10.	Downhill from the habitation site (**AF 29**) to scatter of sherds (**AF 8**) and oval enclosure (**AF 32**)	362
G.11.	Standing field house (**AF 15**)	365
G.12.	Sections of wall (**AF 21**), probably a boundary wall	367
G.13.	View of **AF 21** from above	367
G.14.	Section of wall (**AF 21**)	368
G.15.	Section of boundary wall (**AF 27**)	370
G.16.	General view of the habitation site (**AF 29**)	371

G.17.	The habitation site (**AF 29**) showing the east side	371
G.18.	The main entrance to the habitation site (**AF 29**)	371
G.19.	Opening of a well (**AF 39**)	374
G.20.	The modern road (**AF 40**)	375
H.1.	Minoan pottery from the surface, **AF 17** to **AF 32**	379
H.2.	Minoan pottery from the surface, **AF 32**	383
H.3.	Minoan pottery from the surface, **AF 32**	384
H.4.	Minoan pottery from the surface, **AF 32**, **AF 33**	386
I.1.	Location of the beehive fragment	392
I.2.	The beehive fragment	392
J.1.	Byzantine to Ottoman pottery from the surface of the territory (J-1 to J-5)	395
J.2.	Ottoman pottery from the surface of the territory (J-6 to J-15)	396
K.1.	Cave **AF 9**	400
K.2.	The cave's entrance	400
K.3.	Terrace group **AF 22**	401
K.4.	Terrace **AF 22b**	401
L.1.	Analytical data for agricultural terrace **AF 22b**	407
L.2.	The sediment deposit at Lakkos Ambeliou	408
L.3.	Analytical data for the road exposure	409
L.4.	Locations of the two sets of analyzed soil samples	409

TABLES

2.1.	Seasonal variations at Hagios Nikolaos, Crete	20
5.1.	Final Neolithic sherds from the slag pile	76
5.2.	Early Minoan I–IIA sherd from the slag pile	76
5.3.	Early Minoan IIB sherd from the slag pile	77
5.4.	Early Minoan II–III sherd from the slag pile	77
5.5.	Early Minoan III–Middle Minoan IA sherds from the slag pile	87
5.6.	Final Neolithic sherd from the apsidal structure	87
5.7.	Early Minoan II sherd from the apsidal structure	88
5.8.	Early Minoan III–Middle Minoan IA sherds from the apsidal structure	97
5.9.	Comparison of Early Minoan III–Middle Minoan IA pottery shapes from the slag pile and the apsidal structure	97
12.1.	Plant impressions in chimney fragments from the metallurgical workshop	154
B.1.	SEM/EDAX analyses of copper prills arranged by increasing arsenic content	293
B.2.	SEM/EDAX analyses of metallurgical remains	294–295
C.1.	Chemical analyses of Chrysokamino slag	306–307
C.2.	Results of applying the Bachmann computations to the analyzed slags from Chrysokamino	308
C.3.	Lead isotope analyses of copper ores and galenas from Crete	311
C.4.	Lead isotope ratios of pieces of Chrysokamino slag	313
C.5.	Cycladic ores and copper slags isotopically consistent with slags from Chrysokamino	314

C.6.	Early Minoan copper-based artifacts isotopically consistent with the slags from Chrysokamino	318–319
E.1.	CHR 55, Points 1–5	325
E.2.	CHR 55, Points 6, 7	326
E.3.	CHR 48, Points 1–4	326
E.4.	CHR 50, Points 1, 2	327
E.5.	CHR 50, Point 3	327
E.6.	CHR 8, Points 1–3	327
E.7.	CHR 8, Points 4, 5	328
E.8.	CHR 74, Points 1, 2	328
E.9.	CHR 74, Point 3	328
F.1.	EDS analysis (average composition) of copper ore and iron ore samples	334
F.2.	EDS analysis of slag pellets	336–337
F.3.	EDS analysis of main components of silicate and glass phases in slag samples	339
F.4.	Results of Mössbauer spectroscopy and calculated Fe^{+2}/Fe^{+3} ratios of slag samples	340
F.5.	EDS analysis of copper prills in slag samples	341
F.6.	EDS analysis of matte inclusions in slag samples	343
L.1.	Analytical data from natural deposits at Chrysokamino	404
M.1.	Descriptions of sherds analyzed for organic residues	415
M.2.	GC-MS instrument parameters	417

PREFACE AND ACKNOWLEDGMENTS

The metallurgical workshop at Chrysokamino posed some interesting challenges. The site was located in a rural part of eastern Crete, away from the large archaeological sites. It had been first visited by archaeologists early in the 20th century, and although its metallurgical nature was apparent to almost everyone from the beginning, its date and the nature of the activities undertaken there were controversial during most of the 20th century. Sometimes regarded as ancient, and other times as Medieval or Early Modern, Chrysokamino lay unexcavated until 1995.

We approached the workshop as a multidisciplinary research project, with a large complement of scientists in different fields. Specialists were invited to contribute information of several types to help explain what we knew would be a complex series of problems. In addition to excavation and a large program of scientific analysis, we decided to use an intensive surface survey to place the small workshop within the context of its immediate territory. We felt that a context was needed in order to address questions that could not be answered by examining the metallurgical site in isolation. Was it possible to support a metallurgical installation at Chrysokamino based on local ores, personnel, fuel needs, and other resources? Would such a workshop have operated independently, or would it have been an appendage of a nearby farmhouse, like herding or agriculture? How large was the territory associated with this small rural site? How would this territory have been organized and managed, and what was Chrysokamino's relationship to similar sites and to the larger region of which it was a part? How would such a rural establishment have changed through time, and how did its situation compare with the Classical, Roman, and later systems used for agriculture and craft production?

We knew from the beginning that these would not be easy questions to address. The results of this study make some progress in resolving them, but they do not provide final answers. Yet even if the evidence does not allow complete explanations at present, it is to be hoped that by posing the questions and trying to address them, we will take an important series of steps toward understanding many of the fundamental issues about this aspect of Minoan life. Small farms and workshops situated away from the Minoan palaces and the dynamic events taking place there must have constituted some of the basic building blocks of Minoan culture. We cannot understand the Cretan Bronze Age without looking at small workshops and farms as well as the larger architectural complexes.

The study uses approaches that have not been applied in the study of Minoan Crete before, and these approaches have led to many new conclusions. Such results have been possible only because of the hard work of a very fine team of students and colleagues who have assembled considerable data on the Bronze Age workshop and its surrounding territory. Because of this team's efforts, we know much more about the beginnings of Minoan metallurgy than we did before the project began. In addition, we can begin to define the Minoan system of agricultural land use, and we can distinguish between gardens and fields using the archaeological data collected from survey. This information provides valuable insights about the small rural establishments that were an important part of Minoan society.

Acknowledgments

This project entailed three field seasons followed by a period of study. Work began in 1995 with a geological investigation and an electronic instrument survey and mapping project. Excavations were conducted in 1996 and 1997, under the direction of Philip P. Betancourt, with the assistance of co-directors James D. Muhly (1996–1997) and Cheryl R. Floyd (assistant director, 1996; co-director 1997). Study seasons were conducted at the INSTAP Study Center for East Crete in 1998 and 1999, when the project was completed except for preparation of manuscripts and laboratory analysis. The project was sponsored by Temple University with the collaboration of the University of Pennsylvania Museum of Archaeology and Anthropology. It was conducted under the auspices of the American School of Classical Studies at Athens, with a permit issued by the Greek Ministry of Culture. Financial support was provided by the Institute for Aegean Prehistory, Temple University, the University of Pennsylvania Museum of Archaeology and Anthropology, Dennis and Janice Verstegen, and other donors.

The directors are grateful to the citizens of the village of Kavousi, who assisted with the project in many ways, as well as to the large staff of colleagues and students who participated in the project.

Staff members included Mary A. Betancourt, apotheki supervisor and registrar (1995–1999); Lada Onyshkevych, supervisor for the electronic survey and mapping team (1995–1998); William B. Hafford, computer specialist (1995–1998); William R. Farrand, geologist (1995–1996); Carola Stearns, geologist (1995); Polymnia Muhly, excavation consultant (1996); Mark Hudson, archaeobotanist (1996); Ann Schofield, archaeobotanist (1997); Lyla Pinch Brock, artist (1996–1997); Susan C. Ferrence, artist (1998); Ann Foster, artist (1996); Stephanie E. Gleit, artist (1996); Laura A. Labriola, artist (1996–1997); Ian Verstegen, artist (1998); Sari K. Uricheck, conservator (1996); Elizabeth Baxter Shank, assistant cataloger (1996), assistant pottery specialist (1998); Gerardo I. Medrano, assistant cataloger

(1997); Jane D. Hickman, assistant cataloger (1998); Gayla Weng, assistant to the director (1998); Doniert Evely, stone tools specialist (1997); David S. Reese, faunal analyst (1997); Eleni Nodarou, soils specialist (1998–1999); and Natalia Poulou-Papadimitriou, Byzantine pottery specialist (1996). Robert Huber assisted with layout (1998).

Electronic survey and mapping team members assisting Lada Onyshkevych and William B. Hafford included Jaime J. Alvarez (1998), Leigh Ann Bingham (1995), Terrence P. Brennan (1998), Katherine May (1995), Robert S. Powell (1996), Stephanie Takaragawa (1996, 1998), and Jonathan Wallis (1995). The team used a Topcon GTS 303 Electronic Total Station interfaced with a Gateway 2000 (486) laptop computer for data collection. In 1995, the software used was PC AMP (Archaeological Mapping Program), provided by the Center for Archaeological Field Training, Pima County Community College, Tucson, AZ. In 1996, Easy Survey Plus was used. Surfer 6 and AutoCAD 13 and 14 were employed for additional processing. To run the computer, a Mercury II Solar Panel (Keep it Simple Systems, 32 S. Ewing, Suite 330, Helena, Montana, 59601) provided electric power generated by solar energy.

Trench supervisors included Eleni A. Armpis (1996), Brigit Crowell (1997), Barbara J. Hayden (1996), Katherine May (1996), Eleni Nodarou (1996–1997), Robert S. Powell (1996), Elizabeth Baxter Shank (1997), Evi Sikla (1997), Suzanne Stichman (1997), Stephanie Takaragawa (1997), and Tanya Yangaki (1996).

Staff members during the 1999 study season, in addition to James D. Muhly, Mary A. Betancourt, and the writer, included Susan C. Ferrence, Alejandra Gimenez, Jane D. Hickman, BethAnn Judas, Elizabeth B. Shank, and Gayla Weng.

Support and technical services were provided by the following personnel at the INSTAP Study Center for East Crete: Thomas M. Brogan, director (1997–1999); Stephania Chlouveraki, chief conservator (1997–1999); Ann N. Brysbaert, conservator (1997); Katherine May, chief photographer (1997–1999); Westley Bernard, assistant photographer (1997); and Eleanor J. Huffman, assistant to the director (1997–1999).

Photography was undertaken by Katherine May, Philip P. Betancourt, Gayla Weng, William R. Farrand, and Lillie Floyd. Donald Haggis took the photographs for Chapter 19. Aerial photography for the site was done by Jan Driessen, using a camera mounted on a kite (1997). The Soviet Sputnik program provided aerial photography for the region. Elizabeth B. Shank completed the plan of the metallurgy site.

Thanks are expressed to the Directors of the American School of Classical Studies at Athens, the late William D. E. Coulson (1987–1997) and James D. Muhly (1997–2002), and to the Director of the 24th Ephorate for East Crete, the late Nikos Papadakis. The representatives from the Ephorate during excavation were Giorgios Charoulis (1996) and Giorgios Katsalis (1996–1997). Help with logistics in Crete was furnished by Manolis and Maria Tsagarakis. The team is grateful to the proprietors of the Tholos Beach Hotel at Kavousi, Aristidis Chalkiadakis, Maria Chalkiadaki, and their son Ioannis Chalkiadakis.

The authors of the chapter on the Theriospelio cave are grateful to the following persons for assistance with various aspects of their research: Douglas Haller, David Romano, Pamela Russell, Cornelius and Emily Vermeule, and Wendy Watson.

Special thanks are extended to our many friends in the village of Kavousi, and especially to the former mayor, Dimitris Kophinakis, who offered invaluable assistance at the beginning of the project, and to Manolis Kasotakis, who acted as excavation foreman and provided help in many ways.

Philip P. Betancourt
2004

PART I: THE CHRYSOKAMINO TERRITORY

CHAPTER 1

INTRODUCTION

by Philip P. Betancourt

The Chrysokamino territory is located along the south side of a hill named Chomatas, on the eastern side of the Gulf of Mirabello in northeast Crete (Figs. 1.1, 1.2). The land adjoining the Gulf of Mirabello was first settled in the Neolithic period,[1] and it had a substantial population during the Bronze Age. Its people were part of the Minoan culture that developed in Crete and gradually became an influential force within the Aegean. Those who lived near Chrysokamino, however, did not participate much in the dynamic political and cultural expansion of Minoan society. Their small territory remained essentially rural, making only modest contributions to the advance of Cretan society.

The largest archaeological site in the Chrysokamino territory is an isolated building constructed on a foundation of megalithic dolomite blocks. The establishment can be regarded as a farmstead as defined by Pettegrew, who applies the name to any "isolated, rural unit of habitation and center of agricultural operations."[2] The Chrysokamino territory contains more than an isolated farmstead, however. The area is especially interesting because of the presence of a metallurgical workshop located a little over half a kilometer from the farmstead, on a cliff overlooking the sea (Fig. 1.2). The project published in this volume studied the metallurgical location and its regional context, undertaking excavations at the workshop and the nearby farmstead, along with an intensive surface survey of the territory around these two archaeological sites.

Several informal visits were made by the author to the area of Chrysokamino in the 1980s and the early 1990s. The information acquired from these visits, in conjunction with a study of the relevant literature, suggested that the location offered an opportunity for the discovery of much new evidence about the area's history. Archaeologists knew of an ancient metallurgy workshop at Chrysokamino even before Harriet Boyd visited the region in 1900,[3] but the site had never been excavated. Surface examination of the location had yielded ambiguous conclusions, and the workshop had been assigned dates ranging from Minoan[4] to modern.[5] Only new work involving excavation could properly date the use of the metallurgy site and elucidate the details of its history.

The metallurgy workshop and the farmstead were not the only archaeological features in the immediate area. An ancient burial cave, stone terrace

1. Betancourt 1999.
2. Pettegrew 2001, p. 189.
3. Boyd 1901, p. 156; Hawes et al. 1908, p. 33.
4. Mosso 1908; 1910, pp. 289–292.
5. Faure 1966, pp. 47–48.

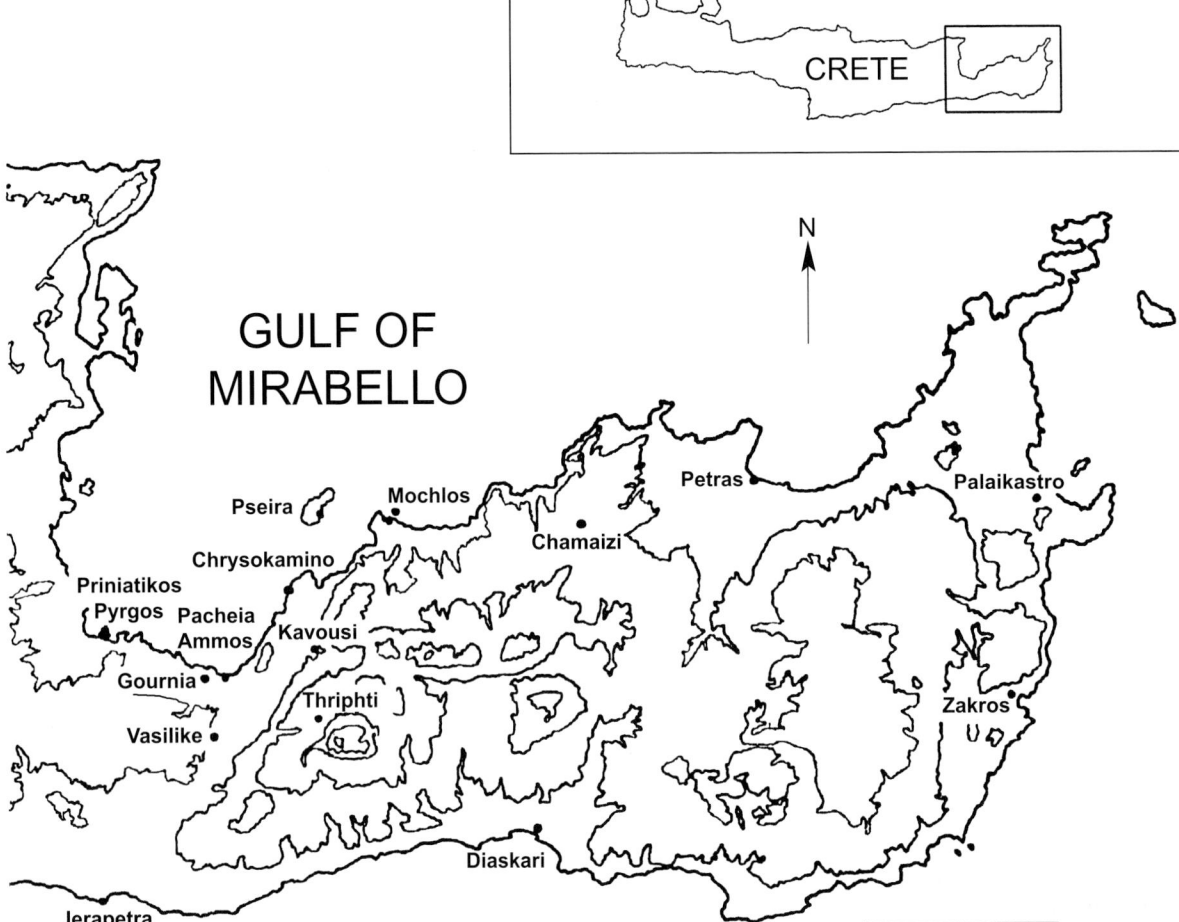

Figure 1.1. Map of eastern Crete

walls, eroded threshing floors, old field walls, abandoned wells, and other landscape features were also observed nearby. Evidence for farming was abundant, and many agricultural terraces could be seen in the immediate territory. An excavation at Chrysokamino would present an opportunity to study the workshop within its local context. A recent intensive survey of the whole region had already been completed as a part of the Kavousi Archaeological Project.[6] The regional survey would be a great advantage to a new project, because it would allow conclusions on the interactions between the tiny territory of Chrysokamino and the larger, regional picture.

LOCATION OF THE CHRYSOKAMINO TERRITORY

The island of Crete has a rugged and mountainous landscape with a complex geological history.[7] The Isthmus of Ierapetra, a break in the mountain chain running across the island from east to west, affords the only easy passage from the north coast to the south. It was probably traversed by an important road in all periods of Crete's human history. The Gulf of Mirabello, at the north of the Isthmus, is a large bay with several harbors

6. Mook and Haggis 1990; Haggis 1992, 1993b, 1995, 1996a, 1996b, 2000, 2005; Haggis and Mook 1993.

7. Creutzburg et al. 1977. For a useful bibliography of the island's geology, see Fassoulas 2001, pp. 93–98.

INTRODUCTION

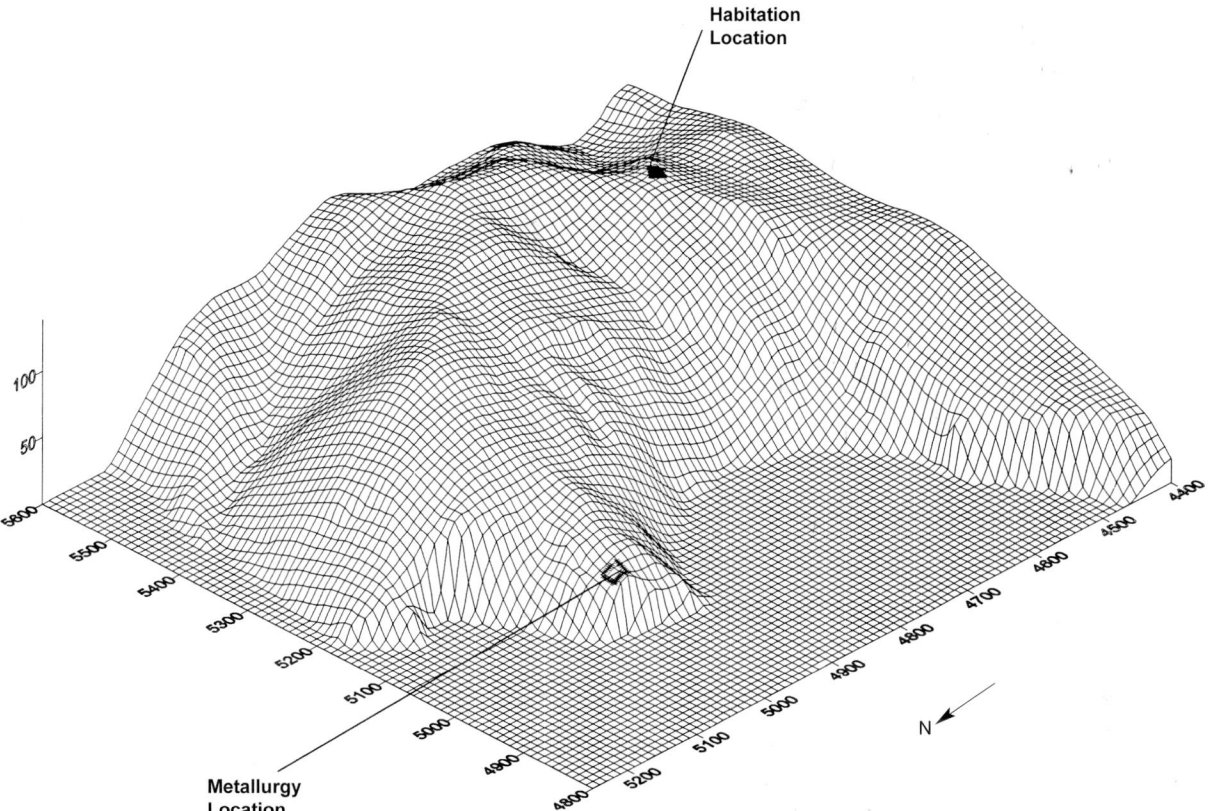

Figure 1.2. Three-dimensional topographic plan of the Chrysokamino metallurgy and habitation sites and the surrounding territory as seen from the sea

that offer shelter for seagoing craft. Traffic and cargo transported across the Isthmus could have connected with several different seaports.

An east–west coastal road meeting the north–south passage at the north of the Isthmus, somewhere near modern Pacheia Ammos, would seem to be a logical result of the topography, but the actual situation is more complex.[8] Before the construction of modern harbor works, the port at Pacheia Ammos was extremely exposed, and it was safe only in good weather.[9] The location was not suitable for the growth of any settlement with year-round maritime interests. In addition, the coastal road could not proceed east for very far beyond the beach and small harbor at Pacheia Ammos. The town at the north of the Isthmus connected easily with places farther west, but the land rises so quickly and becomes so much rougher to the east, that beyond Kavousi it was only a steep path before the modern road system was built. Even in the 19th century, travelers commented on the difficulty of the road leading east from this part of Crete.[10] These geographic limitations probably contributed to the fact that the main Minoan town developed farther west, at Gournia, and the region immediately east of Gournia had smaller settlements. The farmstead at Chrysokamino was one of the smallest of these tiny habitation sites.

Chrysokamino is visible from Pacheia Ammos because it sits on a headland east of the small harbor, on the western slope of a hill called Chomatas (Fig. 1.2). Although it is not far from Pacheia Ammos Bay, it would not have been easy to travel by land from Chrysokamino to the more accessible western sites. Wheeled vehicles would have had difficulties reaching the location before the construction of the modern road, and

8. Betancourt 2004a.
9. Seager 1916, p. 6.
10. Spratt 1865, pp. 157–158.

Figure 1.3. Aerial photograph showing Chrysokamino, the village of Kavousi, and the surrounding region, taken at 6:52 a.m. on June 5, 1988, with a 3.00 × 3.38 km field of view. © 1999 Aerial Images, Inc., and 1999 SovInformSputnik

INTRODUCTION 7

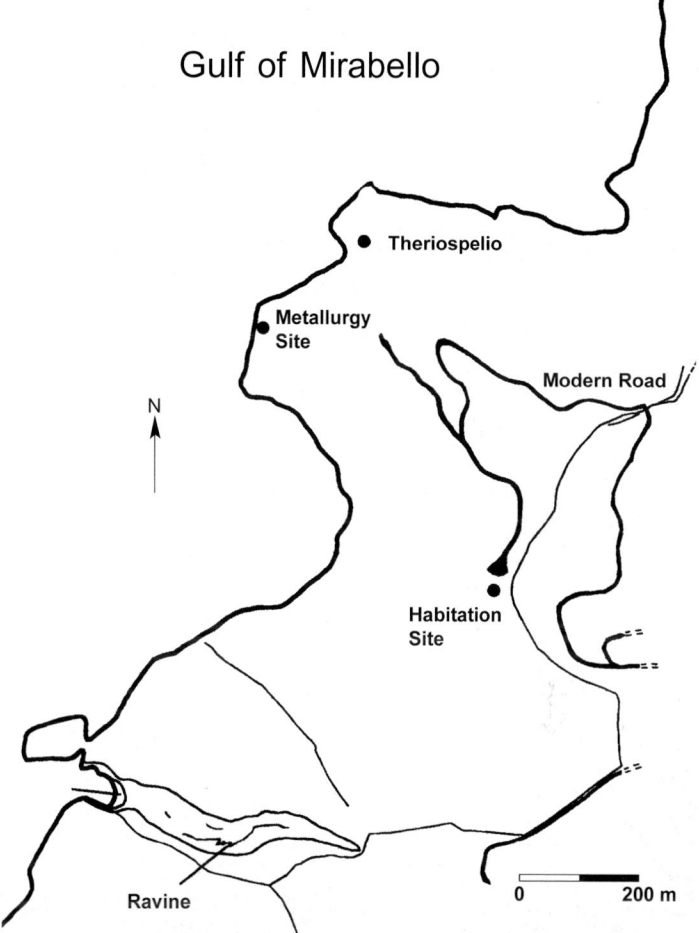

Figure 1.4. Map of the Chryso-kamino territory

ancient traffic probably moved more by sea than by land. The closest modern highway runs through the village of Kavousi, located directly inland from the territory of Chrysokamino (Fig. 1.3).

The part of the Cretan coast where the Chrysokamino territory is located is not particularly fertile farmland, and it is not adjacent to any harbor. The topography is illustrated in the schematic drawing (Fig. 1.2) and in the local map (Fig. 1.4). The farmland consists of a series of low, rounded hills. The spaces between them are sometimes extremely eroded, and small ravines and gullies are common. The soil is partly terra rossa and partly eroded and decomposed phyllite. Steep hillsides and several cliffs demarcate the boundary between land and water. The architecture of the main habitation site, situated on a rocky outcrop of pale gray dolomite, commands a good view of the Gulf of Mirabello, but it is set back more than half a kilometer's walk from the sea. The metallurgy workshop lies on a headland overlooking the Aegean.

The hilly nature of the terrain and the absence of any adjacent boat landing make the location of Chrysokamino less desirable for seafaring than the two good beaches located at Pacheia Ammos to the southwest[11] and Tholos to the northeast.[12] Like other nearby parts of Crete, the rough topography does not allow a coastal road,[13] and the east–west highway is located well inland today, as it was in antiquity.

11. Seager 1916; Harrison 1993, pp. 190–191; Spanakis 1993, p. 618.
12. Haggis 1996a.
13. Betancourt 2004a.

Although both Pacheia Ammos and Tholos had towns by the Late Bronze Age, no large Minoan sites have ever been found along the rough, 8 km long coastal strip that lies between them (see Fig. 1.1). Regional survey has shown that while this region was already inhabited by the Final Neolithic period, before 3000 B.C.,[14] it supported only tiny settlements in the Bronze Age.[15] In this publication we discuss the three main sites that occupy the Chrysokamino territory: a farmstead last inhabited in LM IIIB, a metallurgy site ca. 600 m northwest of it, and a burial cave named Theriospelio found northeast of the metallurgy location (Fig. 1.4). Among these three sites, only the habitation location has evidence for long-term occupation.

MODERN TOPONYMS

Several locations near Chrysokamino have modern names (Fig. 1.5). Some of the labels are used for specific topographical features, while others are applied to geographic areas. In several cases, a name is used both for an area and for the feature that gives the area its name. Interviews with many local residents in the village of Kavousi provided substantial information on nomenclature. The local people are well aware of the variations that exist in regard to the names of several of the locations. They explain that the alternatives stem from different oral traditions.

Agriomandra

A small harbor at the foot of a ravine that leads downhill from Lakkos Ambeliou is called Agriomandra (Αγριόμαντρα). The name is derived from *agrios* (ἄγριος), wild or savage, and *mandra* (μάντρα or μάνδρα), a pen for animals. The word *mandra* is sometimes used in Crete for unfenced locations where animals are gathered together to be loaded on board ships (such a location is found on Pseira Island). The beach at Agriomandra has rock outcrops on both sides, and animals were herded together here easily. The harbor was used until after the middle of the 20th century for shipping animals and other cargo. An alternative, Ayiomandra (from the church of Hagios Ioannis in the ravine uphill from the location), is also used occasionally.

Chalepa

Chalepa (Χαλέπα) is the name of a rugged hill south of the ravine that leads to the beach at Agriomandra. The name is derived from *chalepos* (χαλεπός), meaning difficult. The hill is so rocky and eroded that it cannot be used for any agriculture, and its only modern use is as pasture for sheep and goats.

14. Haggis 1992, p. 269; 1993a; 1996c; Betancourt 1999; Hayden 2003a, 2003b.

15. Mook and Haggis 1990; Haggis 1992, 1993b, 1995, 1996b, 2005.

INTRODUCTION

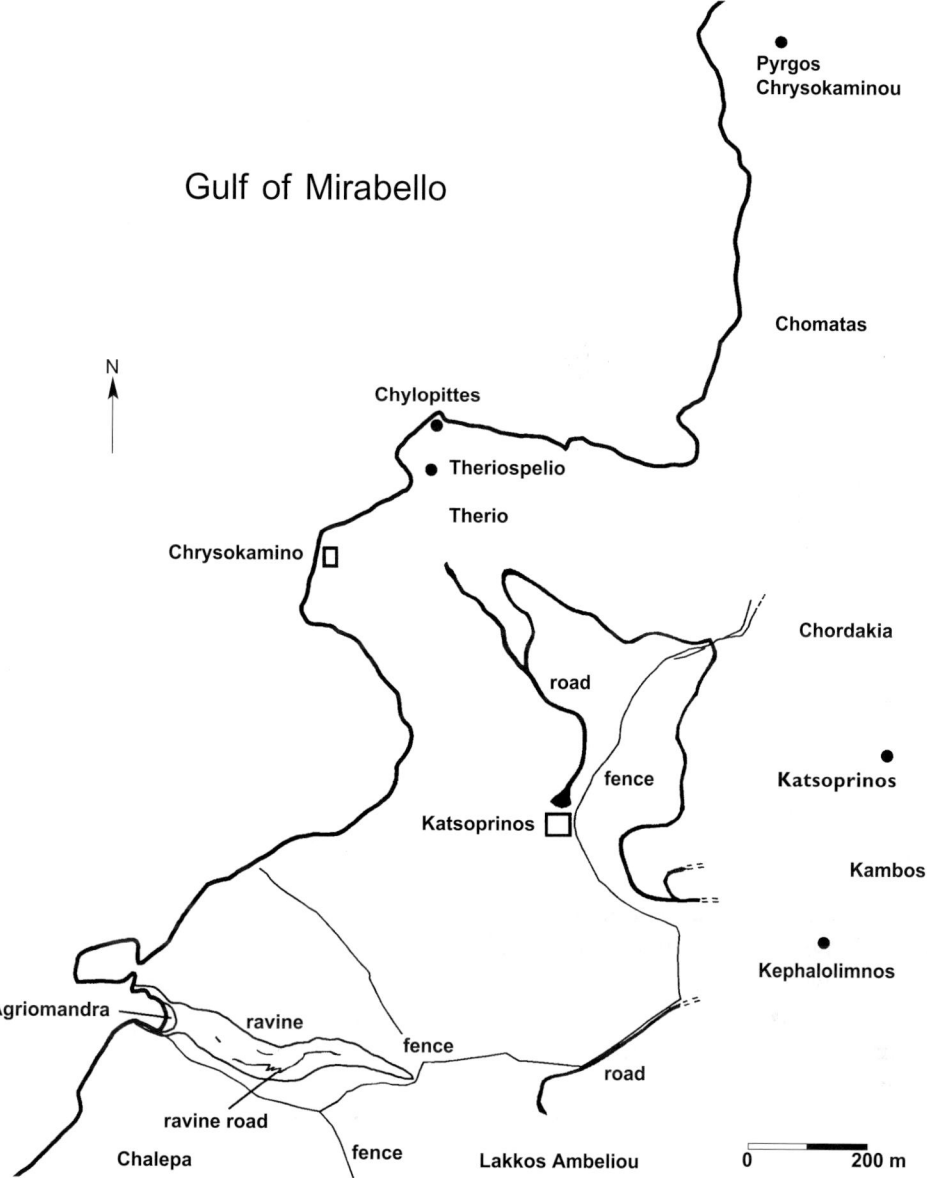

Figure 1.5. Toponyms of the area near Chrysokamino

Chomatas

Chomatas (Χώματας) is the name of a coastal hill (sometimes called a mountain) with two peaks, both of which are northeast of the Chrysokamino territory. The southern part of the hill extends slightly to the east of the location marked Chordakia on Figure 1.5. The area of the excavations, where the settlement site and the metallurgy workshop are located, is on the western slope of the southern part of this hill. The name is applied locally to the hill's crest, a bare area with little vegetation, because the lower slopes have other names. Chomatas is derived from *choma* (χώμα), meaning soil, ground, or dust, and it refers to the dusty, pale-colored, phyllitic soil that covers a substantial part of the hill.

Figure 1.6. Olive trees growing in Chordakia, looking northeast from the habitation site (1999)

Chordakia

The name Chordakia (Χωρδάκια) is used for an area of olive groves on the inland slopes of Mt. Chomatas (Fig. 1.6). The name is used only for the cultivated part of the hill. It is a contraction from Chorioudakia (Χωριουδάκια), from *chorio* (χωριό), meaning village. The term is often used for a place where an old settlement occurs. The reference here is to Kephalolimnos, a Hellenistic, Roman, and Byzantine site with remains from at least as early as Middle Minoan times. Substantial ruins are visible above the ground.[16] Choriodakia (Χωριοδάκια) is a variation recorded by Boyd,[17] and Choridakia is another local variant. Both Khordakia and Khoridakia are used by Kanta[18] and by Haggis.[19] The name is applied to the region from the edge of the Kambos to the modern fence that sets off the public land at the seacoast, not just to the archaeological sites. The sharp boundary between the cultivated plots (Chordakia) and the barren landscape at the coast is clearly visible in the satellite photograph shown in Figure 1.3.

Chrysokamino

Chrysokamino (Χρυσοκάμινο) means gold furnace, from *chrysos* (χρυσός), gold, a Semitic loan word that appears in Greek as early as Linear B,[20] and *kaminos* (κάμινος) or *kamini* (καμίνι), a furnace or kiln. It is used locally both for the actual location of the ancient metallurgical activity (Fig. 1.7) and for the part of the hillside on which the metallurgical site is located. A local variation is in the plural, Chrysokamina.[21] The name Chrysokamino is used by the current project for the territory because it is a unique name, while Katsoprinos, the local name for the location of the habitation site, is used for more than one place in this part of Crete, and it is already in the published literature as the name for a different archaeological site (see Fig. 1.5).[22]

16. Boyd 1901, p. 156; Sanders 1982, p. 141; Haggis 1992, pp. 160–161, locus 2; Fotou 1993, p. 102.
17. Boyd 1901, p. 156.
18. Kanta 1980, p. 145.
19. Haggis 1996b, p. 381.
20. Chadwick 1973, pp. 136 and 343.
21. See Lamb 1929, p. 5; Haggis 1992, locus 88; 1993a, p. 28; 1996b, p. 381.
22. Haggis 2005, p. 35, locus 28.

Figure 1.7. The site of the metallurgical activity at Chrysokamino

Figure 1.8. Chylopittes, a travertine formation at the coast near the Theriospelio cave

Chylopittes

A travertine formation clearly visible from the sea (Fig. 1.8) is named Chylopittes (Χυλόπιττες). It is north of Theriospelio cave, on a cliff above the sea. The name is derived from *chylos* (χυλός), a cooked mass of ground cereal, and *pitta* (πήττα or πίττα), a pastry or pie. It refers to the pale brown stalactitic formations that look like *chylopitta* (χυλόπιττα), a pastalike food.

Kambos

A large valley that extends inland from the coast at Tholos Bay is called the Kambos, from the word *kambos* (κάμπος), which means plain or flat field (Fig. 1.9). The term is borrowed from the Italian word *campo*, derived from

Figure 1.9. The Kambos as seen from Kavousi, with the coastal hills behind it

the Latin *campus*. The valley is inland from Chordakia, and it is outside the Chrysokamino territory as defined in this volume. An intermittent stream, the Platys River (called the Kavousianos River by Faure),[23] flows through the Kambos in the spring and empties into Tholos Bay.

Katsoprinos

Katsoprinos (κατσόπρινος), a local variant of *katsoprini* (κατσόπρινι), the holly oak, is also called *katsiprinos* (κατσίπρινος) in Crete.[24] A local variant is *katsoprinios*. This evergreen shrub, *Quercus coccifera*, is a member of the *Fagaceae* family, and it grows widely in Crete on the lower slopes of hills.[25] It has dark green spiky leaves and seldom grows taller than a few meters. The shrub gives its name to the dolomite outcrop where the habitation site is located because several of the shrubs grew there before excavation began. The name is not used exclusively for this farmstead, as it is also applied to a spot lower on the hill's slope to the east where other Minoan remains can be seen above ground.[26]

Lakkos Ambeliou

Lakkos Ambeliou (Λάκκος Αμπελιού) is south of Katsoprinos. It is a low valley (a part of the Kambos) where a substantial deposit of red sediment has accumulated from the erosion of the nearby hills (Fig. 1.10). The deposit, which is a pedon, is most likely a result of alluvial action that deposited the sediment in a lake.[27] The sinkhole has been used as a clay source by a local company engaged in making brick, and residents of the village of Kavousi regularly take the red sediment to use it as soil for their gardens and vineyards. The removal of the sediment has resulted in a substantial excavated area. The location is appropriately called Lakkos Ambeliou. *Lakkos* (λάκκος) means a hole in the ground, and *ambeliou* is the genitive of *ambelos* (άμπελος), meaning vineyard. The variant Ambeli Lakkos is used occasionally.

23. Faure 1984, p. 47; 1989, p. 320.
24. Alibertis 1994, pp. 77–78.
25. Chavakis n.d., p. 187; Sfikas 1987, pp. 32–33.
26. Haggis 1996b, p. 400; 2005, p. 35, locus 28.
27. Morris 2002, chap. 3.

INTRODUCTION

Figure 1.10. Lakkos Ambeliou, in the Kambos, looking south from the habitation site (1999)

Figure 1.11. The burial cave called Theriospelio

Therio

Therio (Θεριό) is a rugged, uncultivated, barren hillside northeast of the metallurgy location. The name, which means wild beast, is a reference to the rugged nature of the landscape. The word is recent, as the ancient word is spelled with an *eta*.

Theriospelio

Theriospelio (Θεριόσπηλιο) is the preferred term for a cave northeast of the metallurgy site (Fig. 1.11). It has several names. Theriospelio (also Theriospelaio) is a reference to the wild nature of the surrounding landscape, from *therio* (θεριό), wild beast, and *spelio* (σπήλιο), cave. The variation Theriospelos (Θεριόσπηλος) is given by Zois, and Haggis uses Thergiospilio.[28] A synonym, Theriotrypa (Θεριότρυπα), is recorded by Faure.[29] Agriospelio

28. Zois 1993, p. 340; Haggis 1992, p. 171.
29. Faure 1964, p. 227; 1966, p. 48.

(Αγριοσπήλιο) was used by a few residents of Kavousi in the mid-1980s, from *agrios* (άγριος), wild or savage, a synonym for *therio*. Kolonospilios (Κολωνόσπηλιος) is given by Faure,[30] and the neuter gender variation Kolonospilio (Κολωνόσπηλιο) is added by Haggis.[31] These names are from *kolona* (κολώνα), one of many Italian loan words in use in Crete (from *colonna*, a column, a reference to the cave's stalagmite formations).

Discussion

Terms for the land features near Kavousi belong to several classes. Most of the modern names for the territory near Chrysokamino are Greek. Among the exceptions are the word *kolona* in Kolonospilios, derived from Italian, and the word *kambos* from the same language; they are presumably from the Venetian period of Crete. The *chrysos* in Chrysokamino was borrowed from Semitic as early as the Linear B documents, and it has been common in Greek since the end of the Bronze Age.

Like most toponyms in Crete, the names used for locations on and near Mt. Chomatas signify something about a place's appearance or about events or features associated with it. None of the names necessarily derives from the Roman period or from any earlier time. The only designations that allude to antiquities are Chrysokamino and Chordakia. Chrysokamino refers to the wrong metal, and Chordakia is not the name of an actual town, but an indication of an early settlement in the vicinity. All of the names go back at least to the 19th century, and some of them are probably much older, but the dates when they were first used are not known. They have been handed down in an oral tradition, so it is not surprising that more than a single variation or name exists for some places.

EARLIER WORK IN THE CHRYSOKAMINO TERRITORY

During the second half of the 20th century, the Chrysokamino metallurgy location and its nearby territory has been one of the most controversial ancient sites in Greece. Archaeologists visited the region between modern Kavousi and the sea shortly before 1900, and the region is well known from the archaeological literature. The metallurgical workshop's period of use has been particularly problematic, and it has been assigned dates ranging from Minoan to modern. The location has even been considered as a site for lime production as an alternative to a metallurgical installation. The industrial ceramics on the surface have been regarded as pieces of crucibles, furnaces, covers for furnaces, and the floor of a large kiln.

Archaeologists knew about Chrysokamino before 1900. Arthur Evans visited the region in the 1890s. Harriet Boyd excavated Roman remains at Chordakia (east of the Chrysokamino territory) in 1900, and she walked over much of the region. In an article about her 1900 excavations at Kavousi, she reported the Roman discoveries at Chordakia and also noted that "early architecture" (a term she employed for Minoan to Iron Age remains) was on top of the coastal hill.[32] The architecture is probably the hilltop site

30. Faure 1964, pp. 32, 227; 1966, p. 48.
31. Haggis 1992, p. 171.
32. Boyd 1901, p. 156.

recorded here as Pyrgos Chrysokaminou (Fig. 1.5), but it could also refer to the farmstead excavated by this project.

The metallurgy location at Chrysokamino was first discussed in the final publication of the excavations at Gournia.[33] Harriet Boyd Hawes reported that in 1900 the local residents showed her a site called "Golden Furnace," where she collected a few pieces of copper. She said that "rock obtained from the adjacent cliff" contained traces of copper. Her report noted that the ground was strewn with fragments of "an ancient furnace." She did no work there personally, and a later report simply summarized the earlier one.[34]

Angelo Mosso discussed the site in more detail.[35] He reported that Joseph Hazzidakis (who was then Ephor-General of Crete) visited the site twice, once in 1906 and a second time either before or later. Hazzidakis collected a piece of metal and specimens of slag from the metallurgy location, and he also visited the Theriospelio cave. Mosso, who may not have actually visited the cave, did not realize it was completely natural, and he regarded it as the mine for the ore that was smelted nearby. He reported Early Minoan pottery from the cave, but no sherds aside from what he regarded as sections of "crucibles" were found at the metallurgy location itself. Pieces of "scoria" were also discovered in the cave. An illustration of one of the "crucible" fragments from Chrysokamino (reproduced here in Figure 1.12) shows that the artifacts were the ceramic pieces with pierced holes and a glassy deposit adhering to one side that still cover all unexcavated portions of the site. Mosso analyzed the piece of metal and reported that it contained 45% copper.

Mosso also analyzed the "scoria" collected from Chrysokamino. He reported that it contained a trace of copper, and as this result complemented his other researches, he concluded the site was a workshop for extracting copper from its ore. He reported finding pieces of similar "scoria" on the beach at Pacheia Ammos, suggesting another site where copper may have been smelted.

Mosso believed Chrysokamino was a Bronze Age installation. He did not publish any pottery from the metallurgy workshop, but he did describe ceramics from the nearby cave. He reported that Hazzidakis visited the cavern and collected sherds from "the primitive period" as well as from Early Minoan II. One vase from the cave had decoration in white paint on a dark ground, so it was either from EM III or from the Middle Minoan period. A material that contained copper was also collected from the cave, but its exact identification is difficult to comprehend from Mosso's brief description.

Edith Hall and Richard Seager investigated the location more than once, and a visit they made in 1910 was recorded in unpublished letters.[36] They approached the headland by sea after visiting Pseira. Hall called the site "the place where they smelted their bronze [sic] in Minoan times." They visited the cave on the same trip, and Hall returned for two days of excavation there the following week. The limited information concerning her excavation is reported in this volume.

Fritz Schachermeyr visited the territory in the 1930s and made a ground survey of the region.[37] He visited the metallurgy location as a part of his exploration. While he was there, he picked up a sherd of Vasiliki Ware (the

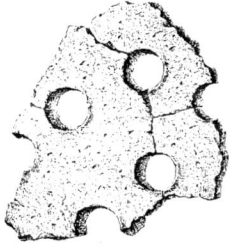

Figure 1.12. A fragment of industrial pottery published by Angelo Mosso. After Mosso 1910, fig. 164. Scale 1:4

33. Hawes et al. 1908, p. 33.
34. Hawes and Boyd Hawes 1909, p. 38.
35. Mosso 1908; 1910, pp. 289–292 and fig. 164.
36. Unpublished letter of April 24, 1910 by Edith Hall in the archives of the University of Pennsylvania Museum of Archaeology and Anthropology; see also Becker and Betancourt 1997, p. 109.
37. Schachermeyr 1938, pp. 472–473.

definitive pottery for EM IIB) at the metallurgical location, confirming the Early Minoan date of the site.

A few discussions of the metallurgy workshop appeared in various scholarly writings in the years after Hall worked in the area. Until the 1960s, Chrysokamino was occasionally mentioned as a mining, smelting, or copper-working location.[38] These reports were all based on the research conducted before 1910 and on Schachermeyr's 1938 report of his visit to the small promontory.

New work began in the 1960s. LM IIIB:1 tombs at Chordakia were excavated by Davaras in 1962.[39] Chordakia is the region just inland from Chrysokamino, and the tombs from LM III were probably related either to the settlement at Katsoprinos or to the Chrysokamino habitation site.

Paul Faure surveyed the area is the early 1960s.[40] He explored the Theriospelio cave, and he also examined the metallurgy site. For the latter, he suggested that the ceramic fragments on the site were not from crucibles but from a single large furnace used for smelting chalcopyrite ore, although he provided no evidence for this assumption. Based on the state of vitrification of the glassy waste products, the absence of soil accumulation, and the fact that he believed the ore reduced was a sulfide, he regarded the site as "relativement récente." He suggested it represented the remains of a smelting operation for copper using either charcoal or coal as fuel. He correctly rejected the nearby cave as a mine.

Keith Branigan also visited the metallurgical location. He regarded the deposit of "cinders, ashes, slag, and crucible fragments" as an indication of copper remelting (not smelting), and he gave the site a different date.[41] Based on an analysis by Geoscan Electron Microprobe Analysis, which showed that a temperature of 1150°F or more had been reached in the slag and that calcium was present, he suggested the site was later than Roman. He used literary sources to propose the 12th century A.D. as the most likely date. Branigan correctly realized that the cave was not a mine but a natural cavern in limestone, with no visible trace of nonferrous metals in the rock.

After 60 years of surface collecting by archaeologists, diagnostic sherds were no longer present on the surface of the metallurgy site when Faure and Branigan visited this part of Crete in the 1960s. They based their dating on the characteristics of the slag, and they cannot be criticized for their conclusions because the methodology they used, which was based on research in Cyprus, was widely regarded as valid at the time. This methodology rested on the mistaken belief that before the Roman period, early metallurgists could never achieve high temperatures with their primitive technological knowledge. Therefore, any slag with evidence for substantial vitrification had to be more recent.[42]

The metallurgy site, like the rest of the territory around it, did not figure prominently in the scholarship on Minoan Crete after the

38. Fimmen 1921, pp. 17 and 120; Glotz 1923, p. 39; Lamb 1929, p. 5; Davies 1932, p. 987; 1935, pp. 7, 113, 264, 270; Pendlebury, Pendlebury, and Money-Coutts 1935–1936, p. 104; Forbes 1950, p. 364; Hutchinson 1962, pp. 40, 247.

39. Kanta 1980, pp. 144–145, fig. 56, nos. 6 and 8.

40. Faure 1964, p. 32; 1966, pp. 47–48.

41. Branigan 1968, pp. 50–51.

42. Koucky and Steinberg nicely summarize the system (1982a, p. 156 and table 1).

1960s. Discussions of early Aegean metallurgy written after this time and before the beginning of the modern project often omitted the site completely.[43] Several writers accepted the theory that the site was Medieval.[44] A general history of metallurgy written by Tylecote suggested the site was used for the remelting of copper.[45] An exception to the general rule, a study of daggers by Nakou, considered the site to be from the Early Bronze Age.[46]

Karen Foster published a catalogue of the collection of Minoan objects in the Mount Holyoke College Art Museum in 1978, including pieces from the Theriospelio cave excavated by Edith Hall.[47] Other pieces from the same excavation were included in a catalogue of the University of Pennsylvania Museum of Archaeology and Anthropology.[48]

In 1983, Noel Gale and Zofia Stos-Gale of the Nuclear Physics Laboratory, Oxford University, collected samples from Chrysokamino as a part of a larger archaeometric survey of metallurgy in the Aegean.[49] They collected slag and pieces of the industrial ceramics for scientific analyses of various types. The results of their analyses showed that the site was extremely important for the early history of metallurgy in the Aegean Bronze Age. They used thermoluminescence to establish the 3rd millennium B.C. date of the industrial ceramics. Their lead isotope studies indicated that the copper might have originated in Lavrion or Kythnos. The Chrysokamino territory was included in the regional survey of Donald C. Haggis. This was a doctoral dissertation project undertaken in conjunction with the University of Minnesota archaeological excavations at Kavousi.[50] The survey by Haggis laid the foundation for the more detailed investigation reported in this volume, inasmuch as it established the regional pattern of settlement in which Chrysokamino functioned.

Two detailed studies of the soils of the area around Chrysokamino were made in connection with the Kavousi Archaeological Expedition. Michael Timpson investigated Late Quaternary alluvial sediments from the vicinity of Pacheia Ammos and Kavousi.[51] Michael Morris studied a series of soil pedons, including the one at Lakkos Ambeliou just south of the Chrysokamino habitation site.[52] The two complementary investigations gathered considerable information on the soils of the region.

The most recent investigation of the area before the current project began was carried out by Antonios Zois. He explored the cave, where he collected Final Neolithic and Early Minoan pottery, and he also surveyed the metallurgy workshop area. He raised the possibility that the latter site was a modern limekiln.[53]

In 1995, the present project began with a preliminary season of geological survey and topographical mapping.

43. A. C. Renfrew 1972; Branigan 1974; Healy 1978; McGeehan-Liritzis 1996.
44. Stos-Gale and Gale 1984, p. 59; Haggis 1992, p. 170; Evely 2000, p. 341.
45. Tylecote 1976, p. 19, table 19.
46. Nakou 1995.
47. Foster 1978.
48. Betancourt 1983.
49. Gale, Stos-Gale, and Gilmore 1985; Gale and Stos-Gale 1989; Stos-Gale 1989; 1993, p. 124; 1998, pp. 720–721.
50. Mook and Haggis 1990; Haggis 1992, Loci 50 and 88; 1995, pp. 373–379, fig. 7; 1996a; 1996b, pp. 380–381 and 401–403; 2000; 2002; see also Haggis and Mook 1993, pp. 287–288.
51. Timpson 1992.
52. Morris 1994, 2002.
53. Zois 1993, pp. 340–341.

GOALS OF THE CHRYSOKAMINO PROJECT

Although substantial information was known about Minoan towns in this part of Crete before this project began thanks to the excavations at Gournia, Pseira, and Kavousi, much less was known about the tiny rural farmsteads that consisted of isolated architectural complexes. No Minoan farmstead had ever been studied with the intent of ascertaining the strategies employed for the use of its agricultural territory. A Minoan smelting site had not been excavated previously.

The goals of the project included a better understanding of human habitation in the Chrysokamino territory, with emphasis on the metallurgical workshop and its relationship to the surrounding territory. The project planned to address several separate items: the date and nature of the metallurgy site, the technology used by the metallurgists, the types of metallurgical products produced at the workshop, the relation of the metallurgy site to the nearby habitation site, the size and characteristics of the local territory, the prevailing subsistence strategy, the diachronic history of the territory, and its relation to the larger region and Crete in general.

The archaeological plan that was developed to address these goals was formulated as an interdisciplinary study involving researchers in several fields. It was based on the nature of the landscape to be investigated, the questions that needed addressing, and the available resources of staff, time, and finances. Because it was anticipated that much of the material that would be discovered would require scientific analysis for its proper understanding, metallurgical specialists were invited to participate both on-site during the excavation that uncovered the evidence and as laboratory researchers after the material was excavated. The plan included the following aspects:

1. Excavation at the metallurgy site
2. Excavation at the habitation site
3. Small-scale excavations at other selected locations
4. An electronic instrument survey and mapping project combined with an intensive surface survey to record visible anthropogenic features
5. A program of laboratory analysis for metallurgical remains, ceramics, and other materials
6. A series of corollary studies including geology, geomorphology, topography, soil studies, examination of land use, toponyms, and other investigations
7. Study and publication

The research plan was devised in 1994. Excavations were planned for the farmstead and the metallurgy location, along with additional programs of small excavations at locations to be decided later. Geological and geomorphological studies, topographic mapping, soils science, and other specialized studies were planned. A large supporting program of laboratory analysis was assembled. Study of the excavated materials was conducted at the Kavousi schoolhouse until the completion of the INSTAP Study Center for East Crete, in Pacheia Ammos. Several preliminary reports have been published.[54]

54. Betancourt 1997; Betancourt, Floyd, and Muhly 1997; Betancourt, Muhly, and Floyd 1998; Betancourt et al. 1999; Muhly 1999, p. 17; 2002, pp. 79–80; Betancourt and Floyd 2000–2001. Preliminary reports on the analyses published in this volume are also included in Gale et al. 1985; Stos-Gale 1989, 1998; Gale and Stos-Gale 1989; Betancourt and Myer 1999.

CHAPTER 2

THE NATURAL ENVIRONMENT

by Philip P. Betancourt and William R. Farrand[1]

CLIMATE AND ITS RELATION TO CHRYSOKAMINO

The climate of eastern Crete is one of the region's most important ecological factors. Like soils and topography, it affects the choice of plant communities and the agricultural methods necessary for successful farming. Eastern Crete has a typical Mediterranean climate, and many authors have discussed its precipitation, temperature, and winds.[2] The general picture is important, but some specifically local conditions must be considered as well because local factors seem to have played a decisive role in the situation at Chrysokamino.

As a whole, mean temperatures and precipitation are moderate, but the crucial factor is not the mean but the seasonal variation. Table 2.1 illustrates the seasonal differences that make east Cretan agriculture a specialized endeavor in which many crops will not thrive without substantial care.

The eastern part of Crete receives less moisture than the western and central parts of the island. Precipitation, mostly in the form of rain, is largely confined to the period between September and May, with the heaviest amounts in December and January. Modern records show that the mean rainfall for January is 96.3 mm, while for July it is only 0.3 mm (Table 2.1). Rain is extremely rare during the hot summer months. It is wetter during the winter, when dry continental air masses move southeast from Europe and cross over the warmer Ionian Sea, with evaporation and atmospheric instability leading to precipitation over the island. Eastern Crete nearly always receives less rain than western Crete because the air masses move from west to east, and the mountains that run along the length of the island except for the Isthmus of Ierapetra act as a shield and cause a majority of the precipitation to fall in the west.

Temperatures are moderate. As the chart in Table 2.1 demonstrates, the temperature in January is substantially colder than it is in July, but it is seldom cold enough to kill hardy plant life. The mean temperature for eastern Crete (as recorded at Ierapetra on the south coast) is 13.2°C for January and 27.2°C for August.[3]

During the summer, fall, and winter, the prevailing winds are from the northwest. They can be very strong, especially during the winter, and

---

1. The climate and natural resources sections of this chapter were written by P. P. Betancourt. The geology section was written by W. R. Farrand.

2. See esp. Allbaugh 1953; Mariolopoulos 1961; Zohary and Orshan 1965; Gat and Magaritz 1980.

3. Zohary and Orshan 1965, suppl.

TABLE 2.1. SEASONAL VARIATIONS AT HAGIOS NIKOLAOS, CRETE

	January	April	July	October
Mean Temperature	12.4°C	16.9°C	25.8°C	20.9°C
Mean Number of Cloudy Days	6.8/month	4.2/month	0.6/month	4.5/month
Mean Rainfall	96.3 mm	27.0 mm	0.3 mm	53.4 mm

After Mariolopoulos 1961.

storms that wash salt spray inland, damaging land and crops, are not uncommon at low coastal sites.[4] In late summer, the north wind *(meltemi)* blows almost every day, and days with wind speeds of up to 7 Beaufort (i.e., 32–38 mph) are not unusual. The force of the breezes varies with the time of day. Winds are strongest in the early afternoon because of a well-known meteorological situation in which the land warms up more quickly during the day than the sea does, causing air over the land to rise, which increases the force of the sea breeze. The result is that the wind is strongest during the hottest part of the day. For the human residents, this situation is an advantage. It makes the coast more comfortable than inland areas because "the sea breeze plays an important role in moderating the temperature of narrow strips of land along the seacoasts," offering "a welcome relief from the summer heat for residents who live near the shore."[5] It is one of the reasons that residents founding new settlements in warm parts of the world (like eastern Crete) establish a large number of coastal towns.

In the Chrysokamino territory, which faces north and is elevated well above sea level, these conditions are more extreme than would be the case with a more protected landscape. The land is so windy and dry that in modern times it has remained unfarmed, and it is used only for sheep and goats. The topography at the metallurgy site, which consists of a troughlike depression between higher exposures of bedrock, is always the windiest part of the hillside when the wind blows from the north. The trough funnels the breezes across the site, and few plants can survive at this dry and barren location. Similar topography was chosen for Bronze Age smelting operations on Kythnos[6] and Seriphos,[7] suggesting that the early metalworkers deliberately chose a particularly windy location for their furnaces.

Winds are not constant throughout the year. In the spring, they are much more variable than in the summer or fall, leading to different conditions in different years. The south wind *(sirocco)* is especially warm, and it can bring Saharan dust to the south Aegean, including Crete.[8] The result of these variations is a considerable seasonal difference in the amount of moisture in the soil. The same wind that contributes to the comfort of the human population is damaging to some types of crops, because the soil cannot retain its moisture during the long, rainless period when the weather is warm and the winds are fairly constant. Planting and harvesting must be arranged according to this seasonal schedule.

The local climate encourages the cultivation of crops that can withstand months with many dry, hot days and are not inhibited by periods of dry soils. The triad of olives, grapes, and grains is often regarded as the main group of such crops.[9] A convincing case has been made for the growing of legumes as a fourth staple,[10] but this is a crop that requires much more care

4. Seager 1916, p. 6.
5. Miller and Anthes 1980, p. 94.
6. Hadjianastasiou and MacGillivray 1988.
7. Gale and Stos-Gale 1989, pp. 24–25.
8. Pye 1992.
9. Vickery 1936.
10. Sarpaki 1992.

than grains, olives, or grapes, so it would have required a different type of agricultural strategy, with extra care.

Evidence exists for a somewhat different climate in the Bronze Age in comparison with the situation in the 20th century. East Cretan searches for pollen suitable for analysis have not been successful,[11] but several studies from western Crete suggest a wetter period in the Neolithic to Early Bronze Age, with gradual change leading to the current climate by the 1st millennium B.C.[12] These conditions may have prevailed in eastern Crete as well. Identification of wood charcoal from cooking fires on the offshore island of Pseira, mostly from LM I contexts, shows that pine was present in addition to the expected trees.[13] Pine does not grow on Pseira today because the climate is too dry, although it is present in some of the ravines and other shaded places in this part of Crete. It is probable, therefore, that the FN–LM I climate was very much like the present, but with more rain in the FN–LM I periods, and with a gradual drying by the end of the Bronze Age. The Bronze Age countryside has been aptly described by Moody, who envisions a setting of rolling hills composed of patches of woodland and maquis, with cultivated fields of grain, olives, and other crops.[14]

As the climate became drier, the agricultural potential of the region must have become increasingly strained. The zonation of the ancient plant communities would have been affected by both the natural moisture available from rain and human activities that altered the environment, including the construction of terraces and the addition of organic matter to the soil to help it retain moisture.

These factors must have contributed to long-term ecological changes taking place in the Chrysokamino region. Elsewhere in this volume, the evidence for substantial care of the Chrysokamino gardens, in the form of chemical traces in the soil and the deposition of small, isolated sherds associated with manuring practices, is presented. The climatic conditions suggest that plants requiring extra attention would have been harder to grow on the coast of eastern Crete than in the western part of the island or inland areas. Local conditions would have required increasing human effort to make the landscape productive.

Another factor to be considered is the gradual rise in relative sea level. In the part of Crete near Chrysokamino, the sea level may be as much as 3.5 to 5 m higher than it was in the Bronze Age.[15] This condition will not have affected Chrysokamino as much as sites located on the seacoast, but it must have been a factor in the viability of the harbor at Agriomandra, used by the local residents.

The abandonment of coastal settlements at the end of the Bronze Age has usually been attributed to warlike raiding and other unsettled political conditions.[16] This explanation may be partly correct, but additional factors must be invoked to explain why a part of the coast that had supported permanent residents for two millennia was not resettled in the more stable Classical and Roman periods. The new Dorian farming methods using serf labor on large estates may have been one of these factors.[17] It is also likely that the climate had become too dry to support a local population on the coastal strip at Chrysokamino. As a result of climatic change, the land, which had always been less than desirable, may now have been so poor that it did not merit permanent settlement.

11. Coring was done by G. Rapp; see Hayden, Moody, and Rackham 1992, p. 307.

12. Bottema 1980; 1994, pp. 57–58; Moody 1987a, chap. 2; Moody, Rackham, and Rapp 1996.

13. Betancourt and Hope Simpson 1992.

14. Moody 1987a, p. 120.

15. Fairbridge 1972; Flemming 1972; Dermitzakis, Karakitsios, and Lagios 1995; see also comments by Farrand below.

16. Desborough 1964, 1972; Drews 1993.

17. Willetts 1955, pp. 46–51; 1990; Jameson 1992.

GEOLOGY OF THE CHRYSOKAMINO AREA

This report results from two visits to the Chrysokamino area from July 21 to 28, 1995 and July 23 to August 2, 1996. Detailed field observations extended somewhat beyond the limits of the map in Figure 2.1 and included mapping of outcrops, recording and sampling road cuts and excavation trenches, and discussions with the topographic survey team. Reconnaissance observations extended still farther into the hills above Kavousi, the Tholos beach area, and into the Ierapetra graben. Detailed studies of the soils of the immediate area and of the surrounding region, reported in two dissertations from the University of Tennessee,[18] proved very helpful.

This section is modified slightly from our previous report of the geology of Pseira[19] because Pseira and the Chrysokamino-Kavousi area are very similar in their relations with respect to their general geological setting in eastern Crete. They are separated by only 2.5 km of water, and the distance from the center of Pseira to the metallurgical site is only 6 km.

The geology of eastern Crete is complex in detail but relatively simple in its broad outlines. For our purposes, the broad outlines will suffice for an understanding of the regional relations of the major bedrock units of the Chrysokamino area. The following summary is based on a spate of fieldwork, mostly in the 1970s, dealing with the tectonic setting of Crete and of the Aegean region as a whole.[20]

On the broadest scale, Crete constitutes the central part of the Hellenic Arc, a major tectonic feature that curves from the Peloponnese through Crete to the Taurus Mountains of southwestern Turkey. The Hellenic Arc marks the collisional junction of the African and European lithospheric plates where portions of the Earth's crust, formerly situated in what is now the central Aegean, have been thrust southward to override a thick section of marine platform limestones constituting the bulk of the "basement" rocks of Crete. The overriding rocks, in the form of relatively thin sheets called nappes, arrived in successive waves (Figs. 2.2, 2.3).[21]

Bedrock of Eastern Crete

The following comments apply most specifically to the part of Crete east of the Ierapetra-Gulf of Mirabello lowland shown in Figure 2.2. The first overthrust sheet is the "Phyllite" nappe that directly overlies the autochthonous, relatively thinly bedded, cherty limestones known as "Plattenkalk" (Fig. 2.4). The rocks of the Phyllite-Quartzite nappe exhibit low-grade metamorphism, which is most intense in its lower part and fades out upward through the nappe.

Stratigraphically between the Plattenkalk and the overlying phyllites is a rock unit that appeared to us at first to be of a character transitional between those formations. However, upon further examination, it appears that this "transitional" unit is really the upper part of the Plattenkalk, a conclusion that is consistent with the regional interpretation. Wachendorf and colleagues described the top of the Plattenkalk as gray marly schists with some limestone interbeds.[22] Thin section examination of one of these rocks showed both the carbonate mineralogy and the metamorphic

18. Timpson 1992; Morris 1994, 2002.
19. Farrand and Stearns 2004.
20. E.g., Wachendorf, Best, and Gwosdz 1975; Baumann et al. 1976; Baumann, Best, and Wachendorf 1977; Seidel et al. 1981.
21. For a diagrammatic explanation of the emplacement of these nappes, see also Baumann, Best, and Wachendorf 1977, p. 517, fig. 8.
22. Wachendorf et al. 1974.

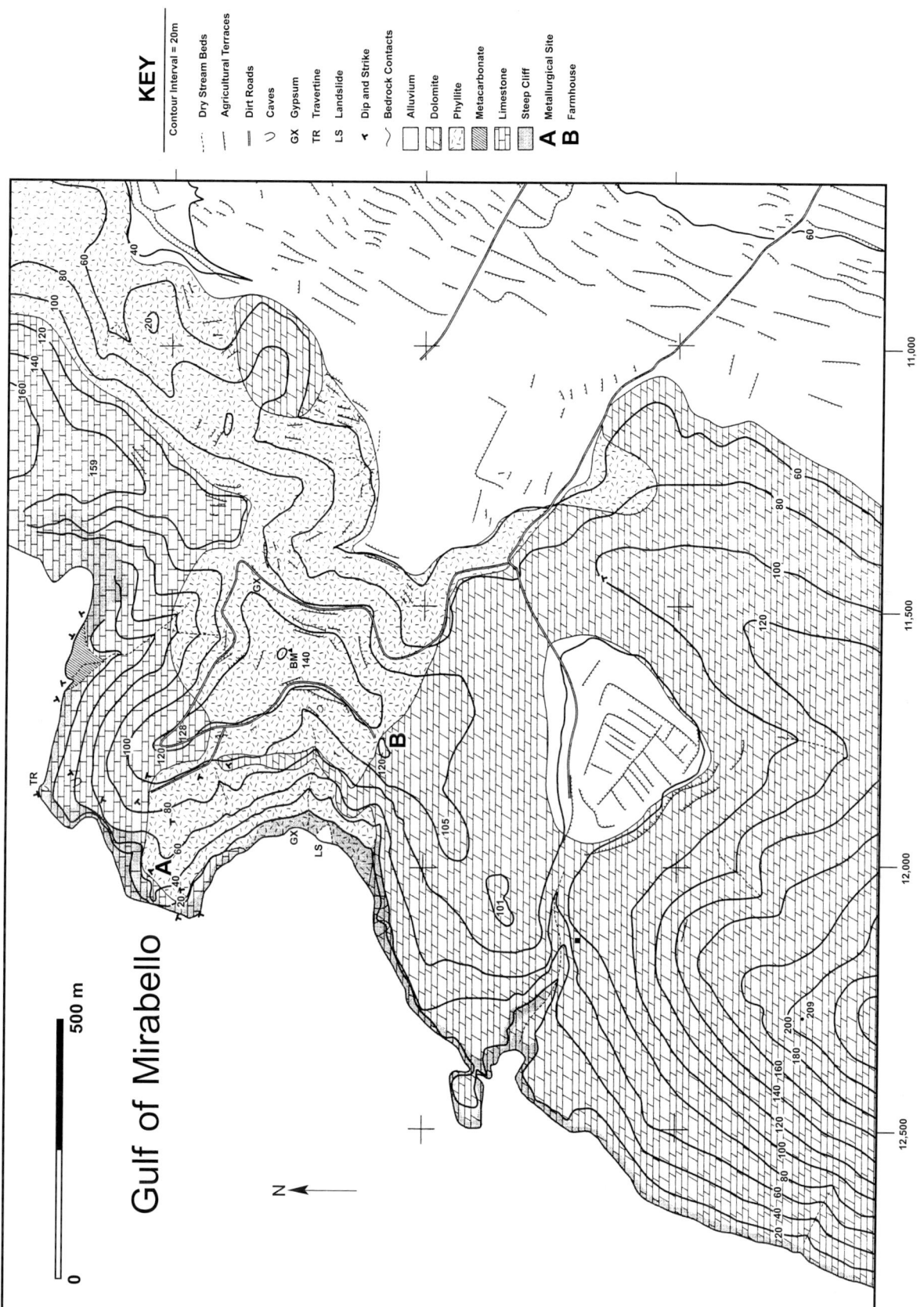

Figure 2.1. Detailed bedrock map of the Chrysokamino area

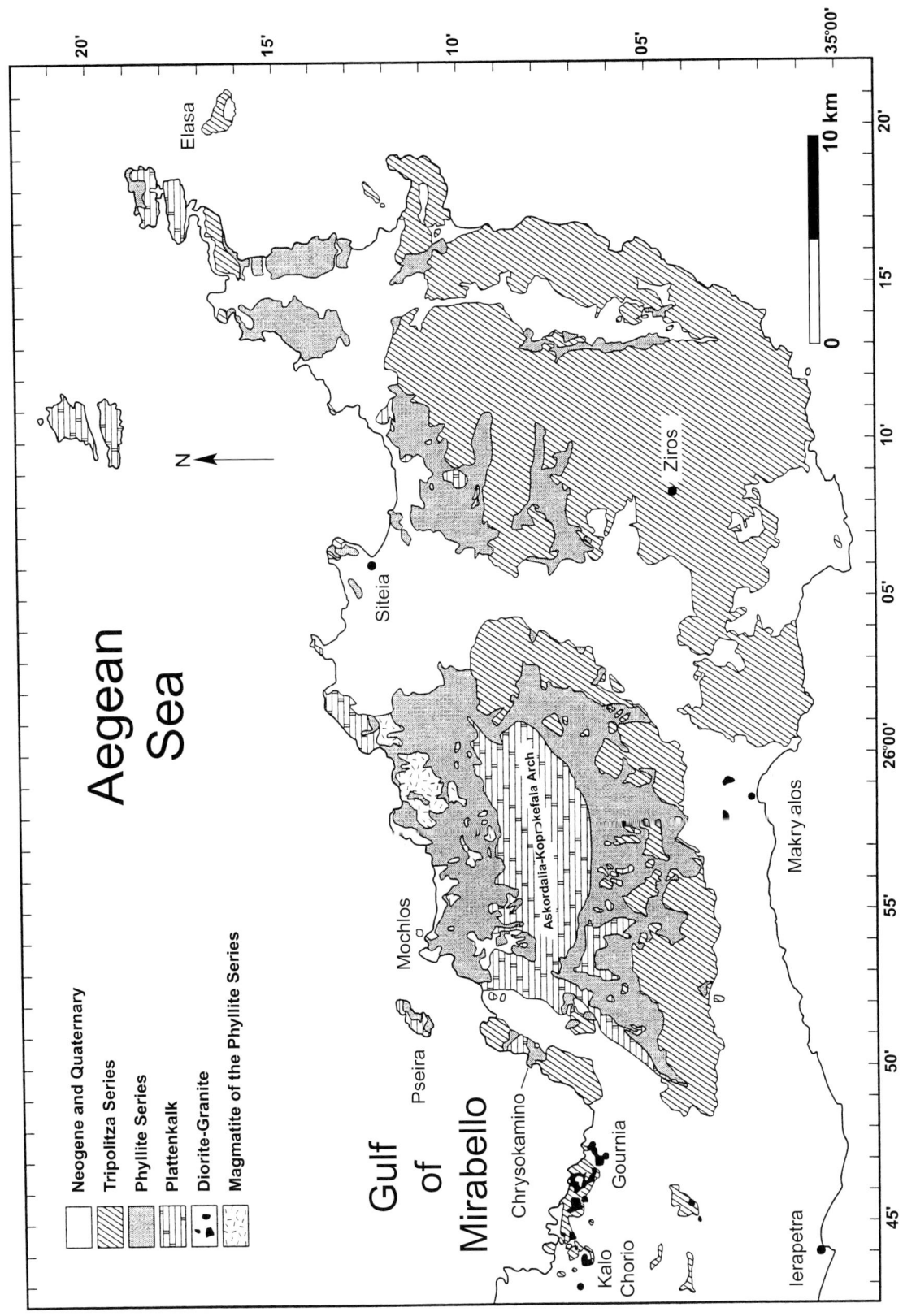

Figure 2.2. Bedrock geology of eastern Crete. After Wachendorf et al. 1974

THE NATURAL ENVIRONMENT

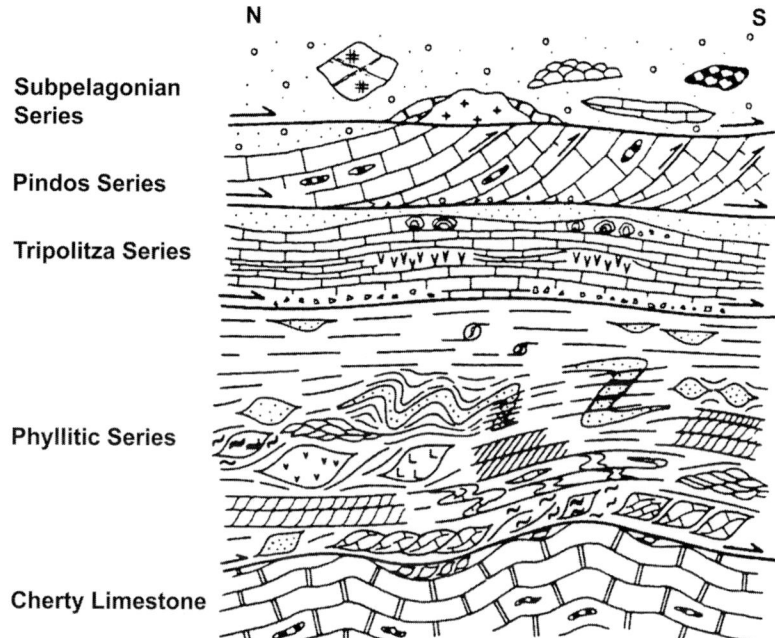

Figure 2.3. Schematic section through the nappe pile of eastern Crete. The cherty limestone (or Plattenkalk) is the basement across which the overlying nappes were thrust. Only the three lowest formations in this figure occur in the Chrysokamino area. The chaotic mélange of the Phyllite series is clearly illustrated, as is the breccia at the base of the Tripolitza (dolomite) series.
After Baumann et al. 1976

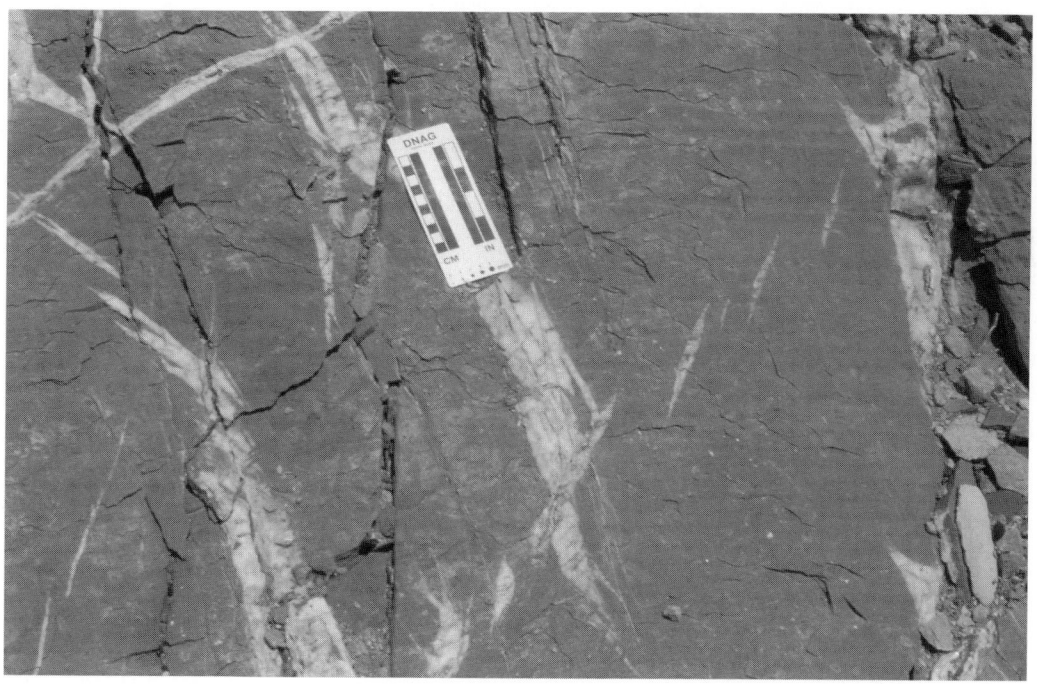

Figure 2.4. Plattenkalk limestone with calcite veins

Figure 2.5. Metacarbonate formation near the metallurgy site

foliation as dominant features, and so we have chosen to call these rocks metacarbonates (Fig. 2.5).

Next above the Phyllite nappe is another carbonate series, the Tripolitza nappe, comprising massively bedded and reefal carbonates and dolomites. Rocks of the Tripolitza nappe are widespread in the far east of Crete, especially south and east of Siteia. Still higher in the pile are the Pindos and Subpelagonian nappes, which are less widely distributed in eastern Crete. Rocks of the Subpelagonian nappe are important, however, at the head of the Gulf of Mirabello near Kalo Chorio, where one finds a chaotic mixture of igneous and metamorphic rocks (lavas, diorite, marble, etc.). Both the igneous activity witnessed by the Subpelagonian rocks and the metamorphism in the Phyllite nappe originated well before those rocks were thrust into their present position, which did not take place until mid-Tertiary times. These rocks have been detached from their deep-seated magmatic (high-temperature) sources, which are to be found well to the north of Crete. Some of them have been uplifted and are now exposed in islands of the central Aegean.

Thus, eastern Crete is characterized by a foundation of Plattenkalk limestone overlain by a succession of nappes that represent displaced terranes of central Aegean origin. During the southward transport of these nappes, the rocks within them were deformed and sheared to various degrees as a function of the physical characteristics of the included rocks.

Such shearing is particularly evident in the Phyllite nappe, owing to the inherent weakness of these thin-bedded, foliated rocks. A conspicuous character of the phyllite series is the occurrence of phacoids: isolated, lenticular blocks of more competent rock types (e.g., limestone) resulting from the fracturing and shearing of original sedimentary beds or lenses.[23] These phacoids range in size from a few centimeters to hundreds of meters, but commonly can be seen as blocks a few meters long. A prominent phacoid occurs about midway up the slope between the metallurgical site and the lower road.

Since the Phyllite nappe is an important element in the Chrysokamino area, as well as on Pseira and in the hills around Mochlos, it is worthwhile to describe its rocks in some detail. Figure 2.6 shows a typical section through this nappe, specifically based on a north–south traverse some 10 km east of Mochlos.[24] Actually, only the lowermost six units described in Figure 2.6 were seen in the Chrysokamino area, from the upper part of the Plattenkalk into the lowermost units of the Phyllite series. These rocks will be described in more detail below. In the middle of the section, relatively thick units of volcanic rocks (especially unit 11, Fig. 2.6) are shown as "Magmatites" in Figure 2.2 and described as intrusive diabase on the Geologic Map of Greece.[25] These volcanic rocks are slightly metamorphosed (to chlorite grade), and they incorporate lenses of phyllite, mica schists, and some large phacoids of marble. Elsewhere in the Phyllite nappe, but not shown in Figure 2.6, are masses of gypsum, notably in the hills southwest of the village of Mochlos where they are being mined extensively.

The geologic ages of the rocks in eastern Crete are not known with great precision because well-preserved fossils are rare. This situation, of course, leads to some differences of opinion in published ages. Some sources suggest that the autochthonous Plattenkalk is probably of Jurassic to Paleogene (Early Tertiary) age. The Phyllite series is of Permian to Triassic age, and the Tripolitza limestones and the Pindos series are both of Jurassic (?) to Eocene age. The igneous complex of the Subpelagonian nappe has been dated radiometrically to 88 to 90 million years (Late Cretaceous).[26] Others have suggested somewhat younger ages, ca. 75 million years (still Cretaceous), for the diorites and granites of the Kalo Chorio area.[27]

Following the mid-Tertiary compressional tectonics that thrust the succession of nappes onto the Plattenkalk basement of eastern Crete, a period of relaxation gave rise to extensional tectonics. This situation led to new structures bounded by high-angle normal faults, oriented roughly northeast–southwest, breaking the island of Crete into a number of compartments. Of particular importance for the Chrysokamino-Pseira area is the Ierapetra graben, a complex downdropped block that includes the Gulf of Mirabello and its extension through the lowland cutting across mainland Crete to Ierapetra on the south shore. Here, multiple northeast–southwest trending faults have lowered the floor of the Gulf of Mirabello in a number of steps or splinters, clearly seen in the present topography as one proceeds westward from the Mochlos vicinity. Pseira and the bedrock ridge leading southwest towards Gournia (Fig. 2.2), which includes Chrysokamino, are horsts, or elevated blocks between downfaulted areas. According to the local boatmen, this faulted topography is expressed on the sea floor on the west side of Pseira, where steep cliffs, presumably part of a fault

23. Durkin and Lister provide a useful description of the origin of phacoids (1983, p. 95, figs. 9 and 10).
24. This is described by Wachendorf et al. 1974.
25. *Geologic Map of Greece.*
26. Baumann et al. 1976; *Geologic Map of Greece.*
27. Seidel et al. 1981.

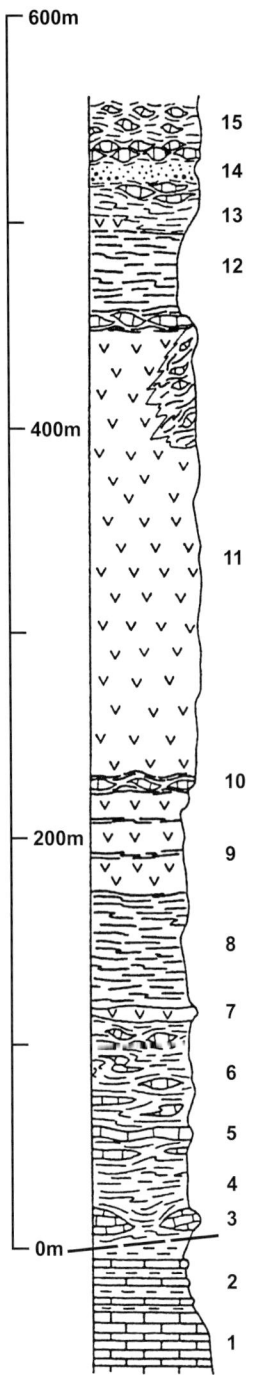

15. Phacoid Series, marble phacoids in a marly matrix; compact carbonate layers at the base, which break up into phacoids upward.

14. Clastics, brownish gray, of variable grain sizes (with volcanic and milky quartz components).

13. Phyllite, olive colored to gray; intercalated with chlorite schist (metavolcanic) at the base and with marble at the top.

12. Chlorite and mica schist with individual cm- to dm- thick marble lenses; at the base marble phacoids in the dm-size range.

11. Basic volcanite, marbled with feldspar, decomposed; toward the east interfoliated with chlorite- and mica schist, as well as large marble phacoids.

10. Marble, laminated.

9. Chlorite schist (metavolcanite), grayish green, with amygdaloid concretions and phyllitic interlayers.

8. Mica schist and phyllite, garnets up to 5 mm diameter; the rock is strongly decomposed.

7. Basic volcanic, dark olive green, decomposed.

6. Marble phacoids, thin-bedded calcphyllite in phyllites.

5. Limestone, dark gray, laminated; thin-bedded at top.

4. Phyllite, red and green, in part with dark gray mica schist; at the top, red phyllite with dm-thick marble beds.

3. Marble, structureless, lens-shaped, resembling phacoids.

2. Marly schist, gray with flaser structures as well as individual Plattenkalk beds.

1. Plattenkalk (slabby limestone), coarsely crystalline, with intercalations of marly red and green limestones.

Figure 2.6. Stratigraphic column through the Phyllite nappe somewhat east of the Chrysokamino-Kavousi area. After Wachendorf et al. 1974

line, plunge downward some 300 m to a submerged bench about 300 m wide, which in turn is terminated by another abrupt dropoff, presumably another fault. East of Chrysokamino, the lowland extending from the village of Kavousi to Tholos Beach is another, smaller graben separating the Chrysokamino horst from the mountains to the east. A small bench in this graben is located immediately east of the Tholos Beach Hotel; it is apparently bedrock (although it is now covered with cemented gravel), and it is probably another tectonic sliver.

Neotectonics

Differential vertical movements related to the tectonic activity described above have continued into Quaternary times up until the present. Former shorelines of late Tertiary and Quaternary date have been mapped at various places on Crete.[28] These studies show that eastern Crete stood relatively high above sea level at the end of the Miocene Epoch, after the thrusting of the various nappes. The area was then submerged at least 450 m early in the Pliocene Epoch, only to emerge again progressively throughout the Quaternary Period, during which time oscillating glacial-interglacial sea-level changes were superimposed on tectonic uplift.

Relative sea-level changes affecting young archaeological sites document tectonic adjustments even within the last 2,000 years. A synthesis of these sea-level changes shows that Crete is not reacting as a single block.[29] Western Crete is rising rather rapidly, at rates of about 4 m per 1,000 years, while central Crete has undergone relatively little change. In eastern Crete, the northeastern corner is being submerged, while the southeast is rising.

The Pseira-Chrysokamino horst is situated within the Ierapetra graben, which has been subsiding at a considerable rate throughout late Tertiary and Quaternary time.[30] The amount of post-Bronze Age subsidence of the area cannot yet be determined. However, Flemming et al. give a value of 0.75 m subsidence in the last 1,900 years for Psyra *[sic]*; the basis for this figure is not given in their article.[31] The same authors list values of 0.25 m and 1.75 m subsidence since 1,900 and 3,600 years ago, respectively, for archaeological features at Mochlos. The evidence for these values presumably comes from the interpretation of the presently submerged isthmus between Mochlos village on the Cretan shore and Mochlos Island, a submerged street on Mochlos Island, and drowned fish tanks *(piscinae)* of presumed Roman age at Mochlos village.[32]

It is possible that the area around Chrysokamino has undergone a greater amount of subsidence than Mochlos, according to the figures cited above.[33] Field evidence of uplifted Quaternary shorelines along the south Cretan shore, just east of Ierapetra, shows a very strong increase in subsidence from east to west across the same eastern boundary fault that passes through the Mochlos area.[34] Therefore, one might suggest that the sea floor off Chrysokamino has subsided perhaps two or three times as much as Mochlos has, or some 3.5 to 5 m, since Minoan times. The effect of these movements on the promontory of the metallurgical site is problematic, however, because we do not know which faults were active at which times. In any case, a difference of some 5 m would not significantly change the altitude of the metallurgical site, which now sits 38 m above the sea.

28. Pirazzoli et al. 1982; Peters, Troelstra, and van Harten 1985.
29. Flemming, Czartoryska, and Hunter 1973.
30. Fortuin (1978) gives details of this structural history.
31. Flemming, Czartoryska, and Hunter 1973, table 1.
32. Soles 1978.
33. Flemming, Czartoryska, and Hunter 1973.
34. Angélier et al. 1976, p. 436, fig. 5.

Figure 2.7. Cliffs north of the metallurgy site, looking toward the island of Pseira

Local Bedrock

The Chrysokamino metallurgical site is on an isolated point of land perched about 38 m above the sea with no immediately adjacent boat landing available. The site is backed by a sharp, but passable ridge rising another 80 m without significant level areas. The point of land where the site sits separates two broad coves. The cove on the north side (Fig. 2.7) is flanked by very steep, limestone cliffs and receives the brunt of heavy waves generated by prevailing winds from the northwest. The south cove (Fig. 2.8) is somewhat more protected and is bordered by less steep slopes on the softer phyllite and related rocks. In times of nonviolent weather, small ships could anchor in the south cove, but offloading would be difficult because of the narrow beach and the steep slope up from the water. However, about 800 m southwest of the site, around a small but prominent rocky peninsula, a major ravine (Fig. 2.9) cuts through the local dolomite bedrock down to a small, but sheltered cove (Agriomandra, Fig. 2.10), ideal for small ships (see Figs. 1.5 and 2.1). The cobble beach at the mouth of this ravine contains many exotic rock types, i.e., rocks not found in the local area (for example, basalt, diorite, and rocks with traces of copper alteration products). It is conceivable that the exotic rocks were dumped here from ship ballast. This ravine provides relatively easy access to flat fields above (near Lakkos Ambeliou), situated south of the farmstead at Katsoprinos (Fig. 1.5). At present, remnants of a well-built road (a Turkish *kalderimi*) are still present through the ravine.

The Plattenkalk limestone forms the structural support of the Chrysokamino headland, as well as forming the impressive, near-vertical cliffs rising some 200 m out of the sea to the northeast of the metallurgical site (Fig. 2.7). It is dark bluish gray, thin-bedded but solid, interlaced with conspicuous white calcite veins (Fig. 2.4), which make it easy to distinguish between the Plattenkalk limestone and the local dolomite. Some of the veins have been dissolved away in the immediate vicinity of the water's edge, leaving open fissures. The Plattenkalk contains a number of small

Figure 2.8. Caves and cliffs south of Chrysokamino

Figure 2.9. Red clay and the entrance to the ravine leading to Agriomandra

Figure 2.10. Agriomandra, a small harbor at the foot of the ravine that connects the coast with the Kambos

Figure 2.11. Soft, powdery phyllite exposed in a trench at the metallurgy workshop

folds produced by deformation during the emplacement of the overriding phyllite nappe. To the southwest of the metallurgical site, the Plattenkalk dips steadily down into the sea at angles of 25 to 35°, but on the north side of the excavated area, the limestone cliff drops precipitously almost 40 m to the water (Fig. 2.7).

Metacarbonate rocks, which sit stratigraphically at the top of the Plattenkalk, are particularly obvious in two areas. Northeast of the metallurgical site, at the base of the rather steep slope that descends from the area of the road gate, the metacarbonate rocks crop out around the mouth of the dry streambed where it meets the sea (Fig. 2.5). The metacarbonate rocks here are strongly foliated so that their bedding planes, although discernible, are less obvious than the planes of foliation. The metacarbonate rocks as well as the adjacent Plattenkalk are clearly folded and offset by small faults in this area. Greenish and reddish metacarbonate rocks can also be seen along the foot of the slope due south of the metallurgical site, just below the phyllite contact and above the upper limit of wave action. However, they were not mapped as a distinct unit in this area because of their gradational relations with the Plattenkalk proper.

The metallurgical site sits on phyllite bedrock, just a few meters away from the contact between the phyllite and the Plattenkalk limestone. The phyllite exposed in the excavation trenches (Fig. 2.11) is mostly the white, very powdery (silty), calcium-carbonate-rich residue of weathering, so typical of most phyllite outcrops in this area (see below). In addition, in several trenches a resistant carbonate ledge, probably a phacoid, was exposed, angling diagonally across the trenches (strike N40–55E, dip 40–50S). In contrast to the orientation of this phacoid, the Plattenkalk just west of the trenches strikes N70W and dips 35S. This discordance in orientation supports the interpretation of the phyllite as an overthrust nappe above the autochthonous Plattenkalk.

In the cove between Chrysokamino and Agriomandra (Figs. 1.5 and 2.1), the complex nature of the phyllite series is abundantly illustrated by the chaotic nature of the included rock types: well-foliated phyllite (Fig. 2.12),

Figure 2.12. Kink fold in phyllite

hard carbonate lenses, strongly sheared limestone masses, and gypsum, along with minor sandstone and pebble conglomerate, all of which may have heavy travertine coatings. Geologically, this is a mélange.[35]

As mentioned above, the phyllite outcrops are commonly very powdery as a result of in situ weathering of the calcite-rich, fine-grained rock. At first glance, the outcrops look very much like loess (windblown silt), but when they are cleaned sufficiently, the foliated structure of the phyllite can be seen. This silty weathering is much more conspicuous here than on Pseira. Victoria Herbert of Rutgers University in Newark conducted X-ray analyses of samples of the phyllite rock and the degraded weathering products. She reported that the rock specimen (phyllite) was composed mainly of chlorite and quartz, with minor K-feldspar and plagioclase feldspar. No 10-Å micalike phase was detected (e.g., biotite or muscovite/illite). The weathered phyllite contained chlorite, quartz, and feldspars, but it also contained a 10-Å micalike phase (probably illite) and calcite. The powdery, very pale soil was chemically similar to the weathered phyllite, confirming its identification as a late stage of the degraded rock. It contained chlorite, illite, quartz, calcite, and minor feldspars

A clay separate (approximately <2 micrometers) of the pale colored soil confirmed that chlorite and illite are the only clay minerals. It is unlikely that the sample contains any kaolinite. In the separate, the illite peak is larger than the chlorite peaks, suggesting it is more abundant. Probably no swelling clay is present.

The southern part of the map area is dominantly dolomite of the Tripolitza nappe (Fig. 2.1). A very clear contact separates the dolomite

35. Wachendorf et al. 1974.

from the adjacent phyllite series. The farmstead building is on the dolomite immediately adjacent to the phyllite contact (in the parking area). The dolomite is predominantly gray to dark gray, but it weathers to light gray, grayish brown, or white. It is massively bedded, with the bedding planes obscured so that it is rarely possible to determine the local dip and strike. Where the dip can be determined, it is commonly rather steep to the south, e.g., as in the large caves across the cove south of the metallurgical site. On the other hand, the dolomite along the contact with the phyllite is highly fractured or brecciated into angular rock fragments. They have been recemented together, indicating brittle fracture during overthrusting. This characteristic can readily be seen along the contact west–northwest from the farmhouse site at Katsoprinos. The dolomite also occurs as a thin cover or as isolated outcrops above the phyllite in the area between the farmhouse and the metallurgical site, especially in the form of a long "finger" in the north-central part of the map (Fig. 2.1). Also, a small, isolated outcrop (or outlier) of dolomite (too small to show on the map) occurs just inside the road gate, and another larger outlier is situated on the east slope of the Chrysokamino hills just above the extensive olive groves in the Kavousi-Tholos Beach valley. Large boulders of dolomite have fallen down to the shoreline as part of a landslide in the middle of the cove south of the metallurgical site.

The dolomite is particularly susceptible to solution weathering, as shown by widespread development of lapies, which are sharp, small-scale ridges and grooves on exposed surfaces. Larger-scale solution features are also present. First, several relatively deep fissures occur in the dolomite a short distance west of the farmhouse site. Four of these were investigated. They are 1 to 2 m wide at the top and up to 4 m deep at least. The bottoms are obscured, being partially filled with colluvium and fallen blocks of dolomite. Although the observed fissures are not strictly parallel, they all have a generally northeast–southwest trend. Undoubtedly, they follow joint patterns in the dolomite. Still larger solution features are the two caverns at sea level on the lower slopes of the dolomite, facing north toward the metallurgical site. These caves are not very extensive horizontally, but they have high arching roofs, which make them very conspicuous when viewed from the metallurgical site. Smaller caves occur on the walls of the sharp ravine leading down to the sheltered cove of Agriomandra, mentioned above. One of these caves has been modified into a small chapel, but the other is in its natural condition. All of these solution features appear to be part of a major karstic system that pervades the dolomite; however, the highly fractured nature of this bedrock probably precludes the development of long cavern systems.

Caves are developed in other bedrock formations as well. In the case of the phyllite, some small caves appear to have been dug by humans or perhaps enlarged by animals (the weathered phyllite is very soft and easy to dig). Several examples can be seen along the phyllite slopes in the area of the large agricultural terraces just northeast of the farmhouse site. The largest of these, discussed further in Appendix K, is large enough to shelter sheep and goats.

The largest cave, Theriospelio (Fig. 1.11), occurs in the Plattenkalk just over 200 m north 50° east of the metallurgical site, directly upslope from the white travertine headland called Chylopittes (see Fig. 1.5), at about 50 masl. It has a small opening, only 1.5 m high at the entrance, but at about 5–6 m inside, the cave opens into a very large room with many stalactites and stalagmites, none of which seemed to be actively dripping. This room continues into a dark cavern that appears to be extensive, but we were not equipped to explore farther. The cave is discussed in Chapter 18 of this volume.

Landslides, Colluvium, and Soils

The bedrock surface in the Chrysokamino area is generally barren of soil cover or unconsolidated deposits, with a few exceptions. Along the shoreline in the middle of the south cove is a small landslide deposit (LS on Fig. 2.1) composed of rubble from the phyllite series and some larger blocks of dolomite fallen from the outcrop some 80 m upslope. Directly upslope from the metallurgical site, on the uphill side of the lower road at about 100 masl, is a fossilized ravine (Fig. 2.13). It is obvious by its reddish brown color, contrasting with the underlying, pale-colored phyllite (Fig. 2.14), and by the fact that it appears to be filling a small ravine cut into the phyllite. Its contents include rock fragments from the Plattenkalk limestone that crops out a bit higher on the slope.

Elsewhere on the phyllite rocks is a thin colluvial cover, usually 20 to 30 cm thick, of reworked phyllite and associated rock types that have been moved downslope by sheet flow and soil creep (Fig. 2.15). This colluvium is easy to detect because it has a very clear contact with the underlying phyllite and is browner than the phyllite. It can be seen at the top of road cuts between the metallurgical and farmhouse sites, as well as along the road on the east side of the hills leading down into the Kavousi valley.

The field of red clay colluvium and soil south of the farmhouse site (at the location called Lakkos Ambeliou) is described below.

Figure 2.13. Paleoravine uphill from the metallurgy site

Figure 2.14. Detail of the paleo-ravine showing the contact between the reddish soil in the ravine and the underlying pale-colored phyllite

Figure 2.15. Colluvium consisting of reworked phyllite and other rocks

Economic Geology

Because of the obvious metallurgical interest of Chrysokamino, a search was made for possible sources of ore, but none was found. In fact, none of the local bedrock types seem obvious environments for mineralization. The veins in the Plattenkalk appear barren of any minerals other than calcite. Extensive coatings (as much as 10 cm thick) and fillings of travertine are associated with outcrops of dolomite, especially in and near the two large caves mentioned above, but no mineralization was seen. Also, a large outcrop of travertine (Fig. 1.8 and TR on Fig. 2.1), most of which is brilliantly white, but with zones that are red, brown, or honey-colored, occurs on the north side of the cape about 300 m north–northeast of the metallurgical site (Fig. 2.1). The deposit is known locally as Chylopittes (Fig. 1.5). It is associated with eroded stalactites and capped by Plattenkalk limestone, all of which plunges below sea level at an angle of 45–50°. This travertine appears to be a filling of an ancient karstic cavity, mostly destroyed by coastal erosion, but no trace of metallic mineralization was obvious.

Other geologic raw materials that are locally available include both gypsum and clay. Two modest outcrops of gypsum were noted (GX on Fig. 2.1), one along the coast about 250–300 m southeast of the metallurgical site and the other high on the ridge about 600 m east–southeast of the site (just outside the road gate, below the stone hut recorded as **AF 15** in App. G), both within the phyllite series.

Abundant red clayey sediment fills an upland basin named Lakkos Ambeliou, which appears to be a large sinkhole, located at the head of the major ravine discussed above, down the slope and almost due south of the farmhouse site. This clay is redeposited terra rossa soil eroded from the surrounding dolomite slopes; it seems to be a reasonable source of clay for ceramics. The deposit is at least 4 m thick in the center and covers some 25,000 to 30,000 m². It comprises two layers of red clay: a younger one, as deep as 1.10 m,[36] with occasional Middle Minoan (MM) sherds on the surface, and an older clay deposit without artifacts. The lower deposit, beginning at 2.1 m below the present surface, is brighter red and harder than the upper one and is characterized by light-colored calcium carbonate veins and nodules and by manganese staining. The upper deposit is at least 4,000 years old, i.e., older than MM, and perhaps much older. Much of the terra rossa in this area is considered to be relict from wetter ("pluvial") conditions that prevailed during times of northern glaciations.

Morris studied the red sediment in detail as a part of his dissertation study of soils related to Minoan sites in eastern Crete.[37] He presents a constellation of laboratory analyses including grain size determination, mineralogical analysis, and chemical analysis. He classifies the soil as a Chromoxerert, meaning that it is a highly colored Vertisol developed under seasonally dry climate. A Vertisol is a clay-rich soil with a strong shrink-swell capacity that causes large cracks to open in the dry season and turbation or mixing of the soil upon sequential wetting and drying. The turbation causes a phenomenon called "slickensides" to appear on the faces of the peds (structural units) within the soil, that is, shiny surfaces with lines or grooves produced as the peds rub against each other during swelling of the soil. This particular soil is unusual in that its clay minerals include only illite and kaolinite; no expandable clays (such as smectite) are present.[38] Morris believes that the strong turbation in this soil is a product of the very strong seasonal contrast between the wet and dry seasons on Crete. From an industrial point of view, the lack of expandable clays would be an advantage in ceramic production because of the reduced likelihood of cracking upon firing.

Morris concludes that "movement of the artifacts through the profile seems to be dependent on the limiting diameter of the artifacts and the morphology of the vertical cracks."[39] He believes that the artifacts below the surface moved there through cracks. A depth of 1.1 m appears to be the maximum depth to which the MM sherds have supposedly moved from the present surface, where they are also found.

36. Morris 2002, p. 57.
37. Morris 2002, chap. 3.
38. However, all clays will shrink and swell to some extent upon wetting and drying.
39. Morris 2002, p. 59.

Summary

The bedrock of the Chrysokamino area consists essentially of three formations. The Plattenkalk limestone crops out immediately to the north and west of the metallurgical site and continues in the high cliffs to the northeast; the Tripolitza dolomite begins at the farmhouse site and continues in the high hills to the south and southwest; and the Phyllite series lies between these two carbonate formations. Whereas limestone and dolomite rocks form uplands of resistant rocks, the phyllite is highly weathered and easily eroded, forming unstable, less steep slopes. The phyllite has weathered to an arable soil and has been exploited for terrace agriculture wherever it crops out.

Although no metallic mineral deposits were found in the area, resources in the form of travertine, gypsum, and clay are present. Carbonate solution features are found in the limestone and the dolomite. A deep cave with Neolithic and later use occurs in the Plattenkalk limestone, and lapies, fissures, and smaller caves are found in the dolomite. The topography of the area suggests that the only safe access from the sea was by means of the small cove of Agriomandra to the southwest of the sites, which connects with a trafficable ravine leading up toward the main habitation site.

Crete is a tectonically active area, but it appears that any geological deformation in the immediate area of Chrysokamino has not changed the local landscape significantly since Bronze Age times. Sea level may have changed by a few meters, but the landscape would hardly have been affected because of the steep slopes along the coast.

NATURAL RESOURCES

Natural resources include rocks, minerals, soils, and other geological features as well as a region's local plants and animals. They can make the difference between a successful economy and a hopeless failure. Rich natural resources may be factors in the original decision to settle an area, and they may significantly affect the lives of those who have them. To achieve the successful maintenance of a settlement, balance must be achieved between population and resources.[40] Someone must take steps to ensure enough food and other supplies to support the population from one harvest to the next, and increases in population can only be accommodated if resources (local or imported) are increased proportionately.

The region near Chrysokamino has relatively few natural resources. Its rocks have little mineral wealth, its soils are poor, and the topography does not support a good road system. Access to the sea is limited, and the nearest stream, the Platys River, does not flow most of the year.[41] Chrysokamino's marginal economic and political character, both in the Bronze Age and in its later history, may be largely attributable to its restricted base of natural resources. Chrysokamino did not grow and become a town because the local landscape could never have supported a large population.

With one notable exception, the natural geological materials available in the territory of Chrysokamino pose few problems in their interpretation.

40. Binford 1968b, p. 327; Smith 1972, p. 7; Wilkinson 1981.
41. Faure 1984, p. 47.

The region has soils of two types and several kinds of rock formations, all of which have been carefully mapped. The other natural resources are also easily discussed. The problem concerns the possible presence of copper ore. With a smelting workshop operating for centuries, the possibility of a local ore source is an important issue.

Soils

Several terraced areas occur in the vicinity of Chrysokamino (Fig. 2.16), showing that the land was once farmed. The basis of this farming, the soil, may be divided into two classes. A red soil, called terra rossa, occurs over the local limestone and dolomite bedrock, and a pale brown phyllitic soil occurs over the phyllite bedrock. Both soils have been thoroughly studied.[42] They have different characteristics.

The terra rossa contains substantial carbonate, kaolinite, and quartz. This type of soil forms over carbonate bedrock. Most researchers feel it consists primarily of the insoluble residue remaining after the dissolution of the parent rock, with the red color resulting from late oxidation of iron minerals.[43] The distribution of terra rossa in the Chrysokamino region corresponds to the presence of carbonate bedrock (Fig. 2.1). An especially large deposit is found downhill from the habitation site, at the low area named Lakkos Ambeliou, and the soil also occurs on the carbonate hillsides overlooking this location. The red soil is considered suitable for raising grapes, and the name of the locality means "vineyard hole," a reference to the excavation of the terra rossa soil for use in local vineyards and gardens. The sediment from Lakkos Ambeliou is particularly important for the Minoan metallurgy workshop because it seems to have been the material used for the furnace chimneys (see App. A).

Eleni Nodarou studied this soil (App. L), and Michael Morris also analyzed it.[44] The analyses show that the terra rossa consists of sediment containing a mixture of several different minerals, including enough clay (kaolinite) to make it plastic enough for pottery.[45] The sediment is red because of the presence of Fe_2O_3, with Munsell colors of 2.5YR 4/6 dry and 2.5YR 3/6 moist. A substantial amount of silica in the analyses comes from quartz as well as kaolinite and other silicates, and the sediment also includes other impurities including calcite and manganese oxides. These studies confirm that the soil is a useful material for making industrial pottery or coarse storage jars or cooking vessels. It is not really suitable, however, for fine ceramics that require a smooth surface.

The pale-colored soil that forms from the decomposition of the local phyllite has very different characteristics. Both Nodarou and Morris also analyzed this soil.[46] It is more calcareous than the red soil, and it is richer in some other elements, such as sodium. In addition, it retains more

42. Timpson 1992; Morris 2002; see also Nodarou, App. L.

43. Nevros and Zvorykin 1936; Danin, Gerson, and Garty 1983.

44. Morris used a Thermo Jarrel Ash Model Plasma Atomic Emission Spectrometer [ICAP-AES] at the University of Tennessee, Knoxville; see Morris 2002, pp. 113, 119–120, Kavousi 3 pedon.

45. In the 20th century, the soil at Lakkos Ambeliou was used as a source of clay for brick making.

46. Nodarou, App. L in this work; Morris 2002, pp. 112 and 119, Kavousi 2 pedon.

Figure 2.16. Terrace system AF 22, located near the habitation site. The small cave discussed in Appendix K (AF 9) is at the upper left, and terrace AF 22b is in front of the cave

moisture than the terra rossa, so that it is more suitable for dry agriculture.[47] It also consolidates more firmly and does not erode as easily. Local conditions have stripped many of the carbonate hills of much of their red soil cover, while the phyllite slopes are still covered with deep deposits. The pale-colored soil is useful for agriculture, but not for the manufacture of pottery.

Rock and Mineral Resources

The geology of the area is described above. Most of the land near the habitation site at Katsoprinos has a covering of soil, and the outcrops of bedrock are fairly small. Rocks include phyllite, dolomite, limestone, and a few small deposits of other rocks. The limestone and the dolomite are good building materials. Where they are available, they are the main stones used for nearby terrace walls. The Chrysokamino habitation site is built over the largest exposure of dolomite on the hill, and it uses dolomite blocks for its walls. The carbonate rocks are also a good source of lime for plaster. This material might have been needed locally in the Minoan period, although no evidence for its ancient use at Chrysokamino has been found. The local sources were definitely used for lime in later periods, and a limekiln from the first half of the 20th century stands near Lakkos Ambeliou. The stone is also a good material for ground stone tools, especially for oval hand tools used in pounding and grinding.

Small amounts of gray to black chert occur as inclusions in the dolomite and limestone. As in some other parts of the southern Aegean, however, the quality is not as good as the chert found in mainland Greece.[48] As a result, many East Cretan chipped tools were made of obsidian imported from Melos. In post-Minoan periods, the residents of northeastern Crete used imported cherts and flints for threshing sledge blades and other purposes.

The Minoans also used many other lithic materials. Although eastern Crete has few deposits of metals, it has a wide variety of rock types. The

47. Morris 2002, p. 76.
48. Jameson, Runnels, and van Andel 1994, pp. 302–303.

Minoans collected many of them for various purposes, including tools, ornaments, stone vessels, religious objects, and other items.[49] The Chrysokamino sites are modest in size and wealth, and many periods are poorly preserved, but the residents of the area certainly used east Cretan resources as materials for querns, other stone tools, stone vases, and other objects.

Small deposits of several specific rocks and minerals occur in the Chrysokamino territory and nearby parts of Crete. Some of them have Minoan and later uses, while others do not. A deposit of coal lies north of Chrysokamino's territory,[50] but the Minoans did not find any use for it. Although central Cretan gypsum was used in Minoan buildings,[51] the local gypsum is too porous for architectural blocks (App. G, **AF 15**). A few other local deposits consist of stones used elsewhere in Crete, but whether the specific deposits near Chrysokamino were ever used is not known. In this latter category are the many small lenses of banded white and brown travertine including the one at Chylopittes (App. G, **AF 34**). They are of the type used for Early Minoan stone vases,[52] but no examples of this material come from the Chrysokamino excavations. Iron oxides and carbonates occur on the hill near the metallurgy site, and they could have been used as fluxes in smelting, but little of the original deposits survives today, and the extent of the original deposits is not known.

On balance, the territory has little mineral wealth. The local phyllites and carbonates could be used as building materials, but the other deposits of useful stones are very small. They were suitable only for domestic tools and a few other items.

Copper Ore

The source of copper ore used at Chrysokamino is particularly important in view of the early date of the FN–EM III metallurgy workshop. Although copper deposits exist in Crete,[53] the modern geological survey of the Chrysokamino territory has recorded no trace of any copper ore. It has been suggested that the rocks of the region probably never had any copper because they have little nonferrous mineralization. Furthermore, none of the modern residents of Kavousi who were questioned know of the presence of any copper in the vicinity.

On the other hand, archaeologists who visited the region shortly after 1900 published very positive statements:

> On a headland about three miles east of this town are fragments of an ancient furnace, called figuratively by the peasants the "Golden Furnace"; and some specimens of rock from an adjacent cliff have shown on analysis a low percentage of copper, not sufficient to induce modern enterprise, but evidently not neglected by the prehistoric inhabitants.[54]

49. The most complete study of the lithic materials used in eastern Crete in Minoan times was undertaken for Pseira; see Betancourt 1994–1996.

50. Faure 1966, p. 48.
51. Chlouveraki 1998.
52. Compare Seager 1912.
53. For lists of copper sources and bibliography on this subject, see Branigan 1968, pp. 51–53; 1974, pp. 59–66; McGeehan-Liritzis 1996, pp. 386–387.

54. Hawes and Boyd Hawes 1909, p. 38.

Several writers have repeated the suggestion that copper ore was found near Chrysokamino.[55] Others, however, have rejected these claims. Branigan, who spent considerable time tracking down references to copper deposits, did not list the site among Cretan copper locations.[56]

In view of the erosion and the other geographical changes since the Bronze Age, including a considerable change in sea level, the question of a local copper source remains open, but the available evidence indicates that a nearby deposit is highly unlikely. Ore deposits that were exploited in the Bronze Age leave more than traces of ore. Ancient miners routinely crushed the ore-bearing rock to remove the high-grade ore bits, leaving behind waste rock in large amounts, as well as broken and discarded stone tools.[57] None of this evidence has been recorded from the Chrysokamino hillside. Instead of an ore deposit, it is possible that the early archaeologists found some stray pieces of ore discarded by the smelting workshop. In the unlikely event that an ore deposit did once exist nearby, it must have been very small. It did not fill the local needs, and foreign ore was certainly brought in by ship, as shown by the presence of copper ore on the small beach at Agriomandra and by the fact that the lead isotope pattern of inclusions in the local slag matches Lavrion and Kythnos (App. C). Smelting at some distance from the ore is, in fact, common worldwide.[58] Furnace operations must adhere to schedules based on factors including the availability of labor and weather conditions that do not affect mining. There is, moreover, no compelling reason for their schedules to coincide with mining activities because mining output is based on other variables, such as intermittent discoveries of rich ores.

Forests and Other Plants

Several authors have discussed the wild trees of ancient Greece in general and those of Crete in particular.[59] They agree that timber played an important role in the ancient world, though they differ in details of interpretation, especially in the extent of ancient Greek forest.[60] The role of trees is nicely expressed by Rackham when he says that trees "are not part of the scenery of the theater of human history; they are actors in the play."[61]

In regard to the eastern part of the Gulf of Mirabello, the ancient vegetation can only be partly reconstructed. Several factors are relevant to Bronze Age conditions. The climate, discussed in the first section of this chapter, is one of the most important of these factors. The Mediterranean climate sets the limits for the type of vegetation and the species that can grow and flourish under local conditions.

The modern vegetation of the strip of coastal land near Chrysokamino also provides helpful information. It is mainly phrygana, which is a low vegetation often found in Crete where the land has been overgrazed and

55. Fimmen 1921, pp. 17 and 120; Glotz 1923, p. 39; Lamb 1929, p. 5; Hutchinson 1962, p. 40; Faure 1966, p. 48.

56. Branigan 1970a, p. 80, fig. 17; 1974, p. 60, fig. 1; 1988, p. 80, fig. 17.

57. D. Gale 1991; Gale and Ottaway 1991.

58. Rothenberg 1972, pp. 208–210; Lechtman 1991; Hauptmann, Bachmann, and Maddin 1994, p. 4; O'Brien 1996.

59. See esp. Meiggs 1982; Perlin 1989; Rackham and Moody 1996; Rackham 2001.

60. See in particular the discussion by Rackham (2001), who suggests the amount of ancient forest has been overestimated.

61. Rackham 2001, p. 5.

taller vegetation has been destroyed by fire. Very few trees grow on the coastal strip where the archaeological remains are located. The trees that are present are low shrubs of the type called maquis, species that would grow taller under favorable conditions but remain stunted if browsed by goats or other animals.[62] Thyme and other undershrubs also grow here. Abandoned agricultural terraces, however, show that the region was once farmland. Excavations in selected terraces found only post-Minoan remains, suggesting that many of the terraces were not built until after the Bronze Age (App. K), even though the survey indicates that this land was used already during the Minoan period (App. G).

How much forest was present? Pollen analysis and other forms of evidence suggest that, like some other parts of Greece, portions of Crete probably had substantial forests in the wet period following the retreat of the last glacier.[63] The date at which these hypothetical forests disappeared from eastern Crete is not clear, however. Certainly, Crete was never completely forested, because most of the endemic Cretan plants (species that do not occur anywhere else) are not forest plants but species that grow on more open land.[64] At Chrysokamino, there was surely no forest in the Late Bronze Age because few wild animal bones are present in the archaeological record, and large numbers of sheep/goat bones indicate extensive grazing. No forest could grow on extremely eroded hills like Chalepa. The erosion that removed soil from this hill and deposited large amounts of terra rossa at Lakkos Ambeliou, which was earlier than Middle Minoan based on the presence of sherds on the surface of the accumulation of soil (App. G), indicates that the date of the erosion was earlier than the 2nd millennium B.C. This erosion must have taken place after the natural ground cover had disappeared.

It is likely that human activities during the Bronze Age curtailed or eliminated any local forest cover still in place before the Final Neolithic. Fields needed to be cleared for agriculture, and grazing would also have inhibited natural plant growth. Grain impressions in the furnace chimneys show that agriculture was already prevalent at the end of the Early Bronze Age (Chap. 12). In addition, activities like making quicklime for plaster[65] or the copper smelting operation itself could have required considerable fuel, and the cutting of trees for charcoal may have been a factor in destroying or reducing any remaining forests.[66] By the Middle Bronze Age, the forest was surely restricted to areas not used for farming.

Even without the presence of extensive local forests, however, wild plants would have been an important resource. Timber from the hills rising inland from the coast could have been exploited. Both maquis and larger trees would have grown in the gullies and ravines in all eras, including periods of extensive agriculture. These plants would have been an important source of charcoal and wood for fuel. Herbs, edible greens, and other useful wild plants would always have been available.

Forests and other uncultivated areas would have provided other useful resources, in addition to the plants themselves. They would have sheltered wild animals, including hares and birds that could be hunted. The bone of a hare found in the faunal remains at the metallurgy site (Chap. 11) can only have come from hunting. Flowers were always available, and they would have supported beekeeping (discussed further in App. I).

62. Rackham 2001, p. 8.
63. Zohary and Orsan 1965; Turner and Grieg 1975; Turner 1978.
64. Rackham 2001, pp. 7–8.
65. Betancourt 2004b.
66. For the concept of deforestation as a result of metallurgy, see Wertime 1983; for an opposing view, see Rackham 2001, pp. 33–34.

Fish and Other Marine Resources

The proximity of the Aegean Sea must also be regarded as a resource. Although Chrysokamino is not a coastal site in the traditional sense, it was close enough to the sea to make some use of its potential. The harbor is within easy walking distance, and trails lead down to the shore near both the settlement and the metallurgy location. Research on fishing at nearby Pseira suggests that the Pseiran exploitation of the sea as a food source consisted only of coastal fishing.[67] The resources available near the coast—both fish and other marine life such as sea urchins and mollusks—would have been available to the residents at Chrysokamino as well, and the archaeological record confirms the exploitation of the sea. The marine environment, a rocky coast with shallow water, can support abundant sea life, and the shells and small fish from here would have formed a useful adjunct to protein from other sources. Salt, an additional resource from the sea, would also have been available.

67. Rose 1996.

PART II: THE METALLURGY WORKSHOP

CHAPTER 3

THE EXCAVATION OF THE METALLURGY WORKSHOP

by Philip P. Betancourt, James D. Muhly, Eleni A. Armpis, Robert S. Powell, Elizabeth B. Shank, Evi Sikla, and Tanya Yangaki

After several informal visits to Chrysokamino beginning in 1985, a formal visit was made to the site in 1994 accompanied by representatives from the village of Kavousi. Plans were made to begin an excavation. Dimitris Kophinakis, the mayor at that time, extended the full support of the local government. This included free access to the site (which is public land), the offer of the Kavousi schoolhouse for use, without charge, as a storage and study facility, and the promise of new roadwork to improve access. Other logistics were arranged, and a research plan was made for the project to begin the following summer.

The goals for the project included a better understanding of the nature and date of the metallurgy location and of its relationship to the local territory. From the beginning, the excavation of the workshop was seen as a part of a regional study that would elucidate its broader context. The nearby habitation site, ca. 600 m away on a dolomite outcrop on the southwest slope of the hill of Chomatas, was excavated as an integral part of the Chrysokamino Project.[1]

Even before excavation began, the extent of the metallurgy location was easily visible because its surface was covered with chimney fragments and slag in such a heavy concentration that almost no soil was visible (Fig. 3.1). Only a small amount of grass and a few other plants grew on this bare and infertile spot. The slag heap was concentrated in a single location, with a sharp line at its edge where the slag stopped and the surrounding soil began. The location was on a promontory overlooking the Aegean, with a good view of the sea (Fig. 3.2).

The position of the metallurgy site within the local landscape and its relationship to the closest habitation site can be seen in the topographic drawing shown in Figure 1.2. The 600 meters of landscape between the two locations is easy to walk, with no major natural barriers between the two places. The slag heap, found on a cliff at the bottom of a hill, sits on a coastal headland with an elevation of ca. 38 m. It lies on a small saddle between outcrops of limestone on the west and phyllite on the east, with the land sloping down both to the north and the south.

The surface bedrock, discussed in Chapter 2, plays a crucial role in the formation of the headland. The western end of the promontory, directly

1. The third archaeological location in the territory, Theriospelio, the burial cave on the slope of the hill north of the metallurgy site, was not investigated in this project. The three locations were all first inhabited in the Final Neolithic period, and they were all in use during the Early Bronze Age and the opening phase of the Middle Bronze Age. The habitation site continued to be used in the Late Bronze Age.

Figure 3.1. The surface of the metallurgy site before excavation

Figure 3.2. The metallurgy site during excavation

above the sea, consists of Plattenkalk limestone, which is hard and durable. It is part of the African tectonic plate.[2] The adjacent bedrock is a much softer phyllite, part of the European tectonic plate that has been thrust over the African one. Because the limestone bedrock at the western tip of the headland is substantially harder and more resistant than the adjacent phyllite, and because the interface between the two formations is especially subject to weathering, the limestone remains as an outcrop that rises higher than the adjacent rock. The result is a small natural saddle that slopes more toward the north than the south (Fig. 3.2). Here, on this natural trough that terminates in cliffs at its two ends, the early metallurgists established their workshop.

The modern road stops well inland and uphill from the metallurgy site (Fig. 1.4). The promontory must be approached by descending a steep slope (for the contour lines, see Figure 2.1). The headland can also be

2. Papastamatiou et al. 1959.

Figure 3.3. Sieving at the metallurgy workshop

approached by boat on the western side when the sea is calm; a steep path leads up from a small landing place to the promontory, ending near the location where the dry sieve is shown in Figure 3.3. The steep hill between the metallurgy site and the modern road above has no agricultural terraces, and the absence of sherds suggests it has had little recent use aside from occasional grazing.

The surface survey began in 1995 with mapping and geological studies, and it continued for two additional years. Excavations at the settlement site and the metallurgy workshop were conducted in 1996 and 1997. This volume describes the results of the survey and the work at the metallurgy site.[3]

The 1995 season lasted only two weeks. It included geological study, instrument survey, and mapping. This season was intended to be a preparation for the excavation scheduled for the next two seasons. No artifacts were collected. Survey points were installed near the metallurgical location, and a topographic grid was established to aid in the mapping of the area. Under the supervision of Lada Onyshkevych, this work was accomplished with a Topcon GTS 303 Electronic Total Station attached to a Gateway 2000 Handbook (486) laptop computer.

Using the grid, stakes were inserted at 2 m intervals on the rectangular border of the area to be excavated. In the two seasons when excavation was conducted (1996 and 1997), string was attached to the stakes to create a series of 2 × 2 m grid-squares. A letter-plus-number system was used to facilitate recording. The site was excavated by grid-square (Fig. 3.4), and the slag pile was sampled in a series of squares that created a long trench north and south, with a more complete coverage in the southwest quadrant (Fig. 3.5). A large part of the slag pile was left unexcavated because something should always be preserved for future archaeological investigation.

Because the boundaries of the site could be easily traced before the excavation began (since the ground was completely covered with slag and small pieces of industrial ceramics), measuring the slag heap's outer border was a simple matter. The deposit extended from north to south for a distance of 34 m from location E5015 N4989 to E5015 N5023, and east to

3. The excavations at the habitation site will be published separately.

Figure 3.4. James Muhly cleaning the upper part of the apsidal building, visible at the bottom of the level with grass roots

west for a distance averaging between 10 and 15 m from location E5007 N4998 to E5019 N4998. The deposit had clearly eroded off the cliffs into the sea at both the north and south sides. The overall area of slag and clay fragments consisted of about 375 to 380 m². No architecture was visible before excavation, and (aside from the fragments of industrial ceramics) only a few sherds were present. None of the sherds was easily identifiable as to date. Many of the ceramic pieces had a white calcareous coating adhering to them, especially on their inner surfaces. Almost all of the slag and clay fragments were under 5 cm in size.

Excavation was conducted in 1996 and 1997 under the supervision of James D. Muhly and Philip P. Betancourt. Trench supervisors included Eleni A. Armpis, Robert S. Powell, Elizabeth B. Shank, Evi Sikla, and Tanya Yangaki. The supervisors worked with teams of three workmen. Manolis Hairetakis was foreman. Objects collected from the modern surface were numbered with their grid-square and the letter S. Excavated units were numbered sequentially beginning with the first pass.[4] Unfortunately, this site did not yield a single artifact with any evidence that it was in its original place of use. All pottery consisted of fragments, all of which had been scattered in antiquity.

Where no stratigraphy was present, excavation was conducted in 10 cm deep units. Natural stratigraphy was followed where it existed. All soil except for water-sieving samples and other soil samples was dry-sieved, and material remaining on the screens was washed with fresh water after sieving to remove dust and facilitate visibility. No soil was dumped without sieving and subsequent examination, both dry and wet. In 1996, all pottery, chimney fragments, faunal remains, rounded stones, unusual stones, and slag fragments above the size of 2 cm were saved; similar collecting strategies were employed in 1997 except that slag was not saved. Samples for water-sieving and other purposes were collected as directed by the supervisors, with samples for water-sieving collected from each level. Water-sieving was conducted onsite (Fig. 3.3), using fresh water brought in by truck to a reservoir on the modern road uphill from the excavation. The water was

4. The catalogues of artifacts in this volume include the find contexts, so that the reader can ascertain what came from the same grid-square and stratum.

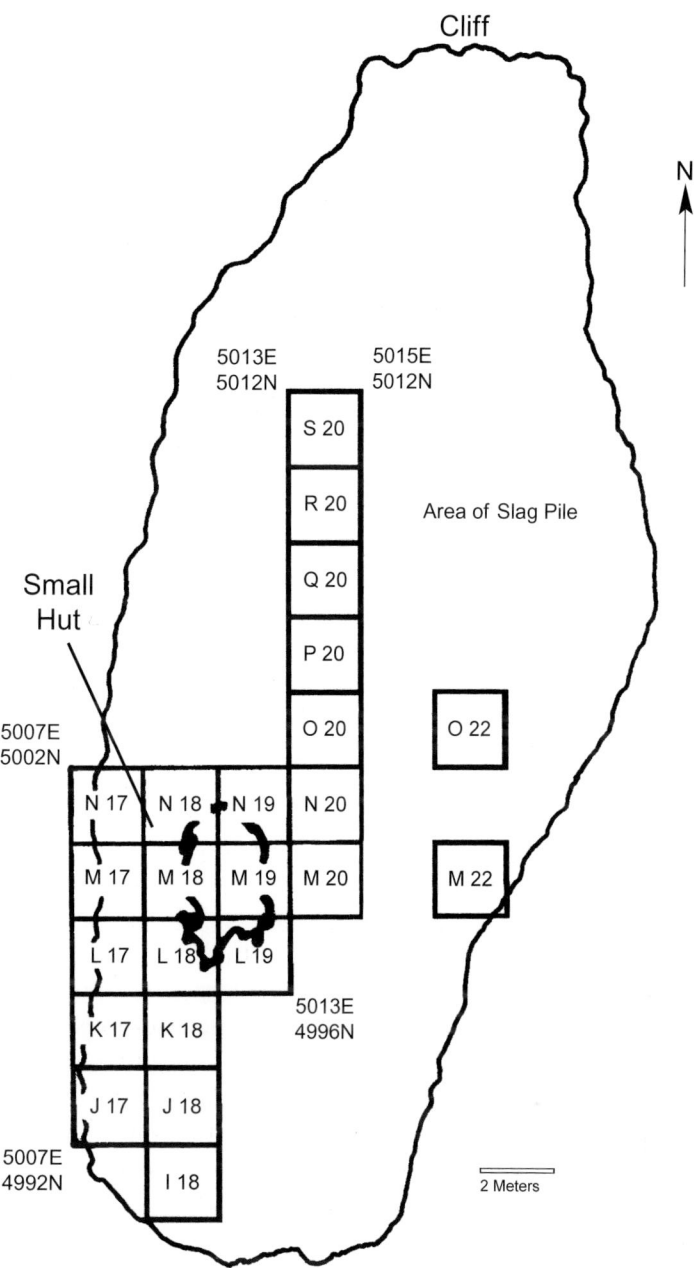

Figure 3.5. Plan of the grid-squares excavated at the metallurgy site.
E. Shank

5. Betancourt 1995.

channelled to the site by irrigation pipe using gravity. Additional details of recovery and recording methodology are documented in a handbook on file in the INSTAP Study Center for East Crete library.[5]

With the exception of the grid squares in which the apsidal building was discovered (discussed in Chapter 4), the same excavation method was repeated in every grid square, with similar results overall. After removal and bagging of any stone tools or sherds found on the surface, the first pass was excavated. It invariably yielded small slag fragments, pieces of furnace chimneys with black vitreous deposits on their interior surfaces, angular pieces of phyllite, and a mixture of soil, ground-up slag, and tiny chimney fragments. The first pass also contained roots from grasses and other plants. Infrequently, the workmen would find a marine shell, a piece of a clay vessel, or some other artifact. Except for the roots, the lower passes were similar. Some passes had no pottery except for the chimney pieces. No stratigraphy was visible anywhere in the slag pile, except for the squares in which the apsidal structure was located.

The methodology was rigorous enough to allow conclusions concerning what materials were not present. As previously noted, all excavated material was dry-sieved. The coarse material remaining on the screen was washed with water in order to remove dust and reveal any green, blue, or red copper ore, but no ore was ever found this way. In spite of two seasons of extensive water-sieving using a water separation machine operated by skilled paleobotanists, not a single seed or fragment of charcoal was recovered. Nor was any piece of partly consumed fuel found in either season of excavation.

During the 1996 season of excavations, the study, conservation, accessioning, and cataloguing of finds were undertaken in the Kavousi schoolhouse. In 1997, work was transferred to the newly constructed INSTAP Study Center for East Crete in Pacheia Ammos where the artifacts were also stored.

Much work was done at these study facilities during the course of the excavation seasons. Trench supervisors studied their own pottery under the supervision of the directors, writing a brief report the day following excavation, after the pottery was washed and dried. They also selected initial objects for accessioning at that time. Accessioned objects were drawn, photographed, and catalogued. Subsequently, additional objects were accessioned. Accessioning and cataloguing strategy was more complete than at some sites. All waterworn stones and all pottery sherds were catalogued except for a few samples retained for possible future destructive analysis. Samples of the fragments of the furnace chimneys were accessioned. All other artifacts were accessioned. The registrar, Mary Betancourt, supervised the cataloguing operation.

Further analytical work on finds from the metallurgy location was done at the following laboratories and institutions:

1. Isotrace Laboratory (Research Laboratory for Archaeology and the History of Art), Oxford University, England
2. Mineralogical Laboratory, Geology Department, Temple University, Philadelphia, Pennsylvania
3. Archaeological Laboratory, Department of Art History, Temple University, Philadelphia, Pennsylvania
4. Bartol Research Institute, Department of Physics, University of Delaware, Newark, Delaware
5. Museum Applied Science Center for Archaeology (MASCA), University of Pennsylvania Museum of Archaeology and Anthropology, Philadelphia, Pennsylvania
6. Materials Characterization Facility, Department of Materials Science and Engineering, University of Pennsylvania, Philadelphia, Pennsylvania

7. INSTAP Study Center for East Crete, Pacheia Ammos, Crete
8. Exhibit Museum, University of Michigan, Ann Arbor, Michigan
9. Department of Archaeology and Prehistory, University of Sheffield, England
10. Center for Materials Research in Archaeology and Ethnology and the Center for Geochemical Analysis, Massachusetts Institute of Technology, Boston, Massachusetts
11. Laboratory for Archaeometry, NCSR "Democritos," Hagia Paraskevi, Attica, Greece
12. Martin Chemical Laboratory, Department of Chemistry, Davidson College, Davidson, North Carolina
13. Amber Research Laboratory, Department of Chemistry, Vassar College, Poughkeepsie, New York

CHAPTER 4

The Apsidal Structure

by Philip P. Betancourt

The project excavated a series of 2 × 2 meter square trenches arranged from north to south on the long axis of the ancient slag pile in order to determine whether any features were preserved within or below the deposit of smelting debris (Fig. 3.5). Only a little vegetation was present on the modern surface of the slag, but a few bushes grew in grid squares L 18, M 18, and N 18. Most of the excavated grid squares showed no change in stratigraphy anywhere between the surface and bedrock. In most places, the deposit consisted of small pieces of black, glassy slag mixed with fragments of industrial pottery containing holes pierced before firing, occasional angular pieces of the same phyllite found in nearby outcrops, and a few other artifacts. The only area that differed lay in the southwest quarter of the deposit where the bushes were growing. It later became apparent that plant roots had taken advantage of ancient lenses of soil within the slag.

A small building was first recognized when an area of pale-colored soil was discovered in Trenches L 18, L 19, N 18, and N 19 (Figs. 3.4–4.4). The pale soil, which contrasted with the darker color of the slag, began to be visible just below the surface, at the level of the roots of the few plants growing in these grid squares. It was first visible at 37.90/37.96 masl. When the area was fully revealed by excavation, it became apparent that the soil consisted of three superimposed lenses, each one with an apsidal shape with its long axis oriented north to south and the rounded end at the north (Figs. 4.1, 4.2).

The lenses of soil were floors within a small structure whose walls were constructed of wood and other perishable materials. The building had three strata, visible as successive apsidal floors. The highest floor was covered with a small amount of slag, and additional slag was found beneath the lowest floor. All of the floors were stratigraphically distinct, and the structure could be traced within six grid squares. A few circular features were present at the margins of the lenses in the lower two strata, and they were most clearly visible at the south and at the east of the lowest floor. They consisted of cylindrical cavities filled with dark, pulverized slag, which had migrated down into the cavities from the sides and from above. The slag-filled cavities, apparently postholes, defined the outer margin of the apsidal area of soil. Eight postholes could be securely identified

56　　　　　　　　　　　　　　　　　　　　CHAPTER 4

Figure 4.1. Aerial view of the apsidal building; squares M 18 and L 18 show the middle floor; other squares in the photograph show the lower floor. Kite photograph by Jan Driessen

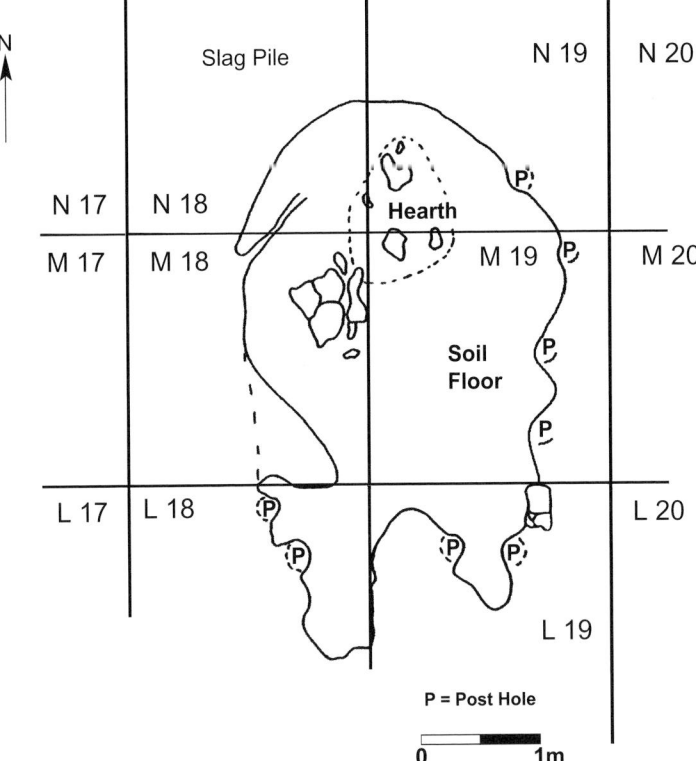

Figure 4.2. Plan of the lower floor

Figure 4.3. The apsidal building

Figure 4.4. Level 4 in grid-square N 19, looking north, showing the lower floor at the eastern side of the apsidal building

(Fig. 4.5), but more were clearly present originally. This structure was the only built feature discovered at the metallurgy site. After the lowest of the three strata was exposed, the building was reburied to preserve it for future generations.

The three superimposed layers of soil show that the floor was renewed twice after the small structure was built (Fig. 4.6). At the north, the outer edges of the superimposed lenses occurred at the same place, indicating that after the wall at the north edge of the soil was built for the first phase of the structure, its position was not moved. At the south, however, the posts were moved for the middle floor. This floor was larger than the original one. The uppermost floor, found just below the modern surface, was very disturbed. It could not be measured at the edges. The soil in all three floors was the same: a pale colored disintegrated phyllite typical of this part of Chomatas Hill. The pale soil contrasted significantly with the darker color of the deposit of slag below and beside it.

Little is known of the upper floor because it was just below the modern surface, and its soil was very disturbed. It appears to have been about the same size as the lower two floors. Its upper surface was irregular and poorly

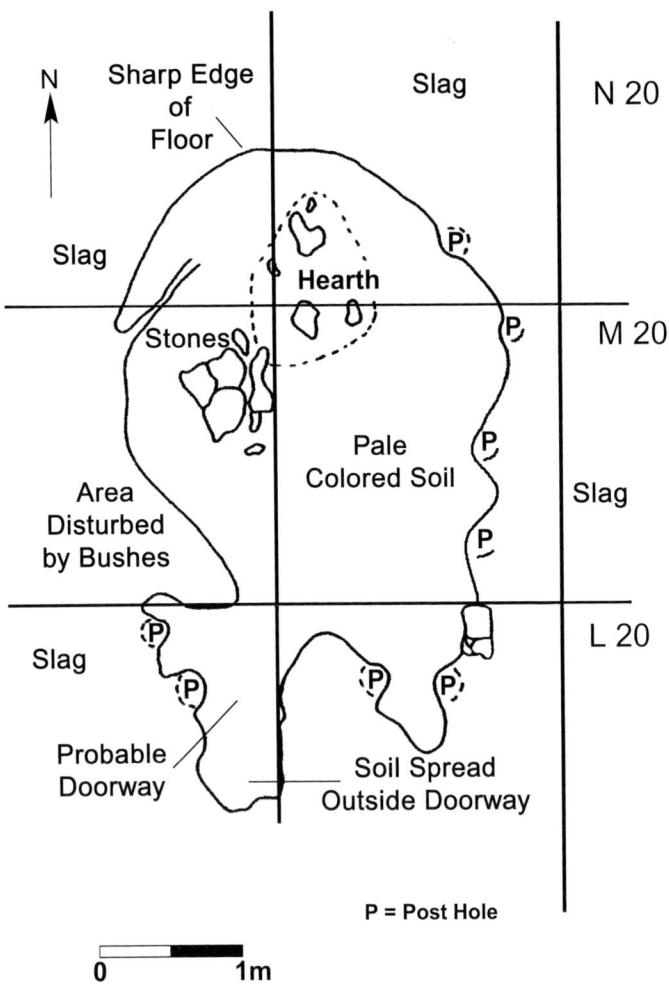

Figure 4.5. Apsidal building, lower floor, showing the location of the doorway

preserved, but it was visible at ca. 37.90 to 37.96 masl (because of the soil color), presumably close to its original surface.

The middle floor (Fig. 4.6), formed of the same pale-colored phyllitic soil as the other two floors, lay at an elevation of ca. 37.91/38.05 masl. At the east, the edge of the middle floor (visible in Fig. 4.1) extended slightly beyond the edge of the lower floor. A lens of burned soil was found at the north, covering an irregular oval area ca. 50 × 65 cm in size. Two large stones resting on the lowest floor projected above this one as well. The soil used to form the floor had many sherds mixed within it, and a few fragments of pottery were found on the surface of the floor (marked on the drawing, Fig. 4.6). None formed complete vessels, and in some cases fragments of the same vessel were both on top of the floor and within its soil, indicating that the pottery was not used in this building. Possibly it was used instead near the place where the soil originated. The southern part of the building was enlarged relative to the previous period of use, and the posts were seated in slightly different locations.

The lowest floor (Figs. 4.4, 4.5, 4.7), at ca. 37.75 to 37.94 masl, sat directly on top of the deposit of pulverized slag and other smelting debris. Between 37 and 45 cm of slag lay below this lowest floor, and bedrock was found

THE APSIDAL STRUCTURE

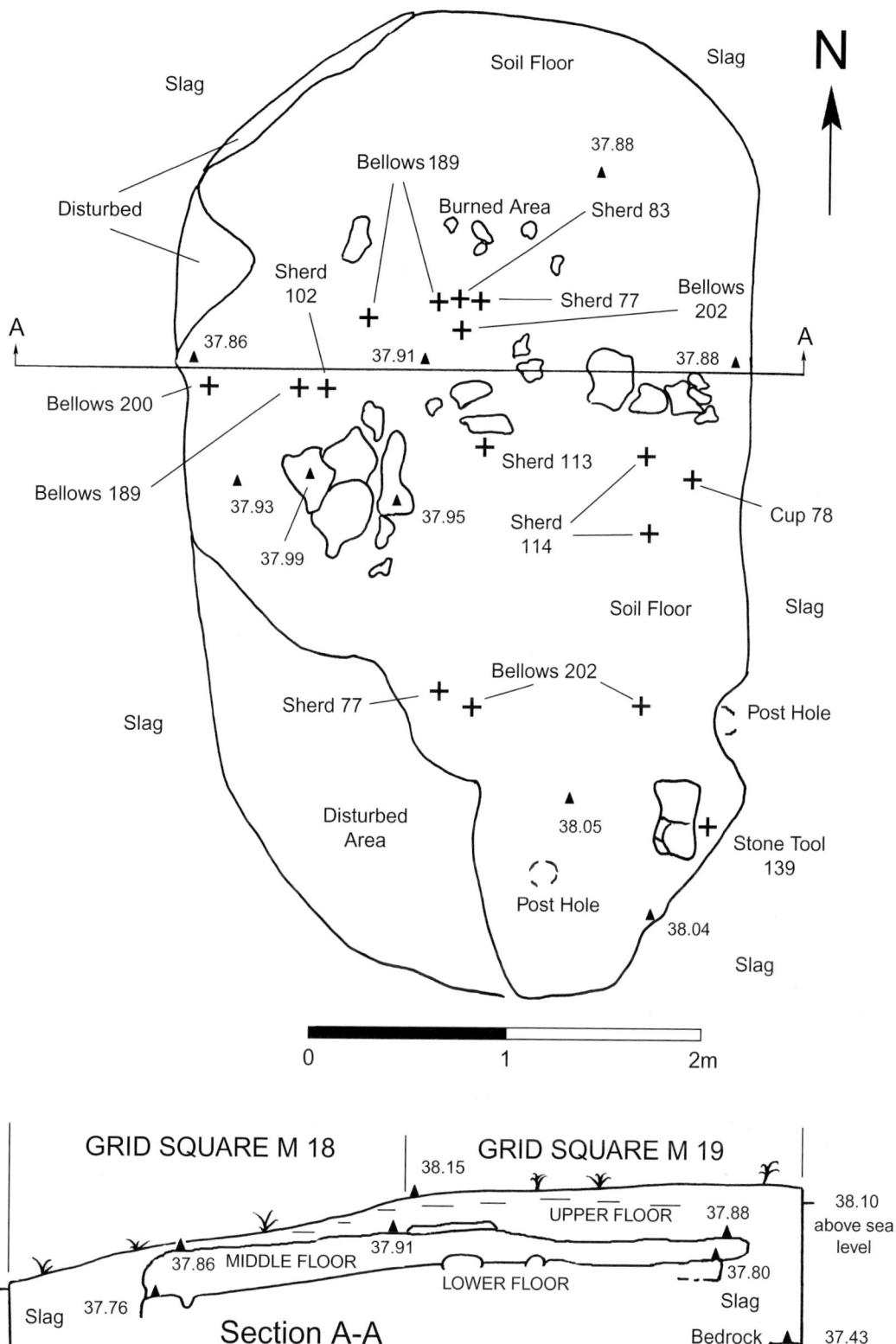

Figure 4.6. Plan of the apsidal building's middle floor and cross section of the building

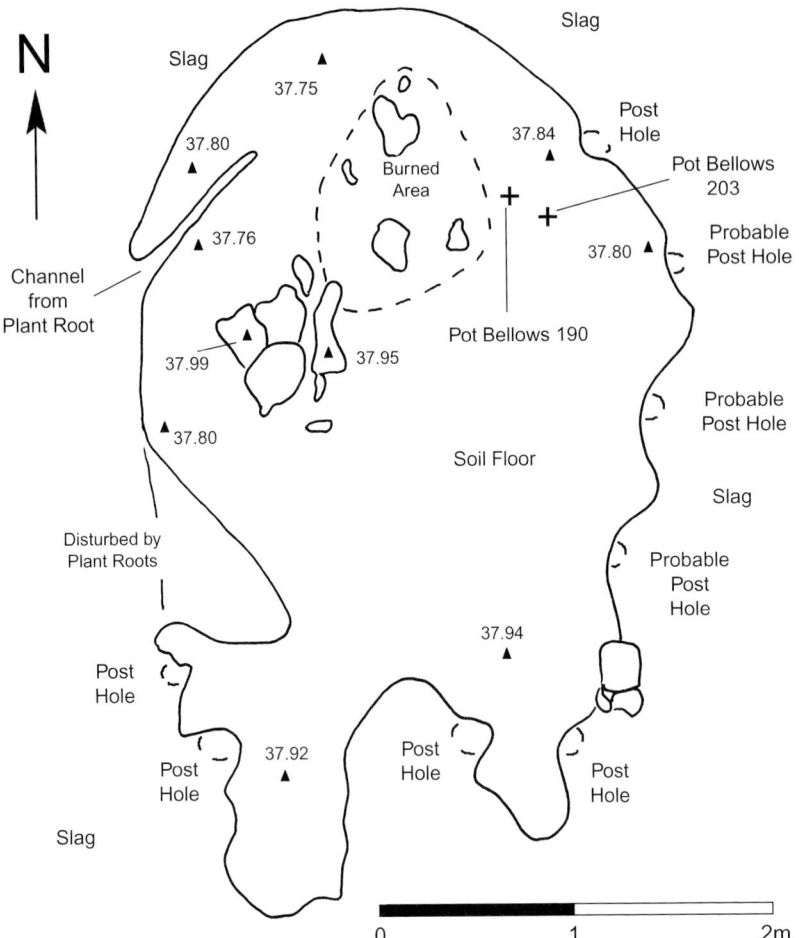

Figure 4.7. Plan of the apsidal building's lower floor

below the slag. The floor was packed and firm, with only two sherds lying upon it (fragments of pot bellows **190** and **203**). A burned area was found at the north. It consisted of about 50 × 65 cm of red soil. Its northern side was approximately 15 cm from the north edge of the lens of soil. Two large stones (neither with any signs of working or use on the upper surface) and several smaller ones rested on this floor (Fig. 4.4, at left of photograph).

The shape of the lowest lens of soil can be best seen in the plan shown in Figure 4.7. The boundaries of the floor could be easily traced by the configuration of the postholes and the edges of the soil floor. Postholes were clearly visible at the south, east, and north. They were less distinct at the west, either because they were never present or because of disturbance from modern plant roots. The soil here (at the west) was extremely disturbed from the roots of bushes still growing at the time of excavation, and the exact outer (western) margin of the lens of soil could not be traced. The south side consisted of a row of three postholes, with a wider distance between the two western ones (forming a doorway). The eastern and northern sides of the building had a scalloped appearance at the edge of the floor, and only a few postholes could be seen. The wall must have been made of some perishable and perhaps flexible material that was attached to the posts. Postholes were about 15–20 cm in diameter.

THE APSIDAL STRUCTURE 61

OBJECTS ASSOCIATED WITH THE BUILDING

A few fragments of vessels and bellows were found both on and within the two upper floors. The lowest floor was almost completely clean on its surface, and it was not removed by excavation. The middle floor, which was ca. 5–10 cm above the lowest one, had the most sherds on it and within it (Fig. 4.6). Most objects were found within the soil, not on top of it. No complete vases were on any of the floors; everything consisted of casual sherds.

The objects are discussed in the other chapters of this volume. They indicate a date in the EM III–MM IA period of eastern Crete, when East Cretan White-on-Dark Ware was in use (see Chap. 5). The three phases of the structure are successive, but they are separated by too little time for any change in pottery style to occur. It must be emphasized that no evidence was found for the actual use of the bellows, jars, shallow bowls, and other ceramic objects in this building itself. The artifacts must be regarded as part of a secondary deposit. Most sherds came from within the soil, not on top of the floors, indicating that the objects must have been used elsewhere, presumably near the location where the soil was excavated to form these successive floors.

THE ARCHITECTURE

Although apsidal buildings are uncommon in Crete, a few structures with curved ends are known from the island.[1] They are more common in eastern Crete than elsewhere. None of the Minoan structures with curved walls provide good parallels for the example from Chrysokamino because all previously discovered apsidal buildings from Minoan Crete are built of stone or mudbrick. The apsidal structure at Chrysokamino is, at present, unique on the island. Apsidal plans are much more common in Anatolia and northern Greece.[2] Although most northern Aegean examples are built on stone foundations, a few buildings of wood are also known.[3]

The use of perishable materials for buildings is also uncommon (or seldom recognized) in Minoan Crete, but the evidence from Chrysokamino is very clear. The presence of posts is well attested, but no traces of clay, plaster, or other weatherproofing materials were present. Postholes were especially clear at the south and east of the lower floor where the remains were well preserved. Although no trace of any wood remained, the outline of the posts could be recognized because the soil from the floor had spread around the wood when the site was in use. The phyllitic soil was not very mobile, so it did not spread into the cavity as the loose slag did when the wooden posts disappeared (perhaps from being pulled up when the site was abandoned). A sharp line of pale-colored soil at the edge of the floor where it had been laid up against a wall or some other barrier could be traced easily, especially at the north. This sharp edge showed that something divided interior from exterior at the edge of the floor; i.e., the wall was built before the soil was spread out in the interior. The soil was only spread outside the structure at the south, where it had been scuffed outside

1. Giannouli 1995.
2. Hood 1986, pp. 41–42; Werner 1993.
3. See Hood 1986, p. 41; Renfrew 1986a, p. 191.

at the doorway. In addition, the position of the sherds and other artifacts indicated a sharp break between interior and exterior everywhere except at the south, helping to confirm that this was not just a set of posts at the edge of a lens of soil but an actual barrier preventing anything inside from casually being kicked out of the structure. Although the interior was strewn with casual debris (Figs. 4.6, 4.7), not a single sherd was immediately north of the edge of the soil.

No evidence survives for a roof. Since the walls were made of perishable materials, it is likely that a roof would have been perishable as well. However, one could also reconstruct the walls as a simple barrier with no roof at all, or with a partial covering of branches providing some shade but no real shelter. The hearth area at the north suggests that probably nothing covered the place where the fire was built.

The architectural tradition that uses posts driven into the ground is called pile-based construction. Several conclusions regarding this practice at Chrysokamino may be inferred from the surviving evidence. Most likely the builders began by driving posts into the deposit of slag to form a U-shaped outline with the entrance at the south or a J-shaped outline with both the south and west open. Some material was then placed either against or beside the posts, creating a solid barrier forming the wall. Finally, soil was placed inside the space to make the floor, and it was spread out evenly to create a relatively flat surface extending as far as the posts and the wall attached to them.

Several aspects of the small hut suggest northern connections: the megaron plan, the pile-based construction, the absence of stone foundations, and the use of wood without mudbricks. Pile-based architecture, in fact, has been present in northern Greece since the Neolithic.[4] Its technology is still known as a folk tradition in the 21st century, and architectural studies of modern pile-based building practices in Thessaly and Thrace demonstrate the techniques used for this class of wooden structures.[5] Several aspects of the modern pile-based buildings help to explain the archaeological evidence at Chrysokamino.

Pile-based buildings use vertical posts with their bases set firmly into the ground to form the core of the walls. The sturdy vertical timbers form the skeleton of the building. Flexible limbs of smaller diameter are often attached horizontally to the posts, with the limbs placed alternately inside and outside the vertical timbers to create the walls. The alternation in the placement of the flexible horizontal branches makes a scalloped wall that is easily visible at the floor. This construction may be the explanation for the scalloped appearance at the outer edge of the soil floor at Chrysokamino (Figs. 4.5–4.7). A combination of rigid vertical piles set into the ground and flexible materials for the walls is a combination that has been regarded as particularly resistant to the lateral forces of fierce winds.[6] It is well suited to the windy conditions on the hill with the workshop.

The framework provides the underlying structure, but a covering is necessary to make the building complete. The walls of modern Thessalian pile-based buildings are regularly smeared with clay or mud to make a daub structure.[7] No traces of daub survive at Chrysokamino, suggesting that if the walls were covered, the builders used leather, fabric, or some other perishable material to make the shelter more windproof.

4. See text, plans, and bibliography (p. 88) in Hiller and Nikolov 1988.
5. Efstratiou 1997; Eres 2003.
6. Rapoport 1969, pp. 112–113.
7. Efstratiou 1997.

THE PURPOSE OF THE STRUCTURE

The location of the apsidal feature on top of the slag pile may seem unusual, but the explanation for its placement is easy to understand because if such a structure were required for the workshop, it would have been more difficult to construct elsewhere in the vicinity. The area immediately around the pile of slag is very compact, consisting of very hard limestone bedrock on the side nearest the sea (the west) and phyllite bedrock with very little soil cover on the inland side. Bedrock is near the surface, with little soil cover. Perhaps driving posts into the slag would have been regarded as easier than finding a place where the soil cover was sufficient to drive the piles deeply enough into the ground to seat them properly.

All the objects in the structure were fragmentary. They included pieces of clay cooking dishes, bowls, cups, and jars in small to large sizes as well as pot bellows fragments. Many of the sherds suggested the possibility of food preparation and consumption. The pot bellows fragments were the only evidence for metallurgical activity aside from a few slag and furnace chimney fragments like those found outside the structure. No crucible fragments, mold fragments, copper prills, copper waste, or other objects suggesting the casting process were present.

Activities inside the small structure cannot have involved more than three or four persons because of the lack of space. The presence of a burned area on all three floors shows that heating played a role in these activities and that the fire continued to be used in the same place through all three stages of the structure's history. A number of possible interpretations are suggested.

OPTION 1. THE HUT WAS USED FOR SMELTING

Watrous has published a suggestion that the lens of soil was used as a place for smelting and that the pot bellows were placed inside what he calls the "so-called postholes."[8] Unfortunately, this idea is not supported by the evidence. The postholes have dimensions of 15 to 20 cm, while the bellows have dimensions of 30 to 50 cm. The bellows were hot enough to have soil fused to them, but neither fused soil nor any other evidence for intense heat was found at the edge of the floor. All the evidence from scientific analysis indicates that the smelting at Chrysokamino generated products that could only have been created inside a bowl furnace excavated into the ground (Apps. C and F), and no such bowl furnace is present in the lens of soil. And finally, since most of the pieces of bellows were found within the floor packing, they were brought to this location along with the soil, so they were used and broken elsewhere.

OPTION 2. THE LOCATION WAS USED FOR SOME OTHER METALLURGICAL ACTIVITY

The hearth area is only a slightly burned area, visible by its reddish color. It is not the base of a bowl furnace, but one might still consider whether it was a location on which something with a metallurgical purpose was heated. Smelting was only one step in ancient metallurgy. The smelting of copper under Bronze Age conditions resulted in very impure metal. After

---

8. Watrous 2001, p. 216.

the metalworkers completed the smelting, they had to purify the copper by remelting it in crucibles before they could pour the molten metal into molds to make useful objects.

According to this theory, the apsidal area would not have been the floor of a building, but a lens of soil used to work on, and the posts driven into the slag would have supported something other than the walls of a building (perhaps a heat barrier?). Although one cannot completely rule out the possibility that metallurgy was practiced inside a closed structure, several points argue against the hypothesis of activity inside an actual building at this location. If the lens of soil were inside a hut, space would have been a major problem with any metallurgical activity, and the high heat would have been a danger to a hut made of perishable materials. If metallurgy were the activity, then the posts would not have supported a roof because of the smoke and sparks that needed venting. In addition, a strong wind plays a role in all metallurgical operations where the heat must be intense, so choosing a very windy site and then placing the location for the activity out of the wind is somewhat illogical, and it would prevent the best results.

If the structure were used for metallurgy, one might envision the posts supporting some type of bracing or framework construction, either to protect the metallurgists or perhaps in connection with the use of the bellows. With this hypothesis, the scallops in the edges of the soil might be regarded as the traces left by the edges of the pots for the pot bellows, which would be used in large numbers to force air into a portable, freestanding furnace or hearth or brazier that would be placed over the burned area. A very serious problem with this reconstruction, however, is that the pots used as bellows seem to have been set into the ground surface and seated with mud, which fused to them during use, and no traces of this fused mud survive on the floors where they would have been with this theory. In addition, if the furnaces or crucibles were placed on the soil surface, pieces of them would have been high up in the slag pile, and they were not present.

In addition, no evidence survives at Chrysokamino for a casting or remelting operation. Not a single crucible fragment comes from the site, and the slag is all smelting slag, rather than the residue from remelting for the manufacture of implements (one obvious difference between smelting slag and slag from remelting is the higher iron content of the former). Little physical evidence supports this possibility.

OPTION 3. THE HUT WAS A DWELLING

The metallurgists had to live somewhere, but the character of this structure indicates that all household tasks could not have been practiced inside it. The hearth takes up so much room that very few people could be inside at once, even sitting down. Insufficient space is present for storage or for sleeping. It is likely that the structure was not very weatherproof, and it would have been cold and uncomfortable in winter. If the hut were a dwelling, one would have to assume that many domestic activities took place outdoors.

In addition, the range of artifacts from the workshop as a whole is not complete enough to represent a habitation site. No ground stone querns are present. The site has neither loomweights nor spindle whorls nor stone vases. Pottery is specialized (mainly dishes, cups, and jars, with no cooking pots, almost no luxury items, and few pouring vessels). Among the White-on-Dark Ware sherds, for example, the corpus from the workshop includes no fragments of the decorated jugs and teapots that are usually found at settlements from this period.[9] Even food remains (animal bones, shells, and other marine remains) are not numerous enough to suggest either permanent or seasonal settlement at the location.

Option 4. The Hut Was a Kitchen

A location for cooking is another possible explanation for the presence of the small hut at the metallurgy workshop. The windbreak need not have been completely weatherproof because it would have been designed to prevent a small cooking fire from burning too rapidly on this windswept hill. It need not have been roofed. Many of the finds (domestic pottery [including cooking dishes], obsidian, animal bones, marine shells, stone tools of small size, and cups and other serving vessels) suggest that at least some food was prepared and consumed in the immediate vicinity. If the lens of soil were the floor of a small building, it may have been a kitchen where a noon meal was cooked, even if the metalworkers returned to their homes at night.

Unanswered questions, however, also exist with the hypothesis that the structure was used as a kitchen. The nature of the hearth area (a flat space discolored by fire) indicates that fires were built over it, not on it; if a fire were made here, the remains were either carefully cleaned away after use or the fire was inside a portable hearth. The only cooking vessel sherds from the site are from a shape called a cooking dish, a large, shallow vessel whose thin walls indicate it was used in place over a fire and not moved. Such a vessel would have been supported by stones or by the edges of a portable hearth, preventing it from tipping over. Fires would have been built around it. However, this configuration was not preserved here, although it may once have been present. The main Minoan cooking vessel of EM III–MM IA is the tripod cooking pot, a vessel that is common at the Chrysokamino habitation site half a kilometer away, but does not occur here. The absence of broken cooking pots, of course, only indicates that no tripod vessel was broken here, offering no evidence as to whether or not the shape was used nearby.

A roof over a cooking area is not desirable because of the smoke that would rise from the hearth. With the building used as a kitchen, one would expect that the posts supported only a screen that was partly roofed to provide a little shade. The structure would have been a temporary shelter rather than a full building.

Option 5. The Hut Was an Apothecary

Analyses of the contents of clay vases from the workshop reveal some interesting results (App. M). Several of the vessels were sufficiently well preserved to yield evidence of their contents. The residues in some of the vases yield evidence for spices and herbs, including saffron, camphor, and

9. Betancourt 1984.

rue. Although some of these herbs may have been used for flavoring food, they would also have had medicinal uses. The absence of normal cooking residues such as plant oils or animal fats in the vessels containing these herbs raises the possibility that some of the mixtures had a specialized use. Some of the herbs have been traditionally employed as analgesics for skin problems such as burns. One should also note that arsenic poisoning causes skin lesions that would have needed treatment. The configuration of the hearth area, suggesting use of a portable hearth rather than a cooking fire, does not contradict this hypothesis. However, because the sherds were found mostly within the floors and not on them, we cannot conclude that anything found in the structure was necessarily used there.

Conclusions on the Hut's Use

In considering all the evidence, a metallurgical process other than smelting, simple cooking, or the preparation of medicines are the most likely activities carried on at the hut. Perhaps the most plausible interpretation for the apsidal structure is that it was a small kitchen associated with the workshop. The metallurgists working in this isolated location may have wanted a cooked meal at noon or perhaps occasionally in the evening, even if they returned to their homes at night, because they could not interrupt the smelting process once it had begun. A windbreak to prevent the cooking fire from burning too quickly would have been absolutely necessary due to the windy conditions on this bare hillside, and a small structure (probably only partly roofed) would have provided for the needs of the cook. The possibility that herbal medicines were also prepared in such a structure cannot be dismissed, particularly given the specialized nature of the pottery from the site and the absence of sherds from tripod cooking pots, which are common in the EM III–MM IA corpus at the nearby habitation site.

CHAPTER 5

THE POTTERY

by Philip P. Betancourt

Relatively few sherds come from the metallurgy workshop. Their dates range from Final Neolithic (FN) to EM III–MM IA, with the largest number of pieces coming from the final period. Relatively few shapes are represented, and the pottery is clearly a specialized assemblage. Even the pottery from the final phase lacks several crucial shapes, including the tripod cooking pot. No complete vessels are present. The sherds come from scattered locations within the slag, from the soil used for the floor of the apsidal structure, and from the ancient ground surface that existed when the apsidal structure was standing.

THE POTTERY PERIODS

In eastern Crete, EM III–MM IA cannot be divided stylistically into two periods. Most of the sherds from before this phase come from the slag pile. Except for one sherd, the FN pieces are all from the slag pile, and some of them come from deep within the deposit (though not at its base). All of the EM I–II fragments are also from the slag deposit. This pattern is extremely important for the dating of the workshop, because it clearly shows that the early pottery was not derived from an earlier settlement existing on the promontory before the workshop was founded. Nor was it brought in with the soil for the floor of the apsidal building, because it occurs within the 35–45 cm deep slag deposit found below the level of the small hut. The FN and EM I–II sherds are definitely a part of the slag heap that existed before the EM III–MM IA phase of the workshop, and the foundation of the site must predate the EM III apsidal structure.

Very little pottery is present from any of the periods before the metallurgists decided to build the apsidal structure. Only 14 sherds, representing 10–14 vessels, predate EM III–MM IA. This assemblage is a limited corpus, but the fact that several different shapes are represented, including a pyxis (**12**) and at least one probable storage vessel (**14**), indicates that the small number of sherds is not necessarily a measure of the level of activity.

Several contrasts can be noted between the pre-EM III pottery and the ceramics from the final phase of the site. First of all, the number of sherds is much larger in the later period. Secondly, the assemblage includes several new classes of pottery, including cooking vessels, more open bowls, and more storage vessels. The EM III–MM IA pottery represents a substantial change in the activities carried on at the site. The workers now spent enough time to build a small apsidal structure that was used long enough for the floor to be renewed twice after its initial construction. The new ceramic classes (especially the cooking vessels and the storage shapes) demonstrate that the preparation of food, cooking, storage, and probably increased food consumption were taking place somewhere in the vicinity. This most likely represents an expansion of the metallurgical activity. The presence of pot bellows sherds in the corpus, as discussed in Chapter 8, supports the same conclusion.

The pottery from the site can best be studied by dividing it into two contexts, the slag pile and the apsidal structure. When the pottery from each of these contexts is examined individually, one can see some clear differences both in date and in character between the two assemblages. These differences are very important for the reconstruction of the history of the small workshop.

THE POTTERY FROM THE SLAG PILE

The bedrock surface below the deposit of slag was reached in all of the trenches excavated within the slag pile except for those that were directly under the apsidal building. In all trenches, the deposit of slag and furnace chimney fragments rested directly on sterile soil or bedrock. No traces of any earlier settlement were encountered anywhere within the trenches that were excavated. The underlying sterile stratum consisted mostly of very pale, disintegrated, highly calcareous phyllite. The soil, which ranged in character from hard and compact to soft and unconsolidated, was similar in character to the local soil found adjacent to the slag pile.

The character of the slag pile did not vary significantly between the surface and its lower limit, except within the trenches that revealed the apsidal structure. Throughout the area investigated, it consisted of small pieces of slag, bits of clay furnace chimneys, small amounts of soil, and a few stones. Only a few sherds of vessels were discovered, and some of the excavated levels (ca. 10 cm deep) yielded no fragments of vessels at all.

The present publication presents all of the sherds noted by the excavators except for a few scraps kept separate for potential destructive analysis in future generations. The context of each catalogued sherd is given along with the other descriptive information, and the lack of clear stratification is evident. Final Neolithic sherds come from the surface as well as from deep within the slag pile. No Final Neolithic sherds come from the sterile surface below the slag, although a concentration of several early sherds in the middle levels in Trenches N 20 and O 20, below the level of the earliest floor in the apsidal structure, indicates a generally early level in this central part of the slag pile (Fig. 5.1). The scattered and mixed nature of the pottery, along with the lack of stratigraphy within the deposit, shows its disturbed nature. Perhaps the metallurgists disturbed the deposit by digging

THE POTTERY

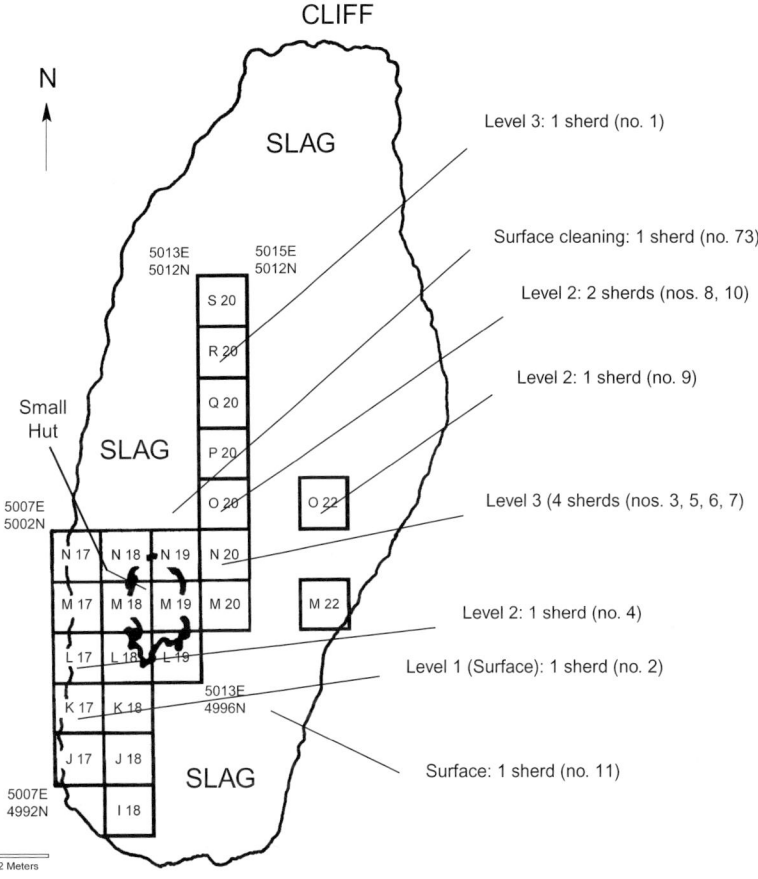

Figure 5.1. Location of Neolithic dark-burnished sherds within the slag pile

into it or moving the material (presumably when the area was leveled for the construction of the small building and when its posts were seated in the ground).

The pottery deposits from the slag pile and the small apsidal structure are organized in separate tables (Tables 5.1–5.5 and 5.6–5.8; see Catalogue below) in order to show their character, following a system first used at Kommos.[1] The tables provide a clear picture of the deposits by recording each of the sherds by ceramic class, vase shape, and position on the vessel, with an additional column at the right of each table giving the minimum number of vases represented by the surviving fragments. If a group of sherds is restorable as a vessel on paper, it is recorded as a single piece in the column called "whole" (no complete vases were found). The list with each table provides references to the catalogue in this chapter where the sherds are described more fully.

THE FINAL NEOLITHIC POTTERY

The FN pottery at the workshop is all made from a single fabric (see App. N). It is coarse, with a brown to dark brown color and a heavily burnished surface. Inclusions, which are abundant, include quartz in several forms, phyllite containing both white and dark mica, and sandstone. These inclusions are not distinctive enough to suggest a place of manufacture,

1. *Kommos* II; *Kommos* III.

but they do not preclude a local source somewhere in eastern Crete. The fabric is substantially different from the main fabric used for the EM III–MM IA ceramics at the site (which is tempered with inclusions in the granodiorite to diorite series), and it can be easily distinguished.

The Final Neolithic vessels have thick walls with a uniform thickness. They are sturdy and well made using hand-forming techniques. Only a few shapes are present: bowls, jars, and a pouring vessel with a globular body and a horn on the one preserved handle. No cups are present, suggesting that local drinking vessels in this period may have been made of gourds or wood.

Coarse, dark-surfaced, burnished pottery is the earliest ceramics from this part of Crete. It occurs at most of the sites in the region, and published finds come from Vasiliki,[2] Mochlos,[3] Sphoungaras,[4] Pseira,[5] Kalo Chorio,[6] Hagios Antonios,[7] the Vrokastro region,[8] and from some of the unexcavated sites recorded in the Kavousi-Thriphti survey.[9] The pottery also comes from elsewhere in eastern Crete.[10] The chronology of this pottery still needs more work. It definitely begins in the Final Neolithic (if not earlier), and at some sites, such as Pseira, the same fabrics persist into EM I, although the harder, well-burnished and pattern-burnished EM I fabrics occur at Mochlos and elsewhere.[11]

Neolithic shapes in the region that includes Chrysokamino consist of jars, bowls, and pouring vessels without spouts. Handles are often wide straps or small, pierced lugs. Decoration is extremely rare. Horned vessels like the jar with the horned handle from Chrysokamino (6) are not known in this part of Crete from any of the published EM I sites, suggesting that the earliest pottery from the metallurgy location is earlier than the beginning of the Bronze Age.

The EM I–IIA Pottery

EM I fabrics in the eastern Gulf of Mirabello region were sometimes thinner and harder than in the Final Neolithic period,[12] New shapes were manufactured during the period, including pouring vessels with spouts, chalices on conical bases, and a series of pyxides of globular to ovoid shape. New decorative techniques, such as pattern burnishing, made their appearance as well.[13]

Pottery was traded over long distances in the Early Minoan I period. Imports into eastern Crete included Hagios Onouphrios I Ware vases from south-central Crete,[14] Lebena Ware from somewhere in the same region,[15] Pyrgos Ware from north-central Crete,[16] and several Cycladic

---

2. Seager 1904–1905, p. 212.
3. Seager 1912, p. 93; Soles and Davaras 1996, pp. 179–180.
4. Hall 1912, pp. 46–48.
5. Betancourt and Davaras 1990, pp. 29–30, tomb 2; *Pseira* VII, pp. 133–134.
6. Haggis 1996c.
7. Haggis 1993a.
8. Hayden 2003a, 2003b.
9. Haggis 1992, 1996b.
10. See Manteli (1992) for Kastelli Phournis; Schlager (2002) for Hagia Triada and Petrokorio (p. 181) and for the Katharades Acropolis (p. 196).
11. On the chronology of EM I, and on the situation in Crete in general at this time, see Hood 1990a, 1990b; Wilson and Day 2000, p. 53; Haggis 1996c. For additional discussion of the Mirabello region, see Betancourt 1999.
12. Compare the situation at Knossos in Hood 1990a.
13. Betancourt 1985, pp. 23–34.
14. Whitelaw et al. 1997.
15. Betancourt 1985, pp. 31–32.
16. Betancourt 1985, pp. 26–29.

shapes.[17] At Chrysokamino, possible imports from outside of the immediate region include a pyxis (**12**) and a vase with basket impression on the base (**14**); the pyxis is from EM I–IIA, while the vessel with the basket impression cannot be dated more closely than EM I–III.

Although some other pieces might be from this period, the only certain EM I–IIA sherd from the metallurgy workshop is a tiny fragment of a pyxis found in the slag pile (**12**). It has a combed decoration incised into the exterior. Such vessels are usually regarded as EM IIA, but the chronology is not absolutely certain. Good parallels come from central Crete,[18] and a similar pyxis comes from a tomb at Gournia.[19]

The EM IIB Pottery

One sherd from a Vasiliki Ware goblet is present from the slag pile (**13**). Vasiliki Ware, the mottled pottery of EM IIB, is one of the most distinctive and easily recognized wares of Minoan Crete.[20] The gray fabric of the sherd from this location is typical of the ware in this part of Crete.[21] The surface is slipped and burnished in the usual fashion.[22]

The EM III–MM IA Pottery

White-on-Dark Ware is the definitive pottery for EM III–MM IA in eastern Crete.[23] It begins before the end of EM IIB,[24] and it persists into a period that is contemporary with the start of MM IA in central Crete.[25] The ware is characterized by linear white decorations on a dark background; it is easily recognizable both by its style and by the type of white paint used because the MM IB white slip in this part of Crete is much whiter than the EM III–MM IA variety.[26] Both shapes and decorations are very distinctive.

Most of the pottery from the slag pile is from EM III–MM IA. Both fine fabrics and coarse fabrics are present in a number of typical shapes. The fine fabrics are the pale, slightly "gritty" clay pastes of the Mirabello Fabric that is typical of this phase in this part of Crete. The typical coarse fabric from this phase is red to yellowish brown. It varies in coarseness from vessel to vessel. Included fragments of stone, in the granodiorite/diorite series, place both fabrics within the definition for Mirabello Fabric.[27] A useful petrographic description of this fabric is published in connection with examples found at Pseira.[28] Mirabello Fabric is also the main local fabric for this region in later phases of the Middle Minoan period.[29]

17. Davaras 1971.
18. Xanthoudides 1924, pp. 34–36; Wilson 1985, p. 305.
19. Soles 1992, p. 32, fig. 13, no. GIII-4, pl. 13:G.
20. See period 2 in Warren 1972; Betancourt 1979.
21. Myer 1979; Myer and Betancourt 1981.

22. Schachermeyr (1938) collected another piece of Vasiliki Ware from the surface of the metallurgy workshop.
23. Zois 1968; Betancourt 1984; 1985, pp. 55–61.
24. See period 2 in Warren 1972.
25. Momigliano 1991, p. 227; 2000, p. 102; Betancourt 2003b.
26. Noll, Holm, and Born 1971,

p. 617; Noll 1982, pp. 190–193; Ferrence, Swann, and Betancourt 2002.
27. Myer 1979; Myer and Betancourt 1981; Day 1991, pp. 91–101; 1997, p. 223.
28. Myer, McIntosh, and Betancourt 1995.
29. Haggis and Mook 1993; Haggis 2000.

THE POTTERY FROM THE APSIDAL STRUCTURE

The pottery from the apsidal structure comes from a level just below the modern surface. It comes only from inside the building and from the small scatter of soil in front of the doorway. Only two periods are represented. One sherd is from the Final Neolithic, and the rest of the pottery is from EM III–MM IA. The date is certain because it is based on several well-dated pieces, including sherds of East Cretan White-on-Dark Ware (Fig. 5.2). Although this ware begins at the end of EM IIB,[30] its latest date is contemporary with MM IA in central Crete.[31] The EM III and MM IA parts of the phase cannot yet be subdivided on stylistic criteria.

The pottery from the apsidal structure (Tables 5.6–5.8) is presented in tables similar to those used for the slag pile. These tables allow the two contexts to be compared and contrasted easily.

Figure 5.2. Sherds of conical cup 75

Final Neolithic

The only pottery fragment from the apsidal structure predating EM III is a body sherd from a bowl of FN date (**73**). Its isolation within a substantially later deposit suggests it is an heirloom.

Early Minoan III–Middle Minoan IA

The pottery from the apsidal structure represents a minimum of 32 vessels. There are many ambiguous pieces whose original shape is uncertain because the sherds are too small for identification. These ceramics are presented in their entirety in the catalogue and charts in this chapter. Where the evidence is sufficient for judgment, small sherds have been combined to form incomplete vessels of nonjoining sherds; in the cases where joining has not been possible, sherds are catalogued individually.

The fact that many of the sherds come from within the soil of the floors in the building proves that the soil, with most or all of its included sherds, was brought into the workshop from a settlement, so that the sherds catalogued here need not have been used originally within the apsidal structure itself. On the other hand, the date, style, and shapes of the EM III–MM IA pottery from inside the apsidal structure and from outside it in the slag pile are so similar that the activities must be reconstructed from both assemblages. This similarity can be seen easily by a comparison between the lists of shapes from the two contexts (Table 5.9; see Catalogue below). These similarities indicate that the two EM III–MM IA contexts are similar parts of a single range of pottery. The EM III–MM IA people were using the entire area of the workshop, and it is not surprising that their pottery is spread over the entire area.

The fragments of cups, pouring vessels, shallow bowls, storage jars with wide mouths, and cooking dishes observed in the sample attest to food preparation activities and the consumption of food and drink. Not a single sherd from a luxury vessel, such as a pyxis that might have contained

30. See period 2 in Warren 1972.
31. Betancourt 1984; 2003b, p. 119; MacGillivray et al. 1992.

perfume or oil, or from any vase that could be called a ritual vessel, is present. A few painted cups and pouring vessels are represented, but the truly elegant painted shapes from the period, such as teapots, are not found. Also missing are transport vessels with narrow mouths (such as amphoras), tripod cooking pots, and all other cooking vessels aside from cooking dishes. The conclusion must be that this is a limited assemblage from a small, specialized site without many of the vessels that would be commonplace within the pottery assemblage of a larger and more diverse settlement.

The EM III–MM IA pottery, like the earlier ceramics, is mostly local. The main fabric is the Mirabello Fabric of the Gournia to Priniatikos Pyrgos region, and the style of the painted vases is typical of this part of Crete. Although the use of an apsidal structure and the possible importation of ore from a region north of Crete may suggest foreign connections, the local pottery demonstrates that the community at Chrysokamino was firmly tied to the local ceramic production system. This implies that it was integrated with other aspects of the local cultural system as well.

CATALOGUE

In the following catalogue, the vessels are arranged by context. Finds from the slag pile are listed first, followed by those from the apsidal structure. Dimensions in this and subsequent catalogues are in centimeters; heights include the handles; the maximum dimension is the length of the longest axis across the sherd. Nomenclature follows the conventions of *The History of Minoan Pottery*.[32] Clay colors are described using the Munsell system.[33]

Pottery from the Slag Pile

Final Neolithic

1 (X 410, from R 20-3) Fig. 5.3

Bowl, rim sherd.
Diam. of rim ca. 24.
A coarse fabric (black, with an unevenly colored surface, mostly dark reddish brown, 5YR 3/3); burnished on interior and exterior.
Open, shallow bowl.

2 (X 1860, from K 17-Surface) Fig. 5.3

Bowl, rim sherd.
Diam. of rim ca. 10–11.
A coarse fabric (unevenly fired, black, with the surface mostly reddish brown, 5YR 5/4); burnished on interior and exterior.

3 (X 694, from N 20-3) Fig. 5.3

Bowl, body sherd.
Max. dim. 4.8.
A coarse fabric (unevenly fired, with the surface mostly dark brown, 7.5YR 3/2); burnished on interior and exterior.

---

32. Betancourt 1985.
33. Kollmorgen Instruments Corporation 1992.

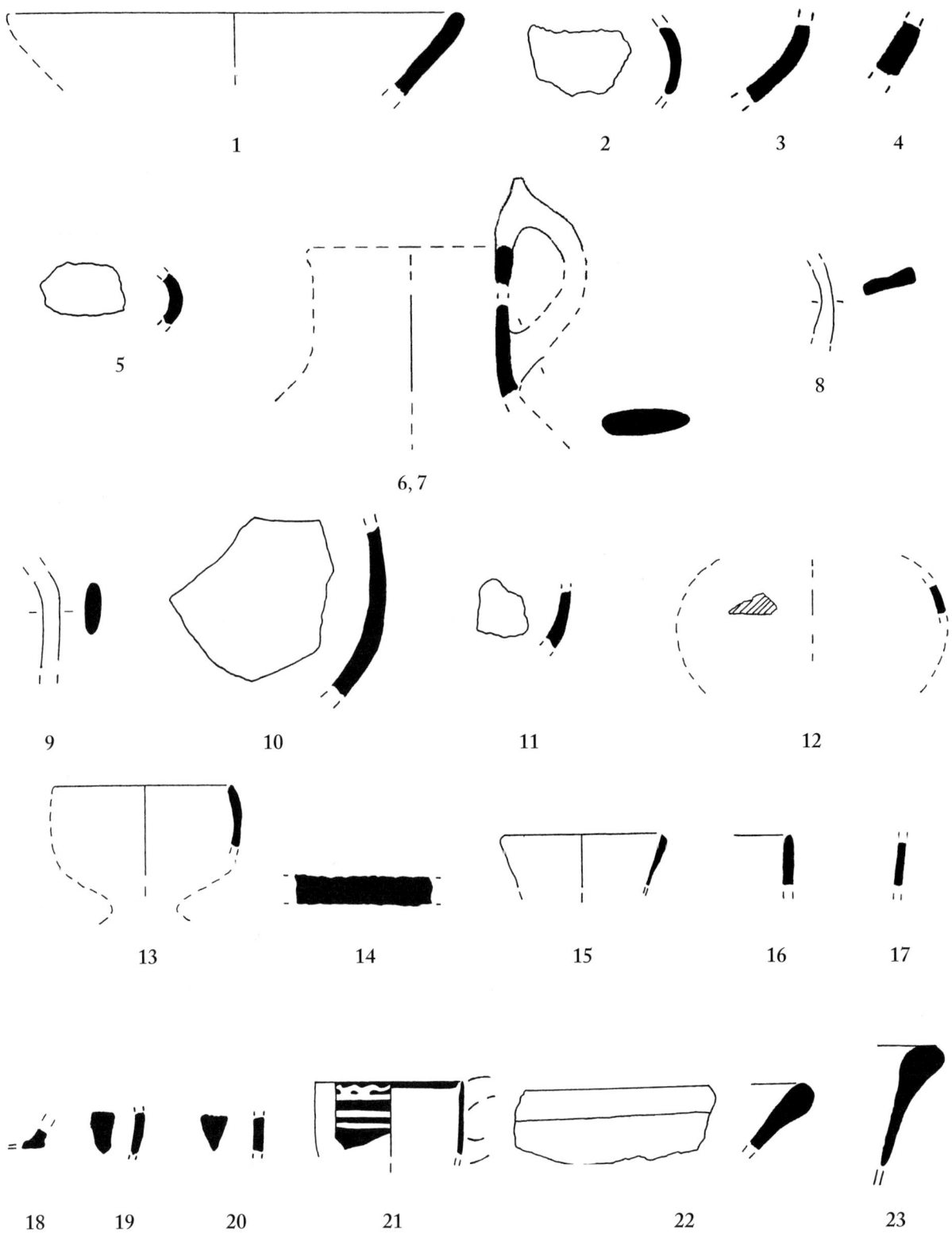

Figure 5.3. Pottery from the slag pile (**1–23**). Scale 1:3

THE POTTERY 75

4   (X 671, from L 17-2)   Fig. 5.3
    Bowl, body sherd.
    Max. dim. 3.6.
    A coarse fabric (unevenly fired, black, with the surface mostly reddish brown, 5YR 5/4); burnished on interior and exterior.

5   (X 718, from N 20-3)   Fig. 5.3
    Carinated open vessel, body sherd.
    Diam. of body ca. 15–16 at the carination.
    A coarse fabric (unevenly colored, mostly dark reddish brown, 2.5YR 3/4, with a reddish brown surface); burnished on interior and exterior.

6   (X 421, from P 20-3)   Fig. 5.3
    Jar, handle sherd with horned rim.
    Restored H. ca. 25.
    A coarse fabric (black, with an unevenly colored surface, mostly reddish brown, 5YR 5/4); burnished on exterior.
    Parallels: For FN closed vessels with horns on top of the handles, see Vagnetti 1975, p. 85, fig. 75, especially no. 8 (Phaistos); Manteli 1992, fig. 8 (Kastelli Phournis). Horned handles are a feature of Balkan ceramics (see, for example, Stefanovich and Bankoff 1998, p. 299, fig. 26:a–d, from Kamenska Cuka).

7   (X 717, from N 20-3)   Fig. 5.3
    Jar, neck sherd with handle.
    Diam. of interior of neck ca. 8.
    A coarse fabric (unevenly colored, mostly dark gray to black, with an unevenly colored surface, mostly light reddish brown, 5YR 5/4); burnished.
    Wide strap handle with small hole pierced in it, made before firing.

8   (X 652, from O 20-2)   Fig. 5.3
    Jar, handle sherd.
    Max. dim. 5.4.
    A coarse fabric (unevenly colored, with a reddish brown surface, mostly light reddish brown, 5YR 5/4); burnished.
    Wide strap handle from a jar.

9   (X 5, from O 22-2)   Fig. 5.3
    Jar, handle sherd.
    Max. dim. 5.4.
    A coarse fabric (unevenly colored with a reddish brown surface, mostly light reddish brown, 5YR 6/3); burnished.
    Wide strap handle from a jar.

10 (X 621, from O 20-2)   Fig. 5.3
    Closed vessel, body sherd.
    Diam. of body ca. 22.
    A coarse fabric (unevenly colored, with an unevenly colored surface, mostly dark reddish brown, 5YR 3/2); burnished.
    Comments: Probably a jar.

11 (X 941, from K 21-Surface)   Fig. 5.3

Closed vessel, body sherd.
Max. dim. 3.1.
A coarse fabric (unevenly colored, mostly black, with the surface mostly dark gray, 5YR 4/1).

TABLE 5.1. FINAL NEOLITHIC SHERDS FROM THE SLAG PILE

	Rim	Handle	Base	Spout	Leg	Body	Whole	Total	Min. No.
1.	2					2		4	1
2.						1		1	1
3.		1						1	1
4.		3						3	2
5.						2		2	2

Coarse fabric sherds: (1) bowl (**1–4**, X 410, X 1860, X 694, X 671); (2) carinated bowl (**5**, X 718); (3) jar with horned rim (**6**, X 421); (4) jar (**7–9**, X 717, X 652, X 5); (5) closed vessel (**10, 11**, X 621, X 941).

EM I–IIA

12 (X 460, from N 17-2)   Fig. 5.3

Pyxis, body sherd.
Max. dim. 2.4.
A fine fabric (brown, 7.5YR 5/4).
Globular pyxis with combed decoration incised into the exterior.
Comments: Small globular pyxides with stamped and incised decoration begin in EM I and continue into EM IIA.
Parallels: Xanthoudides 1924, pp. 34–36 (Koumasa); Wilson 1985, p. 305 (Knossos); Soles 1992, fig. GIII-4, pl. 13:G (Gournia).

TABLE 5.2. EARLY MINOAN I–IIA SHERD FROM THE SLAG PILE

	Rim	Handle	Base	Spout	Leg	Body	Whole	Total	Min. No.
1.						1		1	1

Fine fabric sherd: (1) pyxis (**12**, X 460).

EM IIB

13 (X 750, from N 20-4)   Fig. 5.3

Goblet, rim sherd.
Restored Diam. ca. 8–10.
A fine fabric (pinkish white, 5YR 8/2); slipped and burnished.
Goblet with slightly inturned rim; Vasiliki Ware.
Comments: The small goblet is one of the most common shapes in Vasiliki Ware (Betancourt 1979). It is distributed throughout Crete, but the majority of the examples are from the eastern end of the Gulf of Mirabello region.
Parallels: See Betancourt 1979, pp. 42–45, for a list of other examples.

THE POTTERY

TABLE 5.3. EARLY MINOAN IIB SHERD FROM THE SLAG PILE

	Rim	Handle	Base	Spout	Leg	Body	Whole	Total	Min. No.
1.	1							1	1

Fine fabric sherd: (1) goblet, Vasiliki Ware (**13**, X 750).

EM II–III

14 (X 497, from P 20-3)   Fig. 5.3
    Jar (?), base sherd.
    Max. dim. 10.6.
    A coarse fabric (yellowish red, 5YR 5/8, to very dark gray, 5YR 3/1).
    Base of a closed vessel; basket impression on bottom of base.

TABLE 5.4. EARLY MINOAN II–III SHERD FROM THE SLAG PILE

	Rim	Handle	Base	Spout	Leg	Body	Whole	Total	Min. No.
1.			1					1	1

Very coarse fabric sherd: (1) closed vessel (**14**, X 497).

EM III–MM IA

15 (X 459, from N 17-2)   Fig. 5.3
    Rounded cup, rim sherd.
    Max. dim. 2.6.
    A fine fabric (light red, 10R 6/6).
    Comments: Most of the published examples of this shape are decorated as White-on-Dark Ware, but several examples from Chrysokamino are plain, with only a band on the rim. For the shape, see the list in Betancourt 1984, p. 43, all from eastern Crete.

16 (X 668, from L 17-2)   Fig. 5.3
    Rounded cup, rim sherd.
    Diam. of rim ca. 9–11.
    A fine fabric (light reddish brown, 5YR 6/4). Dark slip: band on rim.

17 (X 698, from L 17-2)   Fig. 5.3
    Rounded cup, body sherd.
    Max. dim. 2.6.
    A fine fabric (light red, 10R 6/6).

18 (X 796, from M 22-4)   Fig. 5.3
    Rounded cup, base sherd.
    Max. dim. 2.0.
    A fine fabric (pink, 7.5YR 7/4).

19 (X 711, from N 20-2)   Fig. 5.3

    Cup (?), body sherd.
    Max. dim. 2.
    Convex profile.
    A fine fabric (reddish yellow, 5YR 6/6). Dark slip on exterior. Added white not preserved.

20 (X 719, from N 20-3)   Fig. 5.3

    Cup (?), body sherd.
    Max. dim. 1.7.
    Convex profile. A fine fabric (reddish yellow, 5YR 7/6). Dark slip on exterior. Added white not preserved.

21 (X 1223, from M 20-2)   Fig. 5.3

    Cup, rim sherd.
    Diam. of rim 8.
    Straight rim.
    A fine fabric (between pink, 5YR 7/4, and reddish yellow, 7.5YR 7/6). Dark slip on inside of rim and on exterior. Added white: bands with zigzag band at rim.

22 (X 3, from M 22-2)   Fig. 5.3

    Shallow bowl, rim sherd.
    Diam. of rim 56.
    A medium coarse fabric (reddish yellow, 7.5YR 6/6).
    Thickened rim.
    Comments: For a list of such open bowls, all from eastern Crete, see Betancourt 1984, pp. 38–39. Complete examples are illustrated by Hall 1912, fig. 20 and Betancourt 1983, nos. 114, 115 (all from Sphoungaras). Both plain and decorated bowls are known. In several cases, the decoration is restricted to the rim. The distinction between the shape of the bowls and the cooking dishes is somewhat arbitrary when only the rim survives, but the bowls are often decorated, while the cooking dishes are left unpainted.

23 (X 1865, from K 20-Surface)   Fig. 5.3

    Bowl, rim sherd.
    Diam. of rim ca. 38.
    Mirabello Fabric (reddish yellow, 7.5YR 6/6).
    Thickened rim. Dark band (red) on the rim.

24 (X 219, from R 20-2)   Fig. 5.4

    Bridge-spouted jar, spout sherd.
    Diam. of rim not measurable.
    Mirabello Fabric (reddish yellow, 5YR 6/6).
    Thickened rim. Dark paint on exterior and on inside of rim. Added white paint not preserved.
    Comments: East Cretan White-on-Dark Ware, based on the typical rim profile with paint on the interior of the rim. For a list of East Cretan examples from EM III–MM IA, see Betancourt 1984, pp. 45–46.

25 (X 607, from O 20-2)   Fig. 5.4

    Vessel, body sherd.

THE POTTERY

Max. dim. 11.6.
A coarse fabric (reddish yellow, 5YR 6/6).

26 (X 6, from O 22-2) — Fig. 5.4
Cooking dish, rim and body sherds.
Diam. of rim 44.
Mirabello Fabric (red, 2.5YR 5/8).

27 (X 448, from P 20-1) — Fig. 5.4
Cooking dish, body sherd.
Max. dim. 2.3.
Mirabello Fabric (red, 2.5YR 5/6, and reddish brown, 5YR 4/3).

28 (X 658, from K 17-1) — Fig. 5.4
Cooking dish, rim of body sherd.
Max. dim. 2.6.
Mirabello Fabric, fine grained (red, 5YR 4/6).

29 (X 672, from L 17-2) — Fig. 5.4
Cooking dish, rim sherd.
Diam. of rim ca. 30.
Mirabello Fabric, fine grained (red, 2.5YR 4/6).
Straight rim.

30 (X 1851, from M 19-Cleaning) — Fig. 5.4
Cooking dish, rim sherd with wide, knoblike handle attached at the bottom of the rim.
Diam. of rim not measurable.
Mirabello Fabric (red, 10R 5/6).
Comments and parallels: For discussion of the cooking dish, a shallow open vessel that appears in Crete as early as EM I and often preserves marks of fire from cooking, see Betancourt 1980. This fabric is typical of EM III and MM cooking dishes in this part of Crete. Compare Banou 1995b, p. 113, no. ADC 42, and p. 116, no. ADC 74 (Pseira).

31 (X 577, from P 20-3) — Fig. 5.4
Shallow bowl, rim sherd.
Diam. of rim ca. 30.
Mirabello Fabric (from red, 2.5YR 5/6, to light reddish brown, 5YR 3/4).

32 (X 1853, from J 17-Cleaning) — Fig. 5.4
Vessel, rim sherd.
Diam. of rim not measurable.
A medium coarse fabric (weak red, 10R 4/4).
Vertical, slightly offset rim with holes pierced in the rim before firing.
Comments: It is possible that this vessel is a Neolithic "cheese pot" from the Cyclades. See Katsarou and Schilardi 2004, pp. 38, 39, nos. 1–8, fig. 10.

33 (X 8, from M 22-Surface) — Fig. 5.4
Bucket jar (or jar or basin?), base and body sherd.
Diam. of base greater than 56.
Mirabello Fabric (reddish yellow, 5YR 6/6).

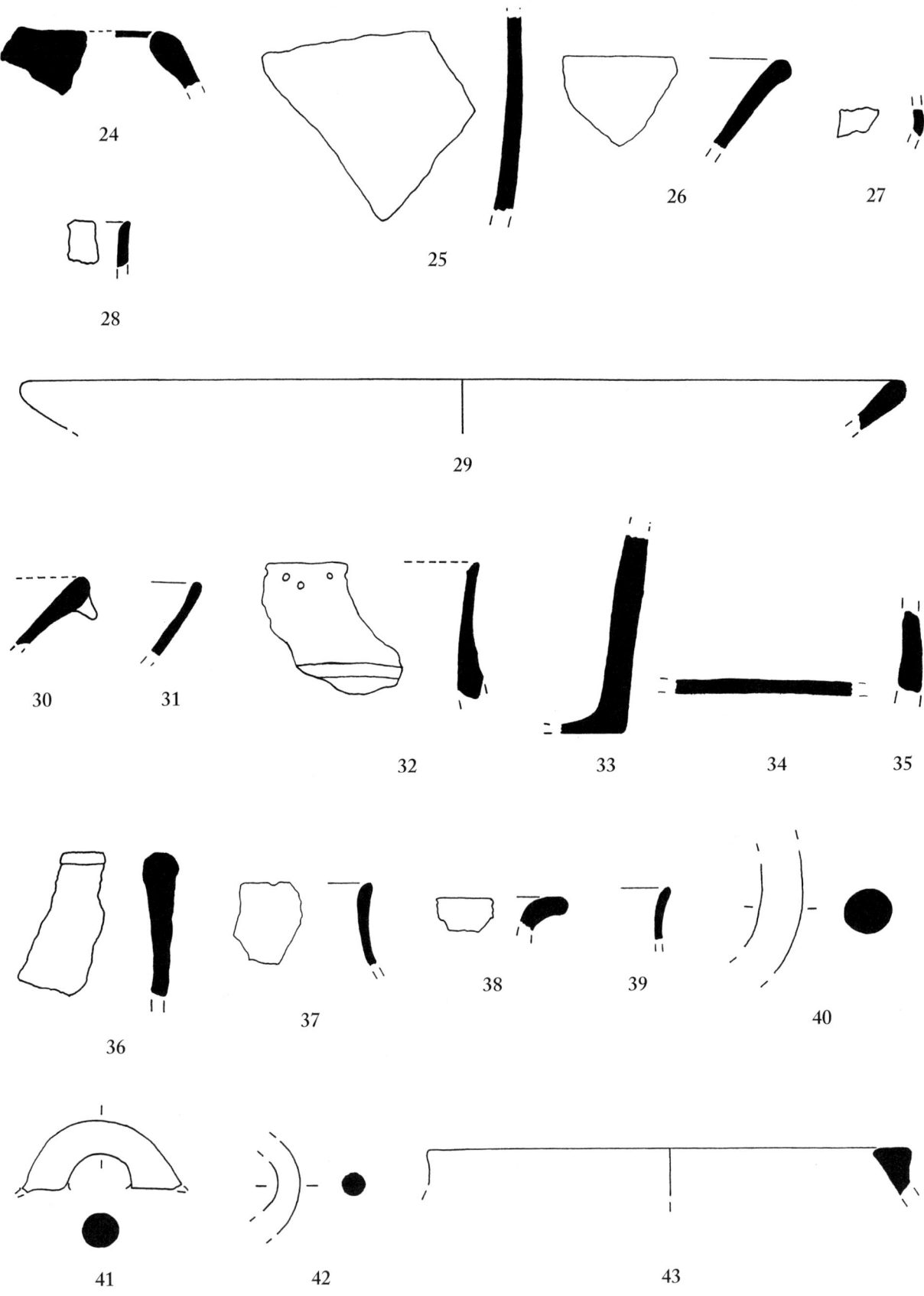

Figure 5.4 *(opposite).* **Pottery from the slag pile (24–43).** Scale 1:3

34 (X 1850, from H 19-Cleaning)
Bucket jar, base sherd.
Max. dim. 13.5.
Mirabello Fabric (reddish yellow, 5YR 6/6).

35 (X 1852, from J 18-Cleaning)
Bucket jar, body sherd.
Max. dim. 6.4.
Mirabello Fabric (reddish yellow, 5YR 6/6).

36 (X 1864, from G 18-Cleaning)
Bucket jar, rim sherd.
Max. dim. 8.1.
Mirabello Fabric (reddish yellow, 5YR 6/6).

37 (X 69, from Q 20-2)
Closed vessel, rim sherd.
Diam. of rim ca. 16.
A coarse fabric (yellowish red, 5YR 4/6).
Probably a jug.

38 (X 99, from S 20-2)
Jar (or jug?), rim sherd.
Diam. of rim not measurable; max. dim. 3.8.
Mirabello Fabric, fine grained (red, 2.5YR 4/6).
Outturned rim.

39 (X 659, from K 17-Surface)
Closed vessel, rim sherd.
Diam. of rim not measurable; max. dim. 3.4.
Mirabello Fabric, fine grained (yellowish red, 5YR 4/6).
Slightly outturned rim.

40 (X 422, from P 20-3)
Jar, handle sherd.
Max. dim. 6.7.
Mirabello Fabric (red, 2.5YR 4/6).

41 (X 661, from K 17-Surface)
Jar, handle sherd.
Max. dim. 8.4.
Mirabello Fabric (light reddish brown, 5YR 6/4).

42 (X 660, from K 17-1)
Jug, handle sherd.
Max. dim. 4.8.
Mirabello Fabric (red, 2.5YR 4/6).

43 (X 405, from R 20-1)
Jar with thickened, flat-topped rim, rim sherd.

Fig. 5.4
Fig. 5.4
Fig. 5.4
Fig. 5.4
Fig. 5.4
Fig. 5.4
Fig. 5.4
Fig. 5.4
Fig. 5.4
Fig. 5.4

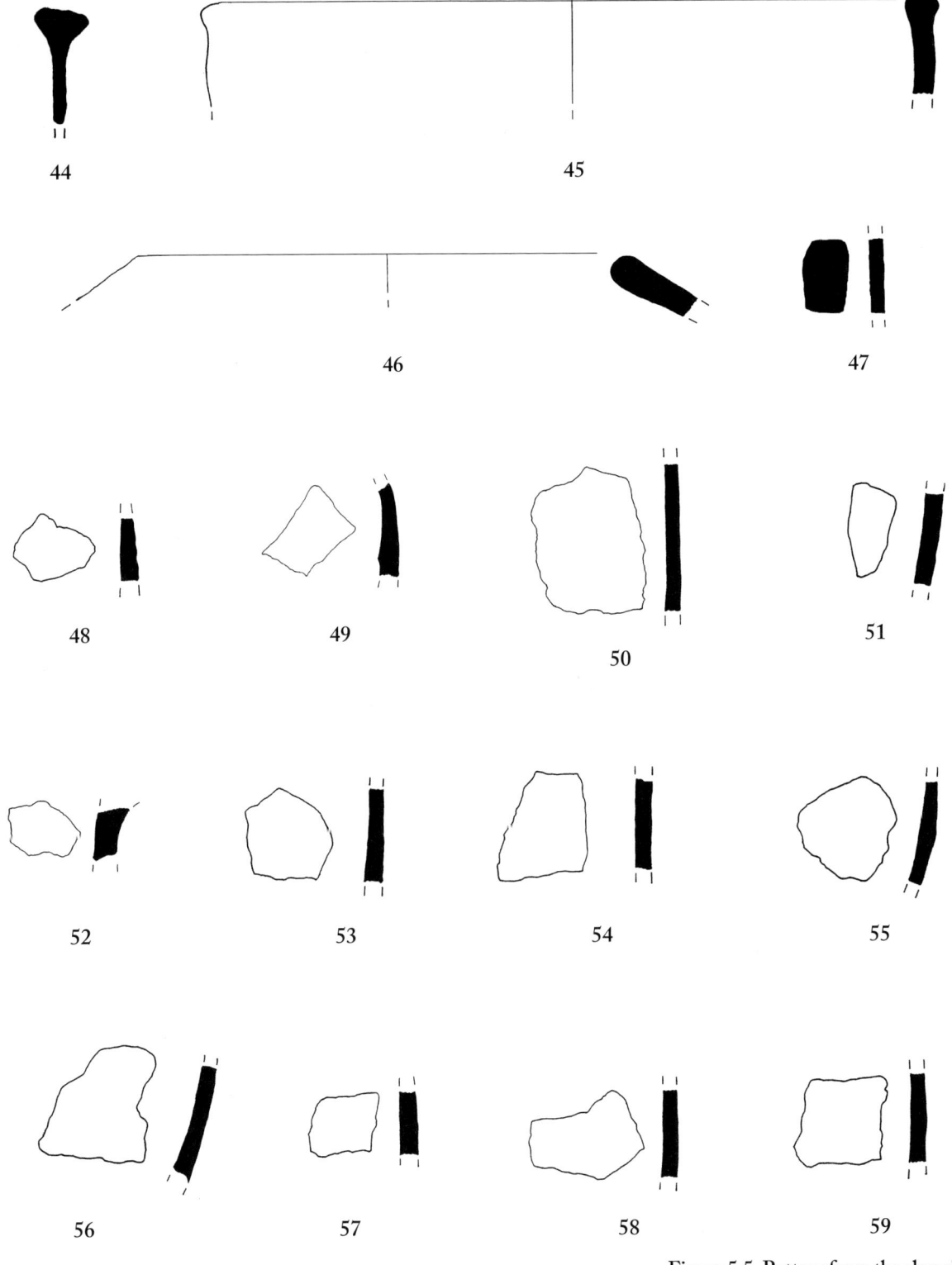

Figure 5.5. Pottery from the slag pile (**44–59**). Scale 1:3

Diam. of rim ca. 32–36.
Mirabello Fabric (red, 2.5YR 4/6).

44 (X 1859, from R 20-Cleaning) Fig. 5.5
Jar with thickened, flat-topped rim, rim sherd.
Max. dim. 8.8.
Mirabello Fabric (red, 10R 5/6).

45 (X 67, from P 21-1) Fig. 5.5
Bridge-spouted jar, rim sherd.
Diam. of rim ca. 28–32.
Mirabello Fabric (light reddish brown, 5YR 6/4).
Thickened rim.

46 (X 664, from K 17-1) Fig. 5.5
Bridge-spouted jar, rim sherd.
Diam. of rim ca. 22.
Mirabello Fabric (light red, 2.5YR 6/6).

47 (X 7, from O 22-2) Fig. 5.5
Vessel, body sherd.
Max. dim. 3.8.
Mirabello Fabric (yellowish red, 4YR 4/6).

48 (X 68, from S 20-2) Fig. 5.5
Vessel, body sherd.
Max. dim. 3.7.
Mirabello Fabric (yellowish red, 5YR 4/6).

49 (X 218, from R 20-2) Fig. 5.5
Vessel, body sherd.
Max. dim. 4.6.
A coarse fabric containing phyllite and other stones (red, 2.5YR 5/8).

50 (X 403, from R 20-Surface) Fig. 5.5
Jar or bellows, body sherd.
Max. dim. 3.6.
Mirabello Fabric (unevenly colored, from red, 2.5YR 5/8, to gray).

51 (X 413, from R 20-Surface) Fig. 5.5
Jar or bellows, body sherd.
Max. dim. 4.3.
Mirabello Fabric (red, 2.5YR 4/8).

52 (X 457, from N 17-2) Fig. 5.5
Jar or bellows, body sherd.
Max. dim. 3.6.
A coarse fabric containing phyllite and other stones (red, 2.5YR 3/6).

53 (X 458, from M 17-2) Fig. 5.5
Jar or bellows, body sherd.

Max. dim. 5.
Mirabello Fabric (unevenly colored, from reddish yellow, 5YR 6/8, to yellowish red, 5YR 5/6).

54 (X 462, from O 18-Surface)   Fig. 5.5
Jar or bellows, body sherd.
Max. dim. 6.4.
Mirabello Fabric (unevenly colored, from red, 2.5YR 4/6, on the exterior, to yellowish brown, 10YR 5/4, on the interior).

55 (X 467, from M 17-Surface)   Fig. 5.5
Jar or bellows, body sherd.
Max. dim. 5.2.
Mirabello Fabric (light red, 2.5YR 6/8, with a gray core).

56 (X 470, from M 17-Surface)   Fig. 5.5
Jar or bellows, body sherd.
Max. dim. 7.2.
A coarse fabric containing phyllite and other stones (unevenly colored, from red, 10R 5/6, to gray).
Comments: Discolored by partial reduction.

57 (X 633, from R 20-2)   Fig. 5.5
Jar or bellows, body sherd.
Max. dim. 4.2.
Mirabello Fabric (red, 2.5YR 5/8).

58 (X 663, from K 17-1)   Fig. 5.5
Shallow bowl or cooking dish, body sherd.
Max. dim. 5.5.
Mirabello Fabric (red, 2.5YR 5/6).

59 (X 666, from K 17-1)   Fig. 5.5
Jar or bellows, body sherd.
Max. dim. 5.8.
Mirabello Fabric (yellowish red, 5YR 5/6, and dark reddish gray, 5YR 4/2).

60 (X 674, from L 17-2)   Fig. 5.6
Open vessel, body sherd.
Max. dim. 3.7.
Mirabello Fabric (yellowish red, 5YR 4/6).

61 (X 675, from L 17-2)   Fig. 5.6
Jar or bellows, body sherd.
Max. dim. 6.
Mirabello Fabric (red, 10R 4/6).

62 (X 676, from L 17-2)   Fig. 5.6
Jar or bellows, body sherd.
Max. dim. 4.3.
Mirabello Fabric (dark gray, 10YR 4/1).

Figure 5.6 *(opposite)*. Pottery from the slag pile (60–72) and the apsidal structure (73–82). Scale 1:3

THE POTTERY

85

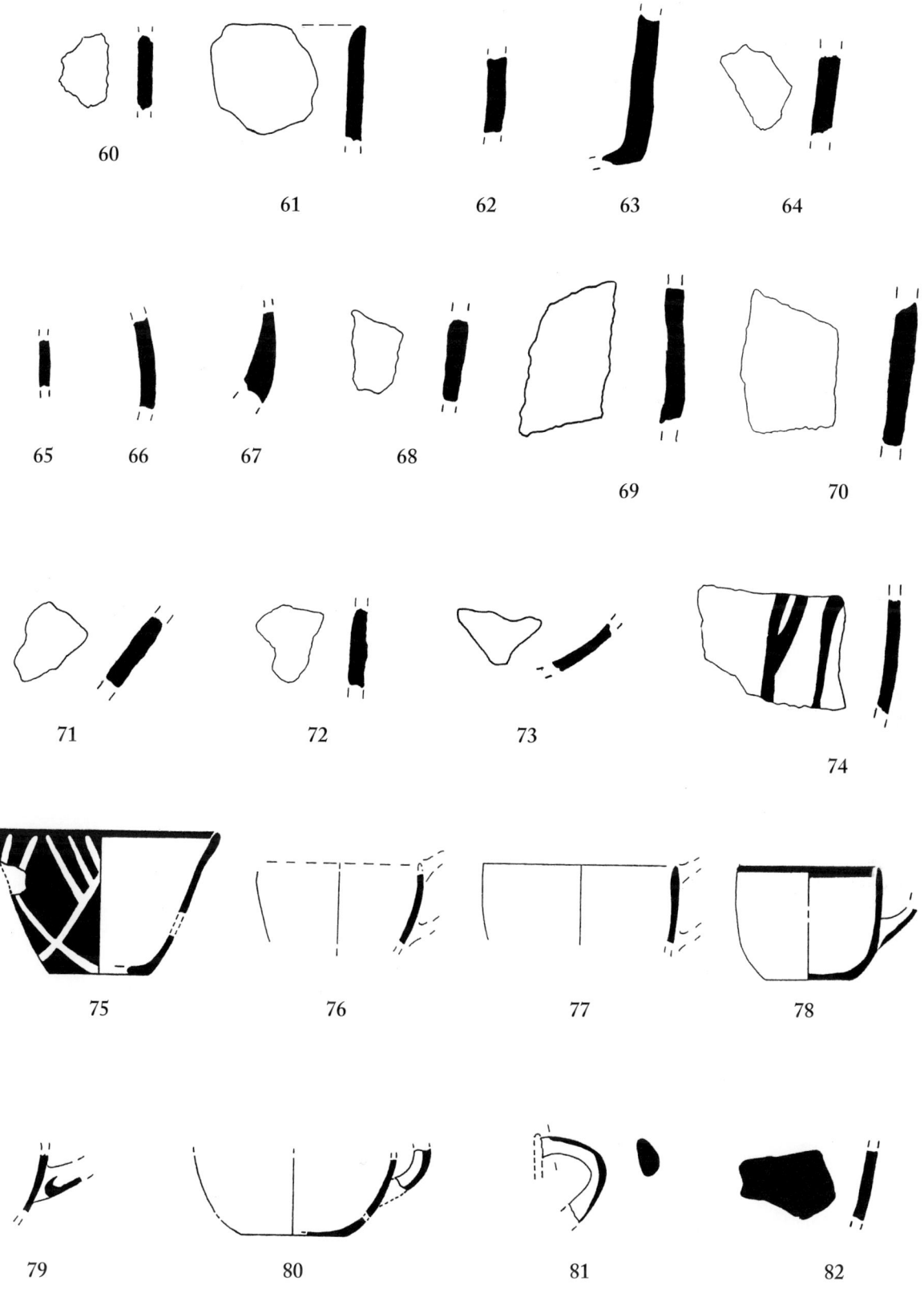

63 (X 685, from N 21-Surface) Fig. 5.6
   Jar (or bellows?), base sherd.
   Diam. of base ca. 38.
   A coarse fabric (unevenly colored, from red, 2.5YR 4/8, to light red, 2.5YR 6/6).

64 (X 689, from L 17-2) Fig. 5.6
   Jar or bellows, body sherd.
   Max. dim. 4.4.
   Mirabello Fabric (red, 2.5YR 4/8).

65 (X 720, from N 20-3) Fig. 5.6
   Vessel, body sherd.
   Max. dim. 2.4.
   Mirabello Fabric (dark brown, 7.5YR 4/2).

66 (X 1722, from N 17-2) Fig. 5.6
   Jar (?), body sherd.
   Max. dim. 4.7.
   Mirabello Fabric (red, 2.5YR 5/6).

67 (X 1855, from O 19-Cleaning) Fig. 5.6
   Vessel, body sherd.
   Max. dim. 5.4.
   Mirabello Fabric (red, 10R 5/6).

68 (X 1857, from O 19-Cleaning) Fig. 5.6
   Vessel, body sherd.
   Max. dim. 4.3.
   Mirabello Fabric (exterior surface weak red, 10R 5/4; dark interior and core).

69 (X 1858, from O 19-Cleaning) Fig. 5.6
   Vessel, body sherd.
   Max. dim. 9.0.
   Mirabello Fabric (red, 10R 5/6).

70 (X 1862, from I 19-Cleaning) Fig. 5.6
   Vessel, body sherd.
   Max. dim. 7.9.
   Mirabello Fabric (red, 10R 5/6).

71 (X 1863, from Q 19-Cleaning) Fig. 5.6
   Vessel, body sherd.
   Max. dim. 4.2.
   Mirabello Fabric (red, 10R 5/6).

72 (X 1867, from N 20-Cleaning) Fig. 5.6
   Vessel, body sherd.
   Max. dim. 3.9.
   Mirabello Fabric (red, 10R 5/6).

THE POTTERY

TABLE 5.5. EARLY MINOAN III–MIDDLE MINOAN IA SHERDS FROM THE SLAG PILE

	Rim	Handle	Base	Spout	Leg	Body	Whole	Total	Min. No.
1.	2		1			1		4	4
2.	1					2		3	3
3.	1							1	1
4.	1					1		2	1
5.	2 or 3					2 or 3		6	2
6.	1							1	1
7.	1							1	1
8.	1		2			2		5	1
9.	3							3	3
10.		2						2	2
11.		1						1	1
12.	2							2	2
13.	2			1				3	2
14.			2			25		27	10

Fine fabric sherds: (1) rounded cup (**15–18**, X 459, X 668, X 698, X 796); (2) cup, painted (**19–21**, X 711, X 719, X 1223). Medium coarse fabric sherds: (3) shallow bowl (**22**, X 3); (4) vessel (**25, 32**, X 607, X 1853). Mostly Mirabello Fabric sherds: (5) cooking dish (**26–30**, X 6, X 448, X 658, X 672, X 1851); (6) bowl (**23**, X 1865); (7) shallow bowl (**31**, X 577); (8) bucket jar (**33–36**, X 8, X 1850, X 1852, X 1864); (9) jar or jug (**37–39**, X 69, X 99, X 659); (10) jar (**40, 41**, X 422, X 661); (11) jug (**42**, X 660); (12) jar with thickened, flat-topped rim (**43, 44**, X 405, X 1859); (13) bridge-spouted jar (**24, 45, 46**, X 219, X 67, X 664); (14) vessel (**47–72**, X 7, X 68, X 218, X 403, X 413, X 457, X 458, X 462, X 467, X 470, X 633, X 663, X 666, X 674, X 675, X 676, X 685, X 689, X 720, X 1722, X 1855, X 1857, X 1858, X 1862, X 1863, X 1867).

Pottery from the Apsidal Structure

Final Neolithic

73 (X 1870, from O 19-Cleaning)   Fig. 5.6

Bowl, body sherd.
Max. dim. 4.3.

A coarse fabric (unevenly colored, with an unevenly colored surface, brown to reddish brown, 2.5YR 4/4).

TABLE 5.6. FINAL NEOLITHIC SHERD FROM THE APSIDAL STRUCTURE

	Rim	Handle	Base	Spout	Leg	Body	Whole	Total	Min. No.
1.						1		1	1

Coarse, dark, burnished fabric sherd: (1) bowl (**73**, X 1870).

EM II

74 (X 66, from S 20-2)   Fig. 5.6

Jar, body sherd.
Max. dim. 9.7.

A medium coarse, gritty, chaff-tempered fabric (light reddish brown, 5YR 6/4).

Large closed vessel, probably a jar; trickle decoration in dark paint.

Comments: This atypical vessel cannot be closely dated. Good parallels for the decoration exist from EM II at Myrtos (Warren 1972). The fabric is not typical for Chrysokamino.

TABLE 5.7. EARLY MINOAN II SHERD FROM THE APSIDAL STRUCTURE

	Rim	Handle	Base	Spout	Leg	Body	Whole	Total	Min. No.
1.						1		1	1

Medium coarse fabric sherd: (1) jar (**74**, X 66).

EM III–MM IA: Fine Fabrics

75 (X 167, from M 18-1, M 18-2, and M 18-3)   Figs. 5.2, 5.6

Conical cup, rim and base sherds.
Restored H. ca. 6.5–7.0; Diam. of rim ca. 10; Diam. of base ca. 5.5–6.0.
A fine fabric (pink, 7.5YR 7/4); burnished. Dark slip on inside of rim and on exterior. Added white paint: pendant triangles with hatched corners.
Comments: East Cretan White-on-Dark Ware. Decorated conical cups, both with and without handles, are common in the ware. The decoration of pendant triangles with hatched corners is one of the standard types.
Parallels: For the shape, see Betancourt 1984, p. 39, shape 2A; for the motif see Betancourt 1984, p. 24, motif 2, nos. 3–5.

76 (X 148, from M 19-3)   Fig. 5.6

Rounded cup, handle and body sherds.
Max. dim. (largest sherd) 4.4.
A fine fabric (light brown, 7.5YR 6/4).
One handle with circular section.
Comments: See **15**.

77 (X 182, from N 19-2, N 19-3, M 19-3, M 19-5)   Fig. 5.6

Rounded cup, rim and body sherds.
Diam. of rim ca. 9–10.
A fine fabric (pink, 7.5YR 7/4). Dark slip; band on rim.

78 (X 210, from M 19-2B, M 19-3, and M 19-4)   Fig. 5.6

Rounded cup, fragmentary.
Diam. of rim ca. 8.
A fine fabric (pink, 5YR 7/4). Band on rim, inside and out; line on outside of handle.
One handle with circular section.
Several sherds burned.

79 (X 415, from N 19-3)   Fig. 5.6

Rounded cup, handle sherd.
Max. dim. 3.8.
A fine fabric (reddish yellow, 7.5YR 6/6). Dark slip; line on outside of handle.
Handle with circular section.

80 (X 534, from M 18-3)   Fig. 5.6

Rounded cup, body, base, and handle sherds.

Diam. of rim ca. 9–10.
A fine fabric (pink, 7.5YR 7/4). Dark slip; line on outside of handle.

81 (X 636, from N 19-3) Fig. 5.6
Rounded cup (?), handle sherd.
Max. dim. 4.5.
A fine fabric (light reddish brown, 5YR 6/4). Dark slip: line on outside of handle. Handle with circular section.

82 (X 149, from M 19-3) Fig. 5.6
Closed vessel, body sherd.
Max. dim. 5.2.
A fine fabric (pink, 5YR 7/4). Dark slip on exterior; no added white paint preserved.
Comments: From the thickness of the sherd and the quality of the dark slip, this vessel is most likely a bridge-spouted jar of East Cretan White-on-Dark Ware with the white paint missing (Betancourt 1984, pp. 45–46).

83 (X 181, from N 18-2 and N 19-2) Fig. 5.7
Closed vessel, body sherds.
Max. dim. of largest sherd 8.5.
A fine fabric (unevenly colored, mostly light brown, 7.5YR 6/4). Dark slip covering exterior; added white paint: bands, spirals, diagonal lines.
Large closed vessel, most likely a bridge-spouted jar.
Comments: East Cretan White-on-Dark Ware. Large bridge-spouted jars in this ware are often decorated with spirals and other motifs. They are more common than jugs or teapots. This sherd helps provide a secure EM III–MM IA date for the deposit.
Parallels: See Betancourt 1984, pp. 45–46.

84 (X 271, from N 18-2) Fig. 5.7
Closed vessel, body sherd.
Diam. of body of vessel above 32.
A fine fabric (light brown, 7.5YR 6/4). Dark slip: covering exterior; added white paint: diagonal lines.
Large closed vessel, most likely a bridge-spouted jar.
Comments: See **83**.

85 (X 466, from M 17-Surface) Fig. 5.7
Closed vessel, body sherd.
Max. dim. 3.2.
A fine fabric (reddish yellow, 5YR 7/6). Dark slip: covering exterior; added white paint: traces.
Large closed vessel, most likely a bridge-spouted jar.
Comments: See **83**.

86 (X 1675, from N 18-2) Fig. 5.7
Closed vessel, body sherd.
Max. dim. 5.1.
A fine fabric (reddish yellow, 5YR 7/6). Dark slip covering exterior; traces of added white paint.
Large closed vessel, most likely a bridge-spouted jar.
Comments: See **83**.

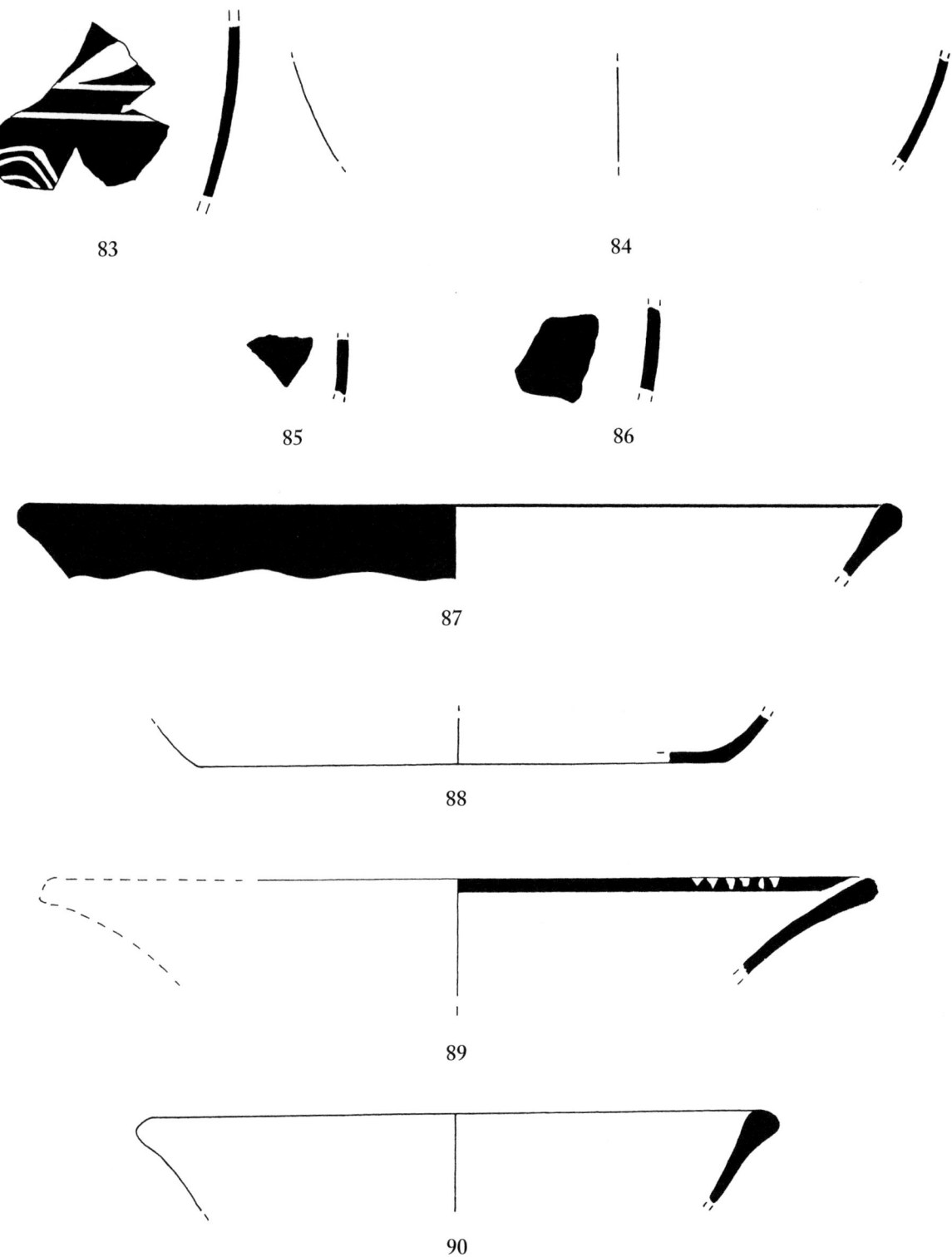

Figure 5.7. Pottery from the apsidal structure (**83–90**). Scale 1:3

EM III–MM IA: Medium Coarse to Coarse Fabrics

87 (X 145, from M 19-3)   Fig. 5.7

    Shallow bowl, rim sherds.
    Diam. of rim ca. 30–35.
    A medium coarse fabric (mostly pink, 5YR 7/4). Dark slip: band on the rim, inside and out.
    Straight, thickened rim.
    Comments: This is a common shape; see **22**.

88 (X 146, from M 19-3)   Fig. 5.7

    Shallow bowl, base sherd with handle.
    Diam. of base 19.
    A medium coarse fabric (reddish yellow, 7.5YR 6/6).
    Comments: See **22**.

89 (X 168, from M 18-2 and M 18-3)   Fig. 5.7

    Shallow bowl, rim sherd.
    Diam. of rim ca. 39–41.
    A medium coarse fabric (reddish yellow, 7.5YR 6/6, with a darker surface on the exterior and a paler surface on the interior). Dark slip on rim. Added white paint: groups of short lines on rim.
    Straight rim.
    Comments: See **22**. Decoration is often only on the rim, as on this example.

90 (X 274, from N 18-2)   Fig. 5.7

    Shallow bowl, rim sherd.
    Diam. of rim ca. 30.
    A medium coarse fabric (reddish yellow, 7.5YR 6/6). Band on rim.
    Straight rim.
    Comments: See **22**.

91 (X 1662, from M 19-4)   Fig. 5.8

    Shallow bowl, rim sherd.
    Diam. of rim ca. 30–32.
    A medium coarse Mirabello Fabric (light reddish brown, 5YR 6/6).
    Straight rim.
    Comments: See **22**.

92 (X 150, from M 19-3)   Fig. 5.8

    Shallow bowl or cooking dish, rim sherd.
    Diam. of rim ca. 25.
    Mirabello Fabric, fine-grained (red, 2.5YR 4/6).
    Straight rim.

93 (X 212, from M 18-3 and M 19-4)   Fig. 5.8

    Shallow bowl, rim sherds.
    Diam. of rim ca. 20.
    Mirabello Fabric, fine-grained (reddish brown, 2.5YR 5/4). Shallow bowl with straight rim. Band in dark slip on rim, inside and out.

94 (X 270, from N 18-2)   Fig. 5.8

    Shallow bowl, rim and body sherd.

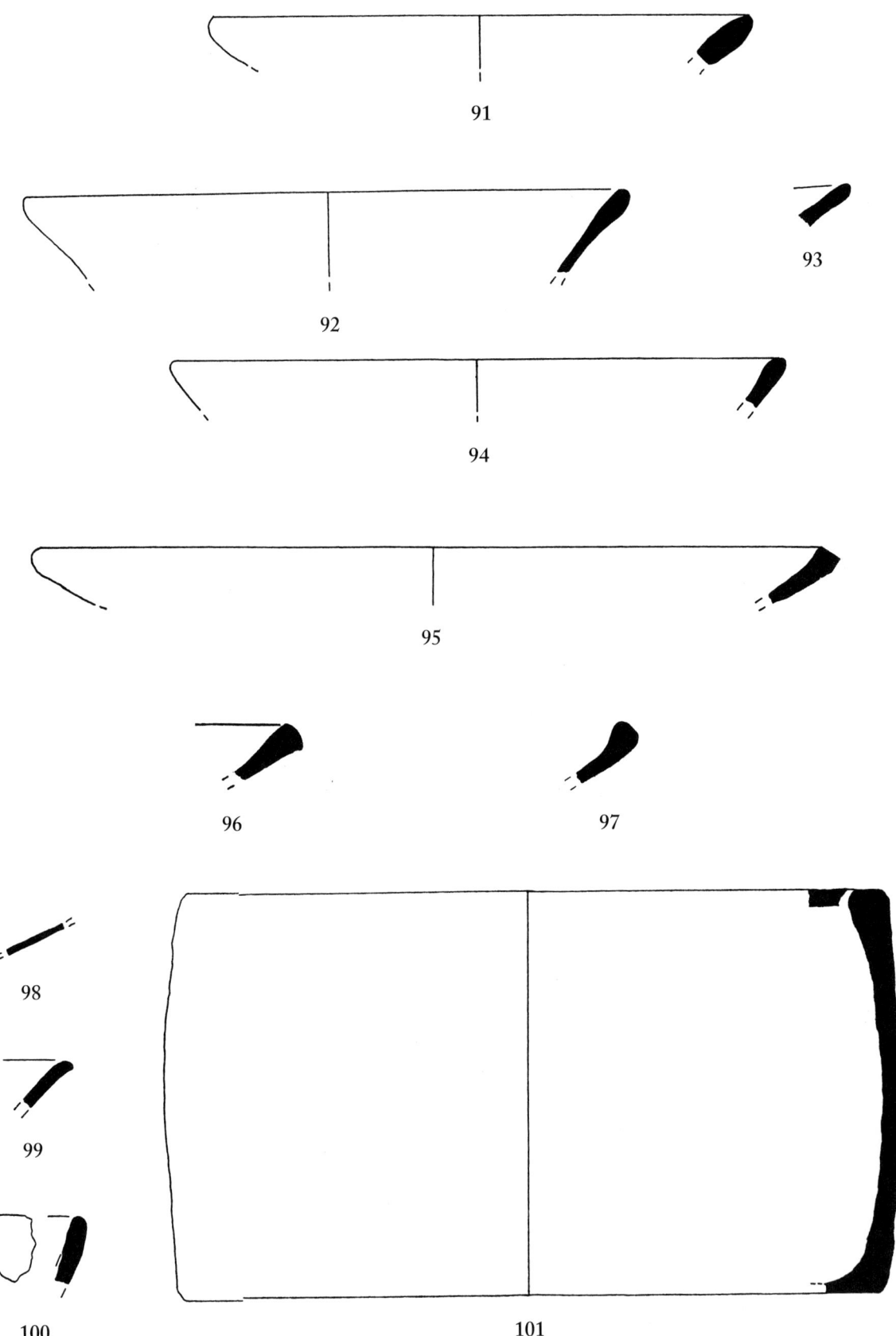

Figure 5.8 *(opposite)*. **Pottery from the apsidal structure (91–101).** Scale 1:3

Diam. of rim ca. 28; max. dim. 3.2.
Mirabello Fabric (unevenly colored, from red, 2.5YR 5/6, to light reddish brown, 5YR 6/4).

95 (X 275 (from N 18-2, M 18-2) Fig. 5.8
Cooking dish, rim sherd.
Diam. of rim ca. 55–60.
Mirabello Fabric, fine-grained (yellowish red, 5YR 4–5/6, with a browner surface).
Straight, thickened rim.
Comments: See **25**.

96 (X 411, from N 18-2) Fig. 5.8
Cooking dish, rim sherd.
Diam. of rim ca. 55–60.
Mirabello Fabric, fine-grained (brown, 10YR 4/3).
Straight, thickened rim.

97 (X 532, from M 18-3) Fig. 5.8
Cooking dish, rim sherd.
Diam. of rim ca. 45–55.
Mirabello Fabric, fine-grained (red, 2.5YR 4/6, with a darker exterior surface).
Straight, thickened rim.

98 (X 1661, from N 19-3) Fig. 5.8
Cooking dish, body sherd.
Max. dim. 3.3.
Mirabello Fabric (yellowish red, 5YR 4/6). Burnished in the exterior.

99 (X 641, from N 19-4) Fig. 5.8
Bowl or cooking dish, rim sherd.
Diam. of rim not measurable.
Mirabello Fabric (reddish yellow, 5YR 7/8).

100 (X 2, from L 17-2, L 18-2, M 18-1, M 20-2, N 18-2) Fig. 5.8
Bucket jar, rim and body sherds.
Max. dim. of largest sherd 11.6.
Mirabello Fabric (light brown, 7.5YR 6/4, to light reddish brown, 5YR 6/4).

101 (X 169, from M 18-1, M 18-2) Fig. 5.8
Bucket jar, complete profile.
H. 20.4; Diam. of rim 35–36; Diam. of base 35–37.
Mirabello Fabric (light brown, 7.5YR 6/4). Band on rim, extending inside the rim.[34]

102 (X 144, from M 19-3) Fig. 5.9
Jar, rim sherd.
Diam. of rim ca. 32.
A coarse fabric containing quartz and other stones (reddish brown, 2.5YR 5/4).
Thickened, flattened rim.

34. The sherds identified by Haggis (2005, p. 114) as larnax fragments are probably from bucket jars or basins.

103 (X 211, from M 19-4, N 18-3) Fig. 5.9

   Jar, rim sherd.
   Diam. of rim ca. 40–42.
   Mirabello Fabric (from red, 2.5YR 5/8, to black).
   Thickened, flattened rim.

104 (X 291, from N 18-2) Fig. 5.9

   Jar, rim sherd.
   Diam. of rim ca. 40–42.
   Mirabello Fabric (red, 2.5YR 5/6).
   Thickened, flattened rim.

105 (X 1659, from N 19-2) Fig. 5.9

   Jar, rim sherd.
   Diam. of rim not measurable.
   Mirabello Fabric (yellowish red, 5YR 5/6).
   Thickened, flattened rim.

106 (X 1663, from M 18-1) Fig. 5.9

   Jar, rim sherd.
   Diam. of rim ca. 30–32.
   Mirabello Fabric (between yellowish red, 5YR 5/8, and dark reddish brown, 5YR 3/4).

107 (X 530, from M 18-3) Fig. 5.9

   Bridge-spouted jar, rim sherd.
   Diam. of rim ca. 30–32.
   Mirabello Fabric (yellowish red, 5YR 4/6).

108 (X 414, from N 19-3) Fig. 5.9

   Closed vessel, probably a jar, handle sherd.
   Max. dim. 10.4.
   Mirabello Fabric (light red, 2.5YR 6/6), with a darker core (between pale red, 2.5YR 6/2, and weak red, 2.5YR 5/2).

109 (X 1664, from N 19-2) Fig. 5.9

   Jar, handle sherd.
   Max. dim. 6.
   Mirabello Fabric (red, 2.5YR 5/6).

110 (X 1658, from N 19-2) Fig. 5.9

   Closed vessel, probably a jar, base sherd.
   Diam. of base ca. 17–20.
   Mirabello Fabric (pink, 7.5YR 7/4).

111 (X 525, from M 8-3, M 19-5) Fig. 5.9

   Jar or basin, body sherds.
   Max. dim. of largest sherd 5.7.
   Mirabello Fabric (between light red, 2.5YR 6/8, and red, 2.5YR 5/8).

112 (X 568, from L 18-2) Fig. 5.9

   Jar or bellows, body sherd.

THE POTTERY

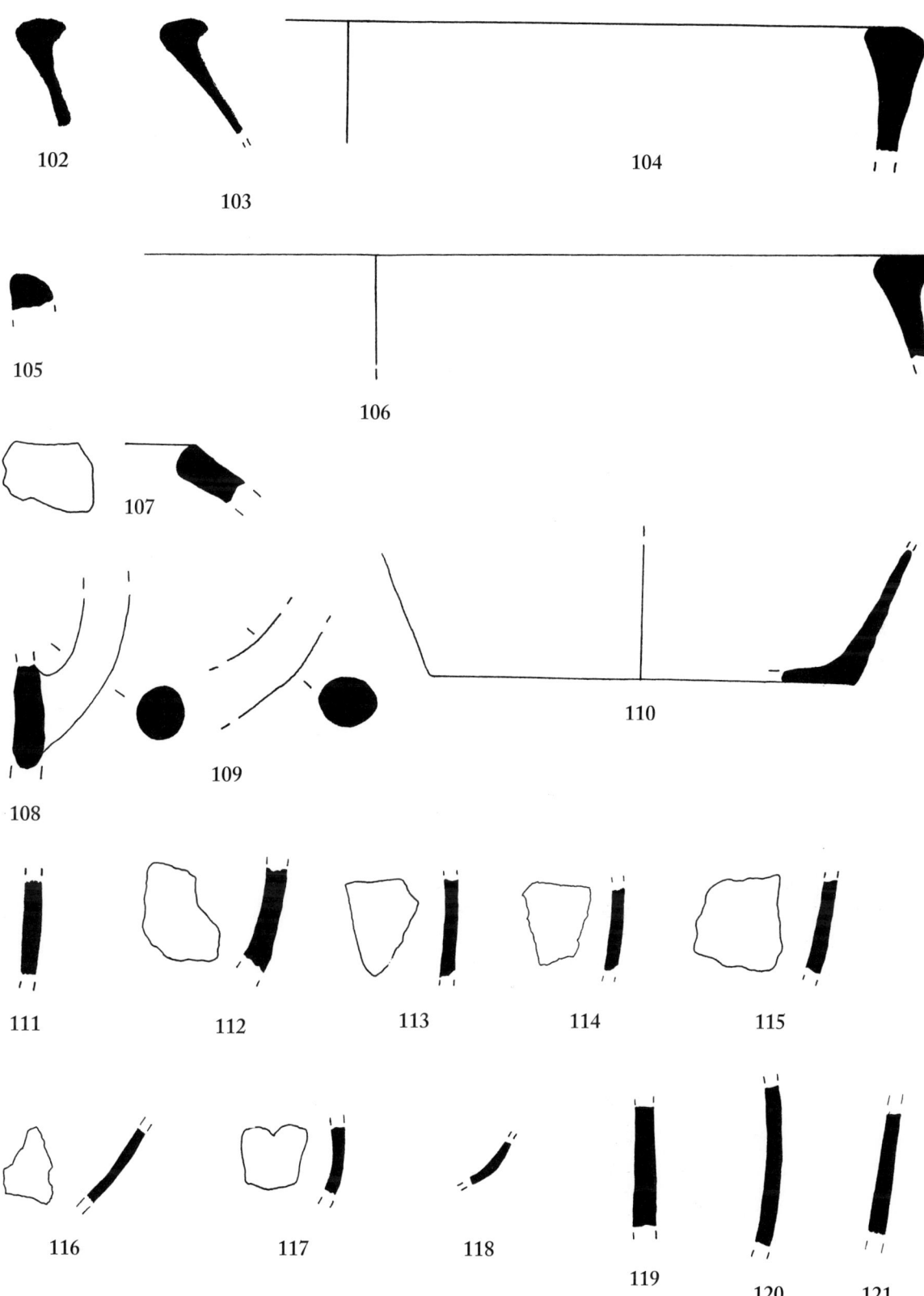

Figure 5.9. Pottery from the apsidal structure (**102–121**). Scale 1:3

Max. dim. 5.7.
Mirabello Fabric (reddish brown, 2.5YR 5/4).

113 (X 601, from M 19-3)   Fig. 5.9

Jar or bellows, body sherd.
Max. dim. 4.8.
Mirabello Fabric (unevenly colored, from red, 2.5YR 4/6, to dark reddish gray, 5YR 4/2).

114 (X 602, from M 19-3)   Fig. 5.9

Jar or bellows, body sherd.
Max. dim. 4.6.
Mirabello Fabric (reddish brown, 2.5YR 4/4, with a gray surface).

115 (X 610, from M 19-5)   Fig. 5.9

Jar or bellows, body sherd.
Max. dim. 5.6.
Mirabello Fabric (reddish brown, 2.5YR 4/4, with a gray surface).

116 (X 643, from N 19-4)   Fig. 5.9

Open vessel, body sherd.
Max. dim. 4.3.
Mirabello Fabric (yellowish red, 5YR 5/6).

117 (X 816, from M 19-3)   Fig. 5.9

Vessel, body sherds.
Max. dim. 3.2 and 3.8.
Mirabello Fabric (brown, 7.5YR 5/4) and (reddish yellow, 5YR 6/8).

118 (X 1660, from N 19-3)   Fig. 5.9

Vessel, body sherd.
Max. dim. ca. 2.8.
Mirabello Fabric (yellowish red, 5YR 4/6).

119 (X 1721, from N 19-2)   Fig. 5.9

Jar (?), body sherd.
Max. dim. 7.3.
Mirabello Fabric (yellowish red, 5YR 5/6).

120 (X 1737, from M 19-Surface)   Fig. 5.9

Vessel, body sherd.
Max. dim. 7.8.
Mirabello Fabric (dark reddish brown, 5YR 3/4, with yellowish red, 5YR 5/6, and dark gray core).

121 (X 1781, from N 19-3)   Fig. 5.9

Closed vessel, base sherd.
Max. dim. 5.9.
Mirabello Fabric (yellowish red, 5YR 5/6).

TABLE 5.8. EARLY MINOAN III–MIDDLE MINOAN IA SHERDS FROM THE APSIDAL STRUCTURE

	Rim	Handle	Base	Spout	Leg	Body	Whole	Total	Min. No.
1.							1	1	1
2.	6	4	1			2	1	14	5
3.						7		7	1
4.	6	1						7	4
5.	8					4		12	5
6.	1							1	1
7.	2		4			2	1	9	2
8.	5		4			2		11	4
9.	1							1	1
10.		2						2	2
11.			1					1	1
12.						16		16	5
Totals	29	7	10			33	3	82	32

Fine fabric sherds: (1) conical cup, painted (**75**, X 167); (2) rounded cup (**76–81**, X 148, X 182, X 210, X 415, X 534, X 636); (3) closed vessel, painted white on dark (**82–86**, X 149, X 181, X 271, X 466, X 1675). Medium coarse fabric sherds: (4) shallow bowl (**87–91**, X 145, X 146, X 168, X 274, X 1662). Mirabello Fabric sherds: (5) cooking dish (**92–98**, X 150, X 212, X 270, X 275, X 411, X 532, X 1661); (6) bowl (**99**, X 641); (7) bucket jar (**100, 101**, X 2, X 169); (8) jar, flat thickened rim (**102–106**, X 144, X 211, X 291, X 1659, X 1663); (9) bridge-spouted jar (**107**, X 530); (10) jar (**108, 109**, X 414, X 1664); (11) closed vessel (**110**, X 1658); (12) vessel (**111–121**, X 525, X 568, X 601, X 602, X 610, X 643, X 816, X 1660, X 1721, X 1737, X 1781).

TABLE 5.9. COMPARISON OF EARLY MINOAN III–MIDDLE MINOAN IA POTTERY SHAPES FROM THE SLAG PILE AND THE APSIDAL STRUCTURE

Shapes	Slag Pile	Apsidal Structure
Conical cup	?	yes
Rounded cup	yes	yes
Shallow bowl	yes	yes
Bowl	yes	yes
Bucket jar	yes	yes
Bridge-spouted jar	yes	yes
Jar with thickened, flat-topped rim	yes	yes
Miscellaneous closed vessel	yes	yes
Jug	yes	no
Cooking dish	yes	yes

CHAPTER 6

The Stone Tools

by Doniert Evely

Only a few stone tools were found in the workshop, in spite of a collecting strategy that saved 100% of the excavated material (Figs. 6.1–6.4). From the ca. 40–50 m³ of deposits excavated (ca. 20%–25% of the slag pile), comprised mostly of small pieces of slag and parts of demolished furnace chimneys, only 23 stone tools were recovered. The corpus includes three pieces of obsidian, one potential surface for working, and a number of limestone hand tools, along with some cuttings in bedrock (Fig. 6.3). This is a small number of tools for a site where the breaking up of furnace slags to retrieve copper prills was a major part of the working cycle.

The three pieces of obsidian from the workshop (**142**, **143**, and **144**) may be debris from working the material on site, as none of the pieces has either the size or other attributes that would merit identification as a tool. The obsidian, like at least some of the metal ores (see App. C), must have been imported from the Cyclades. It is a useful material, and its presence here is not surprising. So little has been recovered that nothing can be deduced from it except the likelihood that it was worked and used at the site.

The largest class of stone tools consists of water-worn limestone and dolomite cobbles. In terms of size and shape, most examples fall within a small range. They are essentially tools wielded in one hand, either hafted or unhafted. Lengths vary from 10–15 cm, and the weight is normally up to half a kilogram. No uniformity of shape can be recognized among the cuboid, bar-shaped, elliptical, ovoid, and nearly spherical forms. All the stones owe their shapes to the effects of water action, having been tumbled in watercourses and/or on the seashore so that their angles have become rounded and their surfaces worn to a smooth and even texture. The sole exception in the matter of size is **125**, a larger implement whose shape, though softened by the effects of water-erosion, resembles that of a large boulder, split lengthwise. The piece is heavy, and it was probably intended to provide an immobile surface on which some working activity (other than pounding) was carried out.

The tools seem to have been collected as natural cobbles, and none of them appears to have been deliberately worked in order to modify the natural shape. Suitable gathering places for the waterworn cobbles would have been found at the sea's edge, with the nearest readily accessible source being a pebbly beach at Agriomandra, at the mouth of a small ravine, about a kilometer away (west of the Chrysokamino farmhouse). All of the rounded limestone and dolomite pieces at the site, whether they were tools or not, were transported to the workshop by humans; they do not occur naturally among the angular, slab limestones and layered phyllites of the promontory.

An identification of tool types is difficult. Many of the pieces cannot be reliably identified as tools, nor can their function be established, because no wear marks are visible on the surface. Eight stones have been excluded from serious consideration, though their basic descriptions are included in the catalogue. At the most generous level of interpretation, five tools could be considered as pounders (**124**, **126**, **129**, **132**, and **133**), two were used for grinding actions (**122**, **123**), and one (**125**) may be regarded as a working surface or support. The uncertainty concerning use stems mainly from the uniformly poor state of the surfaces. This condition has been called "degraded" in the catalogue descriptions, and it is not clear if this state is the result of natural weathering, of the corrosive effects of the debris produced in the metalworking processes, or of a combination of several factors. In any case, the surfaces of the stones are somewhat powdery, and details are blurred. In addition, patinas and carbonate deposits (called concretion in the catalogue) are also present.

Few materials are represented within the corpus. Excluding the three pieces of (presumably Melian) obsidian, the tools are all native limestone and dolomite. The most common kind by far is a dark gray stone similar to carbonate rocks used as materials for tools throughout the Bronze Age in Crete. A single example is made of off-white to yellow limestone (**125**), and one tool is a carbonate with larger crystalline composition (**136**). Analysis by Myer (unpublished) shows that **131** is a limestone breccia. The dark gray carbonates would have made a reasonably suitable material for the percussive work likely to have been carried out at the site, although finer grained stones would have been better. However, sources of fine-grained limestones are rare in the immediate neighborhood (though they are obtainable closer to Gournia). The pale yellow tool (**125**) is made of an unusual limestone, and it might have been used for a different sort of activity from that of the other tools.

The small number of tools can be explained in several ways. Perhaps when tools were broken or damaged, they were slung to one side, an action that would almost certainly ensure their eventual disappearance down the slopes of the saddle into the sea. If any of the tools were hafted, they might have been taken away when work ceased.

Patterns of wear from usage can be seen on only a few of the limestone tools. They can be placed in the following categories:

1. The loss of small surface chips as a result of direct percussive action of the stone tool against an item of considerable hardness (such as slag)
2. The accidental removal of large flakes from the working areas of the tool

3. The creation, as a result of grinding action (possibly combined with a degree of crushing), of a surface texture varying from smooth to polished
4. Occasional scratches cut into the stone tool by the agitation of harder materials in the course of work (such as stray grains of sand or glassy silica-based substances in the slags)

In addition to the tools that have been recovered, one would expect to find some additional items. Apart from more pounders and grinders, the most obviously lacking set of tools are items with a firm surface against which the slag could have been broken up. The single instance of a stone-working surface (**125**) was not used in this way. It is possible that the limestone outcrops on the hillside, with natural slablike patterns of cleavage, would have supplied a ready supply of working surfaces, although no damaged areas were encountered in the excavation. Two interconnected circular depressions (35 cm across and up to 7 cm deep) were found in situ in the virgin bedrock at the site (Figs. 6.2, 6.3), and they could have been used in this way, but one would expect that many more would have been created.

For the rest of the manufacturing processes undertaken at the site, the same pounders, grinders, and (mostly missing) working surfaces would have sufficed for breaking up the charcoal and other fuels and demolishing the furnaces after each smelt. Sharper edges for cutting and slicing actions (obsidian and metal blades) or for chopping actions (axes) have relatively minor roles to play in the craft activities envisaged for Chrysokamino. Cooking and food preparation activities, however, if occurring in and near the small hut, might have required obsidian and other cutting tools as well as pounding and grinding tools. It is assumed that the clay and straw mix consumed in such quantities in forming the furnace chimneys was probably prepared away from the smelting site and brought in ready for immediate use or long-term storage.

CATALOGUE

GROUND STONE TOOLS

122 (X 451, from N 17-2) Figs. 6.1, 6.4

  Hand tool (?), fragment; ends lost, sides broken.
  Max. L. 10.7; W. 4.9; Th. 3.9; Wt. 333 g.
  Limestone (gray); layered texture.
  Rod-shape with somewhat triangular section; flat face smooth to touch, angled (rounded) face slightly smoother.
  No scratches visible.
  Comments: The smooth zones suggest possible grinding or rubbing actions; missing ends might imply pounding (with loss due to fracture).

123 (X 681, from O 10-4) Figs. 6.1, 6.4

  Hand tool (?), complete; surface degraded.
  L. 10.5; W. 6.3; Th. 4.9; Wt. 580 g.
  Dolomite (very pale gray to off-white).

102 CHAPTER 6

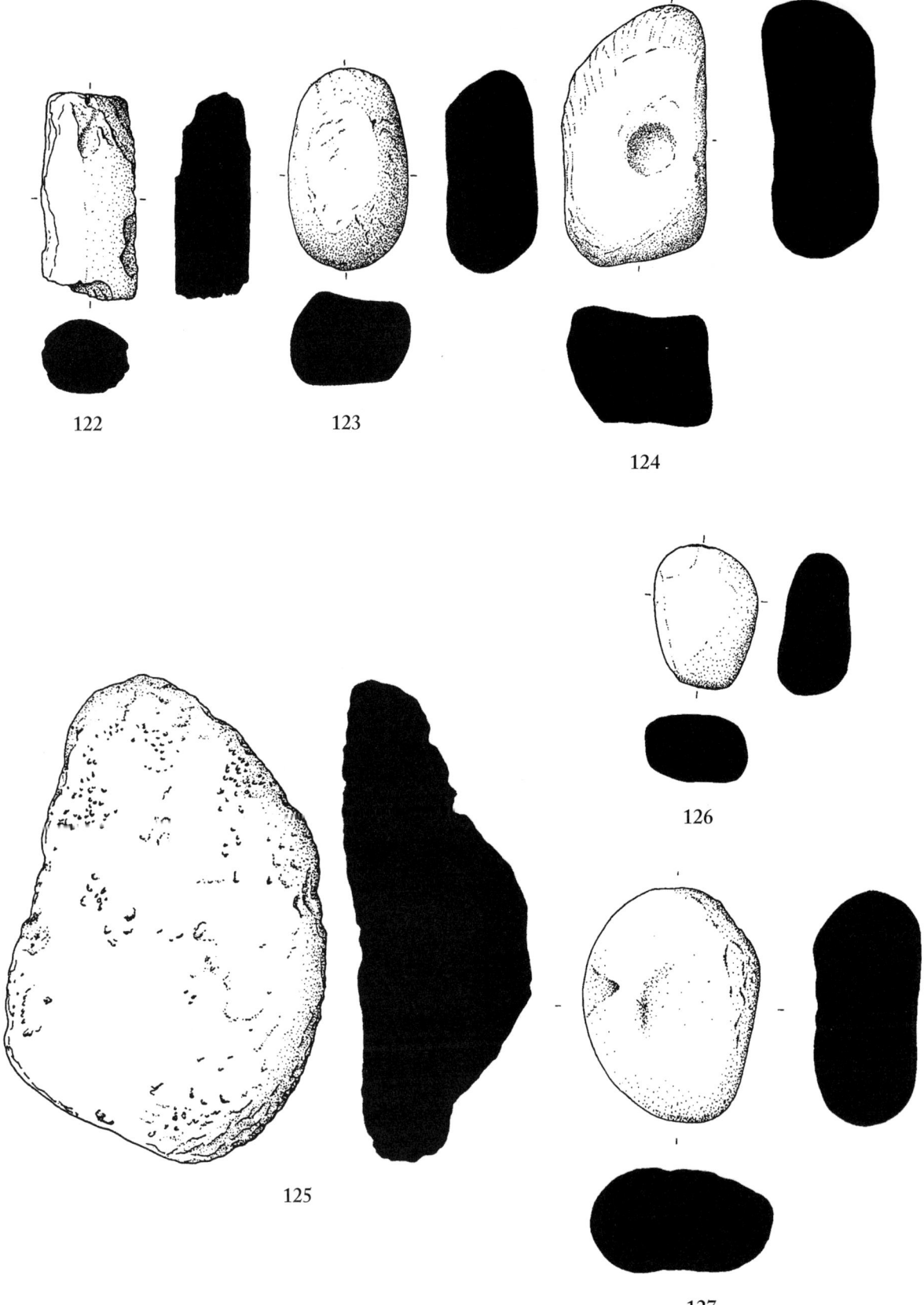

122

123

124

125

126

127

Figure 6.1 *(opposite).* **Stone tools (122–127).** Scale 1:3

Elongated cobble.
Sides and ends show no signs of use. The two main surfaces each have a flat plane, but are not especially smooth; no scratches.
Comments: The flatter zones might argue for grinding or rubbing action.

124   (X 712, from N 20-2)   Figs. 6.1, 6.4

Hand tool, complete; surface degraded.
L. 13.4; W. 7.5; Th. 5.9; Wt. 1.220 kg.
Limestone (pale gray to gray).
Cuboid appearance, with rounded angles; a natural cobble.
Possible ghosts of pounding damage on narrower end. Main faces have slight depression at approximate centers (the more distinct area is 2.5 cm across and 0.3 cm deep).
Comments: Pounder-hammer, perhaps hafted.

125   (X 453, from O 18-1-Surface)   Fig. 6.1

Support (?), complete; stone naturally fractured and damaged by the sea.
L. 23.1; W. 16.1; Th. 7.1; Wt 3.555 kg.
Limestone (off-white and yellow).
Somewhat amorphous in form; natural boulder (split?).
One surface is fairly flat and smooth to the touch, so it was not subjected to percussion, but only to being ground or rubbed.
Comments: The object may have been a passive surface on which work was performed.

126   (X 942, from J 18-Surface)   Fig. 6.1

Hand tool (?), complete; surface degraded.
L. 7.5; W. 5.4; Th. 3.7; Wt. 208 g.
Limestone (gray).
Somewhat wedge-shaped and slightly irregular.
Possible small zone of pounding/bruising activities at either end, more so on smaller end.
Comments: The piece was perhaps a pounder.

127   (X 667, from M 17-2)   Figs. 6.1, 6.4

Hand tool (?), complete; concretion and surface degraded (by heat?).
L. 12.1; W. 9.3; Th. 5.6; Wt. 954 g.
Limestone (pale gray); some breccia adheres to one side.
Ovoid, flattish.
Surface so destroyed that no wear marks can be seen.
Comments: The function is not obvious; the piece would suit as a pounder.

128   (X 217, from P 20-2)   Figs. 6.2, 6.4

Hand tool (?), fragment; end and middle section of larger cobble (approximately a quarter survives).
Max. L. 8.3; max. W. 4.6; Th. 8.0.
Dolomite (pale gray to gray).
From a cuboid (?) cobble.
Possible remnants of a flatter plane on side, either natural or from use.
Comments: The function is not obvious.

129   (X 452, from N 17-2)   Fig. 6.2, 6.4

Hand tool (?), complete.

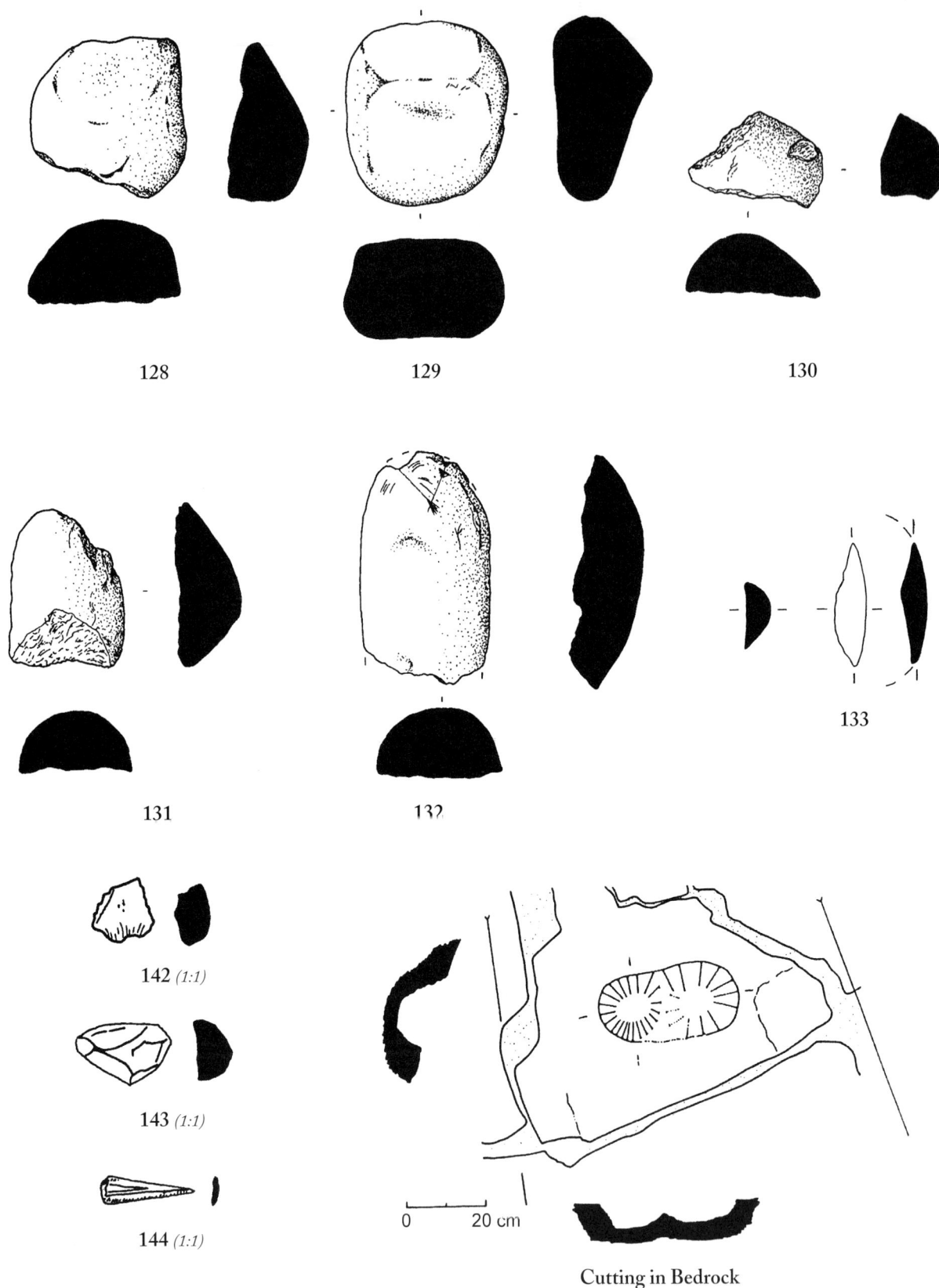

Figure 6.2 *(opposite)*. **Stone tools (128–133), obsidian tools (142–144), and cutting in bedrock.** Scale 1:3 unless otherwise indicated

L. 5.5; W. 8.5; Th. 5.3; Wt. 657 g.
Dolomite (pale gray); soft.
Humped cobble; rounded angles from water action, with one more angular. Perhaps a hint of pounding damage over small portion of one end.
Comments: The function is not obvious.

130   (X 232, from L 21-1)   Fig. 6.2

Hand tool (?), fragment; part of end of a cobble; surface damaged and discolored by exposure to heat.
Dims. 7.1 × 4.6 × 3.0.
Limestone (gray).
No traces of work discernible now.
Comments: The function is not obvious.

131   (X 768, from N 30-3)   Fig. 6.2

Hand tool (?), fragment; part of side of larger cobble.
Dims. 8.1 × 6.0 × 3.2.
Limestone breccia (black and rose elements).
No traces of work discernible.
Comments: The function is not obvious. The material of the tool was analyzed by optical microscopy by George Myer, Department of Geology, Temple University.

132   (X 1753, from T 23-Cleaning)   Fig. 6.2

Hand tool, fragment; part of a larger cobble.
P. L. 12.0.
Dolomite (pale gray).
Rectangular shape.
Peck marks on extant end.
Comments: The piece was possibly broken from use as a pounder.

133   (X 1866, from K 20-Cleaning)   Fig. 6.2

Hand tool, fragment of larger cobble; small piece.
Max. L. 6.3.
Limestone (gray).
Comments: No signs of use, but the fragment may have broken off the cobble in a percussive action.

134   (X 637, from O 20-3)

Hand tool, complete; surface degraded.
L. 11.3.
Limestone (pale gray to gray); partly calcined.
Ovoid, humped cobble.
Comments: No signs of wear, though surface below the calcined (?) crust might indicate pounding. No clear signs of use.

135   (X 455, from Q 21-Surface)

Hand tool, complete; surface cracked and fissured heavily.
L. 11.2.
Limestone (gray).
Ovoid, flattened cobble.
Comments: No signs of use.

Figure 6.3. Cutting in bedrock at the metallurgy workshop

Figure 6.4. Stone tools (**122–124** and **127–129**) from the metallurgy site

136 (X 495, from O 20-2)

Hand tool, nearly complete; large flakes or segments lost.
L. 11.9.
Limestone/calcite (?) (gray and off-white).
Ovoid, flattened cobble; stone fissure throughout.
Comments: No signs of use.

137 (X 496, from O 20-2)

Hand tool, about half missing (split lengthways); surface condition poor in places.
L. 8.7.
Limestone (pale gray).
Subrectangular, cuboid cobble.
Comments: No signs of use.

138 (X 456, from N 17-2)

Hand tool, complete; surface condition poor.
L. 7.6.
Limestone (dark gray).
Humped cobble.
Comments: No signs of use.

139 (X 442, from L 19-2)

Hand tool, complete; concretion on surface.
L. 6.8; W. 2.8.
Limestone (gray to dark gray).
Ovoid pebble.
Comments: No signs of use. Found on the middle floor of the apsidal building (see Fig. 4.6).

140 (X 454, from O 18-1)

Hand tool, complete; concretion (calcined crust) on part of surface.
L. 4.8.
Limestone (gray to pale gray).
Ovoid pebble.
Comments: No signs of use.

141 (X 231, from L 21-1)

Hand tool, fragment of large cobble; ca. one quarter survives; concretion on surface.
Max. L. 3.8.
Limestone (gray).
Ovoid, somewhat spherical.
Comments: No signs of use.

Obsidian

142 (X 945, from J 18-1)   Fig. 6.2

Small chunk/flake, complete.
Dims. 1.0 × 0.9; Th. 0.5; Wt. ca. 1–2.2 g.
Obsidian (black); fairly lustrous.

Removed in core preparation; slight elements of cortex on one side.

Comments: The function is not obvious (too small), but the presence of the flake implies obsidian preparation on the site.

143   (X 944, from J 18-1) Fig. 6.2

Chunk, complete.
Dims. 1.5 × 1.0; Th. 0.65; Wt. 4 g.
Obsidian (black); lustrous.

Comments: The fragment is either a result of core preparation or of accidental severance at the tip of the core. The function of this piece is not obvious, but its presence implies the use of obsidian on the site.

144   (X 1208, from M 18-2) Fig. 6.2

Bladelet, complete.
L. 1.6; W. 0.5; Th. under 0.5 mm; Wt. below 1 g.
Obsidian (black); semilustrous.
Very fine, tapering.

Comments: No signs of edge damage. This piece is probably debris from working a core. The function of the flake is unclear, but its presence shows either the use or the working of this material on the site.

CHAPTER 7

THE FURNACE CHIMNEY FRAGMENTS

by Philip P. Betancourt

Photographs of the metallurgy location taken before it was excavated aptly illustrate the many small fragments of industrial ceramics present at Chrysokamino. Hundreds of thousands of pieces survive in this artifact class, in sizes that range from microscopic to over 10 cm in the longest dimension. Most of the pieces are tiny, and they come from every level and every square meter of the deposit of slag on the site. No stylistic development can be discerned within the class.

Fragments of ceramic vessels in the deposit of slag provide the best clues for dating, but problems exist with this evidence. Pieces of Final Neolithic vases come from the modern surface as well as from deep within the deposit, indicating that the debris is mixed. In addition, most parts of the slag pile contain no vessel sherds at all. Because of these problems with the context, none of the individual industrial ceramics fragments can be assigned a secure date based on the stratigraphic level.

Typical examples of the industrial ceramics are presented in the catalogue at the end of this chapter. The catalogue includes pieces from bases, bodies, and rims of cylindrical clay objects with open tops. The fragments provide a large enough selection to illustrate the range.

The total number of fragments present from the site is enormous. Besides the fragments on the modern land surface, many pieces have eroded into the sea, so that the original dimensions of the deposit are not known. The extant size of the deposit is ca. 200 m^2, with a depth that varies from nothing to ca. 60 cm.

The fragments have a slight curvature, and computation of the degree of arc indicates they are from cylinders with diameters of ca. 16–44 cm. Rim sherds prove that one or both ends of the cylinders were open, with straight to slightly everted rims that were usually slightly rounded at their upper edge. No base sherds from the center of any base are present, but a few angular pieces from the point at which a vertical wall met a base suggest that the cylinders sometimes had small ledges at the bottom to allow them to sit with more stability. Wall thicknesses are usually in the range of 1–2 cm. Most of the bases have diameters that are substantially larger than most of the rims, suggesting that the shapes were usually tapered.

Figure 7.1. Front and back of chimney fragment **180**, showing dark, glassy deposit on the interior (right)

The rim diameters generally measure 20–30 cm, while the bases measure slightly more than 40 cm in diameter.

No decoration or slip occurs on any fragment, but circular holes made before the clay was fired are present on all fragments of more than a few centimeters in width (Fig. 7.1). Diameters of the holes average about 2 cm. In all cases, the configuration of the holes shows that they were made by thrusting something through the wall from the exterior of the cylinder, either at a right angle to the wall or at a slight oblique angle. Experiments with moist clay cylinders of the correct thickness indicate that these holes were almost certainly made by the potters by thrusting their fingers through the clay. Holes are irregularly spaced, typically between 5 and 15 cm apart, and randomly distributed from base to rim.

The fired fabric is red on the exterior of the cylinder and dark gray to black on the interior. It is coarse and porous with many voids from burned-out organic matter. A detailed petrographic description of the fabric is presented in Appendix A. The petrographic analysis and the study of the impressions in the fabric show that the organic material is mostly chaff (short lengths of grain stems remaining as waste material after threshing). Among the molds from burned out chaff are voids in the shape of barley grains, unidentifiable other plant remains, and one olive leaf (Fig. 7.2).[1] Stone fragments in the fabric are not like the igneous rocks found in the Mirabello Fabric, which was used for most pottery at the site. They consist, instead, of carbonate, crystalline and cryptocrystalline quartz, and phyllite fragments, minerals that are typical inclusions in the local soils at Chrysokamino and that also occur in the clay of Lakkos Ambeliou.[2] The similarity with the clay deposited at Lakkos Ambeliou is close enough to suggest that this feature was probably used as the source for the clay to make the cylinders.

Many of the fragments have a coating of dark, glassy, highly vitrified material on the interior side of their curved surfaces (see Chap. 10). This deposit is not usually present on rim sherds. It was once highly viscous, with drips, flow lines, and circular voids attesting to its former liquid state. On some sherds, the coating was once thicker than presently preserved, and it has broken away, while on other sherds, the surface of the vitrified coating is smooth. Small lenses of copper alteration products under 0.5 cm occur occasionally within the glassy coating.

Figure 7.2. Chimney fragment no. X 213, preserving the impression of an olive leaf, compared with a leaf from a modern olive tree

1. For discussion of the botanical remains preserved in the clay of these chimneys, see Chap. 12.
2. Morris 2002, pp. 52–59.

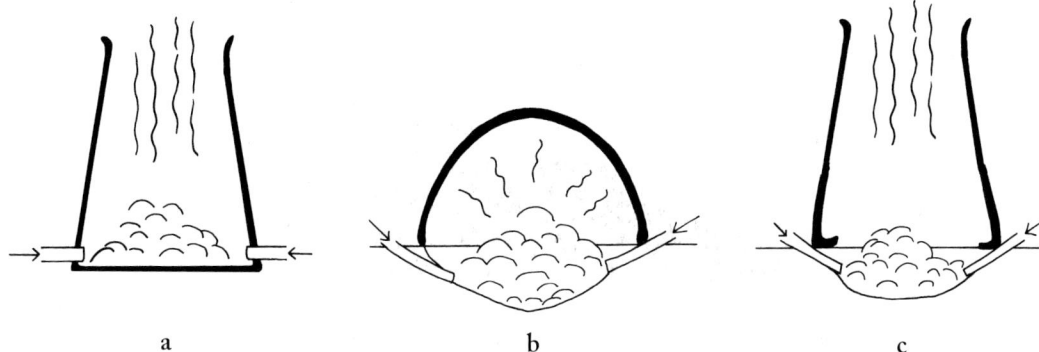

Figure 7.3. Alternative reconstructions of the industrial ceramics fragments: (a) a cylindrical furnace; (b) a cover for a bowl furnace; (c) a chimney for a bowl furnace

THE FUNCTION OF THE INDUSTRIAL CERAMICS

The industrial ceramics fragments differ in several ways from the sherds of clay vessels found at Chrysokamino. Their fabric is distinctive, their shapes are not those of Minoan vessels, and their coating of glassy material is not consistent with household use. The presence of copper alteration products in the vitreous deposit proves the pieces played a role in copper metallurgy.

The fragments were obviously part of some aspect of the smelting process. Previous suggestions include the possibilities that the small fragments were parts of crucibles[3] or that they joined together to form a pierced floor in a single large furnace.[4] The restoration of the fragments as cylinders invalidates these suggestions. The objects were freestanding, tapering cylinders, a shape that considerably narrows the range of possibilities for their use. Alternatives are illustrated in Figure 7.3.

Freestanding, cylindrical smelting furnaces (Fig. 7.3:a) have a long history in metallurgy. They were used in Nigeria until the Medieval period[5] and in Shaba Province, Congo (formerly called Katanga Province, Zaire), until more recently.[6] The furnaces in Nigeria and Congo were used for smelting, and they were only used once before being broken up into pieces.

The possibility that the Chrysokamino cylinders were also freestanding furnaces has been considered,[7] but the evidence indicates that this theory is not tenable. If the cylinders were the furnaces in which the ore was smelted, smelting slag would be found adhering to their inner surfaces, at least in their lower parts. The interiors do not have any such smelting slag attached.

Covers (Fig. 7.3:b) are another alternative. Small bowl furnaces lose heat upward, and a cover to help hold the heat and increase efficiency would be an advantage in maintaining high temperatures. However, since a number of rim fragments can be recognized in the corpus of fragments, the cylinders must have had open tops.

Chimneys are the third alternative (Fig. 7.3:c). They are an important aspect of furnace or kiln design (and also of fireplaces), because proper venting is necessary in order to raise the temperature by increasing the flow of oxygen.[8] In systems using only a natural draft, the exit venting system must be as large as the inlet system because if the outlet flue is restricted, combustion will be slowed, and efficiency will be lowered. If the system adds

3. Mosso 1910, p. 291; Branigan 1968, p. 50.
4. Faure 1966, p. 47–48.
5. Bernus 1983; Bernus and Echard 1985.
6. Forbes 1972, pp. 26–27.
7. Betancourt et al. 1999, p. 362.
8. For the principles involved, see the discussion in Olsen 1973, pp. 36–39.

more oxygen by employing artificial means such as bellows to increase the draft, then the venting system is even more critical. Tall chimneys increase the velocity of the upward draft, driving the temperature higher as long as proper fuel is present and adequate air is forced into the firing chamber. At Chrysokamino, where we know that bellows were used, chimneys would have been a logical part of the furnace system.

The identification of the Chrysokamino cylinders as chimneys is further supported by the presence of the glassy coating on the interior of the fragments. This coating was formed by the condensation of hot vapors that cooled enough to leave a deposit as they moved upward through the cylinders. Such vitreous deposits occur in any situation where the heating process volatilizes mineral-laden materials, and they are common in the upper parts of both lime and pottery kilns. The presence of cuprous inclusions within the coatings at Chrysokamino shows that these particular deposits were made from the intense heating of minerals that included copper.

Clay cylinders used for smelting are not unique to Chrysokamino. In the Aegean, the closest parallels are with ceramic cylinders from smelting sites on Kythnos.[9] The fragments from Kythnos are similar to the ones from Chrysokamino. Like the Cretan pieces, they seem to come from tapering cylinders with open rims (and possibly open bases). They are made of coarse clay, with holes pierced through the walls. The holes, often about 2–3 cm in diameter, are spaced irregularly from rim to base. Evidently the pierced cylinders are part of a regional Aegean technology used for the production of copper.

Reconstruction of the Chimneys

From the curvature of the walls, the clay fragments must be from cylinders with diameters ranging from 16 to 44 cm. Many holes are present in the walls, occurring in a random pattern from just above the base to just below the rim. The holes would have allowed drafts of air to enter and exit. The upper part is open. The bases were probably not closed, because although a few pieces of their angled edges can be identified, no fragments coming from the bottom away from the edge are present. If the cylinders were chimneys, as suggested by the glassy deposits on their interiors, then they had no bases. They must have been used over bowl furnaces built into the ground in locations that have not been discovered.[10]

Some of the details are uncertain. Heights are not known. While it is likely that the chimneys tapered slightly toward the top, with the rims having smaller diameters than the bases, the surviving fragments are all too small to establish their shapes with certainty. If pot bellows were used with the furnaces, the most efficient system would have been for the nozzles to be inserted into apertures of the bowl furnaces, not into their chimneys, because the oxygen should be directed at the fire, not above it. One tuyere was recovered from the excavations, indicating the workshop used the usual method for conveying the draft from the bellows (or an earlier blowpipe) to the fire (see Chap. 9).

The interior diameters of the bowl furnaces may have been about the same as the maximum diameters of the chimneys. The largest diameter,

9. Y. Bassiakos (pers. comm.).
10. A fragment of the furnace lining is discussed in Chap. 9.

A coarse fabric with voids from burned-out organic matter (unevenly colored, mostly black).

Furnace chimney with straight rim; three holes below the rim; rough interior and exterior; vitreous deposit on lower part of the interior.

146   (X 111, from O 22-4)   Fig. 7.4

Furnace chimney, rim fragment.
Diam. of rim ca. 30.
A coarse fabric with voids from burned-out organic matter (unevenly colored, mostly dark reddish brown, 5YR 3/2).
Furnace chimney with outturned rim; rough interior and exterior; no vitreous deposit.

147   (X 112, from O 22-4)   Fig. 7.4

Furnace chimney, rim fragment.
Diam. of rim ca. 36.
A coarse fabric with voids from burned-out organic matter (unevenly colored, mostly yellowish red, 5YR 4/6).
Furnace chimney with straight rim; hole below the rim; rough interior and exterior; vitreous deposit on lower part of the interior.

148   (X 113, from Q 20-2)   Fig. 7.4

Furnace chimney, rim fragment.
Diam. of rim ca. 36.
A coarse fabric with voids from burned-out organic matter (unevenly colored, mostly yellowish red, 5YR 4/6).
Furnace chimney with outturned rim; hole below the rim; rough interior and exterior; no vitreous deposit.

149   (X 198, from Q 20-3)   Fig. 7.4

Furnace chimney, rim fragment.
Diam. of rim ca. 32.
A coarse fabric with voids from burned-out organic matter (unevenly colored, from reddish yellow, 5YR 6/6, to black).
Furnace chimney with straight rim; hole below the rim; rough interior and exterior; no vitreous deposit.

150   (X 222, from R 20-2)   Fig. 7.4

Furnace chimney, rim fragment.
Diam. of rim ca. 18.
A coarse fabric with voids from burned-out organic matter (unevenly colored, mostly very dark brown 10YR 2/2).
Furnace chimney with straight rim; two holes below the rim; rough interior and exterior; no vitreous deposit.

151   (X 299, from M 20-6)   Fig. 7.4

Furnace chimney, rim fragment.
Diam. of rim ca. 54.
A coarse fabric with voids from burned-out organic matter (unevenly colored, mostly yellowish red, 5YR 5/6).
Furnace chimney with straight rim; hole below the rim; rough interior and exterior; no vitreous deposit.

152 (X 319, from R 20-3)   Fig. 7.4

   Furnace chimney, rim fragment.
   Diam. of rim ca. 34.
   A coarse fabric with voids from burned-out organic matter (unevenly colored, mostly black).
   Furnace chimney with straight rim; rough interior and exterior; no vitreous deposit.

153 (X 320, from R 20-3)   Fig. 7.4

   Furnace chimney, rim fragment.
   Diam. of rim not measurable; max. dim. 3.5.
   A coarse fabric with voids from burned-out organic matter (unevenly colored, mostly dark reddish brown, 5YR 3/2).
   Furnace chimney with straight rim; rough interior and exterior; no vitreous deposit.

154 (X 321, from R 20-3)   Fig. 7.4

   Furnace chimney, rim fragment.
   Diam. of rim ca. 24.
   A coarse fabric with voids from burned-out organic matter (unevenly colored, from reddish yellow, 5YR 4/6, to very dark gray).
   Furnace chimney with straight rim; three holes below the rim; rough interior and exterior; no vitreous deposit.

155 (X 426, from N 18-3)   Fig. 7.4

   Furnace chimney, rim fragment.
   Diam. of rim ca. 24.
   A coarse fabric with voids from burned-out organic matter (unevenly colored, mostly dark gray).
   Furnace chimney with straight rim; rough interior and exterior; no vitreous deposit.

156 (X 427, from N 18-3)   Fig. 7.4

   Furnace chimney, rim fragment.
   Diam. of rim ca. 26.
   A coarse fabric with voids from burned-out organic matter (unevenly colored, from red, 2.5YR 5/6, to dark gray).
   Furnace chimney with straight rim; rough interior and exterior; no vitreous deposit.

157 (X 429, from N 18-3)   Fig. 7.4

   Furnace chimney, rim fragment.
   Diam. of rim ca. 26.
   A coarse fabric with voids from burned-out organic matter (unevenly colored, from reddish yellow, 5YR 6/6, to black).
   Furnace chimney with straight rim; rough interior and exterior; no vitreous deposit.

158 (X 491, from P 20-2)   Fig. 7.4

   Furnace chimney, rim fragment.
   Diam. of rim ca. 50 (or from a flat area on a warped or elliptical artifact).
   A coarse fabric with voids from burned-out organic matter (unevenly colored, from yellowish red, 5YR 5/8, to black).

Furnace chimney with straight rim; hole below the rim; rough interior and exterior; no vitreous deposit.

159 (X 500, from L 19-3)   Fig. 7.4

Furnace chimney, rim fragment.
Diam. of rim ca. 28.
A coarse fabric with voids from burned-out organic matter (unevenly colored, from yellowish red, 5YR 4/6, to very dark gray).
Furnace chimney with straight rim; rough interior and exterior; no vitreous deposit.

160 (X 640, from N 19-4)   Fig. 7.4

Furnace chimney, rim fragment.
Diam. of rim ca. 32.
A coarse fabric with voids from burned-out organic matter (unevenly colored, from red, 2.5YR 4/6, to gray).
Furnace chimney with straight rim; rough interior and exterior; no vitreous deposit.

161 (X 752, from N 20-4)   Fig. 7.4

Furnace chimney, rim fragment.
Diam. of rim ca. 28.
A coarse fabric with voids from burned-out organic matter (unevenly colored, from yellowish red, 5YR 5/6, to very dark gray).
Furnace chimney with outturned rim; rough interior and exterior; no vitreous deposit.

162 (X 753, from N 20-4)   Fig. 7.4

Furnace chimney, rim fragment.
Diam. of rim ca. 22.
A coarse fabric with voids from burned-out organic matter (unevenly colored, from yellowish red, 5YR 4/6, to dark reddish brown, 5YR 3/2).
Furnace chimney with outturned rim; rough interior and exterior; no vitreous deposit.

Base Fragments

163 (X 100, from S 20-2)   Fig. 7.5

Furnace chimney, base fragment.
Diam. of base ca. 42.
A coarse fabric with voids from burned-out organic matter (unevenly colored, mostly red, 2.5YR 5/8, with a black core).
Furnace chimney with flat base at bottom of wall; rough interior and exterior; vitreous deposit on the interior.

164 (X 110, from O 22-4)   Fig. 7.5

Furnace chimney, fragment from near base.
Diam. of base ca. 35–40.
A coarse fabric with voids from burned-out organic matter (unevenly colored, mostly red, 2.5YR 5/6).
Rough interior and exterior; vitreous deposit on the interior.

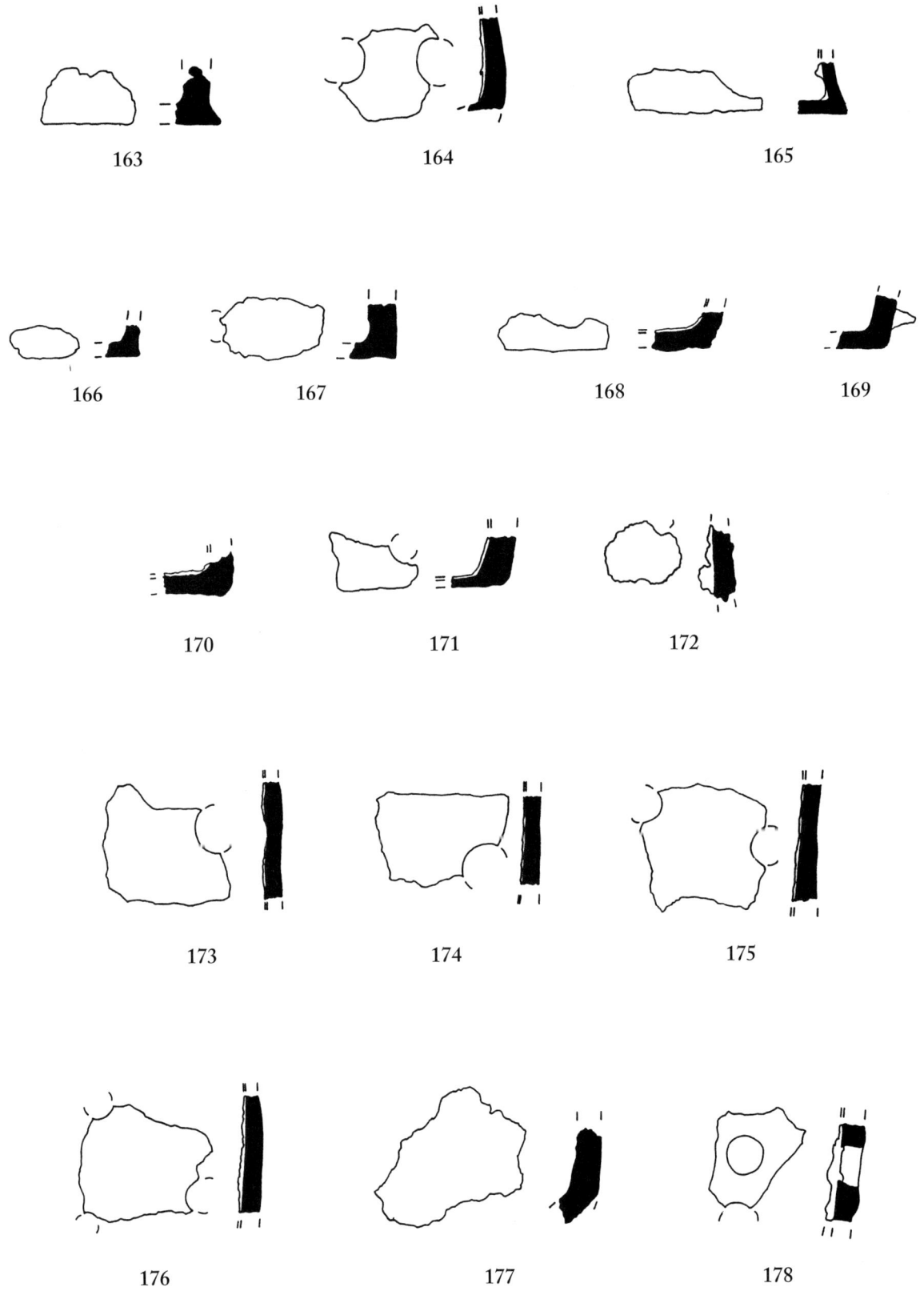

Figure 7.5. Profile drawings of chimney fragments (**163–178**). Scale 1:3

165 (X 114, from Q 20-3) Fig. 7.5

Furnace chimney, base fragment.
Diam. of base ca. 44.
A coarse fabric with voids from burned-out organic matter (unevenly colored, mostly red, 2.5YR 4/6, with a black interior).
Furnace chimney with flat base at bottom of wall; rough interior and exterior; vitreous deposit on the interior.

166 (X 129, from S 20-2) Fig. 7.5

Furnace chimney, base fragment.
Diam. of base not measurable.
A coarse fabric with voids from burned-out organic matter (unevenly colored, mostly red, 2.5YR 4/8, with black interior).
Furnace chimney with flat base at bottom of wall; two holes; rough interior and exterior; vitreous deposit on the interior.

167 (X 205, from O 22-4) Fig. 7.5

Furnace chimney, base fragment.
Diam. of base ca. 22.
A coarse fabric with voids from burned-out organic matter (unevenly colored, mostly very dark gray).
Furnace chimney with flat base at bottom of wall; rough interior and exterior; vitreous deposit on the interior.

168 (X 214, from R 20-2) Fig. 7.5

Furnace chimney, base fragment.
Diam. of base ca. 44.
A coarse fabric with voids from burned-out organic matter (black).
Furnace chimney with flat base at bottom of wall; rough interior and exterior; vitreous deposit on the interior.

169 (X 129, from N 20-2) Fig. 7.5

Furnace chimney, base fragment.
Diam. of base not measurable.
A coarse fabric with voids from burned-out organic matter (unevenly colored, mostly red, 2.5YR 4/8, with a black interior).
Furnace chimney with flat base at bottom of wall; rough interior and exterior; vitreous deposit on the interior and exterior.

170 (X 739, from N 20-3) Fig. 7.5

Furnace chimney, base fragment.
Diam. of base not measurable; max. dim. of sherd 4.3.
A coarse fabric with voids from burned-out organic matter (unevenly colored, from yellowish red, 5YR 5/6, to black).
Furnace chimney with flat base at bottom of wall; rough interior and exterior; vitreous deposit on the interior.

171 (X 188, from O 22-4) Fig. 7.5

Furnace chimney, base fragment.
Diam. of base ca. 28.
A coarse fabric with voids from burned-out organic matter (unevenly colored, mostly dark reddish brown, 5YR 3/2).
Furnace chimney with flat base at bottom of wall; rough interior and exterior; one hole; vitreous deposit on the interior.

Body Fragments

172 (X 1757, from P 20-2) Fig. 7.5

Furnace chimney, body fragment.
Max. dim. 4.1.
A coarse fabric with voids from burned-out organic matter (unevenly colored, mostly red, 2.5YR 4/8).
Furnace chimney with straight wall; one hole; unusually thick vitreous deposit on the interior.

173 (X 105, from O 22-4) Fig. 7.5

Furnace chimney, body fragment.
Max. dim. 7.6.
A coarse fabric with voids from burned-out organic matter (unevenly colored, mostly red, 2.5YR 5/8).
Furnace chimney with straight wall; one hole; vitreous deposit on the interior.

174 (X 106, from O 22-4) Fig. 7.5

Furnace chimney, body fragment.
Max. dim. 7.1.
A coarse fabric with voids from burned-out organic matter (unevenly colored, mostly red, 5YR 4/6).
Furnace chimney with straight wall; one hole; vitreous deposit on the interior.

175 (X 107, from O 22-4) Fig. 7.5

Furnace chimney, body fragment.
Max. dim. 7.3.
A coarse fabric with voids from burned-out organic matter (unevenly colored, mostly red, 2.5YR 4/6, to gray).
Furnace chimney with straight wall; two holes; vitreous deposit on the interior.

176 (X 108, from O 22-4) Fig. 7.5

Furnace chimney, body fragment.
Max. dim. 7.3.
A coarse fabric with voids from burned-out organic matter (unevenly colored, mostly yellowish red, 5YR 4/6).
Furnace chimney with straight wall; three holes; vitreous deposit on the interior.

177 (X 109, from O 22-4) Fig. 7.5

Furnace chimney, body fragment.
Max. dim. 8.4.
A coarse fabric with voids from burned-out organic matter (unevenly colored, mostly very dark gray).
Furnace chimney with straight wall; thickened area; vitreous deposit on the interior.

178 (X 120, from M 19-2) Fig. 7.5

Furnace chimney, body fragment.
Max. dim. 5.9.

A coarse fabric with voids from burned-out organic matter (unevenly colored, mostly yellowish red, 5YR 5/6).

Furnace chimney with straight wall; two holes; unusually thick vitreous deposit on the interior.

179   (X 130, from S 20-2)   Fig. 7.6

Furnace chimney, body fragment.
Max. dim. 4.2.

A coarse fabric with voids from burned-out organic matter (unevenly colored, mostly reddish yellow, 5YR 6/8), with a dark gray interior.

Furnace chimney with straight wall; one hole; vitreous deposit on the interior.

180   (X 140, from S 20-2)   Figs. 7.1, 7.6

Furnace chimney, body fragment.
Diam. ca. 29.

A coarse fabric with voids from burned-out organic matter (unevenly colored, from reddish brown, 2.5YR 5.4, to black).

Furnace chimney with straight wall; six holes; vitreous deposit on the interior.

181   (X 215, from S 20-2)   Fig. 7.6

Furnace chimney, body fragment.
Max. dim. 3.9.

A coarse fabric with voids from burned-out organic matter (unevenly colored, from dark grayish brown, 10YR 4/2 to black; light red, 2.5YR 6/6, on exterior).

Furnace chimney with straight wall; one hole; vitreous deposit on the interior.

182   (X 310, from M 20-3)   Fig. 7.6

Furnace chimney, body fragment.
Diam. not measurable; max. dim. 5.9.

A coarse fabric with voids from burned-out organic matter (unevenly colored, from reddish yellow, 5YR 6/8, to dark gray).

Furnace chimney with straight wall; hole somewhat larger than usual (Diam. 2.3); rough interior and exterior; no vitreous deposit.

183   (X 323, from R 20-3)   Fig. 7.6

Furnace chimney, body fragment.
Diam. ca. 27–30.

A coarse fabric with voids from burned-out organic matter (unevenly colored, from light red, 2.5YR 6/6, to black).

Furnace chimney with straight wall; three holes; vitreous deposit on the interior.

184   (X 324, from R 20-3)   Fig. 7.6

Furnace chimney, body fragment.
Diam. ca. 30.

A coarse fabric with voids from burned-out organic matter (unevenly colored, from light red, 2.5YR 6/6, to black).

Furnace chimney with straight wall; six holes; vitreous deposit on the interior.

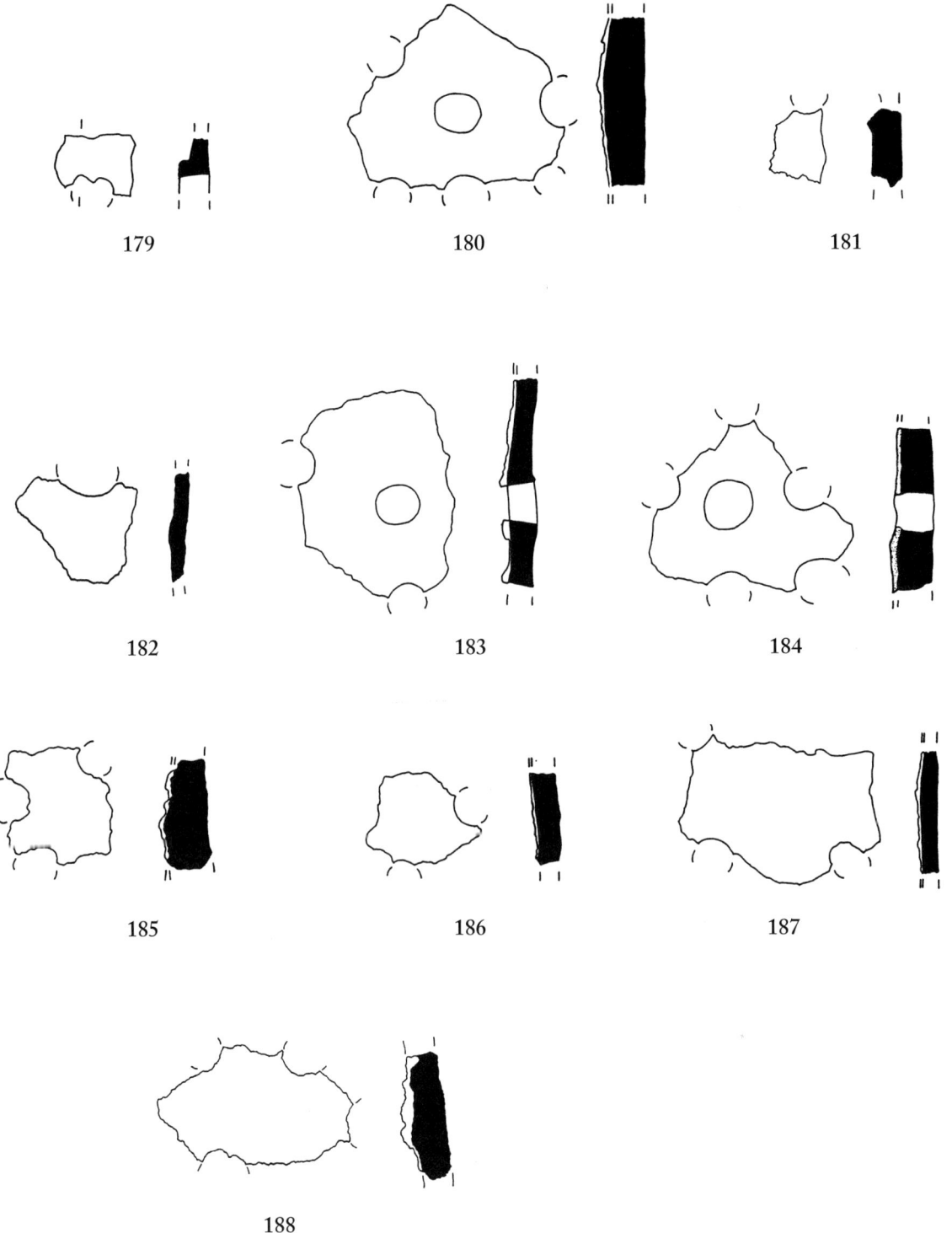

Figure 7.6. Profile drawings of chimney fragments (179–188). Scale 1:3

185 (X 1806, from M 22-2)   Fig. 7.6

Furnace chimney, body fragment.
Max. dim. 6.8.
A coarse fabric with voids from burned-out organic matter (unevenly colored, from red, 2.5YR 5/6–8, to dark gray).
Furnace chimney with straight wall; three holes; vitreous deposit on the interior.

186 (X 101, from S 20-2)   Fig. 7.6

Furnace chimney, body fragment.
Diam. of artifact not measurable; max. dim. 5.8.
A coarse fabric with voids from burned-out organic matter (unevenly colored, mostly red, 2.5YR 5/6).
Furnace chimney with straight wall; two holes; vitreous deposit on the interior.

187 (X 104, from O 22-4)   Fig. 7.6

Furnace chimney, body fragment.
Max. dim. 9.6.
A coarse fabric with voids from burned-out organic matter (unevenly colored, mostly yellowish red, 5YR 5/8).
Furnace chimney with straight wall; three holes; vitreous deposit on the interior.

188 (X 1756, from P 20-2)   Fig. 7.6

Furnace chimney, body fragment.
Max. dim. 9.4.
A coarse fabric with voids from burned out organic matter (unevenly colored, mostly light red, 2.5YR 6/8).
Furnace chimney with straight wall; four holes; unusually thick vitreous deposit on interior.

CHAPTER 8

THE POT BELLOWS

by Philip P. Betancourt and James D. Muhly

The pot bellows is a device used to provide a forced draft for forges, metallurgical furnaces, and other installations in which high temperatures are desired. It was used extensively in the eastern Mediterranean during the Bronze Age, and it has been discussed by several authors.[1] It consists of a cylindrical ceramic vessel fitted with a flexible leather cover having one-way flaps. Moving the cover up and down alternately draws in air through the flaps and forces it out through a nozzle. The device is well known both from illustrations in Egyptian paintings and from actual examples surviving in the archaeological record. Modern examples with similar designs are known from Asia and Africa where they play a role in metalworking and other crafts.[2] Before the discovery of the bellows from Chrysokamino, Cretan examples were only known from later periods.[3]

The earliest use of the pot bellows is not known. Although the bellows themselves do not survive, the device has been suggested for smelting operations as early as the Chalcolithic.[4] Other than the finds from Chrysokamino, the earliest known examples are from the beginning of the Middle Bronze Age.[5]

Figure 8.1 shows a painting of the 18th Dynasty from the Tomb of Rekh-mi-re at Thebes.[6] Metalworkers here use the bellows to melt copper in small crucibles. The bellows are used in pairs so that alternation in their pumping action can produce a more steady draft. The devices in this painting consist of cylindrical vessels with nozzles to allow the attachment of a reed to conduct the air to the fire. A leather cover over each vessel's rim can be lifted with a cord to inflate the bellows and suck in the air, while stepping on the leather forces the air through the nozzle. Four bellows are shown for each small fire.

Fragments of a minimum of nine pot bellows come from Chrysokamino. All examples are presented here (see the catalogue in this chapter). They are somewhat different in design from the bellows known from other sites, although the principle of operation is the same. Perhaps they can be regarded as early, experimental examples of the device.

The bellows from Chrysokamino are drum-shaped clay cylinders with closed tops, with one or more large holes rather crudely cut in the flat upper surfaces. The design of the lower edge (i.e., the open part facing down

1. Davey 1979; Tylecote 1981; Forenhaber 1994; Müller-Karpe 1994, pp. 106–107, fig. 76.
2. Forbes 1950, p. 116.
3. Blitzer 1995, pp. 508–509, from Kommos; Dimopoulou 1997, p. 435, from Poros; Evely 2000, p. 363, from Zakros and Palaikastro.
4. Hegde and Ericson 1985, p. 66; Rothenberg 1985, p. 124.
5. Davey 1979; Tylecote 1981; Müller-Karpe 1994, pp. 106–107, fig. 76.
6. For the context, see Davies 1943, pl. 17.

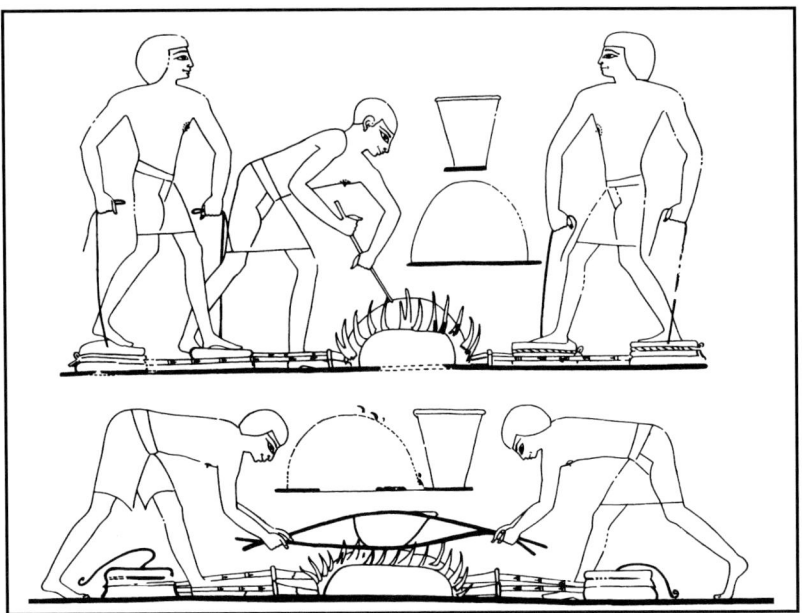

Figure 8.1. Drawing of workmen using bellows, from the Tomb of Rekh-mi-re. After Davies 1943, pl. 23

when the object was in use) is not certain, but sherds with straight rims, such as **191**, may be from this shape. Diameters are between 28 and 53 cm. A nozzle is well preserved on bellows **189** (see Figs. 8.3 and 8.4 below). The nozzle is attached on the upper part of the cylinder, near the hole cut in the upper surface.

The bellows all show signs of darkening from a hot, reducing atmosphere in their interiors. This characteristic is a confirmation of their identification, showing that the devices were used before they were broken. The act of pumping air into a furnace was evidently not so efficient that it drew in only cold air to pump into the fire. Some of the hot air and some sparks seem to have been drawn back out of the furnace and into the bellows, producing a hot, reducing atmosphere inside the bellows and resulting in a dark color in the clay, as its red ferric oxide (Fe_2O_3) was altered to black ferrous oxide (FeO). The process that creates black iron oxides in clay fabrics by the chemical change called reduction is well known.[7] Contact with the hot, oxygen-depleted atmosphere accounts for the condition of the interiors; this heat is one of the reasons why these devices were made of clay instead of a flammable material.

Deposits of fused mud are on the exteriors of several bellows, always away from the end with the holes cut in it. Sherd **191**, which may be a rim from a bellows, is one of the pieces with fused mud on its exterior. It is likely that the bellows were set into the ground when they were in use, and the join with the ground was sealed with mud to prevent air from escaping during the pumping action, so that the mud baked and fused to the vessel by the heat drawn into the bellows from the furnace during use.

All the bellows fragments from Chrysokamino are from EM III–MM IA. The date is based on several considerations, with the most important one being that pieces are found within the soil used for the floor of the small hut, in association with White-on-Dark Ware. No later pottery is found anywhere on the site.

7. Betancourt 1985, p. 6; Noble 1988, p. 80.

Figure 8.2. Reconstruction of the pot bellows in use based on suggestions made by Harriet Blitzer. Drawn by Lyla Pinch Brock

The bellows cannot be earlier in date because they are all made from a type of Mirabello Fabric consisting of a local clay paste with abundant large rock fragments, a fabric not used before EM III–MM IA. Mirabello Fabric uses numerous fragments of angular white stone in the granodiorite/diorite series.[8] It has several subclasses based on the coarseness of the inclusions and their percentage in the fabric, ranging from a fine fabric with just a few inclusions used for Vasiliki Ware[9] to a very coarse class used for storage containers. Deposits of this igneous rock occur at several locations at and near Gournia,[10] and the pottery using it has been analyzed several times.[11] The igneous rock analyzed in the pottery from Pseira is diorite as defined by MacKenzie, Donaldson, and Guilford[12] because it consists of plagioclase, amphibole, and biotite, with very little quartz, while the material analyzed from Mochlos is called granodiorite by Day, Joyner, and Relaki.[13]

Blitzer suggests a reconstruction of the bellows from Chrysokamino as a pot with a leather bag with slits and flaps attached at the top, and the "rim" upside down and buried in the earth, sealed with mud to prevent air from escaping during use and to prevent the bellows from dislodging during the pumping operation (Fig. 8.2).[14] Working the leather bag up and down would pull in air through the one-way flaps and force it through the nozzle and into the furnace.

The widely distributed bamboo called *kalami*, which grows in many parts of Crete, has stems with a hollow interior that would fit the size of the aperture of the tuyere found at the site (see Chap. 9). The presence of the plant on Minoan pottery and wall paintings shows that it grew in Crete during the Bronze Age.[15] It would have made a suitable conduit to allow air from the bellows to flow to the furnace, where the clay tuyere would protect the hollow stem from the heat.

8. Haggis and Mook 1993, p. 273, fabric 2; Myer, McIntosh, and Betancourt 1995, pp. 144–145.

9. Myer 1979; Whitelaw et al. 1997, p. 268.

10. Papastamatiou et al. 1959.

11. Myer 1979; Day 1991, pp. 91–101; Myer, McIntosh and Betancourt 1995, pp. 144–145; Day, Joyner, and Relaki 2003, pp. 17–18.

12. MacKenzie, Donaldson, and Guilford 1982, p. 103.

13. Day, Joyner, and Relaki 2003, pp. 17–18.

14. H. Blitzer (pers. comm.).

15. For pottery, see Popham 1967, pl. 76:a and c; for wall painting, see Platon 2002, pl. 48:a.

CATALOGUE

The catalogue lists all bellows and probable bellows fragments found at the site, grouped into 15 catalogue entries combining pieces similar in thickness, fabric, shape, and general appearance. In many cases one cannot tell which sherds belong together, but a minimum of nine bellows is estimated from this assemblage.

189 (X 143, from L 17-3, L 18-2, L 19-2, M 18-2, M 18-3, M 19-3, M 19-5, N 19-2, N 19-3, P 19-1, R 20-1, locus E5050 N5016)   Figs. 8.3, 8.4

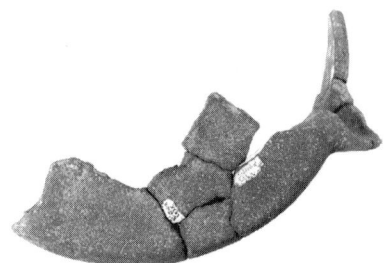

Figure 8.3. Bellows fragment 189

Pot bellows, large part of side and closed top and many nonjoining sherds.
Diam. of top ca. 37.
Mirabello Fabric (unevenly colored, red to black, with the interior darkened from a reducing atmosphere).
Cylindrical vessel with almost straight walls and closed top (lower part of walls missing); nozzle at side, near closed top; rectilinear hole in closed top.

190 (X 609, from M 19-5)   Fig. 8.4

Pot bellows, sherd from closed top.
Max. dim. of sherd 18.9.
Mirabello Fabric (red, 2.5YR 4/8 on interior; dark brown, 7.5YR 4/2, on exterior).
Sherd from the closed top, from near the center; the edges of two long holes survive.
Comments: This sherd demonstrates that the upper part sometimes had more than a single opening cut in it.

191 (X 420, from N 19-3, N 20-3)   Fig. 8.4

Jar or pot bellows, rim sherd.
Diam. of rim ca. 38; Diam. of body ca. 47–50.
Mirabello Fabric (reddish yellow, 5YR 6/6).
Straight rim.
Comments: Burned soil fused to the body on the exterior. The fused soil on the exterior of the vessel suggests this artifact may have been a pot bellows rather than a simple jar.

192 (X 317, from M 18-1, M 18-2, M 18-3, M 19-2, and M 19-5)   Fig. 8.4

Pot bellows, body sherds.
Diam. of body ca. 47–48.
Mirabello Fabric (unevenly colored, red, 2.5YR 4/8, to reddish brown, 2.5YR 5/4, to black; dark in the interior from a reducing atmosphere).
Cylindrical vessel with almost straight walls.
Comments: Burned soil fused to the body, on the exterior.

193 (X 531, from M 18-3)   Fig. 8.4

Pot bellows (?), body sherd.
Diam. of body ca. 40.
Mirabello Fabric (gray core; red surface, 2.5YR 5/6).

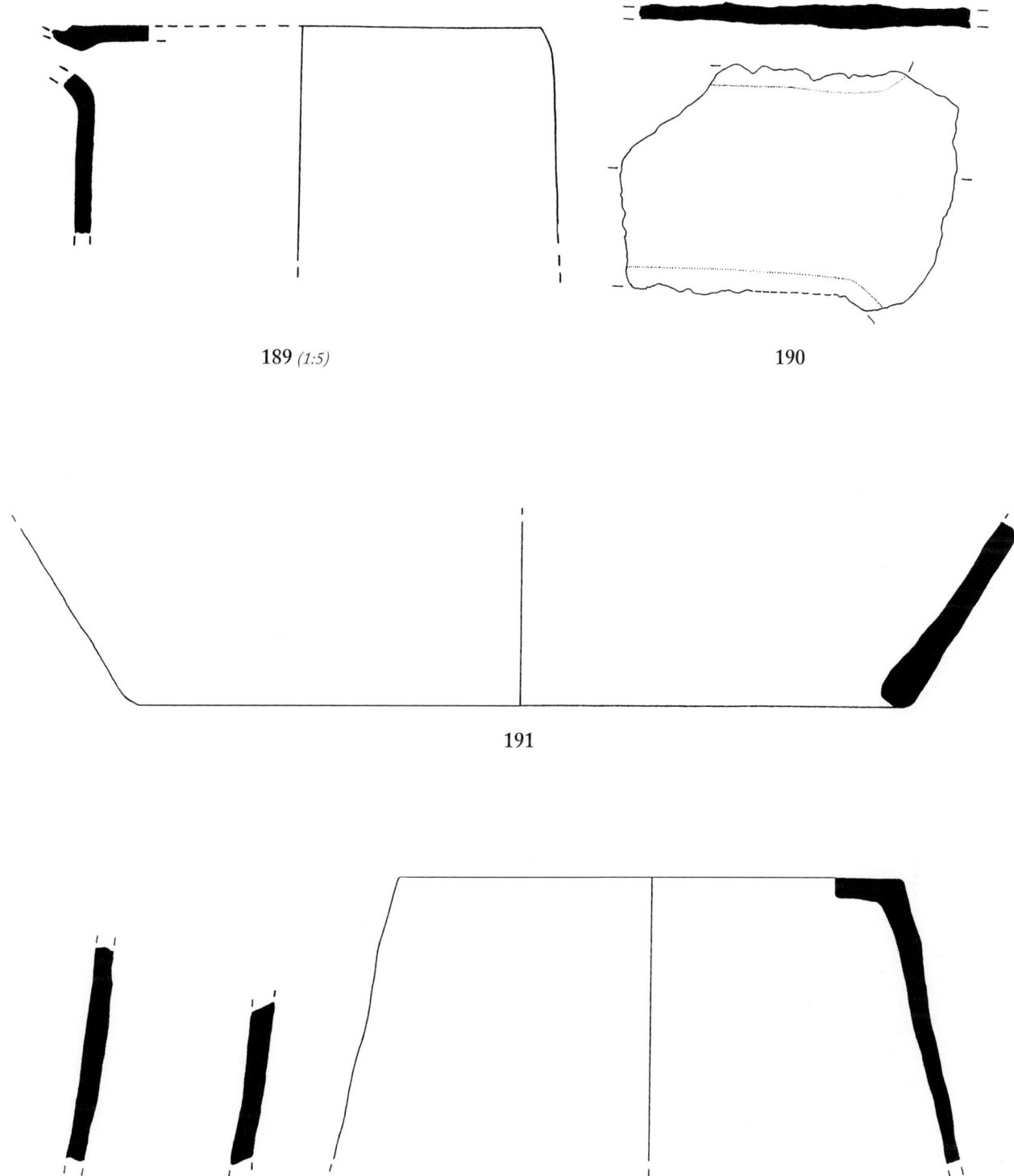

Figure 8.4. Profile drawings of pot bellows (189–194). Scale 1:3 unless otherwise indicated

194 (X 418, from M 18-3, M 19-3, N 18-1, N 18-2, N 19-2, N 19-3)　　　　　　　　　　　Fig. 8.4

　Pot bellows, many sherds from walls and closed top.
　Diam. of body ca. 28.
　Mirabello Fabric (black to red, 2.5YR 4/8; extremely dark in the interior from a reducing atmosphere).
　Cylindrical vessel with almost straight walls; rectilinear hole in closed top.

195 (X 294, from N 18-2, N 19-2)　　　　　　　　　Fig. 8.5

　Pot bellows (?), body sherd.
　Diam. of body ca. 48.
　Mirabello Fabric (red, 2.5YR 5/8).

196 (X 293, from M 18-2, M 18-3)　　　　　　　　　Fig. 8.5

　Pot bellows (?), body sherd.
　Diam. of body ca. 45.
　Mirabello Fabric (very dark gray, N3, to almost completely black from a reducing atmosphere, especially in the interior).

197 (X 483, from P 20-2)　　　　　　　　　　　　　Fig. 8.5

　Pot bellows, sherd from the upper part.
　Diam. of closed end ca. 36.
　Mirabello Fabric (brown, 7.5YR 5/4; dark in the interior from a reducing atmosphere).
　Cylindrical vessel with almost straight walls; closed top with hole in it.

198 (X 416, from K 17-1, L 19-2, M 19-2b, M 19-3, M 19-5, N 18-2, N 10-2, N 19-3)　　　　Fig. 8.5

　Two or more pot bellows made of similar clay, rim and body sherds.
　Diam. of rim ca. 20; restored Diam. of body ca. 52–53.
　Mirabello Fabric (unevenly colored, from dark red, 2.5YR 3/6, to reddish brown, 2.5YR 5/4, to dark brown, 7.5YR 4/2, to light red, ? 5YR 6/6).
　Comments: Burned soil fused on exterior; one sherd with rim, heavily reduced, with a diameter very different from the body sherds. The ridge below the mouth would have helped seat the bellows in the ground.

199 (X 1712, from M 19-3)　　　　　　　　　　　　Fig. 8.5

　Pot bellows, body sherds.
　Max. dim. of largest sherd 6.2.
　Mirabello Fabric (red, 2.5YR 4/6).
　Comments: Burned soil fused on exterior.

200 (X 142, from L 17-3, L 19-2, M 18-3, M 19-2b, M 19-3, M 19-4, M 19-5, N 18-2, N 19-2, N 19-3)　　Fig. 8.5

　Pot bellows, rim, base, and body sherds.
　Diam. of rim ca. 18; Diam. of base ca. 20; restored Diam. of body ca. 36.
　Mirabello Fabric (unevenly colored, from red, 2.5YR 5/6, to reddish brown, 2.5YR 5/4, to dark reddish gray, 5YR 4/2).
　Comments: Hole cut in top; darkened from reduction in the interior; burned soil fused on exterior.

201 (X 1713, from N 19-3)　　　　　　　　　　　　Fig. 8.5

　Pot bellows, rim sherd.

Figure 8.5 *(opposite)*. **Profile drawings of pot bellows (195–203).** Scale 1:3

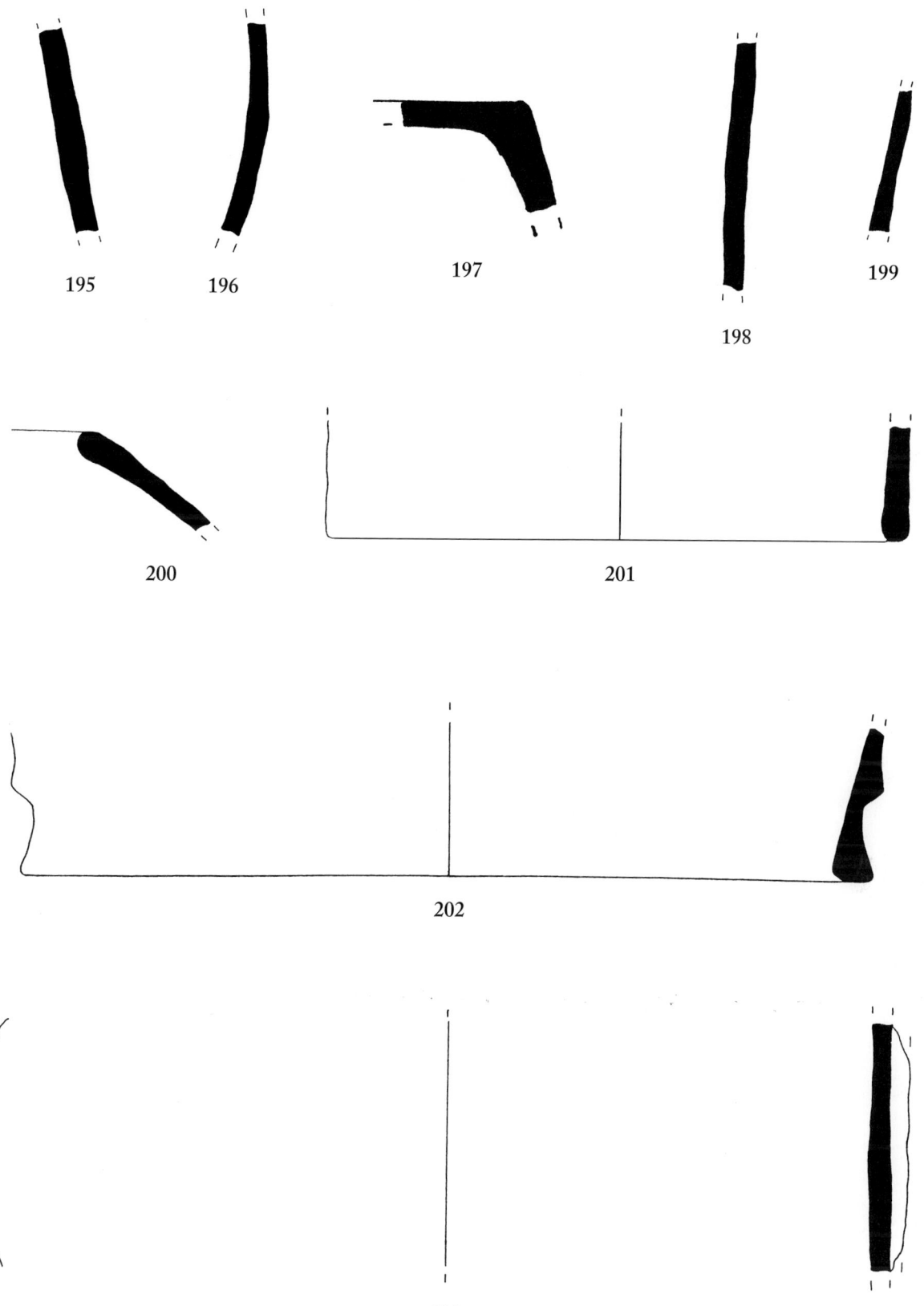

Diam. of rim ca. 28.
Mirabello Fabric (dark brown, 10YR 3/4).
Comments: Burned soil fused on exterior.

202   (X 183, from N 19-2)   Fig. 8.5

Pot bellows, rim sherds.
Diam. of rim ca. 48.
A coarse fabric containing phyllite and other stones (red, 2.5YR 5/8).
Horizontal molding near the thickened rim.
Comments: Burned soil fused on the exterior. This object is made from an atypical fabric, and it is an atypical shape in comparison with the other sherds regarded as bellows fragments. The sherd suggests that more than one ceramic shape was used as a bellows. The molding on the exterior would have helped seat the bellows in the ground during use.

203   (X 1666, from N 19-2)   Fig. 8.5

Pot bellows, body sherd.
Max. dim. 13.9.
Mirabello Fabric (yellowish red, 5YR 5/6, dark on interior and core).
Comments: Burned soil fused on the exterior.

Miscellaneous Ceramic Artifacts

by Susan C. Ferrence and Byron Koukaras

Several ceramic objects from Chrysokamino do not fit easily into the other categories discussed in this volume. Two items were found within the hut, while the others were discovered in the slag area. None of the objects were in their original place of use; they were all fragments found in casual debris. They belong to various categories (see Fig. 9.1 below).

Two of the catalogued items, a tuyere and a piece of furnace lining, provide important evidence for the reconstruction of the metallurgical activities at the site. They make a significant contribution to our knowledge of the workshop. Some of the other objects, however, have unusual characteristics that do not correspond with any of the pottery groups or industrial classes found at the site, and they are difficult to explain. The objects are all from the latest stratum (an EM III–MM IA context), but the poorly fashioned, handmade object **204**, possibly a piece of an unusual vessel, is very crude for the latest phase of the Early Bronze Age. It may be an heirloom produced long before it was deposited in its archaeological context.

OBJECTS FROM THE APSIDAL HUT

The fragmentary pieces of **204** (X 272) consist of rim and body sherds belonging to a vessel with a maximum rim diameter of 22 cm. The most distinctive characteristic of the vase is its chaff-tempered, fine-textured fabric. It is a soft, pinkish white fabric that erodes easily into a very fine, chalky powder. The vessel also has an extremely irregular, lumpy surface. It is handmade in the coil method, as demonstrated by its obvious breakage along coil lines. The object is probably too porous to hold liquid.

This fabric is unique for the site, and it is difficult to determine what function the vessel served at this industrial location. The shape cannot be reconstructed. Both the fabric and the technology of manufacture differ completely from the other artifacts at the workshop and also from other pottery known from this part of Crete. This piece should surely be regarded as an import.

Another unusual item is **205** (X 1667), consisting of two pieces of an object with a cylindrical body approximately 47–48 cm in diameter. The item has holes in its sides (like the furnace chimneys), but it is made in a hard, dense clay fabric, not the porous, soft fabric filled with chaff used for the usual chimneys. It is well fired to a uniform red color, but it shows no evidence that it was ever used as a chimney. It displays no signs of fire or burning.

OBJECTS FROM THE SLAG PILE

Objects found away from the hut may have had some function in the workshop activities. Two of the artifacts, a fragment of furnace lining and a piece of a tuyere, were certainly used in the smelting. The other objects are a sherd possibly fired accidentally and a unique triangular artifact of unknown function.

Number **206** (X 614) is a severely burned piece of Mirabello Fabric, mostly curved but flat on one side. The smooth, bubbly surface of the sherd indicates it may have been near or trapped inside a burning furnace. It could be a sherd that was accidentally fired by one of the furnaces.

The anomalous object **207** (X 940) is a triangular block of well-fired Mirabello Fabric. It comes from the surface of the slag pile. The artifact is 7.7 cm long on its hypotenuse and 1.3 cm thick. The triangular object, which was carefully cut from a slab of clay before being fired to produce a hard, dense object, has no marks from use. Its purpose is unknown.

Number **208** (X 1755) is a tuyere fragment that was found in the slag pile outside the hut. It would have been used as the incombustible end of a hollow reed or piece of bamboo connecting a pot bellows to the furnace. The tuyere, which is only partly preserved, was originally cylindrical in shape with a longitudinal hole in the center. It had an aperture with a diameter of ca. 2.5 cm.

Tylecote discusses tuyeres in detail, and he lists examples from many ancient sites in the Eastern Mediterranean.[1] The use of a tuyere is essential in operations involving pumping air to a furnace using combustible conduits because something must protect the end of the hollow reed from the intense heat of the furnace. The clay tuyere would be fitted to the end of the reed, and only it would come in contact with the furnace. In comparison with the tuyeres illustrated by Tylecote, the example from Chrysokamino has a relatively small exterior diameter and a relatively thin wall, but its hole is the proper size for passing the draft from a bellows or a blowpipe to the furnace. Its 2.5 cm diameter is exactly in the center of the 2–3 cm size range suggested as the optimal dimension for a tuyere's hole in order to pump sufficient air into the fire.

Number **209** (X 1869) is a fragment of a coarse clay object that was probably part of the lining for one of the bowl furnaces. It has no finished exterior sides, and its thickness (over 4 cm) indicates that it cannot have been part of any Minoan vessel. The piece has been fired extensively, and it has a black exterior color from complete reduction, a feature that would be expected if it was buried in the ground, lining a bowl furnace, and was not in contact with oxygen while it was heated. It is broken roughly on all sides.

1. Tylecote 1981.

MISCELLANEOUS CERAMIC ARTIFACTS

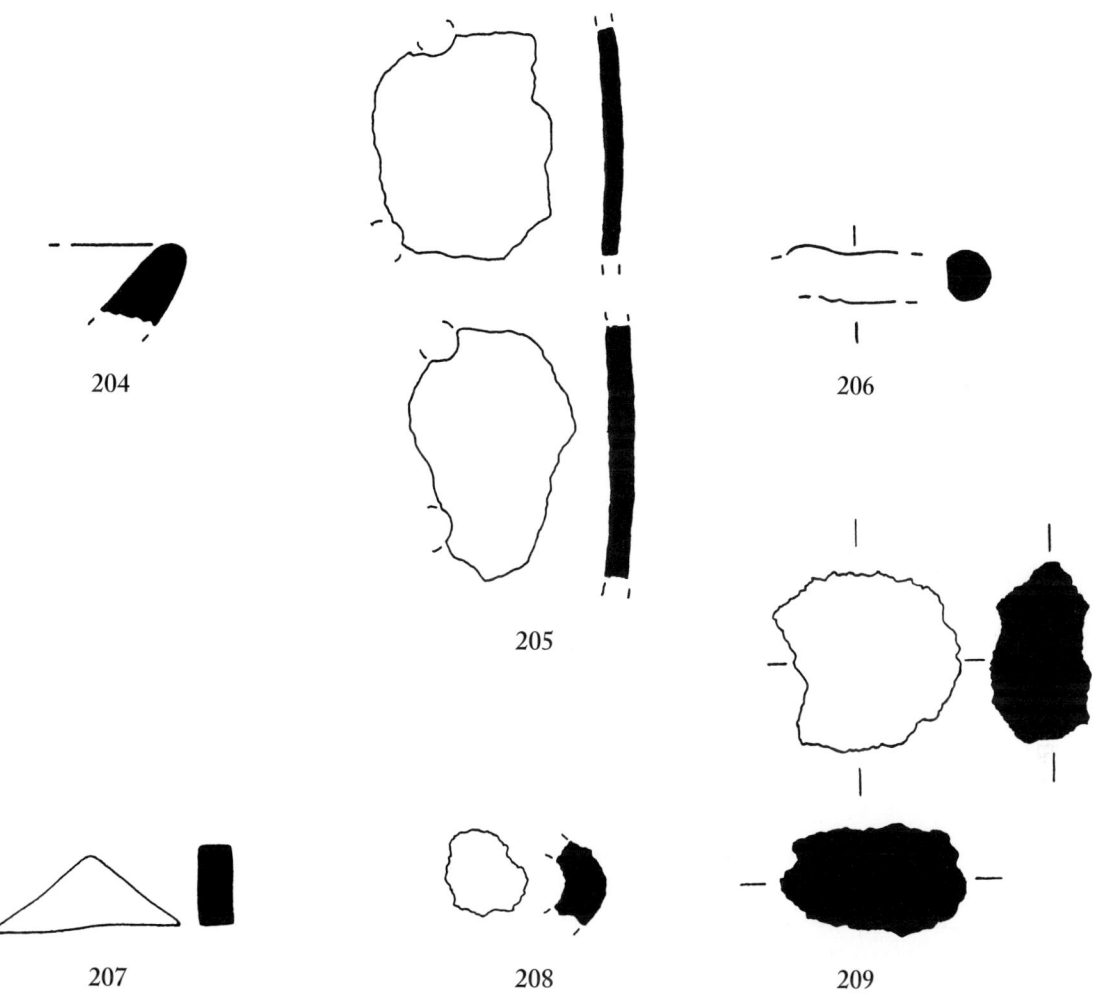

Figure 9.1. Miscellaneous artifacts (204–209). Scale 1:3

CATALOGUE

Objects from the Apsidal Hut

204 (X 272, from N 18-2, N 18-3, N 19-2, N 19-3) Fig. 9.1

Vessel (?), rim and body sherds.
Diam. of rim 22.
A chaff-tempered, fine fabric (pinkish white, 5YR 8/2).
Perhaps a shallow bowl.
Comments: Very crudely made by hand and broken at the places where coils joined.

205 (X 1667, from N 18-1, N 19-3) Fig. 9.1

Unidentified vessel or object, two body sherds.
Diam. of body ca. 47–48.
A coarse, hard fabric (red, 2.5YR 4/8).
Cylindrical form with holes thrust through the body before firing.
Comments: The artifact is similar to the furnace chimney fragments except that the fabric is very different and the item was never subjected to fire after it was made.

Objects from the Slag Pile

206 (X 614, from L 17-1)   Fig. 9.1
   Elongated curved lump of clay with a flat side.
   Max. dim. 5.1.
   Mirabello Fabric (dark gray, 2.5YR 4/0).
   Comments: Severely burned and vitrified.

207 (X 940, from J 17-Surface)   Fig. 9.1
   Triangular block, complete.
   Max. dim. 5.0 × 4.5 × 7.7; thickness 1.4.
   Mirabello Fabric (red, 2.5YR 5/6).
   Triangular object with a slightly convex profile.

208 (X 1755, from P 20-2)   Fig. 9.1
   Tuyere fragment.
   Diam. of exterior ca. 4; Diam. of hole ca. 2.5; original length unknown (broken at both ends).
   A coarse fabric (red, 2.5YR 5/6).
   Cylindrical profile.

209 (X 1869, from P 20-Cleaning)   Fig. 9.1
   Lining from a bowl furnace, fragment, broken on all sides.
   Max. p. dim. 7.8.
   A coarse fabric, highly fired in a reducing atmosphere (black).

CHAPTER 10

OTHER METALLURGICAL MATERIALS

by Philip P. Betancourt

Many different materials discovered at Chrysokamino provide evidence for the activities carried out at the site. The excavation strategy was designed to collect a wide range of manmade artifacts as well as geological samples and other items that could contribute information on the history of the workshop and the technology used there. This chapter presents the uncatalogued metallurgical materials, including slag, ore fragments, and copper prills.

THE SLAG

The word slag has been applied to a wide variety of vitreous materials found in the archaeological record.[1] It has been used for glassy waste products from several different processes including the production of lime, the firing of ceramics, the making of faience, and metallurgy. Within the field of metallurgy, the term has been applied to the residue from smelting, remelting in crucibles, and condensation on various surfaces.

Vitreous materials resulting from smelting and other metallurgical practices have sometimes been divided between furnace slag (formed in the furnace), tap slag (formed when the furnace is tapped to remove molten metal), and crucible slag (formed in a crucible). Although all of these materials share the characteristic of being vitreous, they can differ from one another in their composition and microscopic texture. Description by formation process based on visual examination and on the archaeological context is useful for understanding and describing the technology, but it can be uncertain or impossible when only tiny fragments survive or when the context of the slag is unknown. An understanding of the original composition and the processes that led to the final products can best be done by analyses of various types that demonstrate both the chemistry and the microstructure.

The slag resulting from smelting operations to obtain copper is regularly dark in color and fairly heavy from inclusions of iron oxides, such as magnetite. Archaeologically, it usually occurs in larger quantities than slag resulting from remelting of copper in crucibles. Estimates have suggested

---

1. Bachmann 1982, pp. 1–6; Craddock 1995, p. 20.

a b c

Figure 10.1. Lumps of slag from unit M 22-3

that smelting results in quantities of slag that are between 10 and 15 times greater than the amount of metal that is produced.[2]

The slags at Chrysokamino fall into two broad classes. Some pieces are loose, thick chunks that may or may not preserve an original smooth surface. Other examples are surface coatings on the interiors of furnace chimney fragments (Fig. 7.1, right). Both classes are dark and vitreous, and they sometimes have prills of copper in them. The surface coatings on the chimneys are thin (usually under 0.5 cm in thickness), and they have a smooth surface. The loose chunks, on the other hand, are thicker, solid pieces of glassy material (Fig. 10.1). It is likely that the solid pieces of slag resulted from the melted charge inside the furnace, and the coatings inside the chimneys formed partly from the condensation of vapors and partly from small liquid droplets and solid particles carried upward and deposited inside the chimneys during the smelting process.

The Chrysokamino Slag Deposit

Before the excavation began, a deposit consisting mainly of slag was on the surface of most of the saddle that forms the metallurgy site at Chrysokamino (Fig. 3.2). Although the slag formed the largest component, the deposit also included clay chimney fragments, pieces of natural phyllite, the soil resulting from the phyllite's disintegration, a few bits of ore, and various artifacts (pottery, stone tools, animal bones, shells, and other items discussed in this volume). The small fragments of slag were densely packed, and tiny bits of pulverized slag were also present. Presumably, most of the slag was broken by human activities. Freeze/thaw cycles during winter weather might also have broken some of the vitreous material, but breakage from the freeze/thaw cycles would have been most intense at the surface, so the fact that deeply buried parts of the deposit consisted of the same small fragments as the surface suggests that natural breakage was minimal.

The deposit was low and wide, extending from sterile bedrock at the east and west of the deposit to a maximum height of ca. 60 cm near the center of the saddle (Fig. 10.2). It covered a north–south depression in the landscape between higher exposures of limestone on the west and phyllite on the east. The upper surface of the deposit was level before excavation (i.e., it did not form a mound, but consisted instead of a flat area between the two higher exposures of bedrock). The original size of the slag pile could not be determined because it had partly eroded into the sea at the north and the south where the saddle terminates in cliffs, but it covered an area of ca. 200 m². An estimated 60 to 70 m³ of slag was present.

2. Evely 2000, p. 353.

Figure 10.2. Slag resting on bedrock, looking south

The phyllite on the east was extremely altered, and at the surface much of it had disintegrated into a white calcareous mass that crumbled easily. White powdery material from this phyllite, which had washed downhill over much of the area, looked very much like lime, leading to the understandable but incorrect suggestion that Chrysokamino was once the site of a limekiln.[3] The nearly white soil created a covering of dust that tended to adhere to everything. It made the surface of the slag pile appear paler than it was in actual composition; after the slag was washed, most of it was almost black.

The Vitreous Coatings on the Chimneys

Most of the fragments of clay furnace chimneys had dark, glassy material adhering to their interior surfaces (Fig. 7.1, right). The material seems to have covered most of the inside of the cylindrical chimneys, although it did not reach up to the open upper rims. The thickness of the coating was usually less than 0.5 cm. The coating's surface was smooth, and evidence that it was once liquid was preserved in surface flow structure and drips. The orientation of the drips proved that the cylinders were used in a vertical fashion, with their open rims at the top, as is to be expected in artifacts interpreted as cylindrical chimneys. The vitreous material inside the chimneys seems to have formed mostly as a result of condensation of hot vapors, although liquid droplets and small solid particles carried upward inside the chimneys by the force of the draft probably added to the material. The condensation collected while the smelting operation was in progress because the chimneys were cooler than the furnace, and the coating solidified as soon as its temperature dropped sufficiently.

Analysis indicates that at the microscopic level the material is composed of a glassy matrix containing numerous inclusions of the iron oxide magnetite (Apps. A and B). The magnetite occurs as dendritic crystals that formed when the glassy material cooled quickly. A few very tiny grains of quartz are present as relics of the original charge. A copper chloride recognized in one sample (App. B, CHR 93) can be interpreted as an alteration product of a tiny amount of copper deposited along with the glassy material.

3. See Zois 1993.

The Chunks of Slag

The solid chunks of slag occur mostly in small pieces under 2 cm in size. Much of the material is pulverized. Because the pieces form a continuous series from microscopic particles to fragments about 2–3 cm or larger in their greatest dimension, the total number of fragments cannot be calculated except by an estimate, which would be in the millions. The surfaces have different appearances depending on the degree of weathering. Weathered pieces are dark, vitreous masses with dull, black exteriors. By contrast, bright, highly reflective surfaces can be seen on fresh breaks. When the original exterior surface of the slag is preserved, it is smooth and often rounded (Fig. 10.1). A few rare stalactitic pieces demonstrate that some of the slag reached its melting point during the smelting process (Fig. 10.1:a). Although most of the slag is dense and heavy, a few pieces exhibit considerable variation in physical characteristics. A little lightweight, "frothy" material is present.

The slags have been analyzed by a number of different laboratory techniques (see the individual appendices). In some cases it has been possible to analyze specific samples by more than one methodology, so that the results can complement one another. The typical microstructure consists of a glassy matrix containing crystalline phases. Among these microscopic inclusions, magnetite crystals are especially numerous, as is common in ancient copper smelting slags.[4] The crystals are often dendritic (the branching structure is quite apparent in App. A, CHR 87). This type of inclusion is typical of slags derived from smelting oxidic copper ores, as opposed to those from remelting scrap or ingots of more pure metal, which have lower iron contents.[5] The magnetite forms from the iron oxides that act as a flux during the smelting operation, and it contributes substantially to the black color and heavy weight of the final vitreous material. The composition shows that some of the slag cooled slowly enough for pyroxene group minerals to crystallize, as they are recorded as inclusions within the microstructure (App. A, CHR 96, CHR 99). Also present are fayalite crystals (App. A, CHR 99), another feature of slags from oxidic copper ores.[6] Circular to elliptical voids are present from entrapped gas bubbles. In addition to the other minerals, copper prills are present in the slag as inclusions up to a centimeter across. Analyses of some of the fragments record considerable copper at the microscopic level (App. A, CHR 99). The analyses prove conclusively that the slag at this site results from smelting copper ores.

The final compositions of the slags are the result of several factors. Elements of the final composition include contributions from the ore, the fuel, the flux (whether it was an addition or part of the ore), the air introduced into the furnaces, and potentially even the walls of the furnaces. Some of the crystalline phases are relics of the original charge placed in the furnace, while others formed during the smelting and subsequent cooling of the material. Additional phases formed as weathering products and as secondary depositions during the long period between the use of the workshop and the time of excavation.

A few relics from the original charge can be recognized. These unaltered minerals include phases that were insufficiently affected by the temperatures to alter their composition. Good examples include the quartz and

4. Craig and Vaughan 1994, p. 345, fig. 11.8:d.
5. Bachmann 1982, p. 21, fig. 3:c; Gale et al. 1985, p. 88.
6. Bachmann 1982, p. 12, figs. 3:a, 3:b; Gale et al. 1985, p. 88.

chert grains visible in thin section analysis (App. A, CHR 97, CHR 93). Quartz has also been noted as a relic in comparable slags from the Near East.[7] Tiny traces of the primary minerals that were not oxidized in the secondary ore bodies exploited by the workshop also survived the smelting process. They were detected by Wavelength Dispersive Spectrometry (App. E, CHR 8) and by examination under the polarizing microscope and SEM-EDS (App. F, ORE-1), where they were identified as pyrite, covellite, chalcopyrite, and chalcocite.

Postburial alteration products include several mineral phases. Late depositions of carbonate within voids are occasionally visible (App. A, CHR 93). The weathering has also caused devitrification of the glassy parts of some fragments (App. A, CHR 93), resulting in the formation of new minerals such as hedenbergite (App. B, CHR 93). The devitrification, which is responsible for the dull surfaces on the portions of slag left exposed to weathering in the upper layers of the slag pile, is visible with the unaided eye. One cannot be sure, but some of the cuprite may also be a product of recent weathering (App. A, CHR 99).

Most of the mineral phases in the slags crystallized during the cooling of the material. These minerals are helpful in establishing some of the details of the process of smelting. Fayalite, the iron-rich end member of the olivine group, has been regarded as a characteristic of slags formed under highly reducing conditions. In slags produced under more oxidizing conditions, the formation of the pyroxene hedenbergite would be typical.[8] Olivine also tends to form with high iron content in the melt (see App. F). Fayalite is present in some of the slags produced at Chrysokamino (App. A, CHR 87, CHR 99; App. B, nos. 7, 19, 20, 36), and magnetite and wustite, which also form under reducing conditions with high iron content in the melt, are present as well (App. A, CHR 97, CHR 87, CHR 93; App. B, nos. 6, 13, 23, 28, 33, 38; see also App. F). Reducing conditions were probably usual, but the presence of slags with substantial amounts of pyroxene and cuprite (App. A, CHR 99) means that more oxidizing conditions (and/or conditions with lower iron content in the melt) were also occasionally present. The amount of cuprite is very small in comparison with the amount of cuprite in ancient ingots.[9] The large amount of cuprite in the ingots has been interpreted as a result of the remelting of copper under more oxidizing conditions than those existing in the smelting furnaces. The crystallization of the magnetite and the other phases varies considerably from fragment to fragment, with some pieces having larger and more euhedral crystals, indicating variations in the cooling rate as well as differences in composition.

These variations demonstrate that the smelting practices at Chrysokamino were not completely consistent from one operation to the next. Probably the ore differed in composition from time to time, and the degree of control over the ratio of fuel to ore, the amount of draft, and other elements of the operation were not strictly controlled. The workshop was successful in removing copper from oxidic ores, but repeated operations did not duplicate prior conditions exactly, and the amount of metal remaining in the slag indicates that the process was not always a completely efficient smelting operation.

7. Bachman 1980, pp. 108–110.
8. Hauptmann, Bachmann, and Maddin 1994, p. 6.
9. Hauptmann, Maddin, and Prange 2002, pp. 7–12.

THE ORE

Small amounts of copper ore are present at several locations in Crete,[10] and larger amounts are known from Kythnos and elsewhere in the Cyclades.[11] None of the known Cretan sources are close to Chrysokamino. The sources for copper on Kythnos and Seriphos have lead isotope patterns that are consistent with many of the artifacts found in EM Crete,[12] and these deposits and those at Lavrion were already being mined and smelted in the Neolithic period.[13] Sources north of Crete seem good candidates for the copper used for most of the EBA objects found on the large Minoan island. Analyses of copper prills in the slag from Chrysokamino by lead isotope analysis show that the patterns here also match those of Lavrion and Kythnos,[14] suggesting that some (or all?) of the ore smelted at Chrysokamino originated in the north.

Small pieces of malachite and azurite come from the metallurgy location. These minerals are secondary copper minerals, and they are easier to smelt than sulfide ores. Chrysocolla, a copper silicate that is also easy to smelt under ancient conditions,[15] is also present. The pieces are small (about a centimeter in length), and very few fragments are present in the archaeological record. This situation is not unusual for smelting sites in the Eastern Mediterranean.

In addition to the ore from the smelting site, one piece of malachite also comes from the small harbor at Agriomandra. This small, waterworn pebble, ca. 2 cm in size with small amounts of a bright green mineral in it, was picked up on the beach. Noel Gale confirmed the presence of copper at the Isotrace Laboratory (Research Laboratory for Archaeology and the History of Art), Oxford University. No geological formations that could have produced the malachite are present near the site, and it must have been brought in by ship at some time in the past. The beach is a little over a kilometer from the smelting workshop.

CONSIDERATIONS AFFECTING THE LOCATION OF SMELTING SITES

Contrary to the logic that it would be easier to transport smelted metal instead of the heavier ore, the location of smelting sites away from the mines is a typical practice in the ancient Mediterranean. Similar patterns have been recognized all the way from early mines at Feinan in Jordan[16] and Timna in Palestine[17] to Bronze Age mines in Britain and Ireland.[18] Ores have been found at Shiqmim and Bir Safadi, both of which are located well away from any ore deposits.[19] A similar situation exists elsewhere.[20] In a common metal production model used in many places, only crushing

10. Branigan 1968; McGeehan-Liritzis 1996, p. 387.
11. Gale and Stos-Gale 1989; Stos-Gale 1998.
12. Stos-Gale 1989, 1993; Stos-Gale and Macdonald 1991.
13. Stos-Gale and Gale 2003, p. 87.
14. Stos-Gale 1998; see also App. C.
15. See Tylecote's comments in Koucky and Steinberg 1982b, p. 140.
16. Hauptmann, Bachmann, and Maddin 1994, p. 6; Hauptmann 2000, pp. 167–168; Weisgerber 2003, p. 76.
17. Rothenberg 1972, pp. 208–210.
18. O'Brien 1996; see also comments by Muhly 1997, p. 772.
19. Weisgerber 2003, p. 76.
20. Lechtman 1991.

and sorting occur near the mines, while the ore is transported to another location for smelting.

Good reasons exist for this pattern. The crucial resources that determine the location of the smelting operation involve several factors in addition to the availability of ore. Some of these factors are related to availability of other materials, while others concern markets and social concerns.

Fuel is the first consideration. Wood and charcoal, the most common fuels employed, are very bulky, and bundles of limbs take time to load and transport. After trees are cut down, several years pass before timber is again available from the same region. If the residue from olive oil production was used as fuel for smelting, as is tentatively suggested in this volume, arrangements with settlements capable of providing this product would be required.

Availability of flux can also be a consideration in choosing a smelting site. At Chrysokamino, evidence for a deliberate addition of iron minerals as flux is ambiguous, but enough goethite is available locally for use as flux in copper smelting if this material was desired. At Feinan, no evidence survives for any deliberately added flux until the Late Bronze Age,[21] but the picture at Chrysokamino is apparently different. Iron oxide was obviously present in the furnaces in some fashion, as is proved by the high iron content of the slag. It could have been part of the ore, or it could have been added deliberately as a separate material. If the Chrysokamino smelters needed iron oxide to use as a flux, the proximity of suitable sources could have been one of the influences on the choice of location.

The presence of a suitable topographic situation for smelting may also have been a factor. At Chrysokamino, the windswept cliff chosen for the workshop seems to be a deliberate preference because the wind would aid in providing the draft for the furnaces. A funneling effect associated with the north-facing trough where the workshop is located helps to intensify all northern breezes. Windswept hills were also chosen for smelting on the islands of Seriphos[22] and Kythnos.[23]

An important additional factor is the steady employment of a work force trained in smelting. Ore concentrations in nature are seldom uniformly distributed even within a single deposit, and miners must sometimes spend considerable time excavating worthless rock to get to the next part of the ore body. Smelting can proceed quickly, and it can often handle more ore than can be produced locally. It is not economical to stop the smelting while the workshop waits for additional ore, so smelting sites often use ore from more than one mine. Even a rich tin-mining region like Cornwall has sometimes relied on imported tin ore from South America.[24]

Proximity to markets and to the agricultural and social infrastructure necessary to support a metallurgical workshop may also be important considerations. Mines may be found in remote locations, away from settlements that would be sources of food, water, shelter, and steady supplies of commodities like pottery, tools, or the raw materials to make furnaces. If the smelting work was conducted seasonally, as may have been the case at Chrysokamino, transporting the ore may have been an alternative to transporting the entire workshop and all of its necessities to the sources of copper ore.

21. Hauptmann 2001, p. 426.
22. Gale and Stos-Gale 1989, pp. 24–25.
23. Hadjianastasiou and MacGillivray 1988; Gale and Stos-Gale 1989, pp. 25–27.
24. See Gowland 1930, p. 522.

The preparation of the ore is an additional factor. Before it can be smelted, ore is typically prepared by beneficiation. This process involves the removal of bits of ore from their matrix to create a richer charge and improve the smelting process by giving a higher yield of copper with less slag. In addition, because the ore is more concentrated, less fuel is required during the smelting. The process of beneficiation is simple but time-consuming. The rock containing the ore can be crushed with stone hammers, and the ore can be sorted by color. Because azurite is bright blue, malachite is bright green, and cuprite is red, the sorting operation can be done even by children. If the ore is heated first, it cracks and makes the crushing easier. In some cases where beneficiation has been studied, it has been observed that the ideal size of the ore pieces to be smelted is about 1 cm in length.[25] This is the same size found at Chrysokamino. Because this time-consuming task can be done by unskilled labor, it is not profitable to use metalworkers for it, and it may be preferable not to stop the smelting process for this type of activity.

As a result of all these factors, ancient smelters often seem to have found it more convenient to transport only the beneficiated ore. Mining, sorting, and beneficiating would have been done at the mine itself, before transportation to the smelting site. Transportation, either by land or by sea, would have been easier to accomplish with bags of tiny pieces of beneficiated ore than with large tonnage that included waste rock. Transportation of ore by sea would, in fact, have been very simple, because bags of ore could have been used as the ballast required to keep early ships stable.

The archaeological evidence from Chrysokamino proves that wherever the ore originated, it was not beneficiated at Chrysokamino itself. The absence of large quantities of waste rock and discarded low-grade ore at the site indicates that this process did not occur on the headland. The same evidence shows that the mine was not adjacent to the workshop. Mining removes large quantities of useless rock along with the ore, so the mining site, like the beneficiation site, is easily recognizable by the dumps of waste material. Because the site at Chrysokamino has neither the waste rock nor the discarded tools that are common at such sites,[26] one must conclude that the workshop was not the location of either mining or beneficiation, a conclusion confirmed by the geological survey of the region.[27]

Although the geological survey did not locate any ore deposit at Chrysokamino or any formations that might have once had cuprous metal deposits, the possibility that a small deposit once existed has been proposed previously. The cave immediately north of the metallurgy site, regarded as a mine by Mosso and others,[28] was later properly interpreted as a natural cavern.[29]

Harriet Boyd reported seeing ore on the cliff near Chrysokamino.[30] Whether what she saw was a small amount of ore brought there from somewhere else or a natural geological occurrence, however, is not known. No geologist was present, and she did not describe what she saw in any detail. Local chlorite deposits are green, and someone not trained in geology or mineralogy might mistake them for copper ores. Even if a locality somewhere in the vicinity did once produce a small amount of ore, the deposit was not adjacent to the workshop site, because no signs of beneficiation are present there.

25. Hauptmann, Bachmann, and Maddin 1994, p. 5.
26. D. Gale 1991; Gale and Ottaway 1991.
27. See Chap. 2 in this volume.
28. Mosso 1910, pp. 289–292; Hutchinson 1962, p. 40.
29. Faure 1966, pp. 47–48; Branigan 1968, pp. 50–51.
30. Hawes and Boyd Hawes 1909, p. 38.

The procurement of the ore was just one step in a complex process. Ore acquisition needed to be coordinated with the establishment of a workshop, the training of personnel, the procurement of fuel and possibly flux for the furnaces, and the construction of the furnaces with their clay chimneys. In addition to the general logistics involved with feeding and maintaining the personnel engaged in the work, the workshop also needed a distribution system for the smelted metal.

Ore Sources

The most satisfactory methodology for the identification of sources of prehistoric ore combines lead isotope analysis with the study of elemental compositions.[31] Lead isotope analyses provide a pattern based on the geological date of the deposit, while elemental composition can give some indication of the composition of the original ore. For the prills from Chrysokamino, lead isotope studies have revealed patterns consistent with sources in Kythnos or Lavrion.[32] The high percentage of arsenic in some of the prills and the occasional presence of other metals (see App. B) suggests that at least part of the ore may have been derived from the oxidized upper parts of a deposit containing some arsenic and possibly other metals in addition to copper.[33] Such a deposit has not been identified in either Crete or the Cyclades. The percentages of arsenic from the copper sources on Kythnos are lower than at Chrysokamino, although some arsenic is present there. At Lavrion, the primary ores are iron-manganese deposits and zinc-iron-lead deposits, but many other minerals, including copper, occur in smaller amounts.[34] Malachite is widespread within the region of Lavrion, and minerals containing arsenic (arsenopyrite, arseniosiderite, and many others) occur as well. We have no way of knowing the composition of the upper levels of the deposits at Lavrion because they have been mined away. However, any use of the Lavrion region as a copper source for Chrysokamino must have exploited deposits that were well away from the two main classes of primary ores, because manganese and lead are scarce in the prills at Chrysokamino. The truth is that no known deposit in the Aegean exactly matches both the elemental composition and the lead isotope fingerprint of the prills at Chrysokamino. If Chrysokamino exploited a polymetallic deposit, Lavrion is one geographic region that has both the potential for an elemental general comparison and an isotopic fingerprint that comes close to matching the ores from Chrysokamino. At our present level of knowledge, it cannot be excluded as a source for the ore. Kythnos cannot be excluded either, given the presence of lower levels of arsenic from the region, the match in lead isotope fingerprint, and the large amount of evidence for Cycladic interactions with Crete in other areas during the Early Bronze Age.[35]

Diversity in the slags and copper prills from Chrysokamino, however, as discussed elsewhere in this volume (see esp. Apps. A, B, and C), is a strong cautionary signal suggesting the possibility of a number of different ore sources during the long life of the workshop. If this was the case, and Chrysokamino used more than a single source of ore, then the situation is extremely complicated, and we cannot be sure that any one elemental composition or lead isotope fingerprint will provide a secure identification of the source of ores for the site as a whole.

31. Hauptmann, Begemann, and Schmitt-Strecker 1999.

32. Stos-Gale 1998; see also App. C.

33. See Craig and Vaughan (1994, pp. 247–250) for similar occurrences in central Europe and several other parts of the world.

34. Marinos and Petrascheck 1956; Hanke 1994.

35. Karantzali 1996; Broodbank 2000.

THE COPPER PRILLS

Small prills of copper are present in some of the slag fragments. Their size is always small, ranging from microscopic (under 50 micrometers across) to about a centimeter in size. Although a few prills of especially small size are in the coatings on the interior of the chimney fragments, most of them come from the chunks of slag. Their small size suggests that the metal was extremely valuable, and that larger pieces were carefully removed and saved. If occasional copper lenses or ingots were collected, no remains of them were found in the excavated material. The only prills found at the site were small masses disseminated within the vitreous waste material. They have been examined by a number of analytical techniques, providing different types of information.

Most analyses of prehistoric metals have been directed primarily toward the determination of provenance, but results have been very problematic. Elemental compositions are variable even within a single piece of ore or a single artifact, permitting few conclusions regarding original sources. More encouraging results have come from analyses aimed at understanding ancient technological processes.

Optical thin section analysis indicates that other copper minerals are present in addition to the prills of native copper found in the slag. The occasional presence of tiny crystals of the copper oxide cuprite may be a result of the original crystallization (App. A, CHR 99), but it is also possible that it represents oxidation of small amounts of native copper after burial. The cuprite is always very tiny (under 100 micrometers). The copper and iron oxide delafossite, revealed by X-ray diffraction (App. B, CHR 96), is likely to have formed under oxidizing conditions during the smelting operation.

Most of the recovered prills are extremely altered, and in many cases no pure copper survives. Information on the copper corrosion is provided by PIXE analyses (App. D). This technique was used to examine the surfaces of altered prills, with analyses of several parts of the surfaces. It showed that the chlorine content was enriched on the surface from the formation of copper chlorides. Arsenic content on the surfaces was highly variable, and it was depleted by the onset of corrosion products.

The analysis by SEM/EDAX (App. B) shows that the copper contains small amounts of many elements. Among the metals, iron and arsenic are present up to several percent, with iron occurring up to 4.9%. The arsenic content is extremely variable at the microscopic level. It ranges from a trace that is too small to detect to 26%.

The arsenic content is an important aspect of the copper from this site. A high percentage of arsenic is a feature of copper artifacts from the 4th and 3rd millennia B.C. in a broad geographic area in the eastern Mediterranean,[36] as well as in central and southeastern Europe[37] and the northern Caucasus.[38] It is a characteristic of Early Cycladic artifacts[39] and Early Minoan objects.[40]

Arsenic can be present in copper tools and weapons as a result of several alternative processes. Smelting arsenic-rich copper ores, adding arsenic ores to the smelting charge, or introducing arsenic in subsequent melting opera-

36. Charles 1967.
37. Budd and Ottaway 1991.
38. Ravich and Ryndina 1995.
39. Gale and Stos-Gale 1989, tables 1–3.
40. Gale 1990; Mangou and Ioannou 1998.

tions are all capable of producing arsenical copper. Most ancient arsenical coppers contain amounts ranging from a trace to as much as 7%–10% of the metal. This raises questions concerning the characteristics of the resulting alloys and the possible benefits of arsenical alloys over unalloyed copper. It also raises questions as to whether the ancient metallurgists were aware of these advantages (if they existed) and whether the arsenic was added deliberately.

Several authors have discussed the possible advantages of arsenical copper over pure metal. Favorable characteristics have been suggested both in regard to preventing oxidation during casting and improving annealing.[41] Arsenical copper would also have had a paler color than the pure metal, making it recognizable by sight and, perhaps, aesthetically preferable. An advantage in producing metal for sheets has been proposed for the alloy as well.[42] Cautions have been voiced, however, particularly with respect to the heterogeneous metals achieved by ancient processes.[43] Experiments in casting arsenical copper have shown that it is not necessarily advantageous because too many other factors are involved in creating a superior alloy, including casting temperature, pour rate, and the presence of other impurities.[44] Most researchers have considered that at least 2%–3% of arsenic is required to suggest a deliberate alloy.[45] Percentages of arsenic between 2%–6% yield the best results.[46] Amounts less than 2% have little effect. When percentages exceed 7%, the metal becomes brittle with cold working.

The copper from Chrysokamino does not seem to have had a percentage of arsenic above 5%, and much of the copper surely had less than 3%. Obviously, since the arsenic is present within the prills in the slag, it was included in the charge in some fashion, either from within the copper ore or from the addition of a different ore. The amount of arsenic is so variable that it seems unlikely that the metal was added deliberately. Whether it was deliberate or accidental, it probably resulted from the use of arsenic minerals along with the copper ores. Experiments have shown that if arsenic minerals are included in the charge, concentrations of the metal below 5% can be achieved easily in the smelting process even with temperatures below 1000°C.[47] At Chrysokamino, it is most likely that small amounts of arsenic minerals were present in the charge along with the small pieces of crushed copper ore.

41. Caley 1949; Maréchal 1958; Charles 1967; 1985, p. 25; Gale and Stos-Gale 1989, p. 30.
42. Lechtman 1996.
43. Budd 1991.
44. Budd and Ottaway 1991, p. 135.
45. Hauptmann, Begemann, and Schmitt-Stecker 1999, p. 9.
46. Budd and Ottaway 1991, pp. 138–139.
47. Pollard et al. 1991, pp. 130–131; Pollard, Thomas, and Williams 1991.

CHAPTER II

FAUNAL REMAINS

by David S. Reese

The faunal remains are divided by the two contexts identified at the metallurgy site. These two contexts, the small apsidal structure and the rest of the slag pile, have different dates. The slag pile is a mixed deposit with pottery that ranges from Final Neolithic to Early Minoan III–Middle Minoan IA. The apsidal structure is a closed deposit that comes from EM III–MM IA.

Within these contexts, the finds are organized by grid-square and level. The numbers are presented with whole and restorable individuals before a semicolon and fragments whose complete size is uncertain after a semicolon. Dimensions of the whole individuals are presented in millimeters.[1]

The faunal remains include the following:

Alvania montagui (gastropod, small)
Cerithium vulgatum (cerith, horn shell)
Columbella rustica (dove shell)
Fasciolaria lignaria (tulip shell)
Monodonta turbinata (top shell)
Paracentrotus lividus (sea urchin)
Patella coerulea (limpet)
Pisania maculosa (dove shell)

Lepus europaeus (hare)

Only a few faunal remains come from the pile of slag and industrial ceramics fragments. They include 1 *Lepus* vertebra, 61 *Patella*, and 6 topshells *(Monodonta)*. The shells are all edible species that live in shallow water on rocky shores.

At the time of excavation, the samples were isolated by specific locations, but no pattern of distribution could be recognized because faunal remains were distributed fairly uniformly over the excavated areas, so for this presentation, the remains are given by grid square.

Relatively more marine remains come from the apsidal structure. They include 1 fish bone, 35 *Patella*, 34 *Monodonta*, 3 *Paracentrotus*, 1 crab claw, and 8 other shells of 5 species. Like the shells from the slag pile, the species here are mostly edible species from shallow water on rocky shores. All of the marine invertebrates could have been collected casually from

1. Abbreviations: MNI = Minimum Number of Individuals; ws = the deposit includes material recovered from water sieving.

the seashore in the vicinity of the workshop site. The rarity of mammal and fish bones suggests that meat did not play a significant role in the diet consumed here.

THE SLAG PILE

Mammal Bone

Q20-2 (ws)
    1 *Lepus*-sized vertebra

Marine Invertebrates

K17-2
    3 *Patella*, 25.25; 2 broken, small/medium, 3 MNI

K18-1
    1 *Patella* fragment

M17-2 (ws)
    9/18 *Patella*, 16.75, 18, 19, 20; 5 broken, 2 large, 9 MNI

M20-2 (ws)
    1 *Patella* fragment
    4 *Monodonta* fragments, 1 MNI

M20-3
    1 *Patella*, exterior partly burned black, broken, large

M20-4
    1 *Patella*, broken, medium (?)

M20-5
    2 *Patella*, 26.25, 1 broken, burned, large, 2 MNI

M20-7
    1 *Patella*, 30+, broken, large

N17-2
    31/4 *Patella*, 18.5, 19.5, 19.75 (2), 22, 23, 23.75, 25.5, 25.75,
        26.25, 29.5, 34.5, 1 small; most broken, 31 MNI
    4 *Monodonta* fragments, 2 lips, 2 MNI

N20-4 (ws)
    1/1 *Patella*, broken, large, 1 MNI
    2 *Monodonta* fragments, 1 MNI

O20-4
    1/1 *Patella*, 20 (burned); broken fragment, possibly burned,
        2 MNI

P20-3 (ws)
    1/1 *Patella*, ca. 15, 1 MNI
    1 *Monodonta* fragment, medium/large

Q20-2 (ws)
    1 *Patella* fragment
    1 *Monodonta* fragment

Q20-3
    4 *Patella,* 18.75, 23.25; small/medium; medium/large, 4 MNI

R20-3 (ws)
    2/4 *Patella,* all broken, 2 MNI

THE APSIDAL STRUCTURE

Fish Bone

M18-2 (ws)
    1 fish bone

Marine Invertebrates

L19-2
    2/1 *Patella,* 16, 18.75; internal center, 2 MNI
    25 *Monodonta* fragments, 7 lips, no apices (1 near apex), 7 MNI

L19-3
    1 *Patella,* burned gray, 30.5
    6 *Monodonta* fragments, 5 fresh, 3 lips (3 MNI), 1 waterworn, large

M18-2 (ws)
    8 *Patella* fragments, 3 MNI
    38 *Monodonta* fragments, 4 MNI
    1 *Alvania montagui,* small
    1 *Cerithium vulgatum,* fresh, small, broken lip
    1 *Columbella rustica,* lip fragment
    1 gastropod columella fragment, water-worn
    Paracentrotus, 6 internal pieces, 6 spine fragments, 11 test fragments, 1 MNI

M18-3
    1 *Patella,* burned
    3 *Monodonta* fragments; large, 1 MNI
    1 *Fasciolaria lignaria,* bumpy, open mouth, broken lip, length 35

M19-2 (ws)
    1/1 *Patella,* broken, 1 MNI
    2 *Monodonta* fragments, large, 1 MNI

M19-3 (ws)
    6/1 *Patella,* 27.5, 29.5, 32.75, 34.25; 2 broken, 1 small, 1 medium, 6 MNI
    3/51 *Monodonta,* 6 complete, 6 lips, 11 MNI
    1 *Fasciolaria lignaria,* fresh, broken lip, length 31.75

M19-4 (ws)
- 1/2 *Patella*, medium/large, 1 MNI
- 1/1 *Monodonta*, complete example is small; distal piece is large, with ancient break in half, 2 MNI
- 1 *Fasciolaria lignaria,* body fragment
- *Paracentrotus,* 3 internal pieces, 2 test fragments, 1 MNI

M19-5
- 1 *Monodonta* fragment, body, large, encrusted

N18-2
- 6 *Patella,* 17.25, ca. 20, 24.5; 1 center fragment, 2 broken, medium, 6 MNI
- 1 *Monodonta,* complete, medium

N18-3
- 3 *Patella,* 17.25; 1 small, 1 medium, 3 MNI
- 1 *Monodonta* fragment, burned, body, medium/large
- 1 *Pisani,* fresh, complete, length 18.75

N19-2 (ws)
- 5/9 *Patella,* 18.5, 19, 25.25; 2 broken, 2 medium, 7 MNI
- 1/12 *Monodonta,* 2 MNI
- 1 *Columbella,* complete, fresh
- *Paracentrotus,* 12 internal pieces, 38 spine fragments, 36 test fragments, 1 MNI

N19-3 (ws)
- 1/1 *Patella,* 23.75; broken, 2 MNI
- 1 crab pincer fragment

N19-4
- 3 *Patella,* 2 burned center, broken; 1 unburned, 25, broken, 3 MNI

CHAPTER 12

Evidence for the Use of Threshing Remains at the Early Minoan Metallurgical Workshop

by Glynis Jones and Ann Schofield

During the 1997 excavations of the metallurgical workshop, furnace chimney fragments with evidence of botanical impressions were collected from the surface of the site. Latex casts were made of these impressions, and they were subsequently identified (at ×10 to ×40 magnification) by reference to botanical collections at the Department of Archaeology and Prehistory, The University of Sheffield. These identifications are listed in Table 12.1.

The majority of casts were identified as barley (*Hordeum* L.). The most common inclusions were rachis fragments, and all of them could be identified (either definitely or probably) as two-row barley (*Hordeum distichon* L.) on the basis of the sterile lateral florets characteristic of this species (in six-row barley, all the florets are fertile, i.e., grain-bearing). The shape of the internodes (parallel-sided) was also characteristic of the two-row species (rachis internodes of six-row barley are trapezoidal).

The sterile lateral florets of two-row barley remain attached to the rachis when the ears are threshed to release the grain from the central fertile florets, and so these fragments, lacking the central grain, apparently represent threshing remains. One of the rachis fragments (cast X 991) is from the lower part of the ear and is still attached to the culm (straw). Other grass culm fragments, not attached to rachis, could also represent barley straw resulting from threshing.

A few definite and possible grains of barley (the hulled variety) were also identified, one of which appeared to be straight, as would be expected for two-row barley (six-row barley has both straight central grains and twisted lateral grains). The grains were all rather small—the sort of underdeveloped grains that might remain attached to the rachis during threshing. The most likely interpretation of this material is that it represents the chaff and straw resulting from the threshing of two-row hulled barley, which was used as temper in the construction of the chimneys.

One fragment has been tentatively identified as a glume (chaff fragment) of wheat (*Triticum* L.), which may have been a minor contaminant of the barley crop. A single olive (*Olea* L.) leaf (cast X 213) and a large indeterminate object (cast X 990) are presumably chance inclusions.

Few though they are, these remains provide rare evidence of Early Minoan agriculture. Barley was clearly cultivated at this time, and it

TABLE 12.1. PLANT IMPRESSIONS IN CHIMNEY FRAGMENTS FROM THE METALLURGICAL WORKSHOP

Species		Cast Number																	
		213	977	978	981	982	985	988	989	990	991	993	994	997	998	1000	1001	1002	1003
Hordeum distichon L.	rachis frags.							1			1	1			1		1	1	1
H. cf. *distichon* L.	rachis frags.																1		
H. cf. *distichon* L.	hulled grains				1														
Hordeum sp.	hulled grains												1						
cf. *Hordeum* sp.	grains		1	1															
cf. *Triticum* sp.	glume															1			
indeter. cereal	grain + chaff													1					
indeter. Gramineae	culm. frags.					1	1		1										
Olea sp.	leaf	1																	
indeterminate										1									

has been possible to identify with certainty the two-row hulled type. Impressions of hulled barley were also found in mud plaster at the Early Minoan settlement of Myrtos,[1] but in this case identified, at least in part, as the six-row species (*Hordeum vulgare* L.). The Myrtos impressions also consisted predominantly of chaff and straw and, as at Chrysokamino, it was suggested that they represented sweepings from the threshing floor. At Early Minoan Debla,[2] a few impressions from the coarse fabric of storage bins were identified as barley grains (along with grains of emmer wheat and oat), and at the Middle Minoan settlement of Pseira, a concentration of charred hulled barley grain was tentatively identified to the two-row species.[3] The finds from Chrysokamino and Pseira contrast with those from the rest of Bronze Age Greece, where six-row barley is more usual.[4] The evidence for wheat at Chrysokamino is much less reliable because it is tentatively identified and may be no more than a contaminant of the barley. The olive leaf, too, could be derived from a wild or cultivated tree (Fig. 7.2).

1. J. Renfrew 1972.
2. Greig and Warren 1974.
3. Jones and Smith, forthcoming.
4. Hansen 1988.

CHAPTER 13

Chrysokamino in the History of Early Metallurgy

by James D. Muhly

Archaeological research on Crete since about 1980, especially in the northeastern part of the island, the Bay of Mirabello, has established that the beginnings of Cretan copper metallurgy coincided with the establishment of new settlements in the eastern part of the island in the Final Neolithic period, ca. 4500–3500 B.C.[1] This association of new settlements with incipient copper metallurgy extended to the far east coast (Palaikastro) and possibly as far west as the Mesara and the site of Phaistos.[2] The evidence for FN metallurgy at Phaistos is, however, very inconclusive,[3] and the associations under discussion here are best attested in the eastern part of the island. Inasmuch as metallurgical technology must have been imported to Crete from outside the island, it should be stressed that the Final Neolithic has long been recognized as the first international age in Cretan prehistory.[4]

Prior to 1996, the evidence for incipient copper metallurgy on Crete consisted of isolated finds that would be considered "trinket metallurgy" in a comparable Mesopotamian context.[5] This situation changed with the 1996–1997 excavations at Chrysokamino. Here, for the first time, was an actual metallurgical site, a copper smelting site that seemed to have its main period of activity in the later part of the 3rd millennium B.C. (EM III), but was associated with sherds from all earlier periods, going back to Final Neolithic. It is true that no direct association can be made between metallurgical activities and the FN sherds, but it is hard to imagine any other reason for human presence on this isolated, windswept point of land.

How are we to evaluate this new evidence for the early development of metal technology on the island of Crete? How does it relate to our current understanding of the beginnings of metalworking in the Eastern Mediterranean and the Near East? First of all, it must be recognized that the Final Neolithic does not represent the very beginning of metal usage on Crete. That honor goes to a Late Neolithic flat axe from Knossos, found by Sir Arthur Evans.[6] Evans published only a drawing of this axe; to my knowledge, no photograph exists, and the axe can no longer be located in the Herakleion Museum. Nor, unfortunately, does it come from a secure context.[7] It does, however, seem to be of a common Late Neolithic type, known from sites on the Greek mainland and especially from numerous sites in the Balkans.[8]

1. For the date, see Johnson 1996, p. 271; for a slightly different date, see Johnson 1999.
2. Vagnetti 1975, esp. pp. 94–95.
3. Vagnetti and Belli 1978, p. 131.
4. Betancourt 1985, p. 13; also Vagnetti 1996, pp. 38–39.
5. Moorey 1988, p. 29.
6. Evans 1921–1935, vol. 2, pp. 14–15, fig. 3:f.
7. Evans 1971, p. 115.
8. Phelps, Varoufakis, and Jones 1979.

The Knossos axe is probably best understood as an import in Late Neolithic Crete, most likely from the Balkans, as this type of axe was never produced on Crete itself. A Balkan origin would not be surprising. Throughout the Aegean, the evidence for a Balkan origin of Greek Neolithic metallurgy seems quite overwhelming. Balkan connections exist not only with copper metallurgy, but also with the working of gold, silver, and possibly even lead. I have argued the case for Balkan influence on several occasions,[9] and Balkan connections have also been pointed out by Branigan and Nakou.[10]

EARLY AEGEAN METALLURGY AND THE BALKANS

The quantity of metal finds from the Greek Neolithic increases almost every year. In the terminology of Demoule and Perlès in their recent review of Neolithic Greece, most of these finds come from Phase 5 (Final Neolithic), with a small handful of finds from Phase 4 (Late Neolithic, ca. 5500–4500 B.C.).[11] Only one tiny amorphous lump of copper from level 14 at Dikili Tash is said to be Middle Neolithic, ca. 6000–5500 B.C.[12] Recent finds from excavations at Ftelia, on the north coast of Mykonos, have produced a small collection of copper objects, especially awls and pins, as well as a circular disk of gold with a central perforation.[13] A series of 10 calibrated radiocarbon dates from Ftelia support a date in the first half of the 5th millennium B.C.,[14] or Late Neolithic I in the terminology of the excavator.[15] Yet in the Balkans, gold appears only in what are now called Late Chalcolithic contexts, ca. 4250–4000 B.C.[16] When it appears, however, it is attested in large quantities; some 6 kg of gold have survived from Eneolithic southeastern Europe.[17]

The metal finds from the site of Dimitra, just north of Sitagroi in eastern Macedonia, are currently the subject of some controversy. According to Grammenos, the excavator, five beads of copper come from a Middle Neolithic context in trench II, and four beads of copper, along with two beads of gold, come from a Late Neolithic context in trench I. Metal finds from the site also include a possible fishhook of copper and another of gold in what seem to be unknown contexts.[18] According to Andreou, Fotiadis, and Kotsakis, in their review of Neolithic and Bronze Age northern Greece, the site of Dimitra has three phases, corresponding to Sitagroi I to III, with metal finds, including gold and copper beads, from all phases.[19] Accepting a date for Sitagroi I at 5500–5200 B.C.,[20] this would mean that Dimitra has some of the earliest copper finds from anywhere in Europe, as well as the earliest objects of gold.

9. Muhly 1985a, 1996.
10. Branigan 1974, p. 98; Nakou 1995, pp. 4, 21.
11. Demoule and Perlès 1993; for a summary of the evidence, see Muhly 1996, pp. 78–80; 2002, p. 78.
12. Muhly 1996, p. 76.
13. Maxwell 2002; for the gold disk, see I. Sikka, "Νεολιθικό χρυσάφι στη Μύκονο," *Kathimerini,* December 24, 2002, p. 15.
14. Facorellis and Maniatis 2002, p. 311.
15. Sampson 2002a.
16. Pernicka et al. 1997, p. 51.
17. Pernicka et al. 1997, p. 41.
18. Grammenos 1997, p. 270; for the analysis of these finds, see Mirtsou et al. 1997.
19. Andreou, Fotiadis, and Kotsakis 2001, p. 312.
20. Elster and Renfrew 2003, p. xxvii.

On the island of Andros, copper implements, weapons, and jewelry, along with impressive architecture and marble vases, have been discovered at the fortified settlement of Strophilas. These finds are dated to the middle of the 5th millennium B.C., or the very beginning of the Final Neolithic period.[21] There are also a few Final Neolithic copper objects from the Tharrounia cave in Euboia, principally pins or needles.[22] For Sampson, the Final Neolithic (his Late Neolithic IIA) was the period "when copper objects, always in insignificant numbers, make their appearance at every site of Mainland Greece and the Aegean."[23]

Two aspects of the development of Final Neolithic Aegean metallurgy need to be emphasized. The first is its polymetallic character. The Final Neolithic saw the real development of gold and silver metallurgy in the Aegean world. The number of gold artifacts known to date from this period has greatly increased in recent years, with the discovery of gold pendants from the Theopetra cave in Thessaly and from Anavissos near Yannitsa and Platomagoulia in Magnesia, as well as a gold strip from the Zas cave on the island of Naxos.[24] Silver pendants have been found in the Alepotrypa cave in the Mani peninsula. They were discovered along with a hoard of silver jewelry that included two pairs of bracelets and 168 very small perforated silver beads, now duplicated in unpublished finds from an EM I cemetery at Gournes in central Crete. Other silver pendants have been found in the Amnisos cave on Crete and in the Cave of Euripides on the island of Salamis.[25]

Most of these pendants, both the gold and silver examples, are of the characteristic "ring-idol" design, with a central perforation. Such pendants are a well-known feature of Balkan metallurgy from the time of the Karanovo VI or Gumelnitsa culture, also referred to as the Eneolithic period, roughly contemporary with the Aegean Final Neolithic.[26] The most famous examples come from the cemetery at Varna, on the Bulgarian shore of the Black Sea. The Varna finds have been featured in many books and articles by the late Marija Gimbutas.[27]

From the Aegean world, the most spectacular collection of such ring-idol pendants comes, unfortunately, from the antiquities market. In two sting operations, one at Vouliagmeni (October 1, 1997) and a second in Patras (May 17, 2000), Greek police confiscated a large cache from would-be sellers of looted antiquities, including 67 ring-idol pendants made of thin sheet gold. The most impressive piece, 15.5 cm in height, is, in fact, the largest known pendant of this type.[28] An example in copper is known from Period IX at Emporio on Chios.[29] A stone pendant comes from the Strophilas settlement on Andros.

It would be hard to imagine more convincing evidence for the dominant Balkan character of Aegean Final Neolithic metallurgy. Yet this interpretation is still anathema to many Aegean prehistorians. They want no part of it, for they feel that it reeks of old-fashioned diffusionist archaeology. This brings us to a discussion of the second main feature of Aegean Final Neolithic metallurgy. Whereas earlier periods produced only isolated artifacts, the Final Neolithic period provides us with evidence for the actual production of such artifacts. Now, for the first time, we can speak with assurance of the actual production of metal artifacts on Greek soil. The

21. "Νεολιθικός οικισμός στην Ανδρο," *Kathimerini*, September 7, 2001, p. 15.

22. Andreopoulou-Mangou 1993, pp. 435–437; for more on these objects, apparently made of arsenical copper, see below.

23. Sampson 2002a, p. 164.

24. Muhly 2002, p. 78.

25. Muhly 2002, p. 78.

26. For chronology, see von Görsdorf and Bojadžiev 1996.

27. E.g., Gimbutas 1977.

28. For all this, see Muhly 2002, p. 78; also see the catalogue of a temporary exhibit in the National Museum of Athens (Demakopoulou 1998).

29. This is the earliest metal artifact from that site; for the remarkable chemical composition of this piece, see Muhly 2002, p. 78.

most extensive assemblage of such evidence comes from the site of Sitagroi in Macedonia. The excavators of Sitagroi are also the most outspoken opponents of what can be called the "Balkan Connection."

The site of Sitagroi, also known as Photolivos, was excavated in three seasons (1968–1970) under the joint direction of Renfrew and Gimbutas. The first volume of the final excavation report, published in 1986, dealt with archaeology and chronology, along with specialized studies on the environment and economy of the area. The second volume, dealing with crafts and technology, was published in 2003. At the very beginning of the excavation, Renfrew wrote an extremely influential article on "The Autonomy of the South-East European Copper Age."[30] In this article, he outlined his arguments for the independent invention of metallurgical technology in various parts of the ancient world. He has now returned to this theme in the second Sitagroi volume, in the chapter on "Metal Artifacts and Metallurgy."[31] Meanwhile, a series of studies based on or including discussion of the finds from Sitagroi appeared between 1969 and 2003. Among them, the most important from the perspective of early Aegean metallurgy is a study by Veronica McGeehan-Liritzis and Noel Gale dealing with the analysis of Neolithic and Early Bronze Age metal artifacts from Greece.[32]

This is not the occasion for another evaluation of all this work.[33] What is important here is the evidence for metalworking at Sitagroi. The recently published remarks by Renfrew regarding diffusion[34] seem to have been written in the late 1980s and have no bearing on current discussions. When I talk about diffusion I have in mind such works as the recent book by Oppenheimer, deriving all ancient civilization from Southeast Asia.[35] To argue for a strong Balkan component in Aegean Neolithic metallurgy does not, in my opinion, constitute diffusion.[36]

A small number of copper artifacts is present already in Sitagroi II, a period now regarded as early Late Neolithic, 5200–4800 B.C.[37] Most of the metallurgical materials come from Sitagroi III, a period of uncertain exact date (Sitagroi IV–Va is ca. 3200–2500 B.C., with III covering some or all of the intervening centuries between 4800 and 3200 B.C.).[38] The metal objects from Period III include pins, beads, and a bent strip of copper, as well as a small gold bead.[39] Upon final publication, however, the evidence for actual metalworking has proved quite problematic.

Sitagroi III has 36 sherds with copper incrustation, most likely fragments of crucibles. As the incrustations are found not only on the interiors of the sherds, but also on the exteriors and the rims,[40] the logical conclusion is that the crucibles were used to pour molten metal in some sort of casting operation. The lack of molds is something of a problem in this context, but molds are virtually absent from all metallurgical contexts prior to the Early Bronze Age. Sitagroi III, then, provides evidence only for the crucible melting of metallic copper, perhaps even native copper. This is probably true as well for the isolated finds at other contemporary sites, such as Giali (near Nisyros), the Alepotrypa cave, and Mandalo in western Macedonia.[41]

Copper smelting cannot be documented at Sitagroi or at any other Neolithic site in the Aegean. Nor can the source(s) of the copper ore being

30. Renfrew 1969.
31. Renfrew and Slater 2003.
32. McGeehan-Liritzis and Gale 1988.
33. See Muhly 1991.
34. Esp. Renfrew and Slater 2003, pp. 317–318.
35. Oppenheimer 1999; Terrell 2000.
36. See the important remarks in Diamond 1995.
37. Andreou, Fotiadis, and Kotsakis 2001, p. 308.
38. Andreou, Fotiadis, and Kotsakis 2001, p. 308.
39. Renfrew and Slater 2003, p. 305, fig. 8.1:b–g.
40. Slater 2003, p. 303.
41. Renfrew and Slater 2003, p. 309; for problems of interpretation regarding Sitagroi III, see Muhly 2004a.

exploited in the Aegean Neolithic be identified. The analytical evidence published to date is very unsatisfactory. The artifacts themselves are small; the samples that the analysts were allowed to take were obviously much smaller, and they often contained only corrosion products, not actual metal. Some of these samples, taken from a single artifact, were analyzed by three different analytical techniques.[42] The results vary by over 260%.[43] Obviously, nothing can be said about provenience on the basis of such evidence. The lead isotope evidence is also inconclusive,[44] but it does suggest that sources to the north, perhaps in Bulgaria, were being exploited. Indeed, as Renfrew concludes, "The likelihood of a northern origin for the earliest metallurgical practices at Sitagroi was proposed above on other grounds, and these findings would harmonize with it."[45] And that, in reality, is all I mean when I talk about the role of the Balkans in the development of Aegean metallurgy.

The reasons for turning to the north, to the Balkans in order to understand the beginnings of metallurgical technology in the Aegean have more to do with technological priority than with simple typological comparisons, or with flat axes and ring-idols. What was being done in the Aegean was also being done in the Balkans, but in the Balkans it was being done centuries earlier and on a much larger scale. The earliest copper artifact in the Balkans, an awl from the site of Balomir in Romania, is a massive object, 14.3 cm in length, and it comes from an Early Neolithic context, 7000–6000 B.C.[46] By the second half of the 5th millennium B.C., the Balkan copper industry was producing hundreds of massive shaft-hole axes and axe-adzes.[47] At the same time, Balkan metallurgists were mining copper ores in underground mines with shafts and galleries,[48] and they were capable of smelting the complex secondary sulfide ores from these mines.[49] The number of gold ring-idol pendants known today from the Aegean world is quite impressive, but it does not begin to compare with the quantities from the Balkans. Just four of the richest graves from the Varna cemetery produced a total of 2,200 gold objects.[50]

We now have some perspective for comparing Cretan metallurgical developments with what had gone on to the north, on the Greek mainland and in the Balkans. Metallurgy developed later in Crete and on a much smaller scale, with striking differences. The ubiquitous ring-idol pendant is virtually unknown on Crete. The silver example from a burial near the Amnisos cave seems to be an isolated find. Flat axes never became a part of the Cretan metal assemblage, and the Knossos example remains the only one ever found on the island, although similar axes made of stone were quite common.[51] Crete, as always, seems to have gone its own way, developing a repertoire of types (e.g., the triangular and the long daggers

42. McGeehan-Liritzis and Gale 1988.
43. See Muhly 1991, p. 361.
44. Stos-Gale 2003.
45. Renfrew and Slater 2003, p. 313.
46. Muhly 1996, p. 78; Bognár-Kutzián 1976, p. 70.
47. Todorova 1978; Pernicka et al. 1993.
48. Chernykh 1978; Jovanović 1978, 1995.
49. See Ryndina, Indenbaum, and Kolossova 1999, with detailed analysis of Eneolithic crucibles, pp. 1063–1066.
50. Renfrew 1986b, p. 148. For the Varna finds, see *First Civilization in Europe*.
51. Strasser 2002.

of the EM period) very different from that attested in other regions.[52] Yet even so, there are surprising similarities. The three silver daggers from Early Minoan Koumasa[53] bear a remarkable resemblance to a silver dagger from the Hungarian Middle Bronze Age.[54]

In his 1969 article, Renfrew had very little to say about the Aegean. His goal was to establish the validity of radiocarbon dating and to use the dates it provided to demonstrate the autonomy and cultural priority of European Neolithic cultures. The "Copper Age" of the Balkans could not possibly be derived from Anatolia, especially Troy, as scholars, following Childe, had long believed. This conclusion was necessary because the comparable finds from the Balkans were far earlier than anything known from Troy or any other Anatolian site. Even when early metal-using sites such as Çatal Höyük and Çayönü Tepesi came to light in Anatolia, the tiny bits and pieces of copper from these sites could not possibly compare, it was argued, with the massive copper implements of the Balkan Eneolithic. Prehistorians from Central and Eastern Europe were dismayed, for they had based entire academic careers upon the primacy of Troy. Nevertheless, Renfrew's arguments carried the day, especially after his dramatic presentation of the case in a popular book entitled *Before Civilization: The Radiocarbon Revolution and Prehistoric Europe.*[55]

Renfrew was so sure of the cultural primacy of the Balkans, and of Europe in general, that he was prepared to draw a chronological "fault line" separating Europe from the Aegean, Anatolia, and the Near East. Only on the European side of this fault line did the new radiocarbon dates seem to force drastic, upward revisions of the traditional archaeological chronology. The map showing the fault line appears in the book mentioned above,[56] but it had also been published two years earlier in an extremely influential *Scientific American* article entitled "Carbon 14 and the Prehistory of Europe."[57] This article was reprinted many times, and the fault line map became justly famous. It was not long, however, before it became obvious that the basic premise illustrated by this map was rather misleading. The caption to the map indicated that the fault line only concerned dates after ca. 3000 B.C. At the time, however, the dates after 3000 B.C. from the east of the fault line were based on historical chronologies, while those from the west were based on comparisons with Troy, according to the chronology then accepted. Early radiocarbon dates from the Eastern Mediterranean and the Near East were from contexts after ca. 3000 B.C., and they supported the historical chronologies based on written records from Egypt and Mesopotamia. Early radiocarbon dates from the Balkans came from Neolithic tell sites, and they indicated a chronology much earlier than the accepted "Trojan-based" chronology.

Hence the "fault line," but in reality it did not exist. With the widespread use of calibrated radiocarbon dates, discussed by Renfrew in 1973,[58] revision of traditional chronologies became necessary almost everywhere, especially in the Aegean and Anatolia. The past decade has witnessed a transformation in our understanding of the chronology of the Neolithic period in Greece and Anatolia, and this has had a profound impact on our evaluation of the development of metallurgical technology in both areas. Conceptions of cultural priority and technological autonomy, thought to be secure just 20 years ago, now have to be reformulated.

52. Branigan 1967.
53. Branigan 1968, p. 63.
54. Mozsolics 1967, pl. 45.1; Kovács 1994, pl. 15.
55. Renfrew 1973.
56. Renfrew 1973, p. 105, fig. 21.
57. Renfrew 1971.
58. Renfrew 1973, pp. 69–83.

EARLY METALLURGY IN ANATOLIA

Nowhere has change been more dramatic than in our understanding of the development of metalworking technology in Anatolia. Once thought to be something of a backwater during the Neolithic period, Anatolia has emerged as one of the most dynamic parts of the ancient world, especially during the period once called the Aceramic Neolithic, but more correctly referred to as Prepottery Neolithic A and B, or PPNA and B, ca. 11,000–6200 B.C.[59] The earliest known settlement at Hallan Çemi Tepesi, a PPNA site dating to ca. 10,200 B.C., already had a series of semi-subterranean round houses, stone bowls with incised decoration (one showing a procession of dogs), and the bones of what seem to be domesticated pigs, along with evidence for extensive use of wild plants and animals.[60]

Hallan Çemi is located to the northeast of modern Diyarbakır. To the southwest, in the vicinity of modern Urfa, lie a series of sites in the area known as Upper Mesopotamia. The earliest of these, Göbekli Tepe Ziyaret, has a series of circular or oval-shaped structures with interior T-shaped pillars. The pillars, some 2 m in height, have remarkable relief carvings depicting wild boars, lions, snakes, foxes, and perhaps even wild Asiatic donkeys. These finds are slightly earlier than 9000 B.C.[61]

Just to the north, in the foothills of the Taurus Mountains, the PPNB site of Nevali Çori has five levels of occupation covering the second half of the 9th millennium B.C. Nevali Çori has monumental stone buildings, most likely shrines, with monumental freestanding stone sculptures in the round depicting human figures and a bird in flight. Smaller sculptures depict a veritable bestiary: panthers, lions, wild boars, wild horses, bears, and vultures.[62] A single copper bead from Nevali Çori is one of the earliest metal objects from Anatolia.[63] From the PPNB site of Aşıklı Höyük, 25 km southeast of Aksaray in central Anatolia, a series of beads made of rolled thin sheets of copper provides even more extensive evidence for the early use of copper. Aşıklı represents a real town surrounded by a city wall, the earliest in Anatolia, and it had an extensive obsidian industry. The site is dated to the first half of the 8th millennium B.C.[64]

The best evidence for early copper metallurgy in Anatolia still comes from the site of Çayönü Tepesi, one of the first Aceramic Neolithic settlements to be discovered, and one of the earliest of this period. Excavations underway since 1964 have now uncovered some 7,000 m² in the course of 17 field seasons. Most of the metal artifacts, made of native copper hammered into the shape of objects such as awls and fishhooks, come from the mid-9th millennium B.C. Two calibrated radiocarbon dates suggest the site was occupied much earlier, 10,000–9400 B.C. Çayönü (also known from Nevali Çori) has elaborate buildings with terrazzo floors and the earliest evidence for woven textiles, which were made of linen.[65]

---

59. Bar-Yosef 2001; 2002a, pp. 379–381.

60. Rosenberg 1994, 1998; Rosenberg et al. 1995.

61. Beile-Bohn et al. 1998; Bischoff 2002.

62. Bischoff 2002, with chronological table, p. 30; Aurenche and Kozlowski (1999, p. 68) put Göbekli at ca. 8500 B.C. and Nevali Çori at ca. 8200 B.C.

63. Yalçin 2000, p. 19.

64. Esin 1999; Yalçin and Pernicka 1999; Yalçin 2000, pp. 18–20.

65. Muhly 1989; Özdoğan and Özdoğan 1999; Maddin, Muhly, and Stech 1999. The earliest known cotton, from the site of Mehrgarh in southern Pakistan, dates to the first half of the 6th millennium B.C. (Moulherat et al. 2002).

At the onset of Mellaart's excavations at the Konya basin site of Çatal Höyük in 1961, the scholarly world was totally unprepared for the wealth of finds from a Neolithic site in Anatolia, an area then regarded as provincial and of little significance compared to Syria and Mesopotamia. Mellaart's excavations revealed elaborately decorated shrine buildings, wall paintings of hunt scenes and bizarre funerary rituals with headless corpses and vultures, and clay figurines depicting women in the act of giving birth. These finds came from contexts now dated to the first half of the 7th millennium B.C.[66] There were also small finds of copper, including beads made of rolled-up sheets like those from Aşıklı Höyük. These artifacts, strangely enough, were never published by Mellaart, but they are in the Konya Museum.[67] The small beads from level VIA, identified as lead in the publication, are actually made of the lead sulfide galena.[68]

Çatal Höyük can be seen as the culmination of developments underway in Neolithic Anatolia for over 3,000 years. This was the period known as the Younger Dryas, a time when the drying and cooling of the environment stimulated a shift from a hunter-gatherer way of life to agriculture and settled communities.[69] The Anatolian evidence clearly shows that the introduction of metal and the development of metal technology took place in periods of profound social and economic change, periods that saw the development of specialized crafts, highly sophisticated artistic work in stone and clay, and monumental architecture. Metalworking was but one aspect of a whole series of technological innovations taking place in periods of intense creativity. This remarkable fluorescence of culture in Anatolia came to an end toward the end of the 7th millennium B.C. Severe climatic changes dated to ca. 6200 B.C., as seen in the ice cores from Greenland and the pollen cores from Anatolian lakes, brought about the drought that put an end to PPNB cultures across Anatolia and the Levant.[70] One of the last products of this remarkable age is the massive copper macehead from level 2B at Can Hasan, southeast of Çatal Höyük. This find is dated to ca. 6000 B.C.[71]

There is, not surprisingly, serious opposition among some Aegean and Middle Eastern prehistorians to the whole Younger Dryas climatic change hypothesis and the argument that a sudden reversal of climate at the end of the last Ice Age led to the development of sedentary agricultural societies. In a 1997 monograph entitled *Naissance des divinités, naissance de l'agriculture*, Cauvin shifted the argument from ecology and environment to one based on human mentality and the emergence of new religious ideas and symbols.[72] According to Cauvin, Childe's so-called "Neolithic Revolution" was a fundamental transformation in the world view of the societies of the day, as attested by the elaborate architecture and religious iconography of PPNB societies, especially those of Anatolia.[73]

This issue is of some significance here because most of the scholars who believe that climatic changes brought PPNB cultures to an end ca. 6200 B.C. also believe that another sudden climatic change occurred ca. 2200 B.C., ending the Akkadian empire of Sargon I.[74] According to this theory, contemporary climatic changes across the Eastern Mediterranean coincided with the apparent intensification of copper smelting activity at Chrysokamino in the EM III period.[75]

66. Levels IX–IV; Mellaart 1967.
67. Yener 2000, pp. 23–24.
68. Sperl 1990.
69. Bar-Yosef 2001, pp. 17–26.
70. Bar-Yosef 2001, pp. 26–28; Bar-Yosef 2002b, p. 122.
71. Yalçin 1998.
72. Cauvin 1994.
73. Cauvin's monograph, little noticed in the Anglo-American world of prehistoric archaeology, has now appeared in an English translation by Trevor Watkins (Cauvin 2000) and has been the subject of a review feature in the *Cambridge Archaeological Journal* (Cauvin et al. 2001).
74. Weiss 1997.
75. Weiss and Courty 1993; Courty 1998.

THE USE OF ARSENICAL COPPER

In recent decades, two basic themes have dominated studies of the development of Eastern Mediterranean and Near Eastern metalworking technology ca. 8000–2000 B.C.

The first theme is the development of furnace technology, or the stages whereby metalworkers moved first from heating native copper to melting native copper in crucibles, then to smelting simple oxide and carbonate copper ores in crucibles, and later to smelting simple (and complex?) copper ores in specially built furnaces. Initially, these furnaces may have been nothing more than pits in the ground, with or without crucibles as linings and collars or chimneys on top of the pits to increase the draft. This stage was followed by the use of the fully developed shaft furnace which, on the basis of present evidence, seems to have been a product of the Late Bronze Age. A pioneering attempt at writing the history of these developments has now been made by Craddock.[76]

The other theme concerns the development of alloying technology, a complex of developments integrally related to the development of furnace technology. There is no evidence for the manufacture of any artificial or manmade alloys with native copper as their base. It was only with the smelting of complex copper ores that copper alloy artifacts appeared. The first of these contained ca. 2%–6% arsenic. Later alloys contained 4%–10% tin.

The predominance of arsenical copper alloys during the period from 4000 to 2000 B.C. is one of the most significant discoveries of modern archaeometallurgical research. Contemporary metallurgists, working chiefly in industry, long ago forsook any interest in arsenical copper. Their objective was, contrarily, to remove all trace of arsenic from metallic copper. In the 20th century A.D., copper was of interest chiefly for its ability to conduct electricity, and the presence of even minute traces of arsenic greatly reduced this capability.[77] It was only when the Cambridge metallurgist James Charles examined some of the Early Cycladic daggers in the Ashmolean Museum at Oxford that the importance of arsenical copper alloys in Early Bronze Age metallurgy was rediscovered.[78]

Beginning with a pair of articles published by Charles and Renfrew in 1967, the significance of arsenical copper from the Indus River to the British Isles has been documented in great detail through a series of analytical programs. Much of the technical metallurgical research on the production and the metallurgical properties of arsenical alloys has been carried out on artifacts from the New World, especially from the arsenical copper industries of ancient Peru.[79] Some similar work has also been done on early arsenical alloys from the Old World.[80] It has often been stated that arsenical copper and bronze (the alloy with tin) had more or less equal mechanical properties and that it was only the medical hazards involved in working with arsenic that prompted a shift to tin. This is simply not correct. Bronze is a harder, tougher alloy. The main advantage of the arsenical alloy is its ductility,[81] and the medical hazards have been greatly exaggerated, as all smelting was done out in the open air, not in an enclosed environment. It has also been established in laboratory work, making use of modern alloys with different concentrations of arsenic, that the presence of less that 2.0%

76. Craddock 2001.
77. Lechtman 1996, p. 479.
78. Charles 1967; for his scholarly career, see Charles 2000.
79. Lechtman 1996; Lechtman and Klein 1999; Shimada and Merkel 1991.
80. Northover 1989; Budd and Ottaway 1991.
81. Northover 1989, pp. 112–114.

arsenic has virtually no effect on the properties of the copper artifact.[82] The shift from arsenical copper to bronze was a major technological development of great cultural significance.

The use of arsenical copper in the pre-Bronze Age Aegean must, at present, be regarded as problematic. The presence of arsenical copper is claimed in copper artifacts from Final Neolithic Kephala and the Zas cave, but these analyses are unpublished.[83] The use of arsenical copper in the Thessalian Final Neolithic is said to have been demonstrated by the analysis of two metal artifacts from Dimini,[84] but this cannot be correct. Of the two objects in question,[85] only one, a flat axe, was analyzed. It was found to be 99.7% copper with no trace of arsenic.[86] The two contemporary flat axes from Sesklo[87] were also made of very pure copper, having 99.9% and 99.3% copper.

The metal finds from the Tharrounia cave in Euboia remain enigmatic. From contexts that Sampson regards as LN IIA (late 5th to early 4th millennia B.C.)[88] come four copper artifacts, a dagger and three needles, made of arsenical copper. They have an average of 3.12% arsenic.[89] It is curious that Mangou and Ioannou publish the Tharrounia analyses in their study of prehistoric copper-based artifacts from mainland Greece, but they say nothing about this anomalous use of arsenical copper in a Final Neolithic context. They are only interested in the presence of some 2% nickel in one of the arsenical copper needles from the Tharrounia cave.[90]

It is theoretically possible that complex sulfide and arsenide ores were being smelted in the Aegean already during the Final Neolithic period. As mentioned above, such technological sophistication has already been claimed for the Balkans in the Eneolithic period,[91] based on the analysis of crucible fragments from the site of Durankulak, securely dated to ca. 4300 B.C.[92] Similar claims, involving the use of both copper and arsenic-rich ores, have now been made for contemporary contexts in Anatolia.[93] The analytical evidence from southeastern Europe, however, still supports the reconstruction of an extensive pre-Bronze Age metal industry, one that produced an estimated 4,700 kg of copper, calculated solely on the basis of surviving artifacts. This industry was based on a very pure copper that can only represent the use of melted native copper.[94]

The Aegean documentation for such a copper smelting technology is even more ambiguous than the data from southeastern Europe. The Sitagroi evidence, cited in the past in support of copper smelting activity at that site during Phase III, now seems more compatible with the simple melting of (native?) copper in a crucible, as discussed above. Another crucible from a good late 5th-millennium B.C. context at Mandalo in western Macedonia has extensive vitrification in its interior,[95] but the way in which

82. Northover 1989, p. 113; Budd and Ottaway 1991, pp. 138–139.
83. Sherratt 2000, p. 68.
84. Andreou, Fotiadis, and Kotsakis 2000, p. 267, n. 46.
85. McGeehan-Liritzis and Gale 1988, p. 203, nos. 34, 35.
86. McGeehan-Liritzis and Gale 1988, p. 217.
87. McGeehan-Liritzis and Gale 1988, p. 219, nos. 68, 69.
88. Sampson 1996, p. 73.
89. Andreopoulou-Mangou 1993, p. 436, table 1; Mangou and Ioannou 1999, p. 84, table 1.
90. Mangou and Ioannou 1999, p. 86.
91. Ryndina et al. 1999.
92. Pernicka et al. 1997, p. 127.
93. Yener 2000, p. 35.
94. Pernicka et al. 1997, pp. 41, 121.
95. Papanthimou and Papasteriou 1993, pp. 1209, 1213, fig. 2.

this crucible was used has yet to be determined. Fragments of crucibles also come from the Final Neolithic site of Kephala on the island of Kea;[96] they were found with pieces of slag originally thought to represent a fluid slag that was too advanced for the Final Neolithic period.[97] Now, in light of the Chrysokamino evidence discussed in this volume, the Kephala slags need to be studied in greater detail. Two small amorphous lumps of metallic copper, found with four daggers and other copper artifacts in good Final Neolithic contexts at the Alepotrypa cave,[98] suggest some sort of local metalworking activity, but a more definite interpretation must await future analytical investigation.

Early metal production and usage on Crete is still the subject of much debate and considerable confusion. Certain basic assumptions, eminently reasonable in the early years of the 20th century, still influence our present evaluation of the archaeological evidence. In publishing the metal finds from the Trapeza cave, excavated in 1936, Money-Coutts wrote, "It must be remembered that until the EM II period metal of any kind was practically unknown in Crete, so that the rapid spread of its use and the skill with which it was worked shew the receptiveness and ingenuity of the early inhabitants of the island."[99]

Yet nothing about the deposits within the cave itself necessitated or even supported such a conclusion.[100] Pendlebury's discussion of the nature of these deposits shows that the metal finds, both of precious and of base metal, came from a context that included a jumble of pottery, ranging from Final Neolithic to MM II. Because of the assumption made explicit in the foregoing quotation, none of the metal artifacts was thought to be any earlier than EM II.[101]

The degree to which old assumptions still influence current thinking is even more apparent in the case of the metal finds from the Pyrgos cave, excavated by Xanthoudides in 1918.[102] Recent discussions of the finds from this cave show a consensus that the pottery is EM I, but there is a reluctance to call the associated metalwork EM I, because the metalwork includes a series of long daggers claimed to be no earlier than EM II.[103] Thus, we are left with a situation in which the pottery is EM I but the metalwork is EM II. In the case of Hagia Photia, a large cemetery thought to cover the EM I–II period,[104] there is also a general reluctance to admit that any of the metal finds from the more than 263 tombs could be EM I, even though the bulk (if not all) of the pottery clearly dates to late EM I.[105] As for the Mesara tholoi, which were used for burials during the EM and MM periods, none of the bronzes from these tombs, including the 54 from Hagia Triada, the 56 from Koumasa, and the 95 from Platanos, has ever been thought to be as early as EM I.[106]

The metal finds from Hagia Photia show all too clearly why scholars have been so reluctant to accept the obvious conclusion, one supported by the archaeological evidence. If the metalwork from these tombs is indeed EM I, then one has to accept the use of long daggers, small saws, knives, chisels, and fishhooks at the beginning of the 3rd millennium B.C. Moreover, along with the metalwork, these tombs produced a pair of footed crucibles, displayed in the Hagios Nikolaos museum, of a type known from slightly later contexts at Kommos,[107] as well as from Chalandriani on Syros and

96. *Keos* I, p. 4, pls. 22, 66.
97. The slag was studied by S. R. B. Cooke in the 1970s (*Keos* I, p. 114).
98. Papathanassopoulos 1996, p. 228.
99. Money-Coutts 1935–1936, p. 102.
100. She would probably have been the first to admit this.
101. Pendlebury, Pendlebury, and Money-Coutts 1935–1936.
102. Xanthoudides 1918.
103. Muhly 2004b.
104. Davaras 1971; Davaras and Betancourt 2004.
105. Day, Wilson, and Kiriatzi 1998.
106. For the finds, see Branigan 1984, pp. 30–31.
107. Blitzer 1995, pp. 502–505.

Thermi on Lesbos.[108] The Hagia Photia crucibles were clearly used to cast molten copper, and they would thus seem to testify to the actual working of metal on Crete during the EM I period.[109] Over 30 metal artifacts come from this site, including a silver zoomorphic pendant, a lead bead, and a socketed spearhead of probable LM III date. In the 1980s, the Isotrace Laboratory at Oxford, then under the direction of Noel Gale and Zofia Stos-Gale, analyzed 16 metal artifacts from Hagia Photia. The lead isotope ratios for seven of these objects were published in 1993,[110] but the chemical composition data has only been published in the form of a histogram.[111] The histogram makes it clear that the dominant alloy in use at Hagia Photia was arsenical copper with 1.0%–6.0% arsenic.

Is this the earliest arsenical copper in the Aegean world, apart from the objects found in the Tharrounia cave? Perhaps it is, on the basis of the evidence published to date. However, when published, the site of Poros, the harbor town of Knossos, will take pride of place as the most important Early Minoan I metallurgical site on Crete. Arsenical copper was certainly the alloy being produced and cast at Poros.[112] Arsenical copper was also the dominant alloy at Early Cycladic sites, especially for daggers, but the Cycladic objects have traditionally been dated either to the EC II or the transitional EC II–III period.[113] Noel Gale has further suggested that the triangular daggers from Hagia Triada, also of arsenical copper, could be assigned an EM I–II date,[114] probably slightly later than the copper-based artifacts from Hagia Photia. The electron microprobe analyses of nine unpublished copper-based artifacts from the Thessalian site of Petromagoula, located southwest of Sesklo, show that seven of these objects were made of arsenical copper, averaging 2.9% arsenic.[115] The authors regard Petromagoula as an EB II site, but this does not seem to be correct. Andreou, Fotiadis, and Kotsakis put the material from Petromagoula at the very end of the Final Neolithic or the beginning of Early Helladic I.[116] Johnson assigns the site a similar date, ca. 3700–3300 B.C., in his chronological scheme.[117]

It is hard to see any pattern in these developments on the basis of our present knowledge. The arsenical copper artifacts from Petromagoula seem to be somewhat earlier that those from the so-called Palace hoard of arsenical copper objects found in level VIA at the site of Arslantepe in eastern Turkey.[118] Of the 22 artifacts making up this hoard, 8 have been analyzed using instrumental neutron activation analysis or INAA.[119] All proved to be made of arsenical copper, averaging 4.16% arsenic. The contemporary metal hoard from level XXXIV at Late Chalcolithic Beycesultan in western Anatolia also attests to the use of arsenical copper. The eight analyzed objects averaged 1.33% arsenic.[120] Among Early

108. Day, Wilson, and Kiriatzi 1998, p. 136; for the historical development of the crucible, see Mohen and Walter 1994.

109. For contemporary evidence from Poliochni Azzurro, see further discussion in the text below.

110. Stos-Gale 1993, p. 122, table 11.1.

111. Gale 1990, p. 303, fig. 1. Stos-Gale (1998, p. 725) says that only 10 objects were analyzed; only full publication of the data can clarify this seeming discrepancy.

112. P. Day (lecture, Athens, March 22, 2004).

113. Sherratt 2000.

114. Gale 1990, p. 301.

115. McGeehan-Liritzis and Gale 1988, p. 217, table 4.

116. Andreou, Fotiadis, and Kotsakis 2001, p. 271.

117. Johnson 1999, p. 320.

118. Hauptmann et al. 2002.

119. Hauptmann et al. 2002, pp. 47, 49, table 5.

120. SAM 11774–11781; Muhly and Pernicka 1992, p. 312.

Ninevite 5 contexts of the early 3rd millennium B.C., the site of Hassek Höyük, located just south of the Taurus Mountains in southeastern Anatolia, produced a rich collection of metal artifacts. Of the 76 objects that have been analyzed, not one had any tin, but 52 were made of arsenical copper, averaging 1.90% arsenic.[121] It should be pointed out that this Late Chalcolithic metallurgy in Anatolia also involved the use of silver. The handles of some of the short swords from Arslantepe VIA were decorated with silver inlay, and a silver ring was found in the Beycesultan hoard. This silver is probably just slightly earlier than the silver pendant from Hagia Photia in northeastern Crete.[122]

The preceding evidence shows that arsenical copper was the dominant alloy in use during the period ca. 4000–2500 B.C. This is true for Iran,[123] Mesopotamia,[124] Syria,[125] the Levant,[126] and, of course, Anatolia.[127] It has now been well established that the Early Bronze Age was, for most of its duration, an age of arsenical copper. The actual production of this arsenical alloy, however, still remains a matter of great controversy. It certainly was not made by adding metallic arsenic to molten copper. There is no evidence that arsenic was recognized as a separate metal in ancient times, and no ancient language had a word for "arsenic."

There are, however, various arsenic sulfide minerals that could have been added to the copper or could have been present already in certain types of copper ores that, upon smelting, would have produced an arsenical copper alloy. Recently, it has been proposed that arsenical copper could have been produced by a simple cosmelting operation using copper oxide ores together with either copper sulpharsenide or iron sulpharsenide.[128]

What about the production of arsenical copper in the EBA Aegean? Since the 1980s, it has been argued that the Cycladic island of Kythnos possessed a large and very impressive copper smelting site. This site, known today as Skouries, has massive blocks of copper smelting slag still lying on its surface. Skouries is dated by associated pottery to the Early Cycladic II period, with further support from two radiocarbon dates on pieces of charcoal found in the slag.[129] When calibrated, the radiocarbon dates for the site fall in the first half of the 3rd millennium B.C.[130]

Of even greater potential significance is the discovery, just to the south of the smelting site, of a copper mine said to contain arsenical copper ore. The copper prills in the slag from the Skouries smelting site also proved to have a high arsenic content, with microprobe analyses of six different prills from one Kythnos slag sample averaging 6.3% arsenic.[131] Moreover, the lead isotope analyses of the trace element lead in the ores and slags of Kythnos gave a lead isotope "fingerprint" that seemed to match the "fingerprints" obtained from analyzing the lead in many of the arsenical copper objects from the Early Cycladic period.[132] It also matched the "fingerprints" of

---

121. Schmitt-Strecker, Begemann, and Pernicka 1992; for the metallurgy of the Ninevite 5 period, see Muhly and Stech 2003.
122. Davaras and Betancourt 2004, p. 181. For early silver, see Maran 2000.
123. Thornton et al. 2002.

124. Stech 1999, p. 63.
125. De Ryck, Adriaens, and Adams 2003.
126. Golden, Levy, and Hauptmann 2001.
127. Palmieri, Sertok, and Chernykh 1993.

128. Lechtman and Klein 1999.
129. Hedges et al. 1990, p. 226.
130. Stos-Gale 1998, p. 720.
131. Gale and Stos-Gale 1989, p. 34, table 6.
132. Sherratt 2000, pp. 72–92.

copper-based artifacts from the Minoan Mesara tholoi (including Platanos, Marathokephalo, Hagia Triada, Koumasa, Kalathiana, Hagios Onouphrios, and Porti)[133] and even the copper-based objects from Hagia Photia.[134]

Arsenical copper tools also occur in an Early Cycladic hoard of alleged Kythnian origin that is now in the British Museum. They average 3.2% arsenic.[135] In comparison, the 20 arsenical copper EC artifacts in the Ashmolean Museum analyzed by Gale and Stos-Gale average 4.31% arsenic.[136] Although one can never be sure of the provenience of such material, there seems to be no reason to accept the recent suggestion that the Kythnos hoard actually came from the Zas cave on the island of Naxos.[137] The original information that it came from Kythnos itself is more likely to be correct.[138]

In the 1990s, it looked as if the combination of chemical analysis and lead isotope analysis had solved one of the major problems of prehistoric Aegean metallurgy. The arsenical copper artifacts from the Early Bronze Age Aegean seemed to be made of copper from arsenical copper ores mined and smelted on the island of Kythnos. Unfortunately, the various threads of this argument soon started to unravel. Attempts to replicate the earlier results of Gale and Stos-Gale in subsequent archaeological work at various Kythnian copper smelting sites and in further analyses of Kythnian slags and ores have not yielded the same results.[139] This issue was discussed at the Third Round Table on Aegean Archaeology, devoted to the topic of "Metallurgy in the Early Bronze Age Aegean: New Evidence for Production and Consumption,"[140] and at a colloquium devoted to "Early Copper Metallurgy in the Aegean Islands."[141] Much new analytical work by Bassiakos on ores, slags, and copper prills from copper smelting sites on Kythnos has failed to detect the presence of arsenic. This situation is puzzling, but it raises serious questions regarding the theory that Kythnos was the source of arsenical copper for the Aegean Early Bronze Age.

At the same time, Gale and Stos-Gale are also having serious problems with their definition of a "Kythnian" lead isotope field. It now seems certain that not all the copper ores being smelted on Kythnos actually came from that island. The true picture appears much more complicated, with ores from a number of different islands having been brought to Kythnos for smelting.[142] The Kythnos field itself seems now to consist both of a high and a low field. This leads Stos-Gale to conclude, "We need still more analyses of slags and ores from Kythnos to be finally sure what the term 'Kythnos field' really means."[143] It would seem that we have yet to solve all the problems presented by arsenical copper in the Aegean Early Bronze Age. Obviously, the smelting of arsenical copper ores at Chrysokamino must be seen as but one chapter of a complex story, one that we are still in the process of understanding.

133. Gale 1990, p. 315, fig. 7.
134. Stos-Gale 1998, pp. 720, 725.
135. Craddock 1976, pp. 98, 106.
136. Sherratt 2000, pp. 72–92.
137. Fitton 1989; Sherratt 2000, p. 92.
138. Muhly 1999, p. 17.
139. Work by the Greek Archaeological Service was under the direction of Olga Philaniotou-Hadjianastasiou. The analytical work was undertaken at the Laboratory of Archaeometry at Demokritos, the Greek National Center for Scientific Research, led by Yannis Bassiakos.
140. This conference was held at Sheffield University in January 1998.
141. This colloquium was held at Demokritos on April 28, 2001.
142. Stos-Gale 1998, pp. 723–724.
143. Stos-Gale 1998, p. 726, n. 36.

THE USE OF BRONZE

Arsenical copper, of course, is only half of the technological history of Early Bronze Age copper-based alloys. Even more complex is the development of the copper-tin alloy known as bronze (and the term "bronze" should be used only for this alloy; arsenical bronze is a misnomer, and tin-bronze is redundant). Since my 1985 study of the problem,[144] Pare's magisterial survey has appeared.[145] Nevertheless, the situation in the Aegean still presents serious problems, both archaeological and analytical. At the outset, a distinction needs to be made between the limited use of bronze, the existence of bronze as but one of several competing copper alloys, and the full use of bronze to the exclusion of other alloys. Pare's discussion and his map showing the spread of bronze metallurgy put the full use of bronze in the Aegean at ca. 1600–1400 B.C.,[146] the same date now suggested for Syria and Mesopotamia.[147] Obviously, such a date does not mark the introduction of bronze metallurgy or even the first serious use of bronze.

In recent decades, it has been proposed that bronze may have been used during the second half of the 5th millennium B.C. in Chalcolithic southeastern Europe.[148] These speculations have now been put to rest. In reviewing all the evidence, Pernicka and colleagues conclude, "If we take a closer look at the proposed early finds of tin bronze in southeastern Europe not a single one stands up to close scrutiny."[149]

What needs to be examined in detail here are developments in the Aegean in the mid-3rd millennium B.C. and the early use of bronze in Minoan Crete. It is in the first half of the 3rd millennium, especially in the later part of this period, that we find a substantial quantity of bronze in use in Mesopotamia, at the time of the Royal Cemetery of Ur. In Mesopotamian terms this is Early Dynastic IIIa or 2600–2500 B.C.[150] The tin for this bronze seems to have come from central Asia and Afghanistan,[151] along with gold and lapis lazuli.[152] So-called Royal Treasures and Royal Tombs containing significant amounts of lapis lazuli and gold are known from contemporary or slightly later contexts in Syria, especially from Ebla and Mari.[153]

The texts from Ebla, dating to Early Dynastic IIIB and the following Old Akkadian period, contain so-called recipes for making bronze, adding specified quantities of tin to specified amounts of copper in order to produce a quantity of bronze for a specific purpose.[154] Some of these texts even describe the manufacture of "classic" bronze with 10% tin.[155] The bronze objects from Ebla that have now been analyzed contain up to 13% tin.[156]

In 1986, Stech and Pigott presented a very convincing argument for bronze and tin as status metals in the mid-3rd millennium B.C. They argued, however, that the main trade route for this tin went by sea, from one or more Syrian ports to Troy. It was then distributed from Troy across Anatolia and, presumably, to the north Aegean as well. In exchange, Troy supplied Syria and Mesopotamia with Anatolian silver.[157] Such a trade pattern is, of course, possible, but as I have already argued, it is most unlikely.[158] The situation in Anatolia is ambiguous at best, even if one is willing to discount the possibility of Taurus sources of tin.[159] The lapis lazuli trade route certainly went as far west as Ebla, where blocks of raw lapis from Afghanistan

144. Muhly 1985b.
145. Pare 2000.
146. Pare 2000, p. 26, fig. 1.14.
147. De Ryck, Adriaens, and Adams 2003, p. 579.
148. Schickler 1981.
149. Pernicka et al. 1997, p. 125.
150. Stech 1999, p. 63.
151. Alimov et al. 1998; Boroffka et al. 2002.
152. Muhly 1985b.
153. See the articles in Matthiae, Pinnock, and Scandone Matthiae 1995.
154. Muhly 1985b; for the complex metallurgical terminology of these texts, see Waetzoldt and Bachmann 1984.
155. Palmieri and Hauptmann 2000, p. 1260.
156. Palmieri and Hauptmann 2000, pp. 1261–1264.
157. Stech and Pigott 1986, pp. 55–58.
158. Muhly 1999, pp. 18–20.
159. See Muhly 1993.

have been found in Palace G in late-3rd-millennium B.C. contexts.¹⁶⁰ The inhabitants of EBA Anatolia, however, seem to have had no interest in this semiprecious stone. The only exception is the enigmatic battle-axe of lapis lazuli from Treasure L at Troy.¹⁶¹ One isolated example, in a most unusual context that also included an axe made of jadeite, is hardly enough evidence to support a Troy-Byblos trade route.

What makes things even more complex is our present understanding of the situation at Troy itself. For at least the past 100 years, Troy II and its north Aegean counterpart Poliochni Giallo (V) have been seen as part of an EB II "Age of Gold." In the mid-3rd millennium B.C., this phenomenon spread across the Old World from Mochlos on Crete to Troy II, Alaca Höyük, Eskiyapar, Ebla, Tell Brak, Mari, Ur, and perhaps even as far east as the Harappan sites along the Indus river. The EB II "Age of Gold" hypothesis has, however, been called into question recently by Nakou, who refers to it as "a combination of romantic sensationalism and an overwhelming desire for order in the past" and as part of a "general tendency to historicise the Early Bronze Age."¹⁶² Nakou is certainly correct in calling attention to the fact that the evidence comes from a series of hoards that do not necessarily represent finds from any one period. They are better seen as "palimpsests of material which was current for considerable time, and would probably have escaped deposition were it not for extraordinary circumstances in the late millennium."¹⁶³ In other words, what happenstance has preserved for us is the metalwork, in precious and base metals, that was in circulation during the mid-3rd millennium B.C., exactly the period when bronze became an important prestige alloy throughout the Eastern Mediterranean and the ancient Near East.

What does all this mean for Early Minoan Crete? To answer that question, a proper understanding of comparative archaeological chronology is essential. First of all, the mid-3rd millennium B.C. does not represent the first use of bronze on Crete. I have long argued for an EM I date for the two small daggers from the Krasi tholos, located just outside the northern border of the Lasithi plain and excavated by Marinatos over 70 years ago.¹⁶⁴ Marinatos was a careful excavator, and his detailed report makes it clear that the daggers in question came from the earliest burials in the tomb, associated with the distinctive EM I pattern-burnished pottery known as Pyrgos Ware. Branigan has argued that "the small primitive blades from Krasi might well belong to the very start of the Early Bronze Age in Crete."¹⁶⁵ I would agree.

These two daggers, however, are made of bronze. Analyzed by the SAM Project in the 1960s,¹⁶⁶ they proved to contain 10% and 6% tin.¹⁶⁷ According to current understanding of Early Minoan chronology, these daggers belong to the second half of the 4th millennium B.C.¹⁶⁸ Is the use of bronze at such an early date a realistic possibility? Comparative material from Crete is scarce. The roughly contemporary metalwork from Salami¹⁶⁹ and Pyrgos¹⁷⁰ is all of arsenical copper. The five objects from the Pyrgos cave averaged 1.85% arsenic.¹⁷¹

In wider comparative terms, however, the mid-4th millennium B.C. marked the emergence of a new metallurgical complex, involving new types of objects such as daggers, flat axes, and double-spiral pendants,

160. Pinnock 1988.
161. Antonova, Tolstikov, and Treister 1996, pp. 148–152.
162. Nakou 1997, p. 637.
163. Nakou 1997, p. 637.
164. Marinatos 1929.
165. Branigan 1967, p. 215.
166. See Muhly 1973, pp. 339–342.
167. Marinatos 1929, pp. 119, 131, fig. 13, nos. 30–31; SAM II/3, nos. 9447, 9449.
168. Warren 1980, p. 489.
169. This comprised two daggers: SAM II/3, nos. 9444 and 9445.
170. This comprised three daggers, a needle, and a chisel: SAM II/3, nos. 9366–9370.
171. See Muhly 2004b.

along with new sources of metal. This complex extended over a broad geographic expanse, including southeastern Europe (Usatovo), western Anatolia (Ilıpınar IV), eastern Anatolia (Arslantepe VII and VIA), and Palestine (Nahal Mishmar).[172] The metalwork from Sitagroi IV is part of this complex. The site has artifacts made of bronze, two metal fragments with 3.2% and 5.9% tin respectively.[173] The scholars involved in the publication of these analyses have gone out of their way to urge caution in the interpretation of the analytical data,[174] but these results are not necessarily as anomalous as they once seemed. The most interesting early use of bronze in the EBA Aegean comes not from Krasi but from the site of Kastri on the island of Syros, excavated over 100 years ago by Tsountas. There is no point in belaboring here the confused analytical results from a single set of Kastri metal samples originally taken for the SAM Project.[175] For analytical consistency, it is best to deal instead with results of the SAM Project published by Bossert in 1967.[176] What has always been remarkable about the Kastri analyses is that they seem to document metallurgy involving a combination of arsenical copper and bronze. I would still regard this as representing a transitional period, when metalworkers were adding tin to arsenical copper.[177] But in what sense can the material from Kastri be regarded as transitional?

The pottery from Kastri is now the subject of a major international research project because it seems to document the arrival of Anatolian colonists in the Cyclades, paralleling the EM I arrival of Cycladic colonists in eastern Crete, particularly in the region of Hagia Photia.[178] Central to this discussion is the date of the ceramic assemblage from Kastri. In Sotirakopoulou's recent study of the "Kastri Group," said to include Kastri, Lefkandi, Rafina, and Manika, the pottery is classified as transitional EH II/III,[179] but this study simply ignores the metalwork from all of these sites.

In a 1986 essay, Mellink proposed a radically new understanding of EBA Troy and its role in the Eastern Mediterranean during the 3rd millennium B.C. This reconstruction of traditional archaeological sequences has profound implications for our understanding of the development of 3rd-millennium metallurgical technology and our appreciation of the developments documented by the excavation of Chrysokamino. In Mellink's reading of the pottery, the destructions at the close of the EH II period that put an end to Lerna III and the "House of the Tiles" at that site are now seen as correlating with destructions at the end of Troy Ij, Poliochni Verde (III), and Thermi V. What followed was the Transitional period associated with pottery assemblages from Rafina, Manika, Lefkandi I, and Poliochni Rosso (IV). The distinctive pottery type fossil of this transitional period was a handmade, one-handled, red-polished tankard.[180]

But this tankard is not to be found at Troy because, with level Ij having been destroyed by West Anatolian marauders from Tarsus, the site of Troy lay in ruins during the Transitional period. In the following EB IIIA period Troy was resettled as a "triumphant station built by a successful warrior king of the new breed of West Anatolian navigators and traders."[181] This was Troy II, contemporary with similar settlements at Kastri and Poliochni Giallo (V).[182] With the exception of Lefkandi, all of these sites from the period and from EB IIIA have important metallurgical associations. It must

172. Pernicka et al. 1997, p. 57.
173. Renfrew and Slater 2003, p. 306, table 8.2, nos. 5624, 5625; McGeehan-Liritzis and Gale 1988, p. 216, table 4, nos. 24 and 25.
174. McGeehan-Liritzis and Gale 1988, pp. 220–221; Renfrew and Slater 2003, pp. 313–314; see also Begemann, Schmitt-Strecker, and Pernicka 1992, p. 223.
175. See Muhly 1991, pp. 362–364; 1999, p. 18.
176. Bossert 1967, p. 76, table 1; Muhly 1985a, p. 127.
177. Muhly 1999, p. 18.
178. Betancourt 2003c.
179. Sotirakopoulou 1993.
180. Mellink 1986, pp. 146–148.
181. Mellink 1986, p. 151.
182. For Troy II and Poliochni Giallo as EB IIIA sites, see Efe and Ilası 1997, p. 600, fig. 2.

be remembered, however, that very little of EBA Lefkandi (the Xeropolis site) has been excavated to date.

Rescue excavations conducted by Theocharis at Rafina uncovered two areas of copper smelting activities with apparent remains of a smelting furnace, along with slags, molds, and tuyeres.[183] Unfortunately, no proper study of the material was made at the time of excavation, and any finds brought to the National Museum in Athens can no longer be located.

The metal finds from Manika have been studied in detail by Stos-Gale and Mangou.[184] Of the more than 50 metal artifacts from the site, 20 copper-based objects and 1 of silver were analyzed by atomic absorption spectroscopy (AAS). The copper-based objects contain both arsenic and tin, but in contrast to the situation at Kastri, a clear separation was maintained between the two alloys.[185] With one exception not to be considered here, the objects contained either arsenic or tin, but not both. The analytical results establish very clearly that both arsenical copper and bronze were available to the smiths who made the metal artifacts found at Manika. The 12 objects of arsenical copper (i.e., those with over 1.0% arsenic) averaged 3.5% arsenic. The five objects having over 2.0% tin averaged 5.9% tin. The possible reasons behind the choice of alloy are not at all clear. Pins and knives were made of both alloys, while daggers and chisels were made only of arsenical copper.

The situation at Poliochni is perhaps the most complex of all. In the late 1980s, the Heidelberg-Mainz team analyzed a total of 97 copper-based artifacts. This study used both instrumental neutron activation analysis (INAA) and atomic absorption spectroscopy (AAS), the latter only for lead and bismuth.[186] The samples came from all four metal-using periods at Poliochni, and the number of samples taken (shown in parentheses) gives a good indication of the growth of metal usage at that site: Azzurro (6), Verde (11), Rosso (28), and Giallo (52). In terms of comparative archaeological chronology, these four periods cover almost the entire Early Bronze Age, with Azzurro dating to at least the beginning of Troy I, if not earlier, and Giallo certainly extending to the end of Troy II.

Arsenical copper was already in use in the Azzurro phase, with four of the six analyzed objects averaging 1.96% arsenic. There was no tin, but bronze made a slight appearance in the Verde phase. Of 11 analyzed objects, 6 were of arsenical copper, averaging 2.2% arsenic, and 2 were of bronze, averaging 9.25% tin. The use of bronze increased during the Rosso phase, corresponding to Mellink's period following the destruction of Troy Ij. Of 28 analyzed objects, 17 were of arsenical copper, averaging 2.8% arsenic, and 7 were of bronze, averaging 8.16% tin. A single, very unusual artifact had 19.8% tin and 11.71% arsenic; it is hard to imagine what such an aberrant alloy might represent.

It is remarkable that during the Poliochni Giallo phase we find the use of arsenical copper (with 17 objects averaging 1.94% arsenic), bronze (with 14 objects averaging 9.17% tin), as well as the same mixture of the two alloys found at contemporary Kastri. There are 12 objects with more than 1.0% arsenic and 2.0% tin. Taken together, they average 1.34% arsenic and 6.68% tin.

183. Theocharis 1955.
184. Stos-Gale, Sampson, and Mangou 1996.
185. Stos-Gale, Sampson, and Mangou 1996, p. 51, table 1.
186. Pernicka et al. 1990.

These figures do not take into account the incidence of lead in the Poliochni artifacts. Already in the Azzurro phase, the six analyzed artifacts all had over 1.9% lead, averaging 3.1% lead. This is most unusual; one can only assume that there was a high trace element concentration of lead in the arsenical copper ores being utilized at Poliochni. Metalworking activity of some sort was underway at Poliochni already during the Azzurro phase, contemporary with Thermi I;[187] the German team analyzed the slagged interior of six crucible fragments from this period at Poliochni.[188] Very little lead was found in the objects from the Verde and Rosso phases, or in those of the Giallo, but there are several notable exceptions in the latter phase. Two objects from Poliochni Giallo are made of leaded arsenical copper, averaging 2.01% arsenic and 9.05% lead, and two are of leaded bronze, averaging 9.55% tin and 9.33% lead. At least three objects have small amounts of arsenic, tin, and lead. Clearly the copper-based alloys in use during Poliochni Giallo were more complex than those found in all previous phases at the site.

The site of Thermi on Lesbos has played a significant role in all discussions of early Aegean metallurgy for over 70 years. Lamb, in her excellent final report on excavations at the site,[189] paid careful attention to all the metallurgical evidence uncovered, and she even had some of the metal artifacts analyzed by Desch, the scholar who best deserves the title of "Father of Archaeometallurgy." Consequently, there has been a long discussion in the scholarly literature regarding the crucible fragments from the lowest levels at the site (Thermi I) and, from the same context, a bronze pin, with 13.1% tin, often cited as the earliest object of bronze from the Aegean world.[190] From his detailed analysis of all the relevant pottery, Podzuweit concluded that Thermi I and II were contemporary with Troy Ia and that Thermi III was contemporary with Troy Ib.[191] A bronze artifact of such an early date is unexpected but, as we have seen, not without precedent. The context and analysis of this pin have now been called into question,[192] but it is probably best to accept the original report. Unfortunately, the pin was not included in the new programs of Thermi analyses because it could not be located in the museum.[193]

The recent programs of analyses[194] included 19 objects from Thermi I. They were all made of arsenical copper, averaging 1.91% arsenic. Of the seven artifacts studied from Thermi II, only one was bronze, with 2.4% tin. Five were arsenical copper, averaging 2.36% arsenic.[195] The Oxford

187. Begemann, Schmitt-Strecker, and Pernicka 1992, p. 221.
188. Pernicka et al. 1990, p. 269, table 4.
189. Lamb 1936.
190. Lamb 1936, pp. 214–215, no. 31.64.
191. Podzuweit 1979, pp. 38–40.
192. Gale 1996, p. 118.
193. Begemann, Schmitt-Strecker, and Pernicka 1992, p. 223.

194. The Thermi metallurgical studies present special complications. The metal artifacts from the site were sampled by the Heidelberg-Mainz team with the support of the Greek authorities. The Oxford team lodged a strong protest, with the support of the British School at Athens, claiming its members should have been consulted because Thermi was originally excavated by the British. This resulted in a division of the samples. Both sides issued publications, and each tried to discredit the work of the other. Having no desire to reopen what has been a most unfortunate controversy, I present here what I feel to be a reasonable account of the development of EBA metallurgy at Thermi.

195. Begemann, Pernicka, and Schmitt-Strecker 1995, p. 125, table 1.

analysis of these samples gives somewhat different results. Of 34 analyzed objects from Thermi I and II, 1 was bronze (a punch or drill with 4.12% tin) and 24 were arsenical copper, averaging 2.62% arsenic.[196] Of the 16 objects analyzed from Thermi III, again only 1 was bronze (a pin with 4.2% tin). The 12 objects of arsenical copper averaged 2.22% arsenic. The nine analyses from Thermi IV produced no objects of bronze, and the six objects of arsenical copper averaged 2.43% arsenic. The eight analyses of objects from Thermi V produced two objects of bronze, averaging 10.7% tin, and six of arsenical copper, averaging 3.41% arsenic. In all five periods at Thermi, a clear separation existed between arsenical copper and bronze; no object had a mixture of both metals. No leaded bronzes were analyzed, but two objects were leaded arsenical copper. One artifact from Town III had 2.83% arsenic and 2.64% lead; one from Town IV had 3.8% arsenic and 3.2% lead.

Given the relatively small number of objects involved, little can be said about the comparative use of the two alloys. Single bronze pins came from Thermi I, II, III, and V, but 11 pins were made of arsenical copper: 4 from Thermi III, 4 from Thermi IV and 3 from Thermi V. Arsenical copper was also used to make chisels: 2 from Thermi III and 3 from Thermi V. No great significance should be attributed to these figures. Thermi was clearly a major center of Early Bronze Age Aegean metallurgy. Although Thermi seems to have had one of the earliest examples of bronze from the Aegean, as well as the only tin object from the area,[197] the use of arsenical copper seems to have predominated at the site. This would support the idea that the occupation of Thermi probably ended at a time contemporary with Poliochni Verde,[198] before the real expansion of bronze metallurgy.

This comparative study has shown that arsenical copper, as it was being produced in the Aegean Early Bronze Age, most often had an arsenic content of just under 2.0%. Yet the modern metallurgical studies cited above all conclude that an arsenic content below 2.0% has little effect on the physical properties of the metal being produced. Why, then, did Early Bronze Age metalworkers so consistently produce an alloy that gave them a metal with the same basic properties as unalloyed copper? I have no answer to this question at the present time.

In many respects, all of this research began over 100 years ago in the Troad. From the beginning of his work at Troy in 1870, Schliemann demonstrated an interest in the development of metal technology and a willingness to have metal artifacts examined and analyzed by competent scholars. On the basis of this analytical work, he concluded that there was no use of tin in what is now called Troy I, an opinion supported by later scholarship.[199] However, by the time of his Great Treasure, which Blegen placed in Troy IIg, there was ample evidence for the use of bronze. A total of nine flat axes from this treasure, analyzed for Schliemann by metallurgists in England, France, and Germany, averaged 5.20% tin.[200]

The use of bronze in Troy II has been confirmed by recent analytical work.[201] In 1984, the Heidelberg-Mainz team published the results of its work in northwestern Anatolia.[202] One table gives the analytical results from the study of 50 artifacts contemporaneous with Troy II from north-

196. Stos-Gale 1992, p. 175, app. 3.
197. This was a remarkable tin bracelet from Thermi IVa; see Lamb 1936, pp. 171, 215, no. 30.24, fig. 50, pl. 25.
198. Begemann, Schmitt-Strecker, and Pernicka 1992, p. 224.
199. Muhly 1985b, pp. 283–284; Begemann, Schmitt-Strecker, and Pernicka 1992, p. 223.
200. Muhly and Pernicka 1992, p. 310.
201. Easton 2002, p. 327.
202. Pernicka et al. 1984.

western Anatolia. Of these 50 objects, 30 had over 4.0% tin, averaging 7.95% tin.[203] From Treasure A at Troy itself, 28 objects were analyzed. Results were published only in terms of whether the object was made of copper, arsenical copper (having over 2.0% arsenic), or bronze (having over 1.0% tin).[204] On this basis (and it should be noted that the Heidelberg-Mainz definitions of arsenical copper and bronze are the reverse of those used here), only 1 object was made of arsenical copper; 16 pieces were of bronze. This table also includes the results of analytical work done on a hoard of bronze objects from the Troad, purchased by Istanbul University and published by Bittel.[205] The objects consisted mainly of attachments for a type of metal "teapot," an example of which had been excavated by Schliemann.[206] Of the 22 objects analyzed, 15 were bronze and 6 were arsenical copper. Bittel had already published these analyses in 1959, giving actual percentages.[207] Twelve of these bronzes had the "classic" amount of ca. 10% tin. The six objects of arsenical copper averaged 3.27% arsenic.

In 1985, the Oxford team published its own analytical work, making use of the same samples originally taken for use in the SAM Project. They published neutron activation analyses (NAA) of 15 objects from Troy II.[208] Nine proved to be bronze, averaging 8.23% tin, and four were arsenical copper, averaging 2.45% arsenic. But this hoard of metalwork, designated as the Great Treasure, Treasure A, or Priam's Treasure, is no longer dated to the end of Troy II (Troy IIg). Recent work directed by Korfmann at Troy would now place this metalwork in the middle of the Troy II phase, with an absolute date of ca. 2500 B.C.[209] Korfmann does not accept Mellink's 1986 revision of Trojan stratigraphy. The gold and bronze metalwork from Troy, in other words, is now seen as contemporary with that from the Royal Cemetery of Ur. This same combination of gold jewelry and bronze metallurgy can be seen in central Anatolia, especially at the site of Alaca Höyük, where a group of so-called royal tombs roughly contemporary with the "treasures" of Troy II were excavated.[210] The metalwork from Alaca shows the same combination of arsenical copper and bronze. Of 36 objects from Alaca analyzed by the SAM Project, 12 had over 5% tin, and 13 had over 1.0% arsenic.[211] The very strong Caucasian connections of the finds from Alaca Höyük are of considerable interest.[212]

In all the studies dealing with copper-based artifacts from Anatolia and the Aegean undertaken by the Heidelberg-Mainz and Oxford teams, chemical elemental analyses have been combined with the study of lead isotope ratios. Over the past twenty-some years lead isotope ratios have come to be seen as the best way to study the provenience of copper, silver, and lead. This work has generated an enormous bibliography and a great deal of controversy, especially over Lavrion and Cyprus as sources of copper for the Aegean Bronze Age. Both the Heidelberg-Mainz and the Oxford teams agree that the real introduction of bronze, occurring in the middle

203. Pernicka, et al. 1984, p. 575, table 3.
204. Pernicka et al. 1984, pp. 578–579, table 4.
205. Bittel 1959.
206. Bittel 1959, p. 14, fig. 28.
207. Bittel 1959, p. 34.
208. Gale, Stos-Gale, and Gilmore 1985, p. 148, table 1.
209. Korfmann 2001, p. 380.
210. Mellink 1956.
211. Stech and Pigott 1986, p. 54.
212. Mansfeld 2001.

of the 3rd millennium B.C. along with the use of tin rather than arsenic as the alloying metal, is associated with the appearance of a new, so-called "exotic" lead isotope signature. Bronze artifacts, especially those from Troy and Poliochni, tend to plot right off the lead isotope diagram. That is, they contain a lead that is older than anything known from the Aegean or from Anatolia. This lead could only have come from Precambrian ore deposits 700–900 million years old, and such deposits do not exist in Greece, Turkey, Cyprus, or the Middle East.[213]

What is being studied here, of course, is the geological date of the lead contained as a trace element in copper-based artifacts. As this "exotic" lead does not appear in the arsenical copper artifacts, the logical conclusion would be that it came from the tin. This lead might in turn provide evidence for the long-sought provenience of Bronze Age tin. However, alluvial cassiterite, the dioxide of tin and presumably the source of this "exotic" lead, has been shown to contain virtually no lead.[214] It is now being proposed that the lead came with the copper. In the mid-3rd millennium B.C., therefore, new sources of copper were being exploited by metalworkers, especially those in northwestern Anatolia and the northeastern Aegean.[215] Such copper is even found in objects not made of bronze, as in Thermi I and II.[216] This copper must have come from central Asia, the very area now favored as the source of Bronze Age tin (see above). But how and where was the copper mixed with the tin in order to produce bronze? If this was being done in central Asia itself, then it is possible that Early Bronze Age metalworkers were, in fact, smelting stannite, an ore containing both copper and tin. Such ore is found in central Asia, especially at the site of Mushiston in Tajikistan. This area has deposits of stannite high in copper and tin, but without iron. The smelting of these ores "would result in copper with high tin contents."[217] In Mushiston stannite, the usual iron content was replaced by zinc, and it cannot be accidental that some of the objects from Thermi proved to be so high in zinc that they were considered possible examples of the early production of brass.[218]

Given the distinctive typology of Early Bronze Age artifacts from the Eastern Mediterranean world, it seems reasonable to assume that this metal came into the Aegean in the form of bronze ingots and was turned into artifacts by local metalworkers. The evidence for metalworking at all of the relevant sites supports this conclusion. Metalworking at Poliochni and Thermi has already been discussed; the evidence for Troy II has recently been summarized by Easton.[219] Admittedly, bronze ingots are something of an anomaly, but no other explanation seems feasible at the present time. Recent lead isotope analyses indicate that this metal, with "exotic," very early lead, is also to be found in contemporary contexts in Oman and the United Arab Emirates.[220] The implications of this, obviously, are enormous.

213. For a summary of these arguments, see Muhly and Pernicka 1992, pp. 312–315.
214. Muhly 1978, p. 45, table 1.
215. Pernicka 1987, pp. 702–703; Muhly and Pernicka 1992, pp. 312–313.
216. Stos-Gale 1992, p. 170.
217. Boroffka et al. 2002, p. 141.
218. There are four objects averaging 7.6% zinc; see Stos-Gale 1992, p. 175, app. 3.
219. Easton 2002, p. 328.
220. Weeks 2003.

CHRYSOKAMINO AND THE DEVELOPMENT OF METALLURGICAL TECHNOLOGY

In light of the preceding discussion, the site of Chrysokamino seems to occupy something of a threshold into the Middle Bronze Age. On the one hand, its smelting technology, in terms of what might be called proto-shaft furnaces and the use of pot bellows, seems to be a harbinger of what is to come; on the other hand, its use of arsenical copper reflects the technology of previous centuries. Arsenical copper has already been discussed. It is appropriate, therefore, that this chapter close with a brief discussion of smelting technology. For over a century, scholars have struggled to understand the curious fragments of perforated clay that littered the surface of Chrysokamino. In 1910, Mosso took them to be fragments of crucibles.[221] Our current thinking on this subject is well described by Betancourt in Chapter 14 of this volume. Similar fragments are known from the Early Cycladic II smelting site of Skouries on Kythnos, mentioned above. Although these fragments are largely unpublished, they seem to come from more substantial, larger, and thicker structures than the Chrysokamino examples. However they were used, and the idea of their being some sort of a cover for a bowl furnace is most attractive, they must have been designed to increase the flow of air into what was a wind-powered smelting operation (with additional draft supplied by the pot bellows in the EM III period). Wind-powered smelting technology was probably widely used all over the world. It has now been well documented for iron smelting furnaces in Medieval Sri Lanka.[222] Roughly contemporary with Chrysokamino are the Early Bronze Age examples from the site of Feinan in Jordan.[223]

It is interesting that 20th-century copper smelters were unable to achieve a separation of slag and metal in replication experiments done at Feinan,[224] although such a separation is now being proposed for contemporary Chrysokamino. Perhaps this was achieved with the additional draft provided by the pot bellows, but there is also no evidence for any sort of perforated structure at Feinan. Yet a primitive structure positioned on top of the bowl furnace was already being employed at the Negev site of Shiqmim, dated to the mid-4th millennium B.C.[225] At Shiqmim, a bowl furnace, consisting of nothing more than a pit dug in the ground, possibly lined with a crucible, was surrounded on the surface by a sort of clay collar about 5 cm in thickness, 9 cm in height, and 30 cm in diameter. The original collar was about 80% of a circle, leaving an open space for the draft of air.[226] It is probably not too fanciful to see this Shiqmim collar as the precursor of the structures used at Skouries on Kythnos and at Chrysokamino on Crete a millennium later.

221. Mosso 1910, pp. 290–291, fig. 164.
222. Juleff 1998; Craddock 2001, p. 162, n. 47.
223. Bunk et al. 2002.
224. Bunk et al. 2002, p. 336.
225. Golden, Levy, and Hauptmann 2001.
226. Golden, Levy, and Hauptmann 2001, pp. 956, 958, fig. 8.

CHAPTER 14

DISCUSSION OF THE WORKSHOP AND RECONSTRUCTION OF THE SMELTING PRACTICES

by Philip P. Betancourt

The main period of the smelting workshop at Chrysokamino can be placed in EM III–MM IA, but the quantity of slag beneath the stratum from this period shows that the workshop had a substantial earlier history on the spot before the EM III period. The number of Final Neolithic sherds found stratified within the deposit of slag below the EM III level indicates that the installation was already in operation by this period. As has been demonstrated in the previous chapter, the Final Neolithic of Crete (often equated with the Chalcolithic in western Asia) inherited a long earlier tradition of copper technology. From the lead isotope analysis of Neolithic Aegean copper objects, we know that the ore deposits on Kythnos, Seriphos, and Siphnos were already being exploited during this period.[1] Since these ores do not consist of native copper, this exploitation must have involved smelting; the situation at Chrysokamino cannot be unique.

The presence of specialized metallurgical technology in the Aegean before the beginning of the Bronze Age is not an isolated event, and it fits well with a rapidly growing body of evidence for specialized technology in other crafts as well.[2] Pottery production was becoming more sophisticated, and the earliest Cycladic marble sculptures are from the Neolithic as well.[3]

On the other hand, the exact date of the heavily burnished pottery in this part of Crete is far from secure. It definitely begins during the Neolithic,[4] but its latest date is at least contemporary with EM I, and it may persist even later. The length of time over which the workshop was used cannot be determined at our present level of knowledge.

The metallurgy workshop at Chrysokamino was not the only copper smelting facility in eastern Crete. Other, similar workshops were undoubtedly in operation during the Early Bronze Age, even though no additional examples have been excavated. Two other locations for probable workshops are recorded. In describing the metallurgy at Chrysokamino, Mosso mentioned another place with similar perforated ceramic fragments, somewhere on or near the beach at Pacheia Ammos.[5] The exact location was not recorded. Another location with a similar sherd was found by Haggis

1. Stos-Gale and Gale 2003.
2. Perlès and Vitelli 1999.
3. Papathanassopoulos 1981.
4. Betancourt 1999.
5. Mosso 1910, pp. 289–292.

during his regional survey in the vicinity of Kavousi.[6] The piece came from an excavation within the village of Kavousi, and no other evidence for the workshop was present. It is unlikely that the furnace site at Chrysokamino would have been the only location where a process as important as the extraction of metal from its ores would have been practiced, so that even without the hints of these other locations, one would suppose that other workshops once existed.

TOPOGRAPHY

The topographic situation of the smelting site is an important component of its location. It consists of a limestone outcrop at the west and a phyllite outcrop at the east, with a lower space between them, situated at the top of a north-facing cliff slightly over 38 m above the sea. This topography creates a funnel-like trough that concentrates the wind when it blows from the north. The situation has parallels from elsewhere in the Aegean. Slagheaps at ancient copper smelting locations on Kythnos and Seriphos are found on similarly situated north-facing cliffs high above the sea.[7] In all of these cases, the topographic placement suggests a deliberate choice to maximize the exploitation of the north wind.

The workshop's location was obviously based on factors other than proximity to the ore. The geological survey shows that no copper ore was present in the vicinity of Chrysokamino, and it had to be imported. That transporting ore some distance away from the mines in order to smelt it was not unknown in antiquity is demonstrated by the similar situation at several other places. At Feinan in Jordan, for example, Chalcolithic ores were shipped over 100 km for smelting.[8] Similar examples have been documented from elsewhere in Europe.[9] Ore was especially easy to ship by sea because it would have made good ballast. Smelting sites might have been situated near other necessary resources like fuel and labor, in preference to locations near the ore, because it was easier to ship the ore than it was to transport the fuel.

One presumes that even if the smelters arrived as strangers in a new territory, the local region could provide some needed commodities. Requirements included clay for the making of chimneys and furnaces, water and food for the metalworkers, fuel for smelting, additional laborers, and markets for the final product. For Chrysokamino, however, whatever the situation was when the workshop began, there is no evidence that the smelters were foreigners by the period of the EM III–MM IA activity. By then, the workshop personnel were integrated fully within the local culture, as is shown by their access to local clay sources and chaff (the two ingredients in the manufacture of the chimneys), the presence of local styles of pottery at the workshop site (with no foreign products), and the use of local materials for all the ground stone tools. The abundance of East Cretan pottery and stone tools, in particular, demonstrates that whatever situation existed earlier, by the period of the greatest activity, this workshop was an integral part of eastern Cretan society.

6. Haggis 2005, p. 109, a chimney fragment with two holes found in soil from a site in the village of Kavousi.

7. Gale et al. 1985, pp. 90–91; Hadjianastasiou and MacGillivray 1988; Stos-Gale 1998.

8. Hauptmann 2000, pp. 167–168; 2001, p. 425.

9. O'Brien 1996.

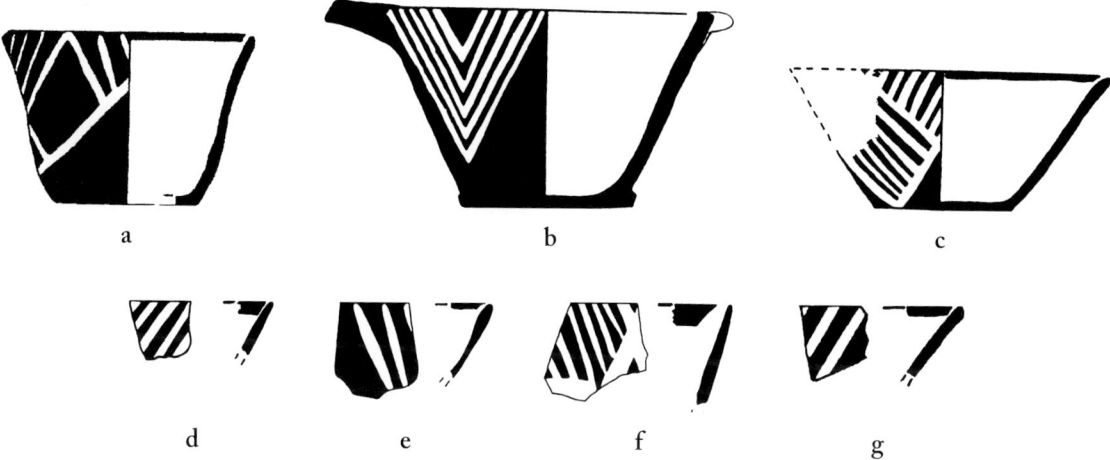

Figure 14.1. East Cretan White-on-Dark Ware from Chrysokamino and nearby sites: (a) Chrysokamino 75; (b) Myrtos Fournou Korifi, EM IIB; (c) Vasiliki, EM III–MM IA; (d–f) Gournia, EM III–MM IA; (g) Pseira, EM III–MM IA.
A, B: after Warren 1972, no. P25; C: after *Pseira* VI, Ill. 13B; D–F: after Betancourt and Silverman 1991, nos. 313–315; G: after *Pseira* VI, Ill. 13A. Scale 1:3

ORGANIZATION OF THE WORKSHOP

Like the Cycladic slag heaps on Kythnos and Seriphos, the deposit of slag at Chrysokamino in Crete is not near any large settlement. The nearest known habitation site is 600 meters away, and the nearest substantial town was located either at Kavousi, at Tholos Beach, at Pacheia Ammos, or at Gournia, all of which were important in later times (but none of which have excavated Final Neolithic habitation levels). The Mirabello Fabric used for the EM III–MM IA pottery in the most important phase of the site suggests a possible relationship with Gournia, where the ware was the main fine pottery in this period.[10] The pottery in this class occurs at a whole series of sites near Chrysokamino (Fig. 14.1).

The EM III–MM IA date for the final and most intense period of activity at the workshop is secure. East Cretan White-on-Dark Ware is the only fine pottery from this phase at the site. The ware begins at the end of EM IIB with simple motifs like pendent chevrons (Fig. 14.1:b), and it develops fully during EM III–MM IA.[11] Conical cups decorated with white hatched triangles (like the one illustrated in Fig. 14.1:a) are among the most common cups from the period. The end of the ceramic phase is contemporary with MM IA in central Crete.[12]

The Chrysokamino workshop could not have existed in isolation. The evidence is very clear on this point: the site has no houses, no complete range of domestic pottery, and no household debris. The metalworkers slept and lived somewhere nearby, not at the workshop itself. The workshop must have been an appendage of some other site (presumably the habitation location situated half a kilometer away) because the workers needed food, water, housing, and other necessities as well as access to local commodities such as pottery, fuel for the furnaces, and clay and chaff to make the chimneys. They also had to have access to markets, and they depended on someone who could coordinate the acquisition of the ore. Since the lead isotope pattern of the copper in the slag matches Kythnos and Lavrion (App. C), the ore must have been shipped in by sea, requiring access to the port at Agriomandra, as well as personnel to unload, transport, and store it.

10. Hall 1904–1905; Betancourt 1984.
11. Betancourt 1984.
12. Betancourt 2003b.

At the regional level, the nature of the workshop would have necessitated some type of directing authority. Importing ore would have required the coordination of a number of different activities, not all of which were performed by the same individuals: prospecting to find sources of metallic ores, operation of the mines, preparing the ore by removing waste rock (beneficiating), transporting the ore to an embarkation point, and shipping by sea over a distance of many kilometers. After the ore was received and smelted at Chrysokamino, the copper was then transported to places where the copper was remelted in crucibles and cast in molds to form useful articles. The metalworking industry in east Crete seems to have been divided into specialized stages, as proposed for other regions by Ottaway.[13] The small workshop at Chrysokamino was most likely only a small part of a large operation with both Cretan and overseas connections.

The presence of other workshops in the area, as shown by the chimney fragments from Pacheia Ammos and Kavousi, suggests a regional production system designed to provide the many villages and towns of the eastern Gulf of Mirabello with an important commodity not available exclusively from local resources. In order to supply these workshops with ore, fuel, and other necessities, some coordinating agency must be envisioned. The implication is that some political or economic authority already existed in the region by EM III–MM IA, and that it was directing the Chrysokamino workshop as well as other aspects of the industry.

The nature of this higher authority, and whether it was palatial or private, is a complicated problem that cannot be resolved with the presently available evidence. In western Asia, private companies could manage comparable international ventures like long-distance trade, and their personnel often had kinship ties.[14] On the other hand, the vast quantity of EM III–MM IA pottery from Gournia[15] suggests that whoever was in charge of this site, which would have a palace by LM I,[16] was already expanding its economic base. The issue may be moot because the modern dichotomy between private and palatial may have had little meaning in Early Minoan society, where kinship may be the cornerstone of both intensification of craftwork and the emerging political hierarchy.[17] The lines between public and private may have been very blurred.

At the local level, the metalworkers must have slept, cooked their meals, and lived their daily lives somewhere in the Chrysokamino territory. The sparse pottery record for all periods before EM III–MM IA indicates that the site was definitely not used continuously during the long period represented by its surviving artifacts; indeed, it is possible to argue that it was used only for very brief episodes, with periods of abandonment in between. Even in years when it was active, the workshop could not have operated year-round. The evidence from the furnace chimneys suggests they were probably all made in the fall because the chaff, an important ingredient in the clay fabric, would only have been available after the threshing that followed the grain harvest. This period also coincides with the windiest season, when there is no rain but the north wind called the *meltemi* blows every day, optimizing the conditions for its exploitation to help provide the draft for the furnaces.

The location of the later farmhouse at Chrysokamino is the best candidate for the metalworkers' place of residence, even though it is 600 m

13. Ottaway 2001.
14. For Karum Kanesh, see Garelli 1963; Veenhof 1972; Kraus 1982; Larsen 1967, 1976; Dercksen 1996; for short summaries, see Foster 1977, 1987.
15. Hall 1904–1905.
16. Soles 1991.
17. Warren 1987, p. 52.

away. No architecture has been found for the Early Bronze Age, but the small settlement has Final Neolithic through EM III–MM IA pottery from later contexts, and the pottery's style exactly matches that found at the workshop.[18] In addition, the early settlement pottery has a more complete range of shapes than exists at the workshop, including cooking pots and cooking dishes, cups and other serving vessels, storage shapes, and even a few luxury pieces. Its pottery convincingly supports the argument that the settlement was occupied throughout the long period represented by the pottery at the workshop. On the other hand, the 600 meters of distance between the two locations raises questions. In particular, the soil for the three successive floors of the small apsidal structure was carried to the workshop from somewhere away from the pile of slag. This soil, with many sherds and pot bellows fragments included in it, must surely have come from wherever the metallurgists lived. Unless the soil had a symbolic meaning (which may very well have been the case), it would have been simpler to bring it from somewhere closer than a spot over a half kilometer away. Wherever the metallurgists lived, the location has not been excavated. Our information about the metallurgists, therefore, must come almost exclusively from the evidence unearthed in the workshop itself.

RECONSTRUCTION OF THE SMELTING PROCESS

In order to form the chemical reactions that separate copper from other elements in secondary ores like malachite and azurite, the ore must be heated in a reducing atmosphere in the presence of fluxing agents.[19] The flux, usually quartz and iron oxides, will eventually form much of the slag. Burning a fuel like wood charcoal in a small, mostly closed space can create the reducing agent, carbon monoxide. The chemical reactions take place in stages. With malachite and azurite, heating to ca. 500°C decomposes the carbonate ores to create copper oxide. On higher heating, the carbon monoxide unites with the oxygen in the copper oxide and forms carbon dioxide, leaving the copper behind. The reaction can be achieved at temperatures under the melting point of copper (1083°C), and lower temperatures can be used if heat is maintained for longer periods of time.

Evidence for the process at Chrysokamino comes from several items found in the archaeological record. The expedition uncovered a fragment of furnace lining, pieces of slag, chimney fragments, bits of ore, a fragment of a tuyere, stone tools, and pieces of pot bellows. This physical evidence provides substantial information for a reconstruction of the smelting processes. Analyses of the slag fragments and their included prills provide information on the operation as well.

Modern experiments in the smelting of copper have demonstrated several of the principles and techniques that must have been present in the operation at Chrysokamino. Bamberger produced good results with a furnace having a diameter of up to 40 cm, preheated for 1.5 hours.[20] He found that air would not penetrate more than approximately 20 cm into the charge, so he suggested ca. 40 cm (which is also the maximum diameter of the Chrysokamino chimneys) as the maximum diameter for a simple

18. Betancourt and Floyd 2000–2001.

19. For useful summaries of the process, see Bachmann 1982, p. 121; Maddin 1996, pp. 9–10.

20. Bamberger 1985.

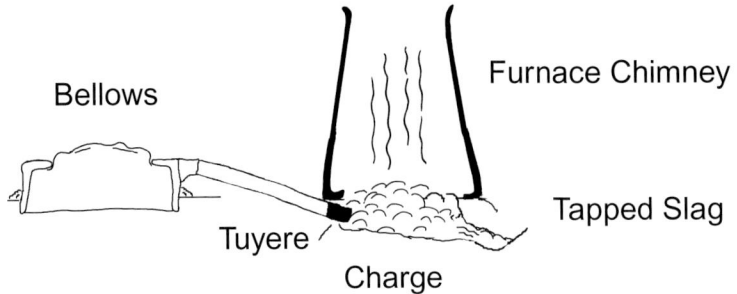

Figure 14.2. Reconstruction of a bowl furnace from Chrysokamino based on fragments found in the excavation

furnace. The ratio of ore-to-fuel was 2:1. Artificial ore was used along with charcoal as fuel, with three air intakes yielding good results.

While a few questions remain, most of the steps in smelting can be reconstructed. The evidence suggests that the technical process in Crete was not very different from other Chalcolithic to Early Bronze Age smelting operations in the eastern Mediterranean.[21] Before the smelting could begin, the workers needed to construct a chimney from coarse clay, build a bowl furnace, and gather together the raw materials and fuel.

Chimneys were made from local clay. A deposit of terra rossa at a nearby sinkhole called Lakkos Ambeliou, about a kilometer away from the workshop, is the closest clay deposit whose mineralogy matches the fabric of the chimneys. The deposit is closer to the habitation site than to the workshop itself. Its sediment includes both abundant kaolinite and small stone fragments, so it (or a similar deposit) would have worked well as a raw material for ceramics without the addition of extra materials, aside from the organic material that would burn away and leave the fabric more porous. At Chrysokamino, this organic material was chaff and straw from the threshing of barley (Chap. 12). The addition of short, cut pieces of the stems of plant material was not unusual for molds and other items such as crucibles used in Cretan metallurgy, because the creation of a porous clay fabric made the ceramics less likely to break under the stress of high temperatures.[22] The chimneys were built by hand to form cylinders that were open at both top and bottom, with slightly smaller diameters at the upper end. Holes were pierced in the cylinders by thrusting fingers through the walls while the clay was still wet and soft. The chimneys were probably not fired before use.

The smelting process would have taken place in a bowl furnace constructed in a small hollow in the ground (Fig. 14.2). The furnace would have been small, 20–40 cm in diameter, and lined with clay. Ore, already broken into small pieces, would have been placed in the furnace along with the fuel and flux to form the charge. The absence of waste rock or of stone tools for preparing the ore indicates it was brought to the site ready for use. Chimneys over the furnaces would have aided in the draft, and, by EM III–MM IA if not earlier, the natural draft would have been augmented by the use of pot bellows.

Three copper minerals, azurite, malachite, and chrysocolla, occur in the archaeological material excavated from the site (Chap. 10). They are all present in very tiny fragments, and no direct evidence survives for their source. Several writers have considered the possibility of Cretan copper

21. See, among others, Bachmann 1980; Rothenberg 1972; 1985; Hauptmann 1989; Hauptmann, Bachmann, and Maddin 1994; Hauptmann 2000.

22. Blitzer 1995, pp. 504–508; Evely 2000, pp. 353–356.

ores. The island does have small copper deposits,[23] and some writers have suggested that they could have supplied Early Minoan needs.[24] Others have discounted some of these sources,[25] but the use of small deposits that are as yet undiscovered or that were depleted long ago remains a possibility. Alternatively, all the ores may have been imported from sources outside of Crete.[26] Isotope analyses of the lead in copper prills trapped within the slag support the argument for a source outside of the island. They indicate that the minute quantities of lead in the prills best matches the isotope pattern of ores found at both Lavrion and Kythnos.[27]

The ores probably arrived at Chrysokamino by sea. In calm weather, the ships could have landed at the foot of the cliff at the metallurgy site, but in windy weather, landing at the small beach of Agriomandra would have been safer. A piece of waterworn stone containing malachite (discussed in App. C) was found on the beach at the small harbor of Agriomandra. This cove, with a small, sheltered beach covered by waterworn stones, is located at the foot of a ravine that begins just downhill from the Chrysokamino farmhouse, the most likely location for the metallurgists' habitation.[28] It is the closest harbor to the Chrysokamino workshop. Malachite is one of the copper ores found at the metallurgy site. Although one cannot determine when the malachite specimen was deposited on the beach, its waterworn condition indicates it is not a recent arrival. It also indicates that copper ore was brought to Agriomandra at some time in the past.

The ores found at the workshop are all secondary copper ores easily recognized by their bright colors (malachite is green, azurite is blue, and chrysocolla is pale blue-green). Even novice metalworkers would have no problem sorting ores or recognizing the difference between higher-grade and lower-grade shipments. These minerals are easier to smelt than copper sulfides, and they would have yielded suitable amounts of copper, but they do not account for all of the mineralogy present in the slags from Chrysokamino. In particular, the high silica and iron contents of the slags are not present in these ores. One must assume that an iron-rich flux was present in the charge along with the copper ores and fuel.

The variable arsenic content also requires an explanation (Apps. B and D). It is not a constituent of any of the three copper ores found at the site. Arsenic is of particular interest because of its presence in arsenical copper artifacts from Crete and elsewhere in the Aegean,[29] as well as in objects from western Asia.[30] The presence of arsenic in Aegean copper artifacts has been discussed many times.[31] The element can occur in nature in brightly colored green and blue-green ores such as clinoclase and pharmacosiderite. Such ores may have been recognized as yielding copper with different

23. Faure 1966; Branigan 1968, pp. 50–52; 1974, pp. 57–63; McGeehan-Liritzis 1996, p. 387.

24. Faure 1966; Branigan 1974, p. 117.

25. Wheeler, Maddin, and Muhly 1975; Becker 1976; Gale 1990, pp. 304–305.

26. Gale and Stos-Gale 1982; Gale 1990.

27. See App. C; for comments on the statistical methods to achieve the probability of a match, see Baxter 2003, chap. 18.

28. For Agriomandra, see Faure 2000, p. 459.

29. Gale and Stos-Gale 1989, tables 1–3; Gale 1990; Mangou and Ioannou 1998.

30. Ravich and Ryndina 1995, p. 4.

31. Charles 1967, 1980; Coghlan 1972; Gale 1990; Zwicker 1991; Tylecote 1991; Katsa 1997; Mangou and Ioannou 1998; see also discussion in Chap. 13.

characteristics from copper smelted without their addition, or they may have been accidentally mistaken for the more common copper ores.

The high arsenic content in some of the prills (App. B) can only mean that arsenic minerals were present in some of the charges placed in the furnaces.[32] At the microscopic level, the arsenic is irregularly present, resulting in one tiny mass that is as high as 26% arsenic (App. B, no. 48), although in most prills, the arsenic content is much lower. The element is not uniformly distributed, and many prills contain only a trace. When only traces occur, the arsenic in ancient metalwork has been regarded as an accidental inclusion.[33] Even if the metallurgists were not fully aware of their process, however, the conclusion must be that in addition to the malachite, azurite, and chrysocolla, the charge sometimes contained small quantities of more complex polymetallic ores with small amounts of several metals, including arsenic, nickel, lead, and (rarely) even gold (App. B, no. 1).

Fluxing agents were also present. The analysis of the slags from Chrysokamino shows that the original charge used for the smelting operation was highly siliceous, with the additional presence of substantial amounts of iron oxides and/or hydroxides (Apps. A and B). The silicates and the iron minerals were either present in the ore or in a separate fluxing agent that needed to be added to the charge. To a certain extent, the point is moot, because both fluxing agents were clearly present in sufficient quantities to produce the desired product.

The smelting operation could begin after the charge of ore and fuel and flux was placed in the small bowl furnace. Fuel will probably have included some charcoal, but additions of high-efficiency fuel like the residue from olive oil extraction cannot be ruled out. We do not know what fuel was used at the site. In contrast with some of the other early smelting operations, no trace of unconsumed fuel comes from the Chrysokamino excavation. Charcoal, however, is not always found along with Bronze Age slag if the waste pile is located away from the furnace itself.[34] This material is usually assumed to be the fuel of choice for early smelting operations, but the high temperatures achieved at Chrysokamino, along with the absence of either ash or charcoal anywhere on the excavated part of the site, suggest that seeds and skins from olives may have been used as an alternative or an addition to the charge. In the centuries before the use of large presses, olive oil extraction would have involved the use of mortars and hand pestles, producing a residue that was a crushed mass of seeds and skins.[35] This oily mixture would have been highly combustible, and it would have burned with an intense heat, accounting for the high temperatures, the absence of unburned fuel, and the lack of unsmelted ore in the slags. It was used as a fuel for pottery kilns at nearby Mochlos.[36] Its efficiency as a fuel is attested at modern limekilns and pottery kilns in Crete, where it is regularly used as a fuel.[37]

After the fire was started, the cylindrical chimney would have been placed above the furnace to aid in the draft, which was created by the natural wind as well as by the use of bellows. The bellows at Chrysokamino were drum-shaped, cylindrical pots with nozzles for the reed to conduct the air toward the fire and holes cut in their upper surfaces to allow for a leather

32. Thompson 1958, p. 4.
33. Hauptmann, Begemann, and Schmitt-Strecker 1999, p. 9.
34. Herdits 2003, p. 71.
35. Soles 2003, p. 24.
36. Sarpaki and Bending 2004, p. 126.
37. Betancourt 2004a.

inflatable pumping mechanism. They were set into the ground and fixed there with mud so they would not move around during use. Several will have been used for each furnace, with a reed that terminated in a tuyere conducting the air from bellows to fire. The bellows were not very efficient at this site, and in addition to the fresh air, they sucked in enough heated air from the furnaces to darken their interiors and to bake the mud that fixed them to the ground sufficiently to turn the mud into pottery. The pumping mechanisms (presumably made of leather and wood) could not have lasted very long.

The technology that produces either a copper ingot at the bottom of a furnace or a series of prills inside the slag is well known, and modern experiments have replicated the process.[38] In small furnaces, ancient metallurgists had to achieve a temperature of ca. 1200°C and maintain it for about four hours so that droplets would flow downward and collect as an ingot in the bottom of the furnace.[39] At lower temperatures, however, even as much as 200°C below the melting point of copper, malachite heated under reducing conditions will form small prills within the slag. These prills stay within the slag and do not flow to the bottom of the furnace, so that their removal requires that the slag be broken into small pieces.

Did the smelting operation at Chrysokamino produce a high enough temperature to cause a mass of copper to collect at the bottom of the furnace? Phase diagrams that show the chemical compositions of slags[40] can demonstrate the liquidus temperature (a reasonable estimate of the lower temperature achieved in the furnace). An alternative method derives temperature estimates from the chemical composition based on the fractional cation composition.[41] By using both of these methods, Stos-Gale and Gale estimate a temperature of near 1200°C for the slags at Chysokamino, and they note that the slags they analyzed fall into two groups, with some examples reaching temperatures approximately 100° higher than others (App. C). These temperatures are sufficient for prills with a diameter of 1 cm to flow downward and collect at the bottom of the furnace, with tiny prills remaining within the slag. The presence of slag fragments with surfaces showing that the material flowed out of the furnace and cooled quickly (Fig. 10.1) provides additional proof that the process resulted in a pool of copper at the base of the furnace. This would have necessitated tapping the furnace to remove the copper, and the tapping process would also have caused the slag to flow out and form rivulets of molten siliceous material (note the flow formations on Fig. 10.1:a and c). No rivulets would have formed if the smelting resulted only in a mass of slag inside the furnace.

The small size of the chunks of slag from the workshop is typical of early copper smelting operations where the slag was broken up to remove the copper.[42] As at Chrysokamino, these roughly contemporary operations produced some or all of the copper as isolated prills within the slag, and crushing was necessary to retrieve the metal. Even if (as seems to have been the case at Chrysokamino) some of the metal collected at the bottom of the furnace, the remainder of the copper was obviously desirable enough to be retrieved even if it consisted only of tiny prills trapped within the slag. The accumulation of the resulting small pieces of broken slag created the slag heap at the site (Fig. 10.2).

38. Tylecote and Merkel 1985.
39. Maddin 1996, p. 14.
40. Bachmann 1980, pp. 120–131.
41. Nathan and Van Kirk 1978.
42. Bachmann 1980, pp. 108–110; 1982, pp. 21–22; Rothenberg 1985, p. 124.

The slags at Chrysokamino indicate that at least some of the prills flowed down to form an ingot. Many pieces of slag from Chrysokamino are a result of consistently higher temperatures than the other known early smelting sites. No partly smelted pieces of ore or intermediate stages between ore and slag are present. The site does not have any of the incompletely smelted material (sometimes called furnace conglomerate) present in Chalcolithic smelting operations at Feinan, Jordan[43] and elsewhere.[44] Also, some of the pieces of slag from Chrysokamino demonstrate the presence of a liquid stage during the smelting. Small drips, stalactite-like formations, and other traces of a melted stage are present among the fragments of slag (Fig. 10.1:a). Flow-banding is also occasionally recognizable at the microscopic level (App. A, CHR 99).

The high temperatures, however, were not always achieved successfully. The differences in the slags at the microscopic level and the varying amounts of arsenic in the copper prills indicate that surviving slag fragments were formed under different conditions. Some of the slag reflects differences in the amount of reduction during smelting (App. A), as is shown by the relative amounts of fayalite (a characteristic of slags formed under highly reducing conditions) and pyroxene group minerals, which indicate a more oxidizing atmosphere.[45] Concerning the arsenic, experiments have shown that incorporation of this element within copper prills in the slag can result if arsenic minerals are present and if the temperature is not raised high enough to result in a molten metal that would be removed by tapping the furnace.[46] The evidence indicates that conditions varied substantially between different firings, and that temperatures were sometimes higher than other times.

The high temperatures that would have been necessary to produce these slags are important because of Chrysokamino's early date. The nature of the slags from this site contradicts an opinion that was once generally held, that slags may be dated on the basis of their physical description, and that "glassy" slags (i.e., highly vitrified pieces indicating a high temperature) are always Roman or later.[47] In fact, this basis for dating, based on slags from Cyprus and regarded incorrectly as a general pattern, was once widespread, and Chrysokamino itself was once assigned a date in the Medieval period or later on this basis.[48]

After the smelting operation was completed, the slag was broken into small pieces to retrieve even the small bits of copper within it. The importance of the availability of water for this process has been emphasized by Herdits.[49] The sea is easily accessible at Chrysokamino because it adjoins the site, and wetting the hot slag with sea water would have shattered it easily and quickly, explaining the relative scarcity of stone hammers at this site (Chap. 6). The composition of the waste pile, which was mostly composed of tiny pieces of slag with angular corners, attests to the final step in the process of smelting (Fig. 10.1:a).

In conclusion, a combination of evidence from scientific analysis along with the physical evidence from the excavation has resulted in a useful reconstruction of the main steps in the smelting process. This reconstruction indicates that the methods at Chrysokamino were successful in achieving

43. Hauptmann, Bachmann, and Maddin 1994, p. 5.
44. Bachmann 1980, p. 108.
45. Hauptmann, Bachmann, and Maddin 1994, p. 6.
46. Pollard et al. 1991, p. 132.
47. The theory is well summarized by Koucky and Steinberg 1982a, p. 156, table 1; see also 1982b.
48. Faure 1966; Branigan 1968.
49. Herdits 2003, pp. 69–70.

good results. Copper from Chrysokamino could have been disseminated to several sites in Crete, to be made into chisels, daggers, awls, and other products. That the resulting metal was a highly valued commodity is proved by the extreme economy practiced at the workshop. Ore was not wasted (the largest piece found at the excavation was a centimeter across). Even small prills were desirable, and the slag was shattered into small pieces to recover the tiny bits of metal. Copper was clearly a highly prized material.

PART III: THE SURFACE SURVEY

Introduction to the Surface Survey

by Philip P. Betancourt

A metallurgy workshop, a small Minoan habitation site, and a burial cave were present in the Chrysokamino region within an area of less than a square kilometer. In order to better understand these sites, an intensive surface survey was planned as one component of the investigation. The main goal was to develop a better diachronic understanding of the local territory with respect to its settlement, economy, and land use. In particular, it was absolutely essential to elucidate the context of the metallurgy workshop in order to interpret its significance. The local context was by no means understood before the project began. Was copper ore present in the vicinity? Did the metallurgy develop as a consequence of local mining and the exploitation of a small ore deposit, as had been claimed by some of the early investigators,[1] or did it exist in isolation from the natural resources necessary for its development? How did the metallurgy site fit into the local economy? What was the relation between the metallurgy site and the small cave claimed at one time as a copper mine?[2] Did the metallurgists live at the habitation site or somewhere else? Did the metallurgy workshop exist within a specific and definable territory, and could one learn anything about the history of that territory during the Bronze Age and later?

In addition to questions about the metallurgy operation, many general questions needed to be addressed. The date of the earliest settlement of the local area had not yet been established. When did the human occupation of the region begin, and how did it develop? Were Final Neolithic sherds found elsewhere as well as at the metallurgy workshop?

The nature of the region's development through time was an important issue. Was the local development continuous, or did interruptions and recessions exist? What was the relation between the habitation site, the cave, and the metallurgy location? Did the habitation site function as a center for an agricultural farm or estate at any time during its history, and if it did, could the boundaries of the local territory be determined? Did the nature and extent of this territory change through time? What were the characteristics (location, description, and distribution) of small individual sites and features in the territory of the main domestic complex, and how were all these locations interrelated?

The exploitation of the land and its natural resources must have been one of the important factors in local history. To answer the many questions

1. Hawes and Boyd Hawes 1909, p. 38.
2. Mosso 1908, pp. 518–521; 1910, pp. 289–293.

posed by the metallurgy site, the project needed to establish the potential of the region, including its geological materials (soils, mineral wealth, stones for building and other purposes) as well as its favorable environmental factors such as access to natural harbors, landing places for ships, and transportation routes. It also needed to establish access to water, the presence of different types of landscape with varying potentials for exploitation, and resources such as forests and wild game. These and other attributes of the local environment that may have contributed to the settlement of the area needed to be defined before we could determine how such resources were used.

Information on Bronze Age land management was a special focus of the survey. An important goal was to define key characteristics of the landscape as a basis for dividing it into different spatial units (i.e., different classes of land), using as many factors as possible: agricultural potential as determined by soil type and presence of mineral nutrients, water retention capability of the soil, use of the land in antiquity as determined by the presence of sherds and other artifacts, proximity to or distance from settlements, and topographic setting (e.g., steepness of the land and other factors). The plan was to use these characteristics to establish different classes of land as part of the evidence for how the landscape may have been managed and used. Agriculture must have been important. Did the exploitation of the farmland change through time? Was the land use of the Bronze Age the foundation for the later agricultural exploitation of the territory, or could differences be recognized in later periods?

Answering even some of these questions would complement the conclusions to be drawn from the data collected by excavation. Placing the metallurgy operation within a historical context and trying to understand something of the economic and social systems within which it operated would hopefully contribute insights into the nature of the local history in general as well as the history of the small metallurgical installation and its role in the local region.

PROCEDURES OF THE INVESTIGATION

A research plan was developed to address the many questions that required answers. It was organized as an intensive surface survey with several components, some of which were extensions of the excavation project and used the same personnel required for the excavation, such as the instrument survey team. Other components required the expertise of specialists who worked only on the intensive surface survey.

The boundaries of the Chrysokamino territory were an important component in determining the extent of the survey. They were defined based on a combination of several factors including topography, natural barriers in the landscape, analogies with territorial boundaries from elsewhere, information from the excavation that showed what resources were being exploited, and the proximity of nearby sites. Because this methodology could not be employed in detail before the project assembled its data, preliminary estimates of the territory were made at the inception of the work, and the plan was modified as work progressed.

An especially close examination of the landscape was required because of the many questions raised by the metallurgy operation. One of the most important of these questions involved the possible use of local copper ores, an issue raised several times in the early 20th century.[3] The presence or absence of former ore beneficiation sites, mining operations, or locations for stockpiling of ore could only be determined by examining the entire landscape. Beneficiation, for example, involves the enrichment of bulk ore by the removal of rock that does not contain any ore, and the practice results in large quantities of waste rock.[4] Even if a small ore deposit is completely exhausted, the piles of waste rock will survive as dumps. Since the excavation of the metallurgy site demonstrated that no beneficiation occurred at the workshop itself, a search for piles of waste rock in the Chrysokamino territory could help demonstrate the presence or absence of former mining or stockpiling sites. Any potential conclusions based on the absence of ore deposits (i.e., using negative evidence) could only be valid after the entire landscape was examined closely enough to be certain that such waste rock was not present.

A specialized surface survey of the Chrysokamino territory was planned. It differed substantially from the traditional intensive walking survey.[5] A traditional survey had already been conducted in this territory,[6] and most of the pottery fragments on the surface had already been collected. In 1983, a team of metallurgists headed by Noel Gale and Zofia Stos-Gale had also examined the territory as a part of their metallurgical survey of smelting and mining sites in the Aegean (App. F). The soils of the region had also already been studied.[7] The new investigation was planned as an extension of the previous work.

The new intensive surface survey included the following components, carried out between 1994 and 2003:

1. Geological Survey. In order to understand the geology of the local territory, William Farrand and Carola Stearns conducted a survey of the geomorphology and the bedrock geology. They walked across the entire territory and recorded the relevant information in 1994 and 1995 (see Chap. 2).
2. Mapping of the Landscape. A team of three to five persons used a Topcon Total Station interfaced with a laptop computer to map the Chrysokamino territory, walking over the landscape and recording points at suitable intervals. The team reconfirmed and accurately mapped sites recorded by the earlier survey, collected additional sherds, and recorded all features. The team set up the survey instrument and recorded points around it at intervals of 5–10 m, noting whatever anthropogenic features were present within the area being measured. The survey station was then moved to record new territory. This aspect of the project took four field seasons (1995–1998). The team

3. Hawes et al. 1908, p. 33; Hawes and Boyd Hawes 1909, p. 38; Mosso 1910, pp. 289–292.

4. D. Gale 1991; Gale and Ottaway 1991.

5. Among many others, see the methodology of Keller and Rupp 1983; Cherry, Gamble, and Shennan 1978; Cherry et al. 1988; Wells, Runnels, and Zanger 1990, pp. 214–216; Cherry, Davis, and Mantzourani 1991, especially Cherry, Davis, Mantzourani, and Whitelaw 1991; Wells 1996.

6. Haggis 1992, 1995, 1996b, 2005.

7. Morris 2002.

divided its time between working on the survey and providing measurements and mapping for the excavations. The electronic mapping project produced computer-generated maps and charts as well as detailed information on the nature of the landscape (see Chap. 16 and App. G).

3. Soils Survey. A survey of the soils in the territory was made by Eleni Nodarou (see Apps. L and N).

4. Analysis Program. Several analytical studies were undertaken to contribute information on the territory (see Apps. A–F, M, and N). They examined artifacts and materials by a series of different techniques, contributing substantial information for the project.

5. Register of Anthropogenic Features. All architectural features (including groups of terrace walls) were mapped, recorded with the survey instrument on a three-dimensional grid in the computer, measured, and described (App. G). The instrument survey team recorded the features as a part of their mapping project. Sherd scatters without architecture were also mapped, and sherds were collected (see Apps. H and J). Given the nature of the evidence as well as the occurrence of a previous survey, the new collection was regarded as a sample possibly not representative of all periods.[8] Therefore, the new information was coordinated with the data obtained from the previous survey.[9]

6. Revisits to Anthropogenic Features. All features and sherd scatters were revisited several times by different members of the staff for additional study, photography, and sherd collection.

7. Selective Excavation. In addition to the excavations at the metallurgy workshop and the habitation site, two anthropogenic features, a small cave and an agricultural terrace, were excavated to obtain additional information (App. K).

8. Study, Analysis, and Additional Revisits. After study and analysis, specific locations, regions, and features were again visited and additional data (including more electronic survey points) were obtained as needed. This extra step allowed computer-generated maps to be checked against the actual topography and preliminary written conclusions to be compared with the physical situation on the ground, so that additional information could be collected as needed. Information from the excavation of the habitation site and the metallurgy workshop was coordinated and used to help interpret the data from the survey.

In order to determine land use, the land was divided into classes based on topography, types of soils and their agricultural potential, archaeological evidence for land use in antiquity (including sherds, terrace walls, and other characteristics), location within the territory, and other factors. These studies resulted in new conclusions on the organization and management of farmland during antiquity.

9. Conclusions and Publication. Conclusions were completed and published both in preliminary reports and in this volume.

8. Read 1986.
9. Haggis 1992, 1995, 1996b, 2002, 2005.

CHAPTER 16

TOPOGRAPHY OF THE CHRYSOKAMINO REGION

by Lada Onyshkevych and William B. Hafford

The metallurgy site at Chrysokamino is located in a region of terraced hillsides and steep cliffs west of modern Kavousi, overlooking the Gulf of Mirabello. Throughout this area, the ground surface is covered with dense, scrubby vegetation (maquis), occasional large thorny bushes, and a few lone olive trees. The bare landscape near the coast is divided from the inland fields by a modern fence. Access to the territory from the village of Kavousi is provided by a modern road, with a gate (to prevent sheep and goats from trespassing) at the point where the fence crosses the road. The road terminates at the Minoan habitation location.

The area surveyed for the maps in this volume includes the Chrysokamino metallurgical workshop and a habitation location with the stone remains of a Late Minoan farmhouse (Fig. 16.1). Anthropogenic features are discussed in Appendix G and shown on Figure 16.2. They include terrace walls, field houses, threshing floors, boundary walls, small wells, and other manmade features. None of these are in use anymore, and their distribution contributes substantial evidence for the past history of land use in the region.

The survey area to the north extends to the coastline beyond the Theriospelio cave, while to the south, it continues to the field called Lakkos Ambeliou and to the lower slopes of Chalepa Hill. The hills north of the territory, where the small unexcavated hilltop tower called Pyrgos Chrysokaminou is located, are beyond steep cliffs that form a natural boundary north of Theriospelio. The coastline, with additional steep cliffs, forms the western boundary. On the east, the surveyed area includes terraced olive groves east of the Minoan habitation site (Fig. 16.3). In terms of the grid created by the survey team in 1995, the area surveyed extends from approximately 4000N to 5250N, and from 4500E to 5600E, with some points measured beyond these lines as well (Fig. 16.2; see also Fig. 16.4 for the locations of the datum and other grid points set in cement).

Maps of the excavated areas, as well as of the larger region linked to them, were produced by electronic survey instrument during the 1995–1998 seasons. Approximately 7,000 points were measured electronically, and coverage averaged every 2–3 m surrounding the excavated locations (Fig. 16.5). In areas farther away from the excavation locations, coverage was sparser,

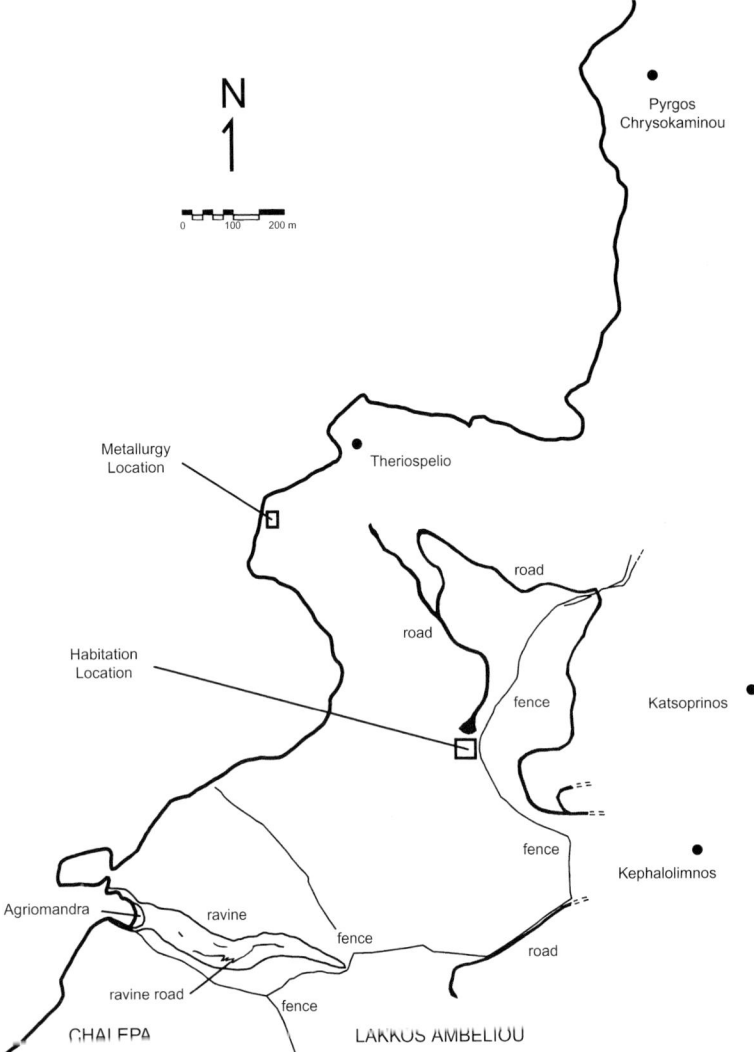

Figure 16.1. Map of the greater region around Chrysokamino, with indications of the habitation and metallurgy locations, Theriospelio cave, Agriomandra beach and the ravine leading to it, the road within the ravine, the road and the fence of the Chrysokamino region, and the neighboring sites of Katsoprinos, Kephalolimnos, and Pyrgos Chrysokaminou

at an average of every 5–6 m, because of the considerable extent of land needing to be surveyed. Very distant areas (such as Chalepa, the hill south of Agriomandra beach) received coverage averaging every 15–20 m. In addition, areas of very steep and treacherous footing (such as the ravine on the northeast side of Therio, or the lower ravine area of Chomatas Hill between the two excavated locations) could in some cases receive only representative coverage: a line of points taken along the highest ridgeline and the lowest line of the ravine. The resulting data were sufficient for the topographic software to generate satisfactory topographic lines for those areas (Fig. 16.6).

This coverage was sufficient for detailed mapping. The work also included an intensive surface survey of the territory, with the recording of all anthropogenic features noted by the team (see App. G). The intent was to produce a series of accurate maps showing the topography and to locate the abandoned architectural remains and other manmade features within the territory, so that both natural landforms and anthropogenic features could be studied in more detail as a part of the overall project.

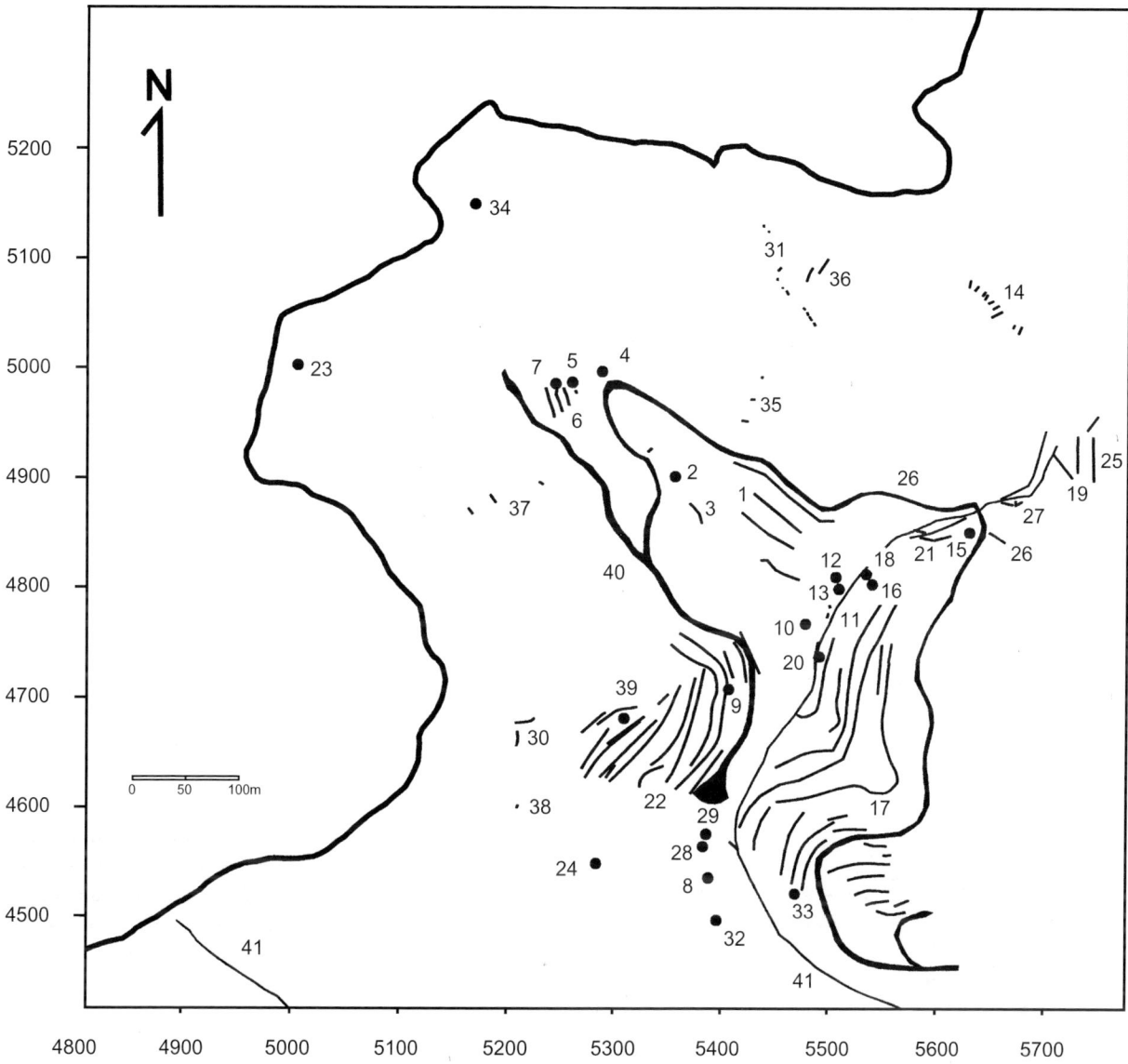

Figure 16.2. Map of anthropogenic features in the Chrysokamino farmstead (see App. G)

Although the survey personnel gathered almost all of the data for the maps in this volume electronically, many of the coastal cliffs at Chrysokamino could not be surveyed directly because of their dangerous slopes and treacherous footing. Therefore, the maps were produced with the aid of a coastline digitized from an old Greek military map and merged with our own electronically surveyed data in Surfer and AutoCAD files. The military map, however, could not be precisely matched to topographic features and landmarks currently visible, due to a number of factors. First, the sea level, and consequently the apparent line of the coast, had undergone changes over the decades since the military map was made. Second, the original purpose of the military map did not require the same level of precision as that yielded by modern surveyed data, and it made use of older technology. Thus, in merging these two bodies of data, it was necessary to make some adjustments in the data recorded on the military map, including a slight rotation of the coastline axis west of north.

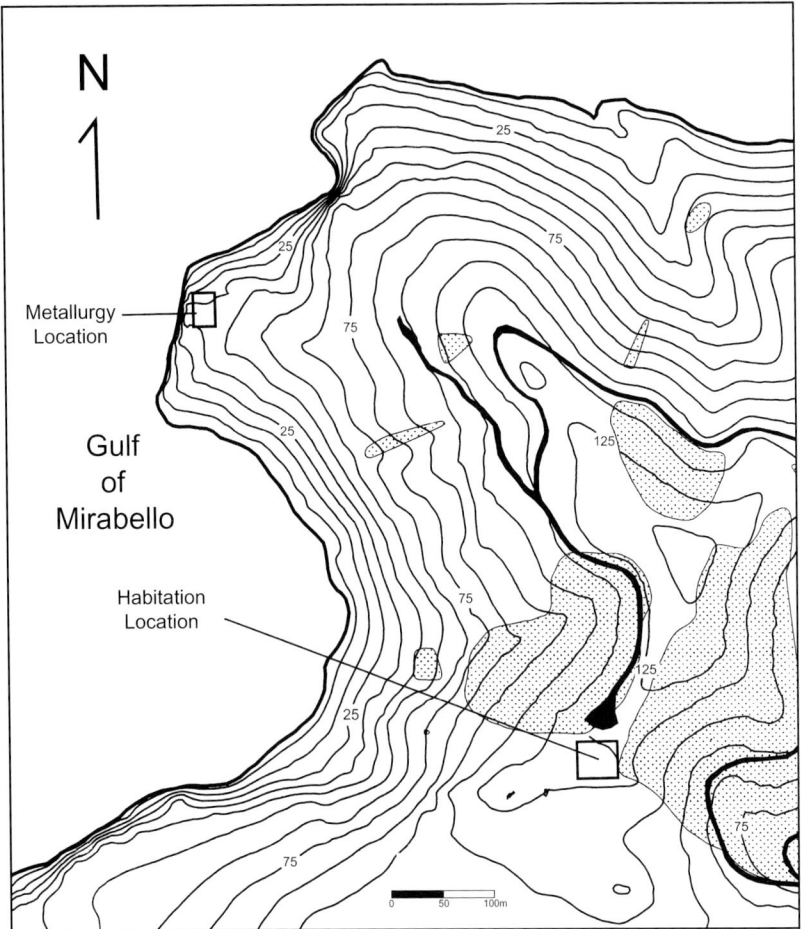

Figure 16.3. Topographic map showing terraced areas (shaded)

A few areas of the coastline and coastal cliffs, however, were accessible for surveying, and they helped determine the correct alignment of the military map's data and the electronically surveyed data. One such area can be found on the coastline immediately southwest of the metallurgical location, curving south and southeast into the bay below the habitation location. A number of coastal cliff areas that descend almost directly into the sea could also be surveyed electronically, and they yielded data consistent with the rotated coastline. Among these areas were Chylopittes, the headland of Therio northeast of the metallurgical location's headland, the cliffs southwest of the habitation location, and the cliffs flanking Agriomandra beach, including parts of Chalepa Hill. Agriomandra, the small beach or harbor at the bottom of the ravine southwest of the metallurgical and habitation locations, also yielded data consistent with the adjusted alignment of the military map's coastline and our own data. Thus, the coastline used in the maps in this volume, while not directly surveyed in most of its extent and not as precise as other surveyed data in this study, can be regarded as reasonably accurate.

The habitation location of Chrysokamino is situated downhill from a small saddle south of the dome of Chomatas Hill, at an average elevation of approximately 120 masl, in an area called Katsoprinos. To the north and east, it is bordered by terrace walls. A modern sheepfold *(mandra)* once

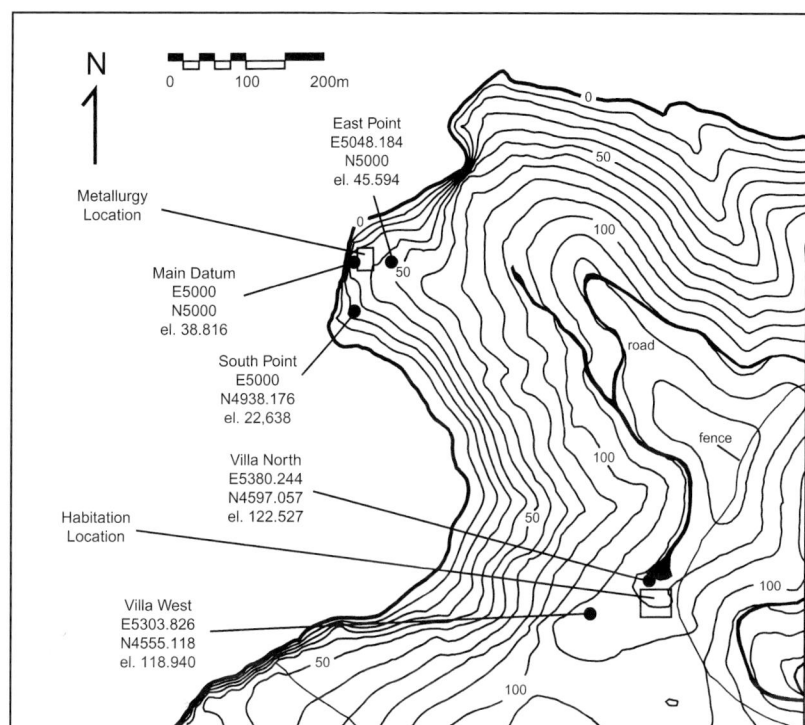

Figure 16.4. Map of grid points for the Chrysokamino farmstead (Main Datum, East Point, South Point, Villa North, Villa West)

bordered it to the south and partly extended over the ancient remains prior to excavation. The architectural complex itself slopes to the southwest at an average gradient of 0.14 (i.e., a distance of 1 m features a change in elevation of 0.14 m). This slope continues down to the southwest, flattening out into a fairly level field (about 106 m in elevation), which contains terra rossa sediment and a large ellipse of rubble wall. This elliptical construction is most likely an old sheepfold, now tumbled down (Fig. 16.2:32; see App. G, **AF 32**, for description and date). The terrain then rises to a small hill of about 110 m in elevation, descends to a lower hill, and continues down again until it reaches Lakkos Ambeliou, a large and fairly flat basin (about 75 m in elevation) filled with more of the terra rossa sediment (Fig. 16.6). West of Lakkos Ambeliou, a steep ravine *(revma)*, with sides dropping down at maximum gradients of approximately 3.0, stretches to the sea, ending at Agriomandra beach. Chalepa Hill rises south of the ravine and Lakkos Ambeliou at an average gradient of 0.4, while another hill rises north of the ravine in two peaks of about 100 m in elevation (with a gradient of about 0.3 from the ravine edge). Northwest of Agriomandra beach is a small, craggy peninsula that is completely inaccessible from land because of a deep crevasse separating it from the cliffs of the north side of the ravine; topographic information for this small peninsula was estimated from the military map and from visual data, and so cannot be regarded as highly accurate.

West of the habitation location is a rocky outcrop, which descends along a knobby ridge and rocky cliffs into a steep ravine, which in turn drops down to the coastline. South of this ridge and ravine, the terrain descends along a gentler incline (with an average gradient of 0.5) to very steep cliffs, which drop 44–50 m into the sea, at a gradient of 3.5. North

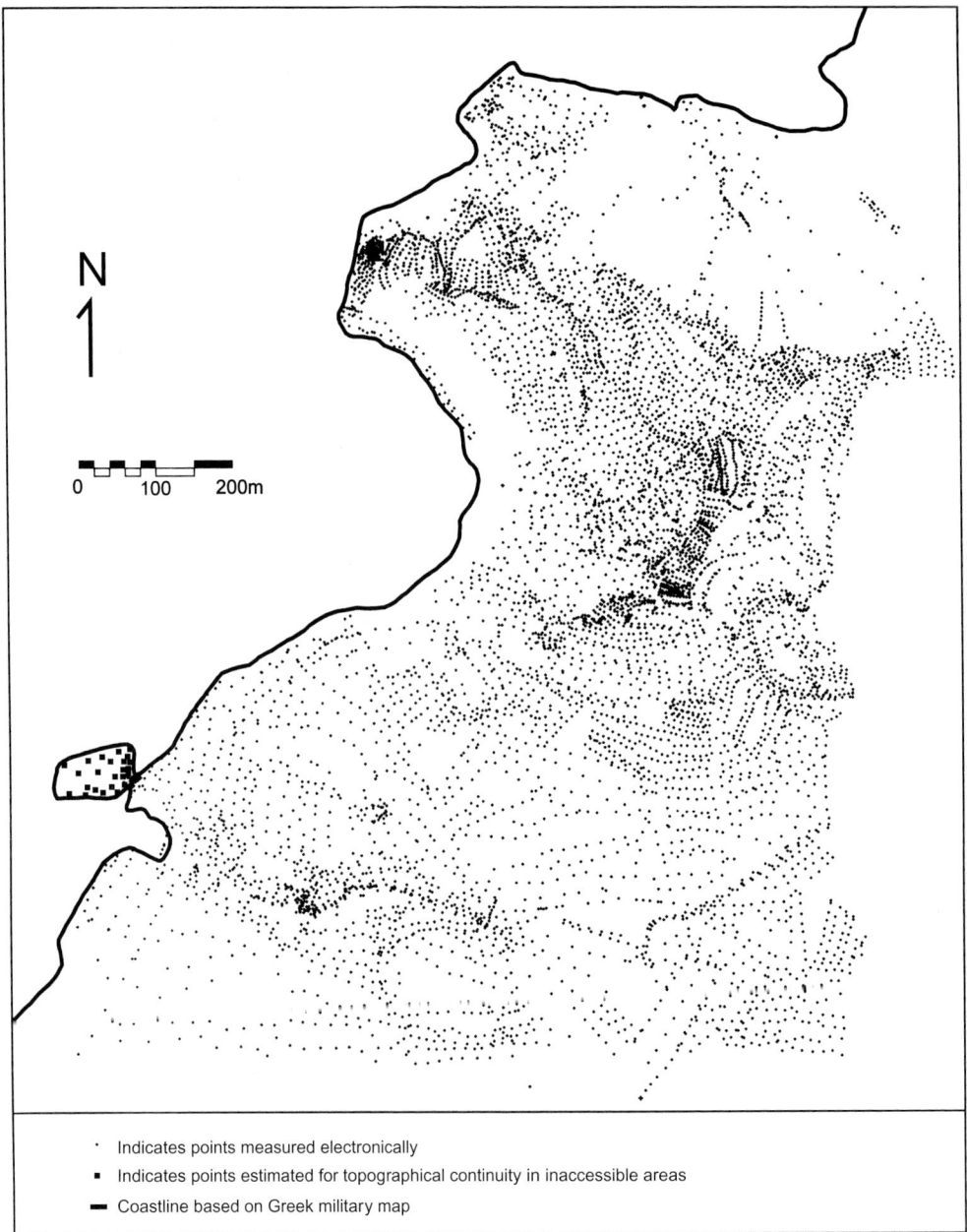

Figure 16.5. Map of the Chrysokamino farmstead, with indications of all points measured electronically, all points taken from the Greek military map, and all points inserted to generate satisfactory topographic contours (see legend)

and west of the habitation location, the terraced hillside steps down along a shallow ravine. Then, at an elevation of approximately 80 m, it drops into a rather steep and very slippery cliff and ravine (with a gradient of 0.53) to the sea coast.

Farther north, the modern dirt road (partly created in the spring of 1996) extends from the habitation location along the terraced hillside onto the southwestern side of the same hill (Chomatas). One branch curves around the dome of the hill to the north side where the gate to the Chrysokamino area is located, and a lower branch descends along the western side of Chomatas Hill and stops abruptly. The southwestern side of the hill descends along a relatively bare and unterraced stepped ridge, down to a very rocky headland above the sea where the Chrysokamino

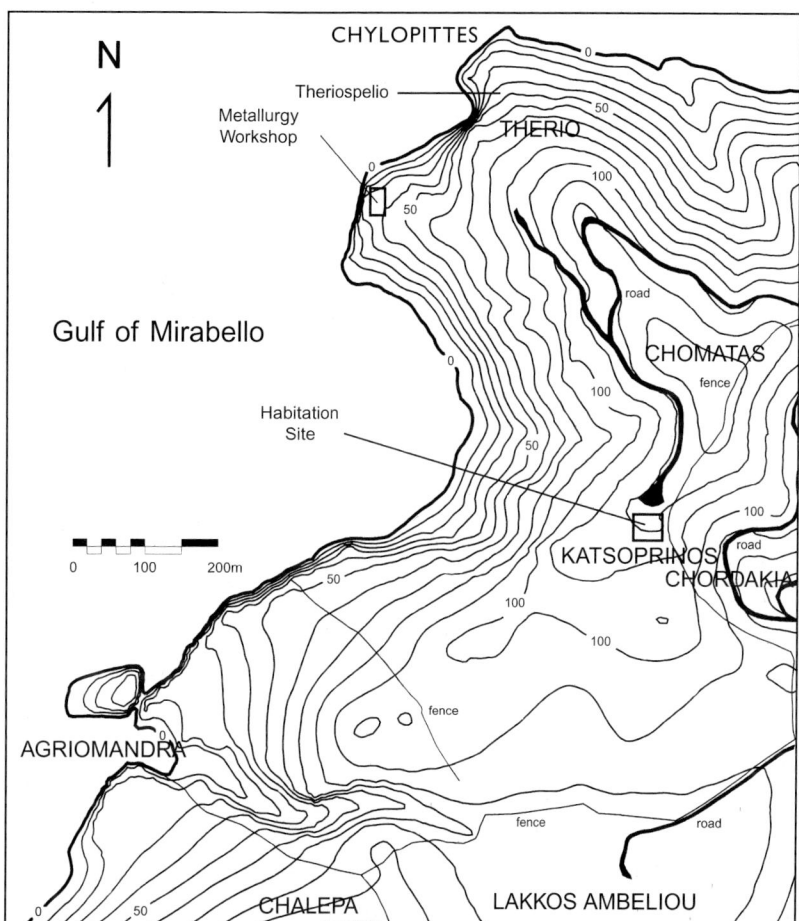

Figure 16.6. Topographic map of the territory of the Chrysokamino farmstead

metallurgy workshop is located. The metallurgical location is on a low saddle on this headland,[1] at an upper elevation of approximately 38 m, dropping sharply to the west at a gradient of 6.5, to the north at a gradient of 1.1, and sloping to the southwest at a gentler gradient of 0.65. The seacoast can be approached with relative ease along the southwestern descent, while the west, northwest, and northeast sides are too steep to allow access to the sea. Located about 222 m below the lower road, the metallurgical area itself requires a walk for ascent or descent (gradient of 0.28). The direct distance between the metallurgical location and the habitation location is ca. 585 m.

Around the curve of the cliff to the north and northeast of the metallurgical location (average gradient 5.5), another headland (Chylopittes, part of the hillside named Therio) protrudes into the bay, descending to a rocky coastline at an average gradient of 0.56 before dropping in stepped cliffs to the water on the west and southwest. The northern corner of this headland, however, steps down directly to sea level, although the wide gaps between its ridges make actual access to the sea difficult here. A cave, called Theriospelio, Agriospelio, or Kolonospelio, is located 44 masl on this headland (Figs. 16.1, 16.2:34, 16.6); this cave has a small entrance, but a large interior (not mapped) with a sloping and slippery floor. The cave is located at a direct distance of ca. 230 m from the metallurgical location and ca. 610 m

1. Haggis 1992, p. 170, locus 88; 1996b, pp. 380–381, esp. n. 22.

from the habitation location. Farther north, the coastline contains another small inlet, from which a ravine ascends to the area of the modern gate (with a gradient of 0.3 along the base, and 0.6 along the sides); a few terrace walls remain along this steep ravine.

The larger Chrysokamino region also contains the remains of a number of features that probably marked the natural boundaries or extent of the territory utilized by the inhabitants in the Bronze Age. The terraced hillsides immediately to the northwest of the habitation location (Fig. 16.3) find parallels on the hills farther north, up to the modern gate to the Chrysokamino area (Fig. 16.2:19, 21, 25–27), and to the east, where an olive grove currently thrives on the inland (eastern) side of the modern fencing, in an area called Chordakia (Figs. 16.2:17, 16.6). Although the terrace walls, built of schist or dolomite rubble, appear to be relatively recent in date, these terraced areas may well have been used for grazing or agricultural purposes in the Bronze Age as well. A number of rubble buildings are also relatively modern.

South of the Minoan farmhouse are two fields with terra rossa-derived sediment (see Fig. 16.2:32 for the upper field and Figs. 16.1, 16.6 for the lower field, named Lakkos Ambeliou). Area residents currently use the larger field in the low basin (Lakkos Ambeliou) as a "clay mine" for brick making and for gardening soil, as well as for numerous apiaries.

Access to the sea for this region was provided from antiquity onwards by means of a substantial ravine, which curves west of Lakkos Ambeliou and north of Chalepa Hill, between two extremely steep cliffs (gradient 3.0), to the small beach of Agriomandra (Figs. 16.1, 16.6). Currently, this ravine is lined with walls of relatively recent date (Venetian or Ottoman?), which support a built path winding from one cliff side to the other for the length of the ravine, often on top of terrace walling constructed especially to support the path. A small church lies in a small cave about two-thirds of the way down this ravine, while the remains of a limekiln can be found at the eastern entrance of the revma.

Assuming that the natural features noted here delineate the outer boundaries of the region (see Chap. 20), a total area of approximately 687,000 m^2 can be ascribed to the inhabitants of the territory of Chrysokamino for agricultural, pastoral, manufacturing, or trade purposes. This territory mostly coincides with the area within the modern fenced land, extending from the northern coastline of Therio at the north, to the northern edge of Lakkos Ambeliou to the south, and from the fence on the east to the sea and cliffs on the north and west.

CHAPTER 17

A Summary of the Habitation Site at Chrysokamino-Chomatas

by Cheryl R. Floyd

The habitation site at Chrysokamino is located on the hill of Chomatas near the coast of northeastern Crete, northwest of the modern village of Kavousi.[1] It is approximately 600 m from the EM metallurgy site. The smelting operations apparently ended in the EM III–MM IA period, while the surviving architectural remains at the habitation site date to the LM IB and LM III periods (Figs. 17.1, 17.2), with potsherds ranging in date from the Final Neolithic to Venetian or Ottoman periods. Although a considerable number of MM potsherds come from the habitation location, no associated architecture has been identified. The preserved architectural remains are mostly from LM IIIA. The latest ceramic material recovered consists of a few Byzantine and Venetian to Ottoman sherds from surface strata.

The site will be published in more detail in a future volume, but a summary of the results is necessary for a proper understanding of the conclusions from the Chrysokamino survey.

THE EARLY PERIODS

The earliest identifiable remains at the habitation site are Final Neolithic to Early Minoan I sherds from strata below the level of the LM I and LM IIIA floors and from fill in Room A, Room 2, and in levels adjacent to Wall 8 at the north. In addition, occasional early sherds were found in various trenches over the site (including Room 6 and the area immediately west of the northern part of Room 2), but none occurred in any concentration.

Larger quantities of EM II–MM pottery were unearthed from floor packing and fill strata over most of the site, indicating a considerable likelihood that EM II–MM structures were located in the immediate area. No permanent structures from before LM IB have been found at this site.

Only the final two passes in Trench 38 (near Wall 8), and the lowest passes in Room A were unmixed, early strata (FN–MM). The early stratum

1. For information on the habitation complex, see Haggis 1996b, pp. 401–403, fig. 13; Betancourt, Floyd, and Muhly 1997; Betancourt, Muhly, and Floyd 1998.

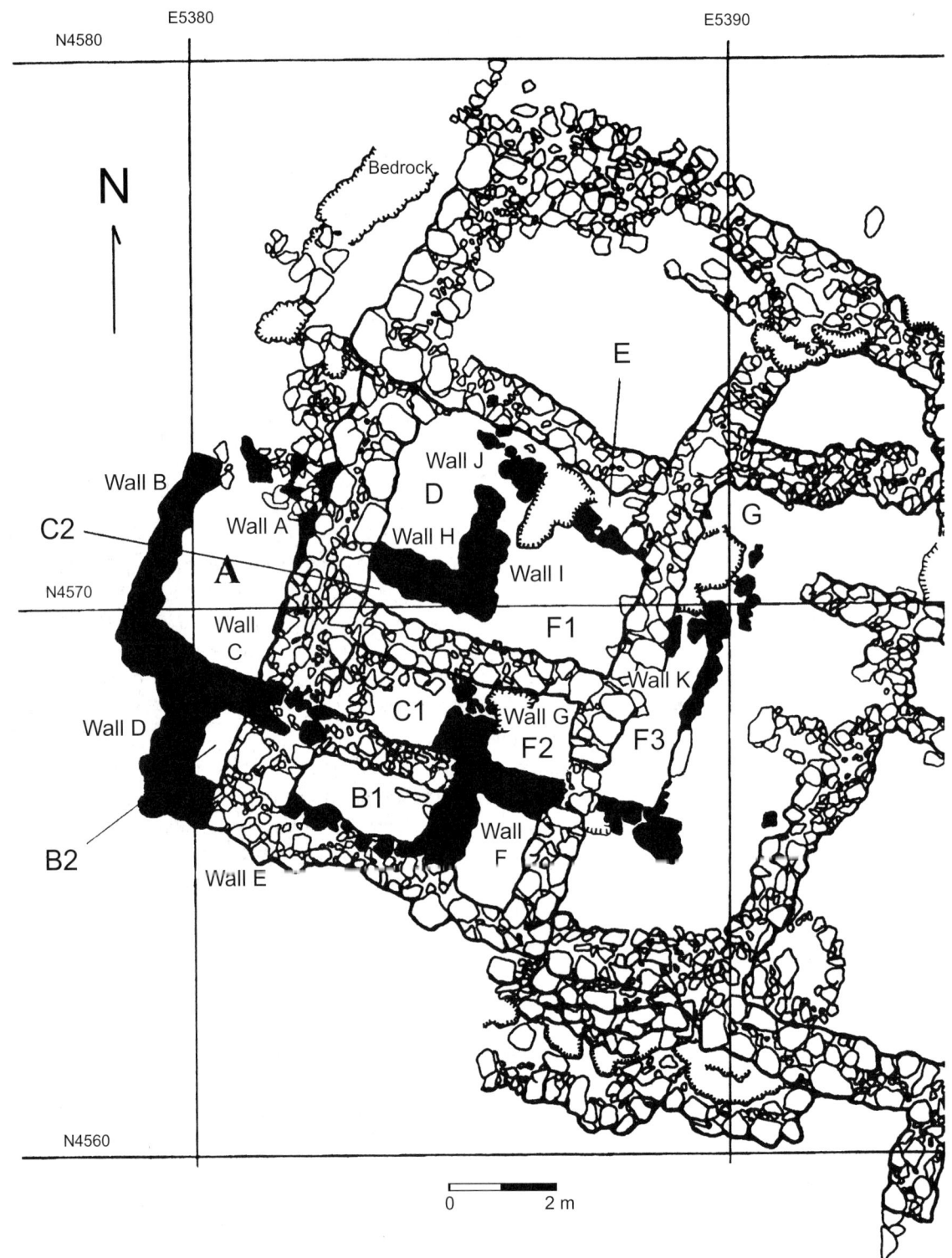

Figure 17.1. Western portion of the plan of the habitation site showing the LM I walls in black. Rooms/spaces are indicated by letters A–G.
L. Labriola, P. Betancourt

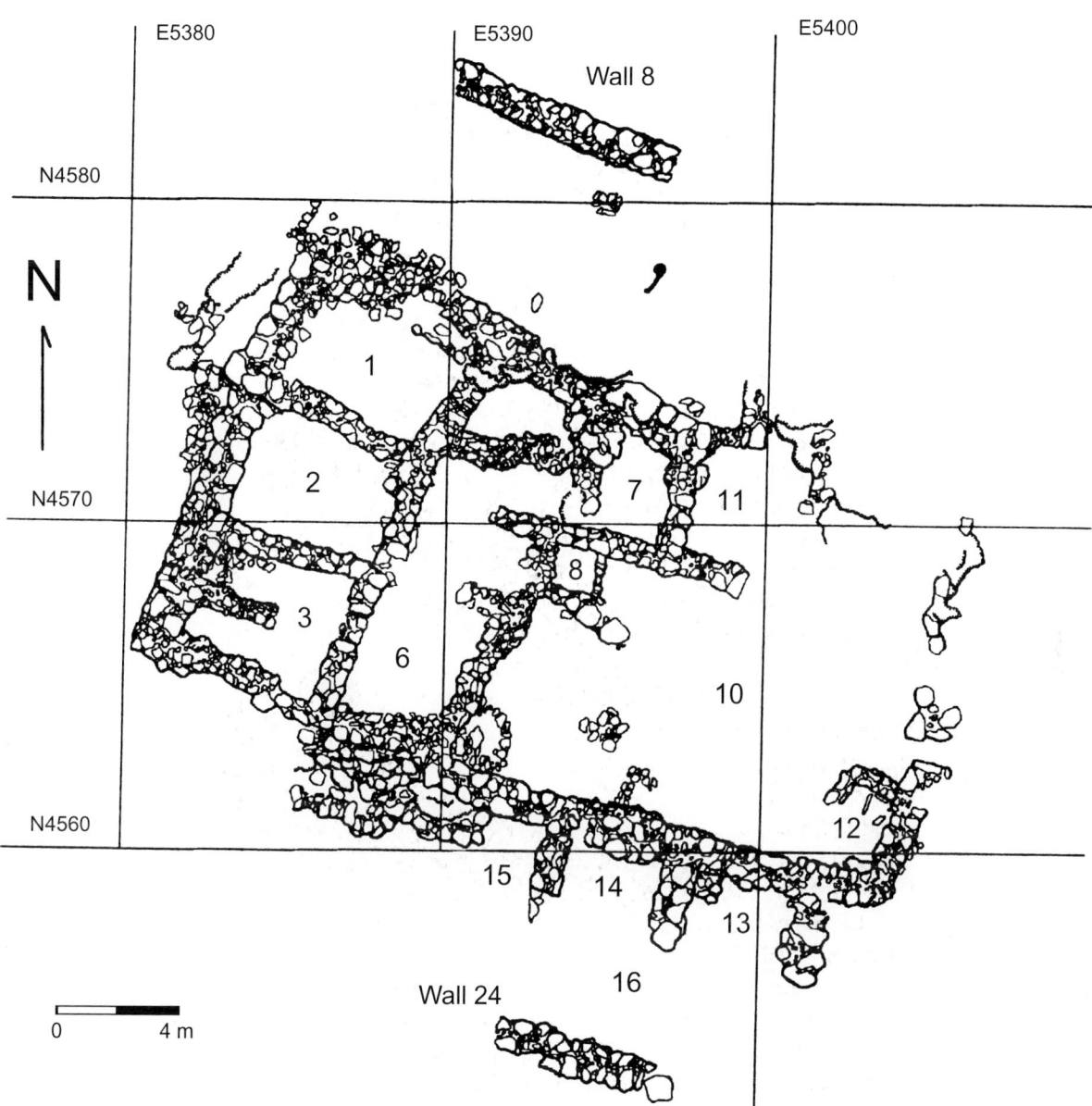

Figure 17.2. State plan of the habitation site showing the LM III architecture. Rooms/spaces are indicated by numbers 1–16. L. Labriola, P. Betancourt

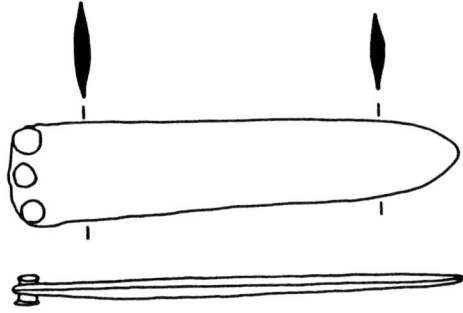

Figure 17.3. The LM I bronze dagger from Room A (X 830)

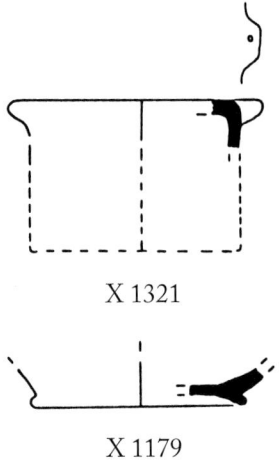

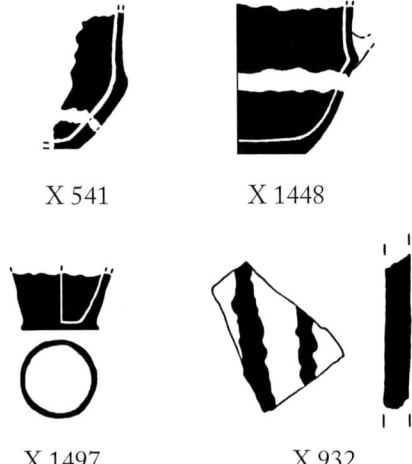

Figure 17.4. Examples of EM ceramics: X 1321, EM I pyxis lid fragment; X 1179, EM IIB Vasiliki Ware closed vessel base. Scale 1:3

Figure 17.5. Examples of MM ceramics: X 541 and X 1448, fragments from carinated cups; X 1497, base of a tumbler; X 932, body sherd of a large, closed vessel. Scale 1:3

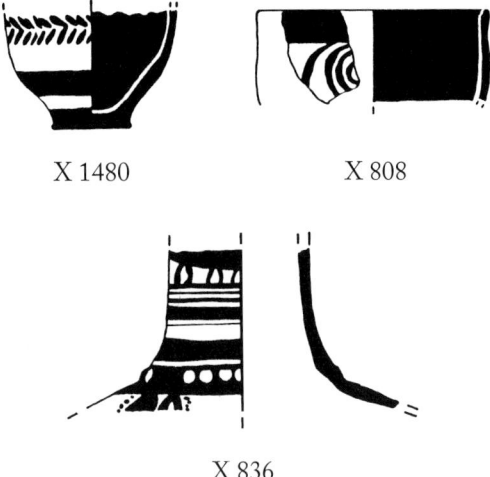

Figure 17.6. Examples of LM IB ceramics: X 1480, base of a bell cup; X 808, cup rim; X 836, neck of a jug. Scale 1:3

in Trench 38 was found at a level below the foundation of Wall 8, however, and it could not be associated with any architecture. In Room A, below the level of the LM I floor with its dagger (Fig. 17.3), a stratum containing FN–MM sherds but no later material was also revealed. FN potsherds were limited in both number and shape; the forms represented included bowls and jars. The vessels were formed from either gritty-textured gray brown fabrics or reddish brown gray fabrics. The walls of the vessels were generally thick, and their surfaces were dark in color and often burnished. Examples of fragmentary EM vases were also recovered, including a sherd in a fine gray fabric from an EM I cylindrical pyxis lid with pierced lugs, and sherds from EM IIB Vasiliki Ware vessels (Fig. 17.4). The range of MM forms was considerably greater; numerous fragmentary cups, bowls, kalathoi, and closed vessels in light-on-dark technique were recovered, as were pieces of MM cooking pots and dishes (Fig. 17.5). This full complement of vessel forms supports the hypothesis that in the MM period a settlement site existed at this location.

THE LM IB PERIOD

The LM IB architectural remains are confined to the western portion of the site (Fig. 17.1). Parts of several rooms from one or more structures have been excavated, revealing a technique of wall construction that differs from the later, LM IIIA technique. Architects in the LM IB period generally built directly atop the dolomite bedrock and used more regularly shaped stones than did their LM IIIA counterparts.

The paucity of LM IB architecture makes it difficult to say much about the nature of the site during this period. LM I pottery recovered from the lower strata in the western part of the site (Fig. 17.6) included vessels associated with general domestic activities, such as eating, drinking, storage, and food preparation. This evidence suggests that the associated structures (or structure) were utilized, at least in part, as domiciles. Other finds from LM IB strata included ground stone tools, a broken obsidian blade, and clay loomweights. LM I luxury goods, as represented by a complete bronze dagger (X 830, Fig. 17.3), a complete stone bowl (X 818, Fig. 17.7), and the rim of another stone vessel, indicate a degree of wealth for the LM IB inhabitants.

 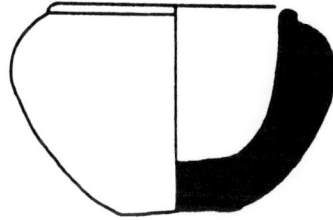 

Figure 17.7. Stone bowl (X 818) from LM I stratum. Scale 1:3

Figure 17.8. LM IIIA hearth above a LM I wall (Wall C), but below a LM IIIA wall (Wall 9)

Figure 17.9. Example of LM IIIA wall construction at the habitation site. Note the seating of larger stones on soil and smaller, chinking/leveling stones

Figure 17.10. East entrance into the LM IIIA complex, delineated with upright slabs

THE POST-LM IB PERIODS

Most of the preserved architecture at the habitation site is post-LM IB in date. Several phases of construction took place after LM IB, the latest of which occurred in LM IIIA. Sometime after the destruction of the LM IB building(s), a hearth was constructed above the level of Room B1 (Fig. 17.8). This hearth was below the LM IIIA wall running east–west as a dividing wall in Room 3 (Fig. 17.2), but above the level of the LM I wall below it (Fig. 17.1). No architecture can be associated with this hearth. Following the phase represented by the hearth, the first recognizable LM IIIA phase can be isolated, represented by the east–west wall within Room 3, which contained LM IIIA sherds in its construction material. This wall, however, is earlier in construction sequence than most of the LM IIIA architecture at the site. In fact, it is actually incorporated into the massive west wall of the latest phase complex (also LM IIIA in date). Many of the trenches excavated on the eastern part of the site had LM IIIA potsherds in their lowest strata, usually above bedrock.

The latest structure is a series of adjacent rooms and open spaces contained within thick enclosing walls. The construction techniques employed contrast with those of the LM IB period on the site. The builders of this complex frequently did not bother to seat their walls on bedrock. More often, they merely built on soil, using small chinking stones to help bed initial irregular courses of larger stones (Fig. 17.9). The type of masonry used was rubble construction with small to large irregularly shaped stones set in a mortar of soil. Two doorways from this phase were articulated using large, upright slabs on either side of their thresholds (Fig. 17.10, the east entrance into the architectural complex). Due to the slope of the site, the builders were required to construct terraces stepping down from north to south. These terraces were formed using buttressing walls in conjunction with natural bedrock shelves (see Fig. 17.2, the walls at the north of Room 1 and at the south of Rooms 3 and 6). Along the south of Room 6, this buttressing system was especially massive and complex, having been built up in stages. The largest enclosed space, Space 10, appears to have been an interior courtyard in which several built features were located. Some rooms or spaces were paved with slabs where the bedrock did not project (for example, see the exterior space north of Room 1 and south of Wall 8, and Room 6). The numerous built features found at the site included two hearths, depressions with channels pecked into the bedrock, and several possible cists or bins.

Not all of the building survives. The northeastern portion of the site is very poorly preserved, and it remains enigmatic. The walls defining Rooms 13, 14, and 15 at the south were truncated at some point after their construction, probably during the construction of a modern *mandra* immediately south of the complex. In addition, Walls 8 and 24 are incompletely preserved and/or excavated, and hence their relationship to the complex is not entirely understood. They may have functioned as further enclosing walls, or they may date to an earlier period.

Most of the pottery from this latest phase was LM IIIA in date (Fig. 17.11). Vessel forms represented in fine fabrics included cups, handled

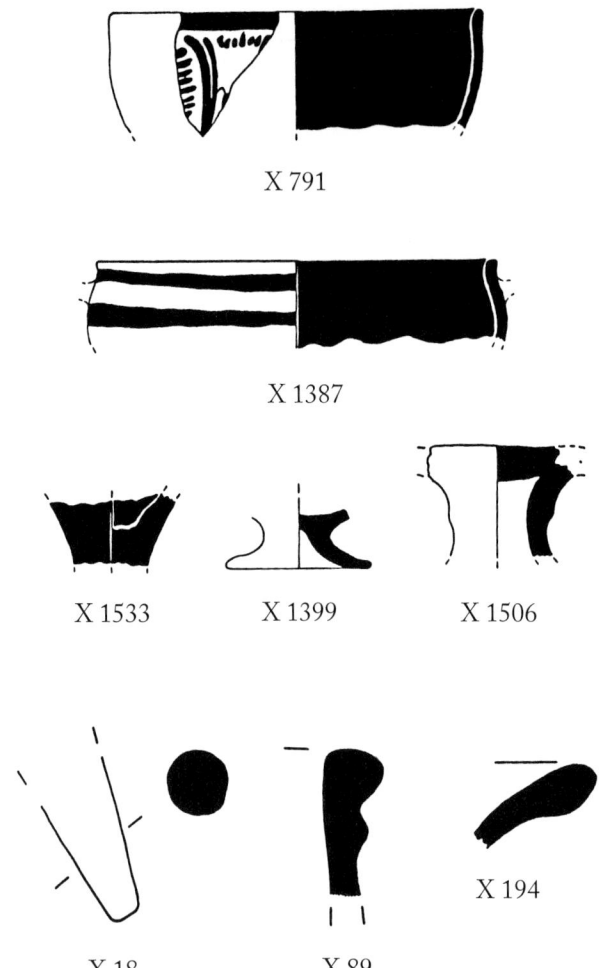

Figure 17.11. Examples of LM IIIA ceramics: X 791, cup rim; X 1387, rim of a bowl; X 1533, kylix stem; X 1399, champagne cup foot; X 1506, stirrup jar neck; X 18, leg of a tripod cooking vessel; X 89, jar rim; X 194, rim of a cooking dish. Scale 1:3

bowls, kylikes, stirrup jars, jars, jugs, and other closed vessels. Many potsherds from cooking and storage class vessels made in tempered fabrics were also present.

Other classes of artifacts were also recovered from the LM IIIA complex. Bronze finds included a chisel (X 1534), a hook (X 1287), a fragmentary knife (X 1025), a buckle (X 1026), and an elaborate, fragmentary jug (X 1149). In addition, several strips of folded lead were found in Room 5, and a sealstone (X 208, Fig. 17.12) came from Room 2.

Ground stone tools at the site included saddle-shaped querns (Fig. 17.13) as well as tools that were probably used as pounders, rubbers, and grinders. While ground stone tools were commonplace at the site, only a couple of pieces of obsidian were found, and these derived from LM IB strata. Quite possibly this scarcity of chipped stone implements and debitage indicates a change or shift in tool production and materials from the preceding LM IB period.

Most trenches at the habitation site yielded some animal bones, which serve as evidence for the diet of the inhabitants, as many of them showed signs of butchering. Although bones of sheep and goats were most common, examples of cattle, hares, weasels, pigs, a shrew, a bird, and an agrimi or

Figure 17.12. Sealstone (X 208) from Room 2. Scale 1:3

Figure 17.13. Examples of ground stone tools from the habitation site

deer were also found. Strata in Room A (LM IB), Room 5/7 (LM IIIA), and Room 11 (LM IIIA) yielded the greatest concentrations of bones on the site.

Of the numerous species of marine shells recovered from the site, limpets *(Patella)* and top shells *(Monodonta)* occurred most frequently. Several whole or nearly complete triton shells *(Charonia)* were also found in LM IIIA contexts. A number of fragmentary examples came from the site.

One of the most interesting deposits was discovered in Room 11. This deposit consisted of two pairs of long goat horns still attached to parts of the skulls of the animals, a complete triton shell, a handle from a ladle of Mycenaean shape, a broken storage jar, a conical cup, and a small ground stone hand tool (Fig. 17.14).

The LM IIIA complex at Chrysokamino presents a series of connected rooms and spaces with built features. The complex was apparently constructed with a concern for defensible architecture. The nature of the architectural features as well as the finds of ceramics, metal implements, stone tools, loomweights, faunal and botanical remains, etc., suggest that the complex functioned, in part, as a residence for individuals engaged in agricultural pursuits, animal husbandry, and small-scale, related industries. These inhabitants, however, also either manufactured or had access to some finer objects such as the bronze jug and the sealstone, and they engaged in certain cult activity, as indicated by the ritual deposit in Room 11. The remains of the final architectural phase at the site attest to a relatively self-sufficient, outlying community that functioned in a variety of capacities for its Bronze Age inhabitants.

Figure 17.14. The ritual deposit in Room 11, in situ, showing the goat horns, triton shell, stone tool, ladle handle, and other ceramic fragments

Edith Hall's Excavations in the Theriospelio Cave

by Philip P. Betancourt and Cheryl R. Floyd

In the spring of 1910, Edith Hall conducted a short excavation in the Theriospelio cave. The results of the excavation were never published. This report is based on a study of pottery from the excavation and records preserved in the archives of the University of Pennsylvania Museum of Archaeology and Anthropology.

The cave is a natural cavern in limestone. It contains many stalactites and stalagmites. It is in the region called Therio, on a steep and almost bare hillside. In direct distance, the cave is 230 m from the metallurgy site and 610 m from the habitation site. It has a small entrance (about 1.15 m high), but it opens approximately 10 m into the earth onto a large room and several smaller ones. From the entrance to the back, the cave is about 60 m in length. Traces of the excavations of Edith Hall or others, including disturbed soil, pieces of coarse, dark-surfaced, burnished pottery, and human bones, are visible on the surface inside the cave.

Few details are known about Edith Hall's work in the cave. She and Richard Seager were excavating at Sphoungaras near Gournia in 1910,[1] and they visited the cave on a Saturday when their workmen had returned home for the weekend. They looked at the cavern on April 23, and Hall returned on April 25 to excavate with six men for two days. Only a few pieces of ceramics are preserved from the excavation.

Most of what is known about the excavations is contained in two letters written by Hall to her parents. The relevant passages are reproduced here. The first letter, written on Sunday, April 24, mentions that the party of visitors had stopped the day before on the way home from a trip by sea to the island of Pseira. They landed on the coast, where they examined

> the place where they smelted their bronze in Minoan times and where there is a big cave like those I dug in Western Crete. I am to go there again tomorrow, with Nikolaos [the foreman from the excavations at Sphoungaras] and six men to clear this cave. It will probably take two days and after that comes Easter holiday when the men won't work.

The second letter was written on May 1st. It provided additional information about the excavation:

1. Hall 1912. Letters of Edith Hall are preserved in the archives of the University of Pennsylvania Museum.

I can't remember whether I had begun to dig the cave when I last wrote you. I was there only two days, Monday and Tuesday of this week. According to the first plan I was to sleep there in a tent guarded by my six workmen but later we decided it was poor ground—a steep, rocky, mountain-side to pitch a tent and that I had better ride back and forth on horseback. That meant three hours a day in the saddle and over [as] steep ground as I had ever taken a horse before but no mishaps. These Cretan ponies are marvelously sure-footed. The cave had a very small opening so that we had to enter on all fours, but once inside it was fairly lofty, supported by splendid stalactite columns. We took in hanging lamps with strong reflectors and suspended them on ropes strung from stalactite to stalactite. The men had candles too. It was a weird sight to see the men tending to their work in dim dark corners with a smoking candle their only light. We got out very good pottery. I hope we may have time for more caves later on.

Sherds from the cave are in two collections in the United States. Six sherds are in the Mount Holyoke College Art Museum, in South Hadley, Massachusetts, and two additional pieces are in the University of Pennsylvania Museum of Archaeology and Anthropology, in Philadelphia. Edith Hall (later Edith Hall Dohan) taught at both institutions during her career, and she arranged for an official gift of sherds from the Candia Museum (predecessor to the Archaeological Museum in Herakleion) to be exported for teaching purposes. This pottery is divided between the two institutions. Sherds in Mount Holyoke were catalogued by Foster.[2] The ceramic pieces in Philadelphia have also been previously described.[3]

The site and its excavation have been mentioned in print several times.[4] Mosso discussed the cavern in 1910 and noted that both he and Hazzidakis had visited the site before Hall excavated there. Hazzidakis collected pottery, including what he called "primitive" sherds (presumably FN or EM I), EM II pieces, and one fragment described by Mosso as "MM III."[5] Haggis is surely correct in suggesting that the "MM III" sherd with white paint on a dark ground is more likely to be EM III.[6] The site has also been briefly mentioned in a few other publications.[7] Zois visited the cave in 1990 and collected several sherds.[8]

THE POTTERY

The pottery in the American collections belongs to three classes.

Coarse Fabric with Burnished, Dark Surface

Two sherds of vessels made in a coarse, dark fabric are in the Mt. Holyoke collection (**H 1, H 2**). One piece is from a bowl with a vertically pierced lug on the exterior, and the other is from a bowl or chalice. The fabric is coarse and gritty. It is heavily burnished and fired in a reducing atmosphere to create a dark brown to black fabric with a dark brown, unevenly colored surface. The fabric is not like the FN–EM I fabrics from Mochlos and

2. Foster 1978.
3. Betancourt 1983, p. 14.
4. The most complete discussion is Haggis 1992, pp. 170–173.
5. Mosso 1910, p. 290.
6. Haggis 1992, p. 171.
7. Becker and Betancourt 1997, p. 114; Haggis 1996b, pp. 380–381, fig. 4.
8. Zois 1993, p. 340.

Pseira, which have phyllite inclusions,[9] but it is visually similar to a fabric from Sphoungaras.[10]

The chronology for this class of coarse, burnished pottery is still uncertain. Similar vessel shapes (not always made in identical fabrics) have been recognized from Kavousi,[11] Alykomouri/Hagios Antonios,[12] Sphoungaras,[13] Pseira,[14] Vasiliki,[15] Mochlos,[16] and several other places.[17] No clear stratigraphic sequence has been found, and chronological synchronisms are based on comparisons between the shapes, surface treatments, and manufacturing and firing technology with Final Neolithic pottery from elsewhere in Crete.[18] The class may persist into EM I, but it does not appear in the late EM I assemblage from Hagia Photia.[19]

Gray Fabric with Pattern Burnishing

Four sherds of pattern-burnished gray pottery (Pyrgos Ware) from the cave are in the American collections. Two pieces are in South Hadley, and two pieces are in Philadelphia. Three pieces are from chalices, and one piece (**H 3**) is either from a bowl or a chalice.[20] This class of pottery, called Pyrgos Ware,[21] has a medium fine, hard fabric with few visible inclusions. The surfaces of the vessels are rubbed with implements when in a leather-hard stage to align the clay particles, causing a shiny, burnished surface. This class sometimes has burnished patterns achieved by rubbing over specific areas. The vessels are fired in a reducing atmosphere, which accounts for their usual gray color.

This ware is particularly characteristic of EM I, although it may persist a little longer. Pattern burnished pottery is widespread in Crete.[22] Analyses would be needed to determine the relation of this pottery (if any) to the later "Fine Gray Ware" of EM I–EM IIA as defined by Wilson and Day.[23]

Several authors have discussed the chalice. The general development was set out by Hood many years ago,[24] and more detailed analyses have been made by others.[25] Based on the fabric and the shapes, the examples here can be assigned to EM I (**H 3–H 5**) and EM I or EM IIA (**H 6**).

Vasiliki Ware

Two pieces of Vasiliki Ware are in the assemblage (**H 7, H 8**). They are made of the fabric typically used for the finest versions of this ware, consisting of a medium fine fabric in the gray to pink color range with a slipped surface

---

9. Banou 1995b, p. 109; 1998, p. 15.
10. Foster 1978, no. 2.
11. Haggis 1992, locus 17/92; 1996b, pp. 389–393.
12. Haggis 1992, locus 58; 1996b, pp. 389–393.
13. Hall 1912, pp. 46–48; Betancourt 1983, p. 46, nos. 110–113.
14. Banou 1995b, p. 109; 1998, p. 15.
15. Seager 1904–1905, p. 212.

16. Seager 1909, p. 279; 1912, opposite p. 82, nos. 29–42, fig. 48; for the FN date, see Vagnetti and Belli 1978, p. 137.
17. Betancourt 1999; Branigan 1999; Hayden 2003a; 2003b; 2004, p. 42.
18. Vagnetti 1973; Vagnetti and Belli 1978; Vagnetti, Christopoulou, and Tzedakis 1989; Hood 1990a; Manteli 1992.

19. Davaras 1971.
20. For the shape, see Haggis 1997.
21. Branigan 1970a, p. 21; 1970b, p. 18; Betancourt 1985, pp. 26–29.
22. See the list in Betancourt 1985, pp. 26–27.
23. Wilson and Day 1994, pp. 4–22.
24. Hood 1971, p. 38, fig. 14.
25. Wilson 1985; Momigliano 1990; Hood 1990a; Haggis 1997.

that was intentionally mottled during the firing process. The sherds are in the collection of the Mount Holyoke College Art Museum.

Vasiliki Ware is the definitive ware for EM IIB.[26] It is first used in this period, and it does not occur in deposits from East Cretan EM III–MM IA. Recent research suggests that this ware was produced by a limited number of East Cretan workshops.[27]

CATALOGUE

H 1 (Mount Holyoke College BAI.3d)   Fig. 18.1

Bowl with vertically pierced lug.

Max. dim. 4.8; Diam. of vessel ca. 14; thick (0.8) wall.

A gritty fabric; core, dark gray (10YR 4/1); exterior surface, reddish brown (2.5YR 4/4); interior surface, very dark gray (10YR 3/1). Slightly convex profile; small lug pierced vertically. Burnished on exterior and interior.

Comments: The fabric is similar to an early fabric from the Mirabello Bay area found at Gournia and Sphoungaras. Marked "Chryso Kamino" in Greek, in pencil. FN (to EM I?).

Bibliography: Foster 1978, no. 5 (incorrectly published as from Knossos).

H 2 (Mount Holyoke College BAI.3g)   Fig. 18.1

Chalice (or bowl?), rim sherd.

Max. dim. 7.9; Diam. of rim ca. 15; thin (ca. 0.5) wall.

A gritty fabric, reddish brown (5YR 4/5); gray areas on exterior surface. Convex profile; straight, rounded rim. Burnished on exterior and interior.

Comments: The fabric is like **H 1**. Marked "Chryso Kamino" in Greek, in pencil. FN (to EM I?). This vessel is more likely to be a bowl than a chalice. Compare Xanthoudides 1918, p. 153, no. 84, fig. 10. If it is a chalice, it is more globular than usual, as is the case with an example from Mochlos in the Museum of Fine Arts in Boston (09.655), a piece from Kanli Kastelli (Alexiou 1951, no. 5, pl. 14, fig. 1), and one from Pyrgos (Xanthoudides 1918, p. 153, no. 76, fig. 10).

Bibliography: Foster 1978, no. 12.

H 3 (University of Pennsylvania Museum 42-34-2-A)   Fig. 18.1

Bowl or chalice, rim sherd.

Diam. of rim ca. 15.5.

A fine gray fabric (gray, 10YR 6/1); heavily burnished. Straight rim. Burnished vertical lines forming panels.

Comments: Pyrgos Ware. The date is late in EM I (based on the presence of the shape at Hagia Photia Siteias, for which see Davaras [1981], fig. 2; 1977, pl. 603; and 1989b, p. 266, pl. 149).

Parallels: For vertical burnished lines, see Haggis 1997, pl. 92, fig. c. For the shape of the rim, see Xanthoudides 1918, p. 150, nos. 42–45, especially 42 and 45, fig. 8; p. 151, nos. 56–59, especially 58, fig. 9; p. 153, nos. 75 and 77, fig. 10, republished by Haggis 1997, pl. 91, fig. 5:a, no. 3, fig. c, lower row nos. 2 and 4, all from the Pyrgos cave; Davaras 1977, pl. 603, from Hagia Photia.

Bibliography: Betancourt 1983, no. 13.

H 4 (University of Pennsylvania Museum 42-34-2-B)   Fig. 18.1

Chalice, rim sherd.

Diam. of rim ca. 15.5.

26. Branigan 1970a, pp. 29–30; Warren 1972, pp. 94, 108–109; Betancourt 1979; 1985, pp. 43, 45–48.

27. Whitelaw et al. 1997; Day, Wilson, and Kiriatzi 1997.

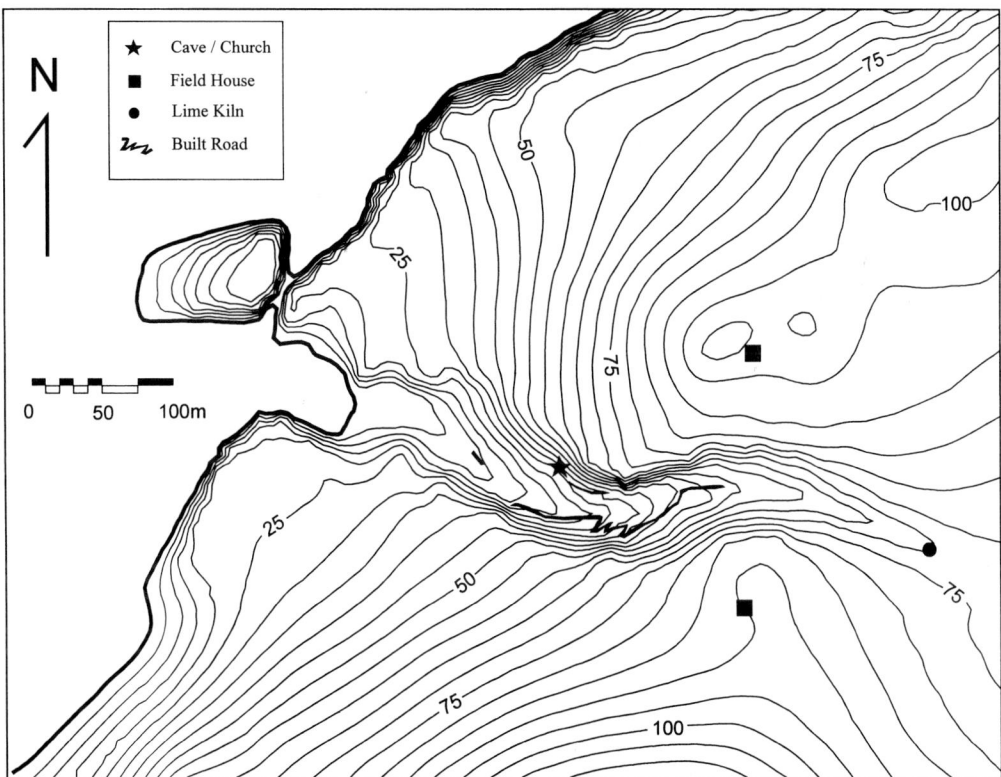

Figure 20.1. Ravine and Agriomandra harbor

Chomatas. The ravine and the bare cliffs and peaks to the north of it form topographic barriers. North and northeast of these physiographic barriers, the closest human outpost on the cliffs is a small walled settlement, referred to here as Pyrgos Chrysokaminou (Fig. 1.5). Its earliest date of use is not known, but it may have served as a lookout post or a refuge site.

To the west and north, the sea forms the border of the territory. Except for the metallurgy location and the burial cave, no Minoan archaeological sites have been found in this direction between the habitation location and the sea. At the southwest, the ravine leading to the harbor of Agriomandra is also a natural border (Fig. 20.1).

No FN–EM sites are located nearby to the south, suggesting the boundary could have extended into the poorer land of the Kambos and to Lakkos Ambeliou or beyond. The bare hill of Chalepa, southwest of Lakkos Ambeliou, cannot have supported agriculture and is a natural barrier itself. It is possible (even likely), of course, that FN and EM sherds will eventually be recognized at additional locations near Chrysokamino, refining our picture of its territorial scope.

In Chapter 19, Haggis suggests that the settlement pattern at the southwestern spur of Chomatas Hill changed markedly in EM III–MM I. Several new sites were founded near the Chrysokamino habitation site, resulting in a cluster of nearby sites. He makes a strong case that these sites were probably related, and that they could have shared resources and managed nearby locations such as the metallurgy site and the harbor at Agriomandra. Their strategic location at the head of the ravine leading to the tiny harbor suggests they monitored traffic between the tiny port and places farther inland. Nevertheless, the presence of a Minoan site at the harbor itself suggests the situation could also be more complex. If the cluster functioned as a unit, as Haggis convincingly suggests, the northern and western territorial limits were probably similar to those of the FN–EM IIB settlement. An expanded population is surely indicated, however, and the southern and southeastern boundaries could have extended somewhat farther.

To the east, Middle Minoan sites existed at Katsoprinos and Kephalolimnos. Because these MM neighbors must have needed farmland of their own, a boundary of some type must have existed between the Chrysokamino territory and the land controlled by these other sites. Although precise boundaries are not known, the proximity of the other settlements suggests an area for the cluster of sites at Chrysokamino and its vicinity that extended across the upper terraced part of the southwest and southeast slope of Chomatas Hill, including part of Chordakia. The territory probably ended near the major ravine north of the modern gate. The outpost to the north (Pyrgos Chrysokaminou) was either an outlying extension or (more likely) the next neighbor to the north.

In MM III–LM I, Haggis suggests that the cluster of sites went out of existence, and a situation arose in which one site dominated the landscape. That site was the Chrysokamino habitation location. This is the first period with excavated architectural remains at the site. One house is partly preserved.

Other communities were nearby. The most important site near here in later times, Kephalolimnos, was already inhabited. A Bronze Age site was still present at Katsoprinos, east of the Chrysokamino habitation location, and Alykomouri lay farther to the east. These LM I settlements must have required their own land, so that the territory of the Chrysokamino farmstead cannot have included all of Chordakia at this period.

For LM III, the boundaries were probably as shown in Figure 20.2, with additional land used for pastures. The northern and western boundaries, established by the geomorphology, are not likely to have changed much. Both Kanta[9] and Haggis[10] have reported LM III pottery from Kephalolimnos, so the residents at Chrysokamino still had the same site as a neighbor to the southeast, suggesting the continuation of a boundary somewhere within the region of Chordakia, on the slopes of Chomatas Hill.

The southern border, near Lakkos Ambeliou, remains a question mark. When some of the tiny MM sites south of the habitation site disappeared, the Chrysokamino estate may have expanded slightly to include more area near Lakkos Ambeliou, but it would probably not have controlled more farmland than the cluster had held jointly. The expansion in territory, absorbing what the cluster had controlled, would be a natural outgrowth of the increase in the size of the architectural complex at the main remaining habitation site, but only if Kephalolimnos did not also absorb population. In the diagram in Figure 20.2, the border is drawn at the bare hillside south of **AF 32** (an oval enclosure), a region with no sherds suggesting LM I–III activity (i.e., it may be a border).[11] Farther south, at Lakkos Ambeliou, LM III and later sherds suggest that activity continued into the 1st millen-

9. Kanta 1980, pp. 144–145.
10. Haggis 2005, pp. 116–117.
11. For the situation and the pottery, see Apps. G, H.

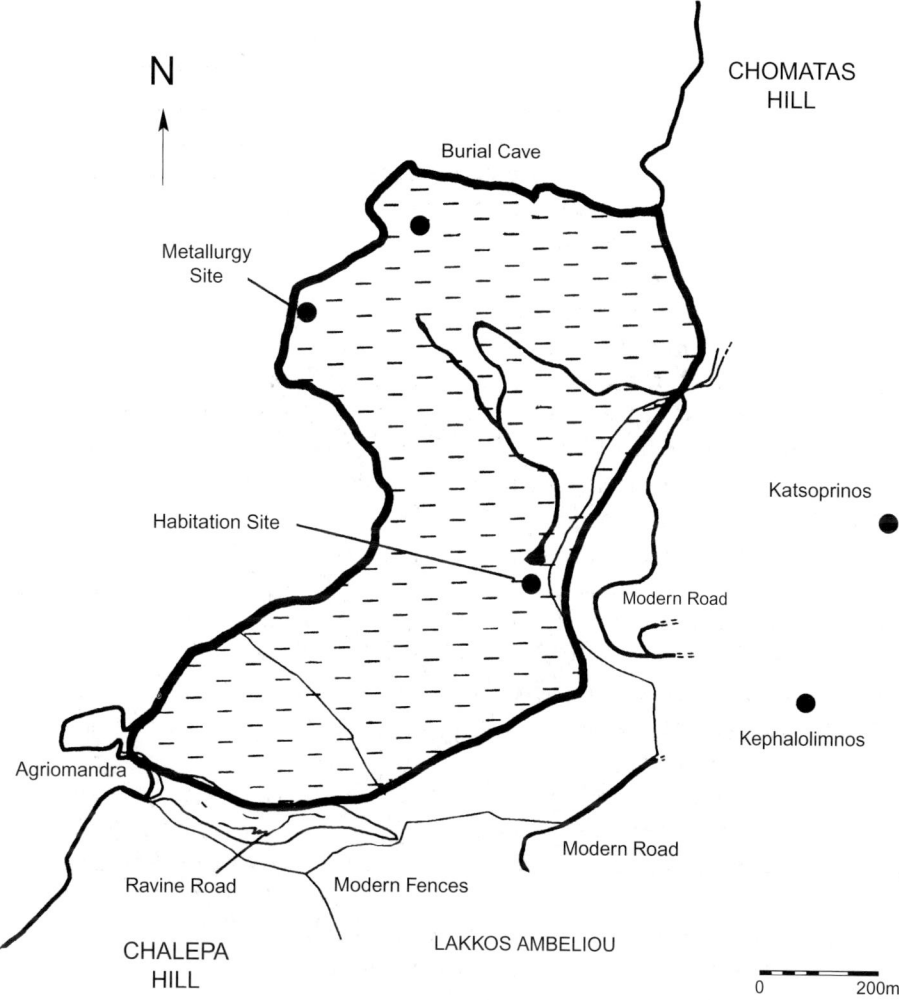

Figure 20.2. Boundaries for the farmstead in LM III

nium B.C., when Chrysokamino's habitation site was not inhabited. This region, therefore, is not included in the LM territory.

By LM I, the estate probably had an eastern border somewhere in Chordakia. The primary territory extended northward to the ravine north of the modern gate, and it probably included the region west to the sea and south to somewhere near Lakkos Ambeliou, where the boundary may have varied somewhat from period to period. This broad region had a shifting border at the south and east, but a stable one at the north, west, and southwest. It always included the locations of the metallurgy workshop and the cave, but both were abandoned before the Late Bronze Age.

Evidence from the excavations contributes additional information. Besides the growing of crops, the residents pursued animal husbandry. Especially in LM I and LM III, the evidence suggests they consumed substantial amounts of meat, which surely means flocks and herds were kept. Access to the bare hillsides south and north of the main farming area would have been essential, whether the pasturelands were shared or owned.[12]

Trade goods were also a part of the archaeological record, suggesting access to a harbor. Many imports were almost certainly brought in by

12. For the practice of herding rather than farming on these hills in modern times, see Haggis 1992, p. 75.

sea through Agriomandra, rather than overland. Because of its regional importance, it is unlikely that any single community ever had exclusive use of this type of harbor; as with grazing land, one must consider the possibility of a shared resource.

In summary, the evidence for the Minoan period suggests a territory like that shown in Figure 20.2. Only one long-lived habitation location has been recognized within this area. It is found on the most stable bedrock outcrop in the region (where most of the land is soft phyllite and its semiconsolidated alteration products). The habitation site is surrounded by a primary territory set off by natural boundaries at the northeast, by the sea at the north and west, and by the proximity of other Minoan sites at the east and south. This primary territory was probably the main farming region controlled by the estate.

Beyond the primary territory are secondary territories that might have been shared. They include a ravine that terminates at a harbor with a small beach, as well as bare hills more suitable for grazing than for farming located both north and south of the ravine. Bare hills are also found north of the territory, where they could have been shared with Kephalolimnos, which would also have surely required pastures. Chrysokamino's access to a primary territory for agriculture and a secondary territory for grazing, along with access to the sea, helps to explain how a small, semi-independent unit was able to subsist for more than a millennium and a half in the face of fundamental changes occurring both locally and throughout the island. A stable relationship between land use and population must have been established early in the region's history, based on constant factors such as the nature of the soils, the topography, crop management strategies that did not deplete the soil, and simple animal husbandry. Metallurgy contributed to the domestic economy until EM III–MM IA, but in later periods, stable factors such as gardening, farming, and herding were more significant in the long run.

CHAPTER 21

LAND USE ON THE CHRYSOKAMINO FARMSTEAD

by Philip P. Betancourt

Agricultural communities were the main economic force in the Classical era,[1] and they would have been at least as important in earlier periods. In spite of commitments to trade and craftwork, Minoan society must have been firmly based upon an agricultural economy. Especially at small sites located away from the large urban centers, subsistence activities would have been focused mainly on growing crops and tending animals.

For the territory considered here, strong evidence indicates an agrarian base. In the first place, accessibility to farmland seems to have been an important factor in the choice of settlement location. The habitation site was neither situated on the coast at the harbor of Agriomandra, which would have accommodated seafaring, nor farther inland where land traffic would have been easier. No major natural resources aside from the land itself are present in the vicinity of the site. It was surely the arable land that attracted the residents. Other evidence for agricultural activities comes from the present excavation. This evidence includes the presence of querns for grinding grain, substantial numbers of animal bones from domestic animals (and very few wild species), pottery evidence for long-term storage of commodities, and the identification of chaff, barley, wheat, and olive in the metallurgical furnace chimney fragments. No doubt exists that the residents here engaged in farming and animal husbandry.

The interaction between human occupation in this part of Crete and the slowly changing conditions of climate, natural resources, and other aspects of the physical environment are crucial factors in the history of the Bronze Age community at Chrysokamino. The natural setting of the Chrysokamino territory is formed by specific types of physical features that can be arranged into separate activity zones. Because these zones differ markedly from one another, they would have afforded different potential economic contributions to the community as a whole. The dynamic ways in which their potentials were used or neglected had an important impact on the history of the region.

The Chrysokamino location provides the setting for a specific way of life based on farming and the keeping of domestic livestock. The pattern of cultural activity must have changed through time, but the need for farmland

1. Barker 1985; Halstead 1987; Wells 1992.

would have been constant. In all periods, it would have been necessary to manage the land with a strategy that was appropriate for the local soils, climate, and crops.

In the methodology developed for this project, the landscape was first divided into different classes. The information used for this analysis came from the intensive surface survey, and the categories of land were distinguished by considering the following three types of data and the relationships between them:

1. Available soils and their characteristics, especially their potential to support agriculture[2]
2. The presence or absence of a substantial scatter of Minoan sherds on the surface[3]
3. The nature of the topography (steep or flat; rocky or covered with sediment; with or without a substantial soil cover)

Two types of soils are present on the land that makes up the Chrysokamino territory. A red soil called terra rossa occurs over carbonate bedrock, and a pale-colored phyllitic soil is present over phyllite. The soils are very clearly distinguished by color. Because they also have very different characteristics, they cannot be farmed in the same way. Any successful agricultural system would have required an understanding of the differences between the two soils and the development of different strategies for their management. The fact that the Chrysokamino territory was farmed for approximately 2,000 years is ample proof that the residents clearly understood their land's capabilities and were able to manage their soil resources successfully.

The presence or absence of a substantial scatter of Minoan and later sherds on the surface can be easily documented by intensive surface survey (see Apps. G, H, I, and J). In the Chrysokamino territory, some parts of the landscape are almost empty of sherds, with only an occasional piece spaced every 25–50 m or more apart. Other portions of the land have many ceramic fragments, with more than 10 pieces in a 10 m^2 area. This circumstance obviously results from a difference in ancient activity, and it is, therefore, a significant factor in defining classes of land use.

The topography around Chrysokamino is extremely varied. It includes areas of steep cliffs, gently rolling hills, a deep ravine, and bare hillsides with almost no soil cover. These different classes of land do not offer the same potential for agricultural use, and they must have been exploited in different ways.

Four classes of land have been recognized based on the three criteria of soil type, presence or absence of substantial numbers of ceramic sherds, and type of terrain. The first three classes are on flat or gently sloping land suitable for farming. Type 1 consists of land with terra rossa soil and a substantial scatter of Minoan sherds on the surface. Type 2 consists of terra rossa soil without the substantial scatter of Minoan sherds. Type 3 consists of phyllitic soil, which does not have a substantial scatter of Minoan sherds on the surface at Chrysokamino. The fourth type consists of land that is too steep for either agriculture or the keeping of livestock. These four classifications, defined initially by simple observations of the geological and

2. For discussions of the importance of this factor, see Bintliff 1977, pp. 99–100; Carothers 1992, pp. 48–49.

3. On the importance of this factor for the presence of subsurface features and for its relation to past human activity, see Cherry et al. 1988, p. 170; for its use in the reconstruction of agricultural practices, see Wilkinson 1982; Bintliff and Snodgrass 1988, pp. 507–508; Betancourt and Hope Simpson 1992 (with substantial additional bibliography).

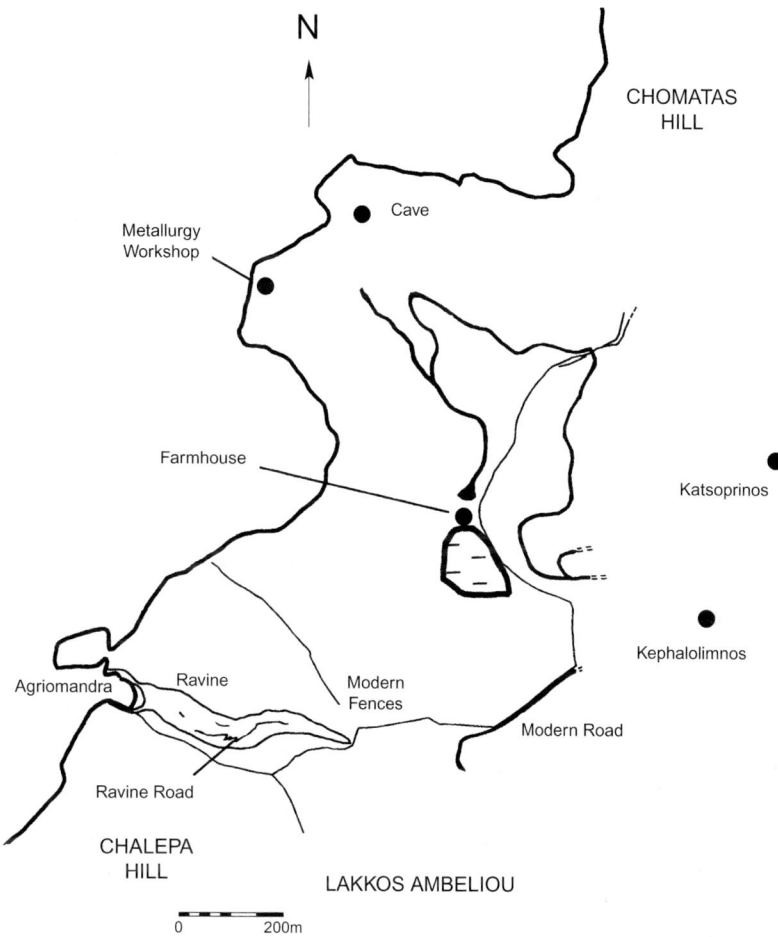

Figure 21.1. Land of Type 1, used for horticulture (terra rossa with substantial scatter of Minoan sherds)

archaeological data, have many additional characteristics that distinguish them from one another as well. The additional characteristics that are unique to each class help confirm the validity of the organizational system as a whole. A recognition of the differences between the categories leads to some significant conclusions on the organization and management of land at the Chrysokamino farmstead.

FARMLAND, TYPE 1: GARDENS

In the classification used in this study, land of Type 1 consists of terra rossa soil with a substantial scatter of Minoan sherds on its surface. The land of Type 1 is fairly restricted in extent (Fig. 21.1). A sloping hillside extends from the habitation site, which has an elevation of 120 masl, to the oval structure called **AF 32** (App. G), which has an elevation of 106 masl. South of **AF 32**, the land rises slightly to a ridgelike hill with an elevation of 110 masl and then descends in a long, relatively evenly sloping hillside all the way to Lakkos Ambeliou (elevation 75 masl). The land of Type 1 is an area south and southeast of the habitation site, extending downhill from the architecture to just over the ridge south of **AF 32**. The roughly circular deposit of sherds extends across an area with a width of ca. 100 m and a length of ca. 110 m. Within this area, the sherds are distributed fairly evenly, occurring as small, worn, pottery fragments on the surface of the ground. The area is too large for the sherds to have been dumped here and later scattered. The soil (red terra rossa) is different from the pale brown soil that covers the settlement, so the material cannot have simply eroded down the hill from the habitation site, and the ridge with a height of 110 masl demonstrates the same conclusion because erosion cannot flow uphill. The even distribution across the land shows that the sherds did not spread out from any single location. The long hillside south of **AF 32** has the same terra rossa soil, but it has no sherds on it until it reaches the vicinity of Lakkos Ambeliou. It is not a part of this class of landscape.

Another thick scatter of sherds occurs at Lakkos Ambeliou. This feature is a pedon (a deposit of sediment) in a sinkhole measuring ca. 280 m north–south by 410 m east–west. Morris studied this deposit in detail.[4] He concluded that it developed as an alluvial deposit from the terra rossa on the surrounding hills, possibly accumulating in a lake during the last glacial period. The surface of the pedon was dry by the Middle Minoan period, when sherds were deposited on it.

Many Minoan sherds occur both within the soil and on the surface of the land of Type 1 between the habitation site and across the ridge south of the farmhouse. Most of them are from the Late Minoan III period, but Middle Minoan and Late Minoan I pieces are present as well.[5] The earliest sherd recognized from the area is from the Final Neolithic. Several sherds are catalogued in the present study as typical examples of the material found here (see **AF 32** in App. G). They are all similar in their physical preservation, in that they are small (mostly under 5 cm in size), worn, and poorly preserved. Little paint survives on them, and they do not join with other pieces to make larger fragments or parts of vessels. Most of their edges are rounded.

The character of the pottery differs from the better-preserved sherds found at the settlement, but it is identical to sherds found away from settlement sites. In comparison with the sherds from the Chrysokamino habitation site, the sherds from Site **AF 32** and its vicinity are smaller and more worn, although they come from the same Minoan periods. In terms of quantity, the number of sherds per square meter was much lower at Site **AF 32** than at the settlement before excavation. Isolated sherds of this type have sometimes been called "offsite scatter."[6] They are commonly found in regional surveys in eastern Crete and elsewhere.

The small, worn sherds also occur within the soil. Morris found Minoan sherds buried to a depth of 110 cm at nearby Lakkos Ambeliou. He demonstrated that the buried sherds were not on an ancient, buried ground surface, and he suggested that they might have moved downward through vertical cracks in the ground.[7] This conclusion may be correct for some of the sherds in the sinkhole he studied, but it is not likely to be correct for all of them because he also identified a buried tillage zone extending from the surface to ca. 30 cm below the surface.

4. Morris 2002, pp. 45–59.
5. For their previous recognition, see Haggis 2005, p. 119, locus 53.
6. Bintliff and Snodgrass 1988; Cherry, Davis, and Mantzourani 1991, chap. 3.
7. Morris 2002, pp. 57–59.

Considering that such vertical movement usually takes place in montmorillonite clays, not in kaolinite clays such as those of this sinkhole, some of these sherds may have been deposited through the practice of manuring. In any case, the sinkhole is not within the region of **AF 32** where the presence of a larger number of Minoan sherds must be explained in terms of Bronze Age activity.

Adding manure to land used for agriculture was a worldwide practice for agriculturalists before the introduction of commercial fertilizers.[8] The value of manuring has been studied in many regions, from Great Britain,[9] to continental Europe,[10] to North America,[11] to Asia.[12] The practice improves the land in several ways. The working of manure and other refuse into the soil renews nutrients, adds organic matter to help retain moisture, aerates the soil, distributes nutrients more uniformly, and breaks up clumps of denser soil to improve root penetration. The practice has often been regarded as a normal part of gardening and farming.

The manuring of fields has also been previously documented archaeologically.[13] It must have been widespread in the ancient Mediterranean because Greek and Roman writers described it as a routine practice.[14] In his manual on farming, for example, Cato regarded it as an essential part of agriculture (*de Agri Cultura.* 5.8, 29.1).

Given its broad distribution elsewhere, it would be surprising if manuring was not also practiced in Bronze Age Crete, and we now have firm proof of its use during the Minoan period on Pseira. The excavation of two agricultural fields on Pseira Island discovered sherds both on the surface of the fields and deeply buried within the soil.[15] The Pseiran field at location G 2 is particularly relevant to the situation at Chrysokamino. The two regions are in the same part of Crete, the fields have contemporary Middle Minoan to Late Minoan deposits, and the sites are both farming locations using poor quality, marginal land composed of terra rossa soil. Sherds in the Pseiran agricultural field, which was located well away from any settlement, were found within the soil all the way from the surface to bedrock. They were small and worn like their counterparts from **AF 32** at Chrysokamino. Small amounts of other refuse occurring with them, including bones, sea shells, scraps of obsidian, broken loomweights, and other items, helped indicate that the material on this and other Pseiran terraces was settlement debris. Scientific analysis by a team of scientists at the University of Bristol has now confirmed that chemical traces of manure are present in the soil from context G 2 on Pseira. There can be no doubt, therefore, about the formation of the deposit.[16]

As on Pseira, the terra rossa at the location near the Chrysokamino habitation site is rather poor terra rossa soil. It has few nutrients, and it has an even greater problem in its inability to retain much water. If this soil was used for horticulture or agriculture over a long period of time, a fertilizer would have helped restore nutrients, and it would have added organic matter for better moisture retention.

Analysis of the soils contributes additional evidence that the terra rossa in this region was used for agrarian purposes. Human influence on soils can be determined by identifying the presence of specific residual chemical

8. Lyon and Buckman 1929, pp. 381–405; Wilkinson 1982; Courty, Goldberg, and Macphail 1989, pp. 133–134.

9. Pitty 1978, pp. 132, 135.

10. Barker 1985, p. 163.

11. Wines 1985, chap. 1.

12. Tamhane et al. 1964, pp. 264–269; Walls 1982, p. 26.

13. Wilkinson 1982, 1989; Bintliff and Snodgrass 1988, pp. 507–508; Bintliff, Howard, and Snodgrass 1999; Bintliff et al. 2002, pp. 260–261.

14. Tilly 1973, p. 73; Garnsey 1992, p. 151.

15. Betancourt and Hope Simpson 1992.

16. Bull, Evershed, and Betancourt 1999.

elements, especially phosphorus.[17] Although the natural surface horizon of soils in this region averages only 15 cm in depth,[18] Morris demonstrated that the red soils from Lakkos Ambeliou have a 20–30 cm deep surface zone containing phosphorus and other chemical relics of agricultural activity.[19] This horizon may be identified as a tillage zone, and the presence of substantial numbers of Minoan sherds shows that this part of the land, not far from **AF 32**, was used in the Bronze Age.

The variation in sherd density is the most important factor in the interpretation of the difference between the Chrysokamino farmland of Type 1 and Types 2–3. At Chrysokamino, the extensive sherd scatter of Minoan pottery exists only on a restricted part of the terra rossa soil. An occasional sherd occurs on the other two types of land in the Chrysokamino territory, but the contrast in the amount of pottery is sufficient to indicate a pronounced difference in land use. The explanation cannot be found in the type of soil because some of the terra rossa (i.e., Type 2 farmland) does not have the extensive sherd scatter.

Other characteristics of the land of Type 1 further help to separate it, albeit less dramatically, from the other two types of farmland at Chrysokamino. All of the Type 1 land is close to the settlement. In fact, it adjoins the habitation site, with no physical barrier in between, while parts of the other two farmland types are up to a kilometer away. It is restricted in size, in contrast with the larger territory covered by the other two types of land. It is close to a deep natural cleft in the bedrock, which would have been a source of water in rainy periods. It has no terraces because it is not very steep.

The evidence at Chrysokamino clearly indicates that local land use was different for various parts of the landscape. This conclusion fits well with a growing body of archaeological evidence for zonation in land use in ancient and more recent periods in Greece.[20] The land available for agriculture was apparently often divided between crops, even though we usually have no clear picture of which crops were on which plots. Variations in agrarian practices have been recognized based on topographic situations[21] and on the suitability of land for grazing and other uses.[22] That selected crops are grown under conditions of intense care (hoeing, weeding, watering, and manuring) can also be recognized from the presence of certain weeds.[23]

The use of different classes of land at Chrysokamino is best understood in relation to research on ancient strategies for raising various types of crops. An excellent summary of the subject is presented by Leach, who notes that a large body of evidence (linguistic, topographic, archaeological, literary, and ethnological, among others) indicates a very different ancient attitude toward gardens and fields.[24] Several Roman writers, especially Cato and Varro, discuss these different attitudes.[25] The different strategies arose as a natural consequence of the types of crops grown; some plants (such as pulses and many vegetables) require constant care, while others (such as grains and olive trees) do not.

Strategies for the two classes of crops were very different in Classical and Roman times. Crops in the fields were handled as large units in terms of tilling, planting, and harvesting, while plants in the gardens were tended individually. Gardens were routinely placed near the settlements where they could receive intensive care, while fields could be located farther

17. Sjoberg 1976; Proudfoot 1976; Eidt 1977.

18. Timpson 1992, p. 137.

19. Morris 2002, pp. 52, 59.

20. Wagstaff and Augustson 1982; Hayden 1995, p. 100; Acheson 1997, p. 175; Halstead 2000, p. 111; Atherdon 2000, p. 63.

21. Wagstaff and Augustson 1982, p. 107.

22. Gamble 1982, pp. 163–164; Forbes 1995; Halstead 2000.

23. Jones 1987, 1992; Jones et al. 1999; Bogaard et al. 2000; Jones et al. 2000.

24. Leach 1997.

25. Tilly 1973, p. 64.

away. References in Homer and in Linear B tablets suggest that gardens for vegetables, fruit trees, and grape vines were situated near the settlements.[26] The basic contrast in attitudes is even reflected in the linguistic origins of the words horticulture, associated with the home, and agriculture, derived from a root meaning wilder areas at some distance from the house. The situation typically resulted in a division of labor along gender lines. Women typically cared for the gardens, which required small amounts of labor on a daily basis. Men were more likely to work in the fields, away from the houses. The fields required longer periods of intensive work at certain stages of the growing cycle, as in the times of pruning, planting, and harvesting.

It is this difference in the amount of care needed for some plants that helps explain the differences in the number of sherds on the land around Chrysokamino. It is likely that the Bronze Age residents of this farmstead were growing some crops that required intensive care, with the addition of substantial amounts of manure, while they also raised grains and olives that could be allowed to grow with much less hoeing, weeding, and manuring. Greater use of manure and rubbish as fertilizer in the gardens near the settlement would have resulted in the higher density of sherds recorded in the archaeological record. A combination of poor soil and plants requiring extra nutrients, more moisture, and more frequent care than other crops is the best explanation for the use of farmland of Type 1.

There are additional reasons for the placement of a garden near the home. Many vegetables are routinely harvested in tiny amounts every day or two to ensure freshness, making frequent trips to the garden necessary. Only small amounts of vegetables are stored, in contrast to crops such as grains and olives that are harvested in large quantities all at once and stored until the next harvest. Planting garden plots close to home means that short trips to the garden to water sprouts, remove weeds, pick fresh produce, or add a small amount of refuse as fertilizer can be interspersed with other household chores.

Plants such as olives, barley, and wheat, all of which are attested at Early Minoan Chrysokamino (Chap. 12), as well as elsewhere in EM Crete,[27] do not require constant care. They can be planted in fields away from a settlement. In addition, many of the tasks associated with their management, including pruning trees and harvesting grain, require full days of very hard labor. Such tasks are not well suited to household members who are also cooking or taking care of very small children.

This distinction in land use is clearly reflected at Chrysokamino. Land of Type 1 contrasts with land of Types 2 and 3, conforming very closely to the distinction between gardens and fields. Land of Type 1 adjoins the habitation site and is easily accessible with no natural barriers such as ravines to be crossed. It has many small sherds on its surface, indicating that this land received more attention than the land of Types 2 or 3, a situation very compatible with garden plots.

A number of crops were probably grown in the gardens. A strong case has been made for pulses as a staple in the Minoan and Cycladic diet.[28] If this hypothesis is correct, it confirms the need for horticulture. Pulses would have to have been raised in gardens near the settlements because the

26. See the discussion in Palmer 1999, pp. 478–480.
27. J. Renfrew 1972; Greig and Warren 1974.
28. Sarpaki 1992.

cultivation of these plants is extremely labor intensive.²⁹ Grapes may also have been grown. Viticulture requires extra care in pruning, staking, fertilizing, and other tasks,³⁰ and grapes thrive on well-drained slopes. The land interpreted here as garden plots would have been well suited to this crop.

Although this pattern, with gardens for vegetables as well as dry farming for olives and grains, has not been previously noted in the Minoan archaeological record, it was probably not unusual. Diversity has long been regarded as an important aspect of Bronze Age farming,³¹ and different crops require different strategies. Distinctions between gardens and fields were routinely made in later periods of antiquity in both Greece and Italy. There seems no reason to suggest that Crete was substantially different from other places where the same crops were raised. The ancient Mediterranean people had limited strategies for solving the problems of agricultural production on marginal land. A division between gardens and fields would probably not have been an unusual situation, but rather an ordinary aspect of Minoan land use.

FARMLAND, TYPE 2: GRAZING LANDS

Areas of terra rossa soil without any substantial scatter of Minoan sherds constitute farmland of Type 2. This category of land is farther away from the habitation site than the land of Type 1. It includes low hills where the underlying bedrock is a carbonate as well as the Kambos where the terra rossa has been extensively deposited. This class of farmland is shown in Figure 21.2.

Chalepa, the hill south of the ravine leading to the harbor, is extremely eroded. Many parts of the hillside are composed of boulders and bedrock rather than soil. The hill does not have any terrace walls on it, and it could not be used very successfully for farming in its present, eroded state. Most of the soil has washed into the valley of the Kambos near Lakkos Ambeliou, an event that seems to have taken place in two stages, which are not precisely dated. Research by Morris suggests that much of the first period of erosion probably occurred during the last glacial period, when the sinkhole at Lakkos Ambeliou was a lake.³² The presence of Middle Minoan sherds on the surface here proves the final stage was earlier than this period. Deforestation and the resulting erosion have often been regarded as anthropogenic processes in the Aegean,³³ though there are strong opinions against the universal application of this hypothesis.³⁴ It is tempting to associate the last period of erosion with the Final Neolithic to Early Minoan clearing of the land for agriculture or with the use of the region for copper smelting, because such operations require substantial fuel, and smelting often leads to deforestation and subsequent erosion, but there is no direct evidence for the association in this case except for the date when the two events occurred.

In spite of their bare condition, the hill of Chalepa and other carbonate hills in the vicinity are used today (1990s to early 2000s) as grazing land for sheep and goats. They support sparse vegetation, making them suitable for grazing even in late summer. No cattle are kept in this vicinity.

29. Barker 1985, p. 73.
30. Hanson 1992.
31. A. C. Renfrew 1972, p. 306; Palmer 1999, pp. 478–480.
32. Morris 2002, p. 58.
33. Bottema 1980; Thirgood 1981, pp. 68, 153–155; van Andel, Runnels, and Pope 1986; van Andel, Zangger, and Demitrack 1990; Jameson, Runnels, and van Andel 1994, p. 325.
34. Rackham and Moody 1996, p. 127; Moody 1997, 2000; Forbes 2000b.

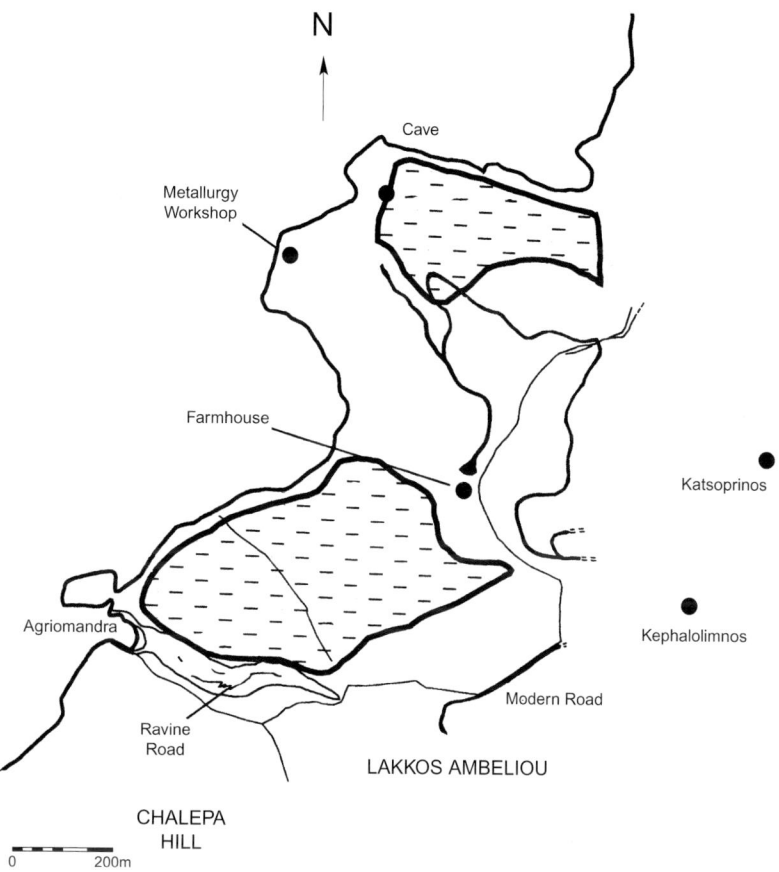

Figure 21.2. Land of Type 2, used for pasture (terra rossa with no surface sherds or with a light scatter of sherds)

Additional hills where the bedrock is a carbonate lie between AF 32 and Lakkos Ambeliou, north of the ravine that leads to Agriomandra, and also north of the main Chrysokamino region. All of these places are mostly bare of vegetation, but they support enough for sheep and goats.

It is suggested here that these hills would have afforded grazing lands for the ancient farmsteads west of modern Kavousi, including Chrysokamino. The Early Bronze Age situation is not clear, and the absence of animal bones from the metallurgy site could suggest either that little meat was in the diet during the Early Bronze Age or that this workshop did not consume much meat while the personnel were working. In other words, the site may have been too specialized to give a true picture of diet in general; the metalworkers may have returned to their homes at night. By the Late Minoan period, however, there is substantial evidence for sheep and/or goats in the LM I–III strata at the habitation site, indicating that meat formed an important part of the diet of the people who lived at the farmhouse in that period. Land of Type 2 would have provided nearby grazing for the animals that they must have managed.

Specialized grazing sites near agricultural communities have also been recognized from later periods of Greece.[35] Animal husbandry, therefore, may often have been more complex than simple transhumance, in which flocks and herds are moved from upland pastures in the summer to lower grazing lands in the winter.[36] It has been suggested that keeping animals locally was very important to small farmsteads, with agricultural sites being

35. Gamble 1982, pp. 163–164; Forbes 1995.
36. Halstead 1981, 1987, 2000.

chosen because of their proximity to grazing lands as well as farmland.[37] This gives us an additional reason to suppose that at least some animals were managed locally at Chrysokamino.

If the situation in the Kambos before the middle of the 20th century A.D. is any indication of ancient times, the low-lying parts of the plain may have suffered from droughts that made them undesirable as farming areas, even though they could have supported flocks and herds, particularly in the summer. This situation would leave the terraced phyllite hills (farmland of Type 3) for raising field crops. A similar pattern of land use, with animals maintained on the upper slopes and cereals and other crops raised on lower hillsides, has also been proposed for other parts of eastern Crete.[38]

FARMLAND, TYPE 3: FIELDS

Evidence is steadily accumulating for olive cultivation in Early Minoan Crete. Blitzer convincingly refutes earlier doubts about its presence,[39] arguing that the olive must have been a significant crop. It was probably also present in the Final Neolithic, although evidence has not yet been discovered for this early phase.

Grains must also have been important. Impressions of chaff, barley grains, a wheat grain, and an olive leaf in the furnace fragments from the metallurgy site show that the Chrysokamino region was already engaged in diversified agriculture at the end of the Early Bronze Age (Chap. 12). Because the impressions in pottery are primarily chaff, it is likely that grain was planted in substantial amounts. Grain and olives may even have been planted on the same farmland, with the gathering of chaff resulting in the casual inclusion of olive leaves.

Olives and grains are crops that can withstand long periods without rain; their cultivation is often called dry farming. They require more land than the small area of farmland of Type 1, and the land of Type 2 is too poor to support them. The large area of Type 3, consisting of the most fertile land in the Chrysokamino territory, must have been the mainstay of dry farming, if for no other reason than that none of the other available land was suitable. The farm could not have been maintained as a semi-independent unit unless the phyllitic soil was exploited.

Land of Type 3 is shown in Figure 21.3. It consists of areas in which the bedrock is phyllite, with the overlying soil being formed by its disintegration. This soil is substantially richer than the terra rossa of the region, and it retains moisture better. The modern residents of the region still prefer it to the red soil for agricultural purposes.[40] These characteristics may explain why it lacks the extensive scatter of sherds associated with manure and other refuse. In this region, an average of one sherd or less is present within an area of 1,000 m². A tiny settlement with limited refuse might prefer to use its manure on the nearby gardens where it was most needed, rather than on the fields where crops grew well.

The area of Type 3 is substantial. The habitation site adjoins land of Type 1, but much of the land of Type 3 lies at some distance from the

37. Forbes 1995.
38. Hayden 1995, p. 100.
39. Blitzer 1993, p. 165; cf. Runnels and Hansen 1986, pp. 305–306.
40. Haggis 1996a, p. 188; 1996b, p. 378.

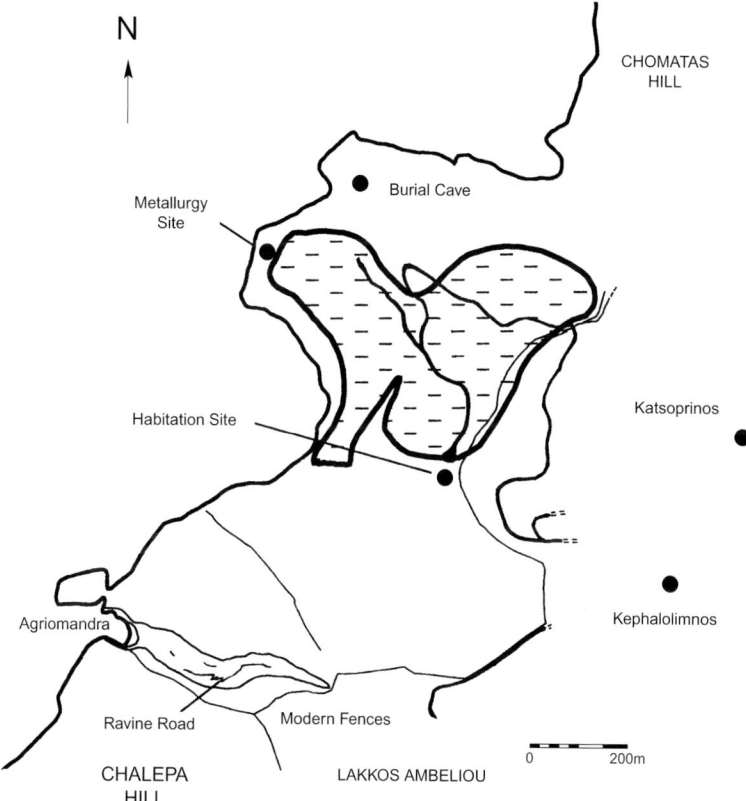

Figure 21.3. Land of Type 3, used for dry farming (phyllitic soil)

settlement. Several areas in this part of the landscape have well-preserved terrace walls. They show that much more of this general region was farmed in the past than today.[41] Terraced areas are shown in Figure 16.3.

Terracing has been a regular practice in many periods of Greek history.[42] The practice, however, was not universal, and it may have been based on local factors, such as the desired crop, the type of soil, and the degree of slope. In northeastern Crete, agricultural terraces have been excavated at two sites in close proximity to one another, revealing very different practices. On Pseira, many terraces were built during the Bronze Age,[43] while at Chrysokamino the only excavated terrace was constructed after the Roman period (App. K).

Rackham and Moody suggest three types of Cretan terraces.[44] Parallel walls form step terraces. Switchbacks at the ends create braided terraces. Small, often semicircular terrace walls create pocket terraces. All three types occur at Chrysokamino (App. G), and a fourth type, in which the walls are built inside a ravine to prevent it from eroding and becoming larger, is present as well (App. G, **AF 30** and **AF 31**).

It is not possible to prove that any of these terrace walls are Minoan from the available data. The only terrace excavated (App. G, **AF 22b**) was Venetian or Ottoman. Occasional Minoan sherds recorded on the land

41. According to local residents, the last farming was abandoned here not long after World War II.
42. Moody and Grove 1990; Foxhall 1996; French and Whitelaw 1999; Rackham and Moody 1992; 1996, pp. 140–145; Krahtopoulou 1997; Frederick and Krahtopoulou 2000.

43. Betancourt and Hope Simpson 1992; *Pseira* IX, p. 287.
44. Rackham and Moody 1992.

indicate it was used in the Bronze Age, although a sherd from the surface does not prove that a terrace was first built in the Bronze Age. The terrace may simply have incorporated earlier soil when it was constructed. Even though the gradual disintegration of included phyllite would have released nutrients gradually into the soil, occasional additions of refuse from animal pens and settlement refuse piles must have occurred.

In most places, the terraces are irregular and somewhat informal, and they do not create a series of even, regularly spaced steps. This type of terracing has been considered most appropriate for olive trees,[45] and it is likely that grains would have been grown under the trees as well. At the Chrysokamino farmstead, no direct evidence survives for the Minoan crops grown on the land that is now terraced.

One of the most important consideration in terrace systems is likely to have been water conservation rather than the creation of level land. Terraces of the "check dam" class, in which successive terraces are built on the slope of a hill, occur throughout the territory. They are a standard method of preserving moisture.

The small buildings found in this terraced zone all seem to have been constructed after the Bronze Age (App. G). Their late date suggests that the Minoan farmers walked out to their fields from the settlement, taking their tools with them. A similar situation exists at nearby Pseira, where all the field houses and threshing floors are Byzantine or later.[46] The gradual drying of the climate in this part of the Aegean[47] may have brought about the need for more terrace walls to retain moisture. This problem may not have been so acute until after the Minoan period. Many of the slopes of the hills are not very steep, and crops could have been grown on them without substantial terracing.

What crops did the residents of the estate grow? The triad of grain, olives, and grapes has been regarded as the foundation of Aegean agriculture for many years.[48] Pulses were probably staples as well.[49] Several pieces of evidence suggest that the main grain grown on poor land in this part of Crete was barley rather than wheat. The molds of cereals preserved in the furnace chimney fragments are mostly two-row barley (Chap. 12). Similarly, at Myrtos, the molds in clay have also been identified as barley,[50] and this grain appears to have been stored in quantity even at Knossos.[51] Barley grows better on poor land than wheat does, and it would have been a logical choice for the main cereal to be grown on the hillsides at Chrysokamino.

Evidence is lacking for other crops, but they surely existed. A strong case has been made for diversification in the small outlying farms of Minoan Crete,[52] and this theory seems to be the best explanation for meager evidence that exists from Chrysokamino. The use of land of more than one type, the presence of more than one species of domestic animal, and the relatively small scale of the operation all argue for a diversified agricultural base.

45. Lohmann 1992.
46. Betancourt and Hope Simpson 1992; *Pseira* IX, pp. 296–300.
47. See Chap. 2, pp. 19–21.
48. Vickery 1936.
49. Sarpaki 1992.
50. J. Renfrew 1972.
51. Evans 1921–1935, vol. 4, p. 622.
52. Halstead 1992.

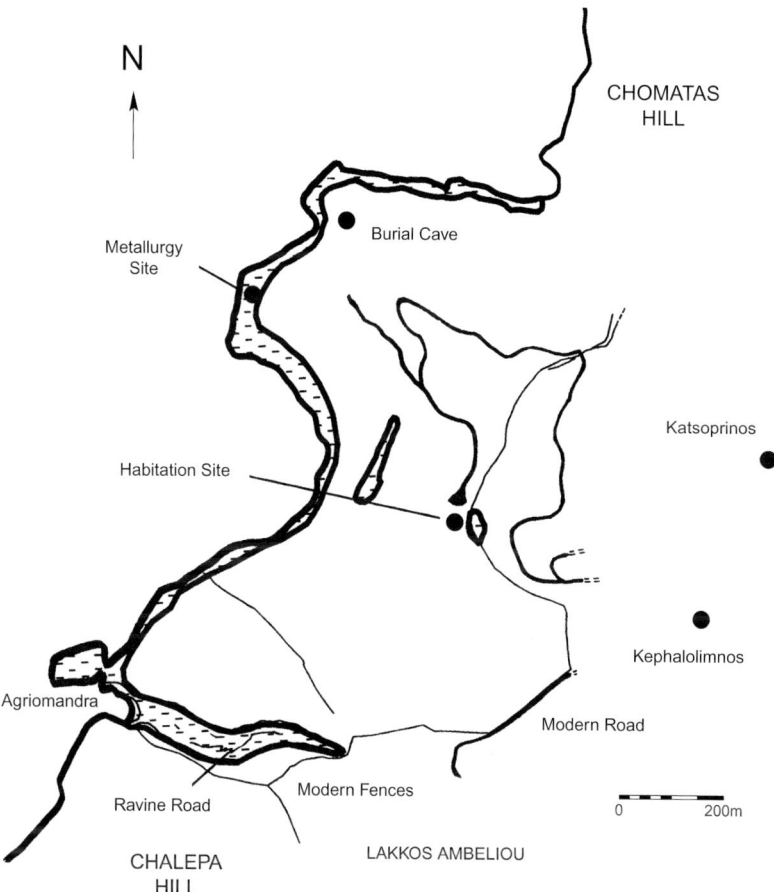

Figure 21.4. Steep or rocky areas unsuitable for farming

RAVINES, CLIFFS, AND OTHER STEEP AREAS

If the land in the vicinity of Chrysokamino is divided into zones, the locations that are too steep for farming make up a very small percentage of the total (Fig. 21.4). They consist only of the sea cliffs, some rocky outcrops, steep areas on the hillside, and a few ravines. Most of the ravines that divide hillsides with phyllitic soil are not very steep, and some of them were terraced in post-Minoan times to conserve water and provide spaces for farming. Numbers **AF 30** and **AF 31** (App. G) are examples of ravines with walls running up and down the slope to prevent the gorges from becoming wider through erosion, with subsequent loss of farmland.

Cliffs and very steep slopes characterize the land at the edge of the sea at this point on the coast. Many of the cliffs are too dangerous even for goats, and the modern shepherd tries to keep his animals away from them (one animal was lost over the cliff in 1996). The steepest gorge is the one that forms the natural barrier at the southwest of the territory. Some of the bluffs at the north of this gorge, bordering on the Chrysokamino territory, are sheer cliffs. In this study, the hectares of steep land are not counted in estimates of the total land area available for agriculture and animal husbandry.

On the seacoast, the steep cliffs at the edges of the hills are not continuous. Numerous steep paths lead down to the sea, and residents of modern Kavousi routinely use them to get down to the sea to fish as a leisure time activity. The ease with which the paths are used suggests that the appearance of isolation from the sea is deceptive. The Minoans probably walked down to the coast fairly often to gather sea urchins and other marine invertebrates as well as to fish.

The gorges were probably once forested, as many ravines are in this part of Crete. Their present bare condition is most likely a result of recent overgrazing (most of their vegetation is stunted from foraging by goats). In the Bronze Age, they could have furnished brush and wood for fuel and other purposes, small amounts of vegetation for animals, and havens for the wild birds, hares, and other creatures that were occasionally hunted in all periods. These resources were extremely important.[53] Hare bones are present in the faunal remains from both the farmhouse and the metallurgy site, suggesting occasional hunting or trapping to augment the local meat diet.

LAND OUTSIDE THE MAIN TERRITORY: AGRIOMANDRA

The ravine that leads to the harbor of Agriomandra is one of the most interesting areas in the region (Fig. 20.1). The access route from the Kambos consists of a steep ravine descending from the area of Lakkos Ambeliou to the harbor. A narrow cobbled road *(kalderimi)*, now with some parts eroded away, testifies to the importance of the route. The road may be originally from the Byzantine or the Venetian period, because it provides access to a small church built into a cave part way down the route, but some of the better-preserved sections may be from the Ottoman period. The church, on the north side of the gorge, is dedicated to Hagios Ioannis.[54] The entrance is a vault, and the interior is covered with concrete and stone *(opus incertum)*.

Beyond the church, the road continues to descend down the gorge until it reaches a tiny beach called Agriomandra. The small harbor was used as a seaport by some of the residents of Kavousi until the middle of the 20th century.[55] Because its small beach faces west, it offers some protection from northern and eastern winds. During the Bronze Age, the beach must have been larger because of the lower relative sea level in this part of Crete.[56] The harbor would have accommodated a series of farms and villages in the territory between the two larger ports at Tholos and Pacheia Ammos. It was surely too important to have been used exclusively by any single small settlement. One assumes that those who lived nearby took care of the harbor's maintenance. Haggis reports Minoan pottery from MM and LM I from a small site at the south side of the harbor,[57] and he also records worn sherds from the beach.[58] A piece of water-tumbled malachite was collected from this beach in 1996 (App. C).

One cannot know if the inhabitants of sites like Chrysokamino had direct or indirect contact with the seafaring that must have entered Crete at the small harbor, but the presence of imported goods at all Minoan sites indicates that access to the sea was a simple and constant part of Minoan

53. Burford 1993, pp. 111–112.
54. Faure 1964, p. 58; 1979, p. 58; 1989, p. 244.
55. Haggis 1992, pp. 169–170; 2005, p. 118.
56. Flemming 1972; Flemming, Czartoryska, and Hunter 1973; Pirazzoli et al. 1982; Pirazzoli 1988.
57. Haggis 1992, pp. 169–170; 2005, p. 118, locus 45.
58. Haggis 2005, p. 118, locus 110.

CHAPTER 22

SURVEY CONCLUSIONS

by Philip P. Betancourt

The inhabitants of the Chrysokamino territory adapted to several changes during their long occupation of this small area in northeastern Crete. The first known inhabitants settled here at the end of the Stone Age. They found an area suitable to their needs and established permanent residence. Some of the territory's features, such as the proximity of the sea and the nature of the soils, remained stable over an extended period. Other aspects of the environment, however, such as climate and vegetation, did not remain constant. The territory's relations with the other settlements on Crete also experienced many deep-seated changes during the region's long history. The residents of Chrysokamino responded to changing conditions, but their choices were made within the strictures of limited local resources. Some aspects of the local economy, such as the copper smelting workshop, developed slowly, reached a successful stage, and then passed completely out of existence; others, such as agricultural practices, lasted much longer. The history of Chrysokamino shows that cultural change, as others have suggested, is usually based on complex forces of formation and dissolution rather than on a strictly linear, gradual development.[1]

FINAL NEOLITHIC TO EM III–MM IA

In the Final Neolithic period, probably well before 3000 B.C., a few settlers arrived on the coastal strip known as Chrysokamino. In western Asia, the period would be called the Chalcolithic, and copper tools were already being made and disbursed in small quantities. The new settlers founded at least one settlement, and at some time within the period, they either established or allowed others to establish a small metallurgical installation for smelting copper ore. Their dead were probably buried in the Theriospelio cave. The residents of the area were farmers, and the evidence they left in the archaeological record indicates their choice for a place to live was dictated more by the availability of low hills where crops could be grown and animals could be grazed than by easy access to the sea or to inland tracts of virgin forest.

1. See Sutton 1994.

Final Neolithic sites have also been recognized at several other places around the Gulf of Mirabello.[2] Settlements and/or cemeteries have been noted at Kavousi,[3] Mochlos,[4] Pseira,[5] Sphoungaras,[6] Pacheia Ammos,[7] Vasiliki,[8] Vrokastro,[9] and elsewhere.[10] The pattern suggests that the settlement of Chrysokamino was part of a general migration that brought numerous residents to this part of Crete.

Chrysokamino was neither the most desirable nor the worst location for a small band of agriculturalists. It was not easily visible from the sea (perhaps an advantage if hostile forces approached), but one could walk down to the water fairly easily. An outcrop of dolomite provided both construction material for houses and a stable bedrock on which to build. Agricultural land was available for both gardens and fields, and a cleft in the dolomite south of the habitation site was almost certainly a source of water in the rainy season. Rain was more abundant than in later times, and the success of the settlement proves the location was favorable. The residents of Chrysokamino established a stable pattern of land management that allowed their territory to support a small group of people for the next two millennia.

The choice to settle in a location that would support diversified agriculture successfully, as opposed to placing the habitation site centrally on only one class of soil, raises questions about a traditional view that Neolithic agriculture was based only on grains.[11] No real evidence for such a restrictive Cretan Neolithic farming practice exists, and although one cannot prove the point, the first residents at Chrysokamino probably required land for almonds, fruit trees, legumes, and vegetables in addition to cereals. The proximity of the habitation site to good garden soil as well as to land for field crops was surely not accidental (for Final Neolithic pottery on the garden area, see App. H, **AF 32.1**).

Agricultural diversification, especially on small, semi-independent farmsteads, was an asset. Failure in one crop from disease, drought, or other factors could be offset by success in another. A varied diet is both healthier and more pleasant than a restricted one. Small labor forces can be maximized with diversification because different crops are planted and harvested at different times of the year, allowing the labor force to move from one crop to another. In regard to storage, some crops complement one another (for example, olive oil and wine or vinegar, made by processing olives and grapes, can be used as preservatives for more easily spoiled harvests). Different soils are suited for different crops. Diversification allows the use of otherwise unusable land (for example, planting fruit trees in gullies or placing grains in the underused spaces that are necessary between

2. Betancourt 1999.
3. Haggis 1992, pp. 173–174, locus 17/92.
4. Seager 1912, p. 93.
5. Betancourt and Davaras 1990; *Pseira* VI and VII.
6. Hall 1912, pp. 46–48; Foster 1978, p. 4, nos. 2–3; Betancourt 1983, p. 46, nos. 110–113.
7. M. Tsipopoulou (pers. comm.).
8. Seager 1904–1905, p. 212.
9. Hayden 2003a, 2003b.
10. Haggis 1992, pp. 53–56; Betancourt 1999.
11. A. C. Renfrew 1972, pp. 280, 304–307.

olive trees). All of these factors suggest the likelihood that several crops were raised at Chrysokamino during the entire history of the territory.

The situation near the Gulf of Mirabello differs from the one in central Crete in some important ways. Knossos had a substantial population much earlier than eastern Crete, and its Neolithic levels were extensive.[12] Neolithic Knossos covered an area of at least 4.5 hectares,[13] and in comparison with other Neolithic sites, it must be regarded as a large, wealthy town or even a small city. The Neolithic levels at Phaistos were also substantial,[14] as were other early settlements in southern Crete.[15]

Although by the Late Neolithic period central Crete had had a large population for millennia, the population living near the southern and eastern shores of the Gulf of Mirabello before the Final Neolithic period must have been much smaller, as it has left few traces. In fact, no large town from any of the Neolithic periods has been found in eastern Crete, and in the Gulf of Mirabello area the occupation seems to have consisted only of tiny sites spread through the region until the late Early Bronze Age development of places like Gournia.[16] The small size and the scattered distribution of the eastern sites must have played a role in their slower development in comparison with the more urban situation in the central part of the island.

Large towns developed in eastern Crete by the end of the Early Bronze Age. Evidence from excavations attests to substantial growth at two places that would later be important: Mochlos[17] and Gournia.[18] But Chrysokamino, like most of the local settlements, remained tiny. The contrast with central Crete in terms of population distribution would have been an important factor in local history. The system of land use for a district with tiny hamlets would have to have been substantially different from that of one dominated by large urban centers.

Settlement patterns appear to have remained stable during the Early Bronze Age. At Chrysokamino, three archaeological locations have substantial evidence for human social/economic/religious activity (Fig. 1.4):

1. A habitation location (App. G, **AF 29**)
2. A metallurgy workshop (App. G, **AF 23**)
3. A cave (App. G, **AF 34**)

Final Neolithic and Early Minoan pottery comes from all three places. The discussion in Chapter 20 demonstrates that the boundaries for this territory, functioning as a single topographic district, were probably formed by natural topographic barriers to the north and west, and the proximity of other sites to the east and south, along with several other factors. The three FN–EM locations are very different types of sites, and they must have made separate contributions to the inhabitants of the territory.

---

12. Evans 1921–1935, vol. 1, pp. 32–55; Furness 1953; Evans 1964, 1971, 1994; Hood and Smyth 1981, p. 6.

13. Evans 1964, p. 132; Broodbank 1992, pp. 40–41.

14. Vagnetti 1975.

15. Vagnetti 1973.

16. For the large size of EM III–MM IA Gournia, see Hall 1904–1905.

17. Seager 1912; Branigan 1991, pp. 101–104; Soles 1992; Soles and Davaras 1994, pp. 394–396; 1996, pp. 178–180; *Mochlos* IA; *Mochlos* IB.

18. Hall 1904–1905; Soles 1992; Watrous and Blitzer 1999.

The Main Habitation Site

The main habitation site's setting is directly related to the geomorphology. It is on a hard, stable, dolomite outcrop, at the highest point on the southwest slope of the hill of Chomatas where the dolomite is visible at the surface. The dolomite provides ample amounts of building material for stone architecture, as well as a stable and relatively flat foundation on which to build. In fact, this bedrock outcrop is the only place within the Chrysokamino territory where these two conditions are present to such a great degree. The location is the only logical choice as a site for substantial stone and mudbrick architecture because the rest of the territory is covered with terra rossa or unstable decomposed phyllite, with only small and highly irregular outcrops of a few other stones.

The site also has other good characteristics. It is near the entrance to the ravine that leads down to the best harbor in the area; it overlooks the depression at Lakkos Ambeliou, where water could be reached easily by digging through soft soil down to the water table, and it has ample land nearby for raising crops and grazing animals. It is also high enough to receive good breezes, alleviating the heat of summer, and it is relatively far away from the mosquitoes that would have troubled people living closer to marshy areas in the Kambos.

Pottery sherds found under the floors of the later farmhouse show that the spot was already settled during the Final Neolithic and Early Bronze Age. The excavation of the farmhouse has produced significant amounts of domestic pottery from these early periods, and the site was probably the controlling factor in the territory, although it is likely that other tiny sites lie buried in the vicinity. The residents of the area used the cave at Theriospelio for burial, and they must have had a relation to the metallurgy workshop. The three locations are so close together that they must have been interrelated. It is significant that Final Neolithic pottery is the earliest pottery at all three places because it implies that the territory was already established as a unit during this early phase of its settlement.

After its establishment in the Final Neolithic, the pattern of land use was stable enough that no major change seems to have taken place until EM III, when several small sites were founded near the previous settlement. This long period of stability must have resulted from a successful balance between the size of the population and the capacity of the land to support it.

The presence of barley, wheat, and olive impressions in the furnace chimneys from the metallurgy workshop shows that agriculture was already being practiced in the region during the Early Bronze Age. An economy based on mixed farming and animal husbandry seems to have been normal for early Aegean communities.[19] In this region, the proximity to the sea allowed access to seafaring as well, and an outlet to the sea would have been a crucial factor in the other important element of the early economy here, a metallurgical installation depending on imported ores.

19. A. C. Renfrew 1972, pp. 304–307; Halstead 1981, pp. 318–320, 326.

The Metallurgy Workshop

Like the main habitation site, the choice of location for the metallurgy site must also be based on topographic factors. The workshop is on a saddle between higher outcrops of bedrock with a cliff at the north. The saddle acts as a funnel that channels the northern winds and intensifies them, so that if the metallurgy needed a natural draft to aid in the smelting process, this spot is the windiest location on the southwest slope of Chomatas. A second factor may be the proximity of the sea. A path leads down to the water, and in calm weather ships could anchor and unload cargo and passengers. If both access to the sea and windy conditions were the factors necessary for the pyrotechnology, the saddle is the most suitable location in the vicinity. A third factor could be the availability of fuel. The territory must have had an original forest cover when the Final Neolithic settlers arrived, and metallurgy needs substantial fuel. It is also possible that local iron ores were used as fluxes.

The metallurgy site is too tiny to have been a complete settlement. The presence of one flimsy hut less than 4 m long does not imply a permanent or even a self-sufficient community. Artifacts are scarce, and they do not include the full range of objects (such as cooking pots and storage vessels) one expects from a permanent settlement. The workshop must have been an appendage of some other site nearby.

The presence of a highly specialized workshop at this isolated promontory calls into question suggestions that early access to specialized technology was concentrated in locations that were centers of socioeconomic power.[20] The workshop at Chrysokamino probably supplied a number of sites with copper, but there is no reason to assume that this technology was established because of any elite presence within the region itself. Instead, the choice is more likely to have been made based on wind conditions, topography, access to the sea, availability of fuel, and proximity of markets. If an elite segment of the population controlled the workshop, which may very well have been true, the control was exercised from a distance.

The Burial Cave at Theriospelio

The third major site in the territory is a natural cave with stalactites and stalagmites. Obviously, such a cavern is a significant and highly specialized topographic feature. The cave is found in Plattenkalk limestone. Although its entrance is fairly small, the space inside is over 40 meters in length, and one can stand up inside the main room.

No evidence for a settlement is adjacent to the cavern. Its entrance is situated on a steep hill, so that construction of houses would have been impossible without extensive terracing. No traces of houses or terraces are present at the location, and the cave appears isolated from any spot used for habitation.

Because human bones are clearly visible on the surface inside the cave (in the disturbed soil either excavated by Edith Hall in 1910 or by others), the cave is interpreted here as a place for burial. The earliest pottery from

20. E.g., Carter 1998.

the location is from the Final Neolithic period. Final Neolithic burials in this part of Crete were associated with a wide variety of tomb designs, including natural caves, manmade caves, and small built chambers.[21] Sometimes more than one tomb type exists in the same cemetery.[22] Choices for tomb design were clearly based on social factors about which we have little information. The Theriospelio cave, the only known natural cavern in the Chrysokamino territory, was apparently the social choice for the local community's burials.

Other burial caves in eastern Crete were regularly used until much later than EM III.[23] It is likely that the Theriospelio cave was also used into the Middle Bronze Age, but without excavation we cannot know the details.

Comments

During the Early Bronze Age, the residents of the Chrysokamino territory made use of three different types of site. If the three locations are considered as a single unit, an agrarian community lived at the habitation site (and possibly elsewhere), buried its dead nearby, and engaged in periodic craftwork including metallurgy. The evidence suggests that the three main archaeological locations at Chrysokamino complemented one another, interacting to form a single socioeconomic unit probably based at the excavated habitation location. This unit seems to have been one of a series of small settlements established in the Kavousi region at the end of the Neolithic.[24]

Very little evidence survives for the Early Minoan agricultural products that must have been the foundation of this stable system. Impressions of grain and chaff in the furnace chimney fragments from the metallurgy site (Chap. 12) show that barley was raised, and the impression of one wheat seed hints at wheat production. An olive leaf gathered up with the chaff that tempered the chimneys suggests that grain and olive trees were growing in close proximity to one another. Organic residues in sherds from the metallurgy site show that wine, flavored with many herbs and spices, was already present (App. M), so grapes were being raised. Grains were already grown in the Cretan Neolithic.[25] By the Early Minoan period, the farmers of Crete raised wheat, barley, and oats along with olives[26] and several other crops.[27] As at these other sites, the agriculture at Chrysokamino was probably diversified.

The economy was supplemented by the smelting of copper. For Final Neolithic and Early Minoan I, copper is known from only a modest amount of evidence in this part of Crete,[28] but the evidence is accumulating for other places (see Chap. 13). The EM I–IIA cemetery at Hagia Photia shows that the metal was already making an impact on Cretan culture early in

21. Betancourt 2003a, pp. 123–128.
22. For Mochlos, see Seager 1912; Soles 1992. For Gournia, see Soles 1992. For Pseira, see *Pseira* VI and VII.
23. Pendlebury, Pendlebury, and Money-Coutts 1935–1936; Branigan 1970a, pp. 153–154; Davaras 1989a, p. 387; Schlager and Dollhofer 1998.
24. For discussion of some of the others, see Haggis 1992, pp. 53–56.
25. Evans 1968, p. 269.
26. J. Renfrew 1972; Blitzer 1993.
27. Greig and Warren 1974.
28. For Mochlos, see Seager 1912, p. 93; Evans 1921–1935, vol. 1, pp. 58, 68.

the 3rd millennium B.C.[29] Lead isotope patterns observed in the copper slag match patterns from Lavrion and Kythnos,[30] suggesting the ore was brought in by ship. The workshop produced only copper, which was presumably taken to neighboring sites to be remelted in crucibles and cast into various objects.

The local smelting operation was only one step in the making of metal objects. The production of copper from ore involved the activities of a series of skilled specialists in addition to the smelters working at Chrysokamino: the prospectors who found the ore, the miners who recovered it from the ground, the workers who beneficiated the ore and packed it for shipment, the ship captains and their crews who provided transportation, and the metalworkers who made the final products. To set up the operation, someone needed to advance the resources to obtain good quality copper ore, arrange for its shipping, mine the rock to be used as flux, establish a workshop run by craftsmen familiar with the smelting process, and secure potential markets for the copper. It would also have been necessary to obtain fuel, food for the workers, and raw materials for manufacturing furnaces and chimneys.

The implications of this complicated production system are very important for our understanding of early Cretan craft organization. Such disparate activities, carried out over an extended period of time in multiple locations, imply the involvement of a series of different persons rather than a small workforce performing every task. One cannot expect that ship captains would also have been prospectors, miners, and smelters. Such a complex endeavor would have required coordination and supervision. The ore did not get transported to the right port or loaded on the right ship automatically. Someone must have accompanied the shipment from mine to destination in order to guarantee its arrival. Behind such an operation was someone with confidence in its successful conclusion and the resources to accomplish a series of tasks whose profits would only be realized at the end of the process. That same manager might also have found it necessary to oversee the intermediate level supervisors, and perhaps to take some action (such as arranging for armed guards or soldiers) to make sure that the potential profits were not stolen somewhere along the way. Such an operation would be fully at home in a bureaucracy like the one reflected on the Linear B tablets. The implication is that some of the supervisory roles that played such an important part in later Cretan administration had a long history in Minoan Crete, emerging well before the Late Bronze Age.

Chrysokamino was probably not unique. Both locally and elsewhere in Crete, successful communities would have needed to develop agricultural systems that were viable within the constraints of locally available land, to secure access to the sea or other modes of transport, and to exploit local and regional resources. Augmentation of the agricultural base with crafts or manufacturing, as in the development of a metallurgical workshop at Chrysokamino, might have been undertaken as a natural response to the leisure time afforded by seasonal farming activities. Patterns of settlement and land use recognized at other locations in Crete[31] are not incompatible with the more detailed conclusions presented here.

29. Davaras 1971; Davaras and Betancourt 2004; see also the discussion in Chap. 13.

30. Stos-Gale 1998, pp. 720–721; see also App. F.

31. E.g., Blackman and Branigan 1977, pp. 69–71.

The culmination of developments in this long period occurred at the end of the Early Bronze Age. EM III–MM IA is defined in eastern Crete by the presence of East Cretan White-on-Dark Ware.[32] The ceramic style begins at the end of EM IIB with the application of simple white lines over Vasiliki Ware surfaces.[33] After the destruction level coinciding with the end of EM IIB, White-on-Dark Ware became more popular, and its range of motifs expanded to include spirals, circle motifs, quirks, and other designs. It used white linear designs on uniformly dark or reddish brown surfaces, without any of the mottling characteristic of the previous period. The phase ended with destructions throughout the region, identifiable in the archaeological record at Mochlos,[34] Pseira,[35] and Gournia.[36] The final stage of the phase overlapped with MM IA at Knossos.[37]

This period is an important one in Crete. Several new developments can be assigned to it at Chrysokamino and elsewhere on the island. The substantial deposits that survive at Knossos suggest this was a prosperous time in north central Crete.[38] In southern Crete, several new sites can be recognized, suggesting an expansion in population there as well.[39] Important pottery from this period is also found at Malia.[40] Gournia, a site that would develop into an important center by LM I, expanded significantly. The large amount of EM III–MM IA pottery at Gournia[41] contrasts dramatically with the tiny amount of Vasiliki Ware from the previous phase.[42]

The period was a stable one in central and eastern Crete, and it was not marked by major destructions until its end. The beginning of polychrome painting on pottery, which should mark the boundary between EM III and MM IA,[43] is actually a rare characteristic that was only used on a small number of vases in this period, even at Knossos.[44] The red pigment was not used during this phase in the pottery workshops at the eastern side of the Gulf of Mirabello.

At Chrysokamino, Haggis has noted a new settlement pattern in which a cluster of sites replaced the single location of the main habitation site (Chap. 19). Most of these new sites are found downhill from the original settlement, at the edge of Lakkos Ambeliou. Other new foundations in the mountains south of Kavousi may signal advances into the higher country inland, perhaps to accommodate seasonal movements of herds and flocks. The new pattern suggests an expansion in population and an exploitation of new resources for intensified animal husbandry. This expansion coincides with the main period of the metallurgy workshop, which would have needed timber for fuel, also available in the higher hills.

This pattern of scattered settlements has been proposed for other parts of Crete as well.[45] It must have been the result of a larger population needing more food. By spreading the population over the land, agricul-

32. Betancourt 1984.
33. Boyd 1904–1905, pl. 25, no. 2; Warren 1972, phase II; Betancourt 1983, p. 69, no. 236.
34. Seager 1909, p. 278; Soles and Davaras 1994, pp. 394–396; 1996, 178–180.
35. Banou 1995a, pp. 19–20.
36. Hall 1904–1905, 1908.
37. Warren 1965, pp. 25–26; Betancourt 2003b.
38. Momigliano 1991.
39. For Patrikies, see Bonacasa 1967–1968; Levi 1976, vol. 1, pp. 747–756; for Drakones, see Xanthoudides 1924, pp. 76–79.
40. Demargne 1945, pp. 1–12; Chapouthier, Demargne, and Dessene 1962.
41. Hall 1904–1905, 1908.
42. Hawes et al. 1908, pl. 6, no. 1.
43. Hood 1971, p. 39.
44. Momigliano 1991, p. 219.
45. Kanta 1983.

tural production might have been increased,[46] but decisions to inhabit dispersed locations are also likely to have been at least partly based on cultural preferences.

An important characteristic of the Early to Middle Minoan system of land use in this region is its open and undefended character. The sites were tiny and exposed, with no traces of fortification walls. This situation is very different from the one in several other parts of Crete, where walled towns and citadels, often with megalithic fortification walls, suggest that small independent communities were fearful of attack.[47] Elsewhere in eastern Crete, fortified road stations may attest to an organized regional defense.[48] The relative safety of the region (perhaps imposed by a higher authority at Gournia or elsewhere?) must have been a crucial factor allowing the dispersal of the population.

All of this activity implies that new social and economic factors were affecting the local area. Several theories have been suggested for the relationship between this part of the island and higher authorities in Crete. For the Middle Minoan period, several authors have made a case for a palace-state that included this region and was centered at Malia.[49] An alternative center was much closer at hand, however. EM III–MM IA was a major period of expansion at Gournia,[50] and this expansion coincided with the spread of the typical Gournia pottery tempered with rock in the granodiorite/diorite series.[51]

This pottery was the majority fabric at Chrysokamino in EM III–MM IA. The small workshop, the habitation site, and the cave all used White-on-Dark Ware and other ceramic products that had the same style and the same fabric as Gournia, Vasiliki, Pseira, Priniatikos Pyrgos, and several other sites in this part of Crete.[52] The ceramics had a regional distribution, replacing a series of other ceramic classes including Vasiliki Ware[53] and other burnished pottery.

Several developments at the transition between the Early and the Middle Bronze Age involved cooperation between sites, indicating a greater awareness of regional goals. Traditional settlement patterns changed. The metallurgy workshop expanded greatly, importing its ore from abroad and sending its copper to locations away from the workshop. New foundations in the inland mountains were surely derived from the coastal population, and they must have shared communal goals. A new, large, and widely distributed class of pottery replaced the Vasiliki Ware of EM IIB. Some of these new regional initiatives would seem to imply a directing organization. The authority must have had strong social (and perhaps religious) power, because there are no local fortifications, in sharp contrast with other parts of eastern Crete.[54]

46. See comments on extensive versus intensive agriculture by Halstead 1992.

47. Watrous 1982, pp. 42–43, 62–63; Brown 1993, p. 65; Nowicki 1996a; 1996b, pp. 35–42; Schlager 1999.

48. Tzedakis et al. 1989; Tzedakis et al. 1990.

49. Cadogan 1995; Knappett 1999.

50. Hall 1904–1905, 1908; Betancourt 1984.

51. Haggis and Mook 1993, p. 273, fabric 2.

52. Betancourt 1984.

53. Betancourt 1979.

54. Alexiou 1979, 1980; Tzedakis et al. 1989; Tzedakis et al. 1990; Manning 1994; Nowicki 1996a; Schlager 1999, 2002.

The cluster of small sites probably managed a territorial region that was not too different from the one used earlier, except that it must have been larger at the south. The land at the south, the Kambos, would have been marshy in the rainy season, but it would have been attractive as pastureland, especially if its use was complemented by exploitation of the inland hills for transhumance. At the same time, the patterns of sherds on the surface, with their evidence for manuring, indicate that agriculture continued to play an important role in local subsistence. The division between gardens and fields suggested in Chapter 21 was probably already in place, and the area adjoining the habitation site (Fig. 21.1) would have been used for vegetables, vines, and other plants requiring periodic care.

The evidence suggests that this period at the end of the Early Bronze Age and the beginning of the Middle Bronze Age was an affluent time, with population growth, expanded craftwork and manufacturing (including at the metallurgy workshop), and the advent of new ways to exploit the larger landscape. The expansion into the hills beyond the coast and the probable use of this territory for transhumance, timber, and other resources, would have contributed to significant advances in the region's economy. If the stability of the earlier system implies a balance between subsistence and population, the expansion of flocks and herds involving transhumance on a larger scale may suggest new surpluses, either for local trade or, more likely, to support a rapidly expanding state centered at Gournia.

MIDDLE MINOAN IB–IIB

After MM IA, outside influences must have increasingly affected the territory of Chrysokamino. The new phase follows a major destruction at Gournia, when large quantities of White-on-Dark Ware were deposited north of the town.[55] New pottery styles, now using the potter's wheel more than previously,[56] signal the beginning of MM IB.

Malia, Knossos, and Phaistos expanded at this time, and magnificent palaces were built. The polychrome pottery developed into a palatial class called Kamares Ware,[57] and the other arts were encouraged as well. Cretan society and economy advanced on many fronts. Palatial art objects had little or no impact on life in the Chrysokamino territory, but the rise of palatial culture must have resulted in a new economic situation for all of Crete.

The metallurgy workshop was abandoned at the end of EM III–MM IA. Its abandonment, whatever the ultimate cause, was part of the broader social and economic changes happening in Crete at this time. With the founding of the Middle Minoan palaces, metals may have been more available, or more centrally controlled, or imported in a smelted condition. In this new era, there was no room for the small independent workshop at Chrysokamino. By MM II, metals were much more available throughout the island than they had been during the Early Bronze Age, and competition from these new sources, surely involving imports from eastern ports, meant that the older ways of obtaining metal would never again be viable.[58] Chrysokamino now depended primarily on agriculture and the tending of animals for its subsistence.

55. Hall 1904–1905, 1908; Betancourt 1984.
56. Warren 1980, p. 492; Betancourt 1985, pp. 77–78.
57. Walberg 1976.
58. Betancourt 1998.

The Site Clusters

During the Middle Minoan period, the Chrysokamino territory and its adjoining region to the south had a cluster of small sites that included the main habitation site as well as some tiny locations near Lakkos Ambeliou (Chap. 19). These small establishments persisted until the end of the MM IB–IIB phase. This scattered settlement pattern seems to have been typical of the period throughout the Kavousi region.[59] The individual house sites are not contiguous, but they are so close together that a relationship between them, possibly kin-based, seems likely. They probably managed the land as a group, with cooperation for large projects affecting the whole region. The cluster was well located both for farming and for controlling the access to the seaport at Agriomandra.

Comments

MM IB–IIB developments in the Chrysokamino region coincide with the growth of the palaces. The more centralized economic system and its success in obtaining imported copper surely led to the abandonment of the local metallurgical installation. The new system of land use and the greater exploitation of the land at Lakkos Ambeliou by the cluster of small sites may have been a response to the demise of local manufacturing activities, forcing an intensification of the agricultural efforts. It might also have been related to a need to produce a surplus for contribution to a larger political unit, one that was based at Kavousi, Gournia, or even Malia.

Several authors have discussed the theory of a large palatial state governed by Malia.[60] The administrative evidence from the Malia hieroglyphic tablets,[61] as well as the size and importance of the Middle Minoan Malia palace, leaves little doubt that the site was supported by a substantial hinterland, but the boundaries of that hinterland have been difficult to establish because of lack of clear evidence. In the case of pottery, for example, one cannot know if connections also imply political control. The most compelling case for a site under the control of Middle Minoan Malia may be found at Pyrgos Myrtou, a town on the southern coast. For eastern Crete, the situation is less clear. Although many of the parallels between pottery from Malia and the Gulf of Mirabello are apparent in vases made somewhere in the Gulf area, imports from the Phaistos area also occur at Malia and elsewhere within the proposed Malia state, and some imports appear in the hinterland but not at Malia itself,[62] suggesting that pottery was not simply sent from a subsidiary political unit to a central one. Evidently, pottery moved routinely across political boundaries, making it difficult to use it for defining those boundaries.

MM IIB ended with widespread destructions in eastern Crete. The final phase was characterized by the use of the thin-walled carinated cup.[63] In the eastern parts of Crete, from Malia to Palaikastro, this cup often has no white paint over its dark slip, and it is decorated only with thin horizontal grooves on the upper part. Although the cup disappears in MM III, hundreds of MM IIB examples have been excavated. The cup is present throughout the region from Malia to the Gulf of Mirabello.[64]

---

59. Haggis 2005.
60. Poursat 1987; Cadogan 1990, 1995; Knappett 1999.
61. Poursat 1990.
62. Knappett 1999, pp. 632–634.
63. For discussions, see Pendlebury, Pendlebury, and Money-Coutts 1935–1936, p. 62; Coldstream and Huxley [1972] 1973, pp. 94–95; Walberg 1976, pp. 150–151, form 47; 1983, pp. 190–192, 234–241, forms 234–241.
64. Hawes et al. 1908, pl. 2, no. 11 (Gournia); Betancourt and Silverman 1991, p. 25, no. 412 (Gournia); Nowicki 2002, p. 34, no. 13, fig. 5 (Katalimata); Demargne 1945, pl. 33, no. 8657 (Malia); Knappett 1999, p. 631, fig. 19 (Malia); Betancourt and Davaras 1999, p. 145, nos. BR 28–30 (Pseira); Knappett 1999, p. 629, fig. 12 (Pyrgos Myrtou); Seager 1904–1905 (Vasiliki); Betancourt 1983, p. 274, nos. 257–260 (Vasiliki).

Nowicki has noted that the end of MM IIB in the Isthmus of Ierapetra coincided with a crisis period when people fled to high refuge sites, including the easily defensible height of Monastiraki Katalimata.[65] The use of such an inaccessible refuge site must surely have been correlated with the widespread architectural destructions in the region. The disruption is especially evident at Vasiliki, where House A was buried with its contents,[66] and at Pseira, where the town was destroyed and the cemetery abandoned.[67] The extensive rebuilding after the period of the carinated cup indicates, however, that the discontinuity occurred throughout the region. The most important site with destructions at this period is the palace site at Malia; the destruction is clearly visible in Quartier Mu.[68]

It is tempting to associate the fundamental changes at the end of this period with the expansion of Knossos into eastern Crete. At Chrysokamino, where no MM architecture has been found, a transformation is evident in the abandonment of the pattern of land use requiring a cluster of sites and the survival of only a single settlement, the habitation site built on the dolomite outcrop at Katsoprinos. A new house was constructed here after the MM IIB period, and it survived until LM IB or early LM II.

MIDDLE MINOAN III–LATE MINOAN I

At the beginning of the Neopalatial period, in MM III–LM I, the cluster of sites southwest of the hill of Chomatas was replaced by a more nucleated settlement pattern. For the remainder of the Bronze Age, until the domestic complex was abandoned in LM IIIB, the habitation site was the only settlement within the Chrysokamino territory. The LM I farmhouse at Chrysokamino must have controlled the whole territory of Chrysokamino, although the absence of sherds in an area south of **AF 32** suggests that the territory included the land in Figure 20.2, but not the region south of it. Several tiny MM sites observed by Haggis south of Chrysokamino (Fig. 19.8) must now have joined the Chrysokamino farmhouse, the village of Katsoprinos, or some other settlement. The consolidation of the scattered sites into larger settlements represents a new system of land use, and it may imply a new bureaucratic management system organized under palatial control.

The LM I architecture at the Chrysokamino farmhouse is not completely preserved, but it seems to have consisted of a single building. It had an entrance on the east and a series of contiguous rooms. Additional interior spaces were at the north (destroyed by the later architecture). No fortifications were present.

Although it seems to have been an independent domestic complex situated in the Minoan countryside, the LM I building at Chrysokamino cannot be regarded as a villa or "country house" as defined by van Effenterre, Cadogan, and others.[69] Minoan villas, a phenomenon of LM I, are large, unfortified architectural units with rich possessions, large amounts of storage, and sophisticated architectural details that imitate the palaces.[70] The architectural imitations of the palaces are, as far as we know, completely

65. Nowicki 2002.
66. Betancourt 1977, p. 346.
67. *Pseira* VI and VII.
68. Poursat 1996.
69. Cadogan 1971, 1997; Hood 1983; Nixon 1987; Betancourt and Marinatos 1997; Effenterre and Effenterre 1997.
70. Moody 1987b.

missing at Chrysokamino. In LM I, the incomplete architecture leaves many questions unanswered, but the finest possessions from this period, including stone vases and bronzes, are certainly not imitations of objects found in the palaces, although they argue for a more complex social station than an impoverished, isolated farmstead. Reused wedges for drilling stone vases (found in LM III levels) suggest an earlier industrial component that may even have produced a surplus beyond its own needs.

Niemeier has convincingly argued that the Minoan villa system of LM I had an earlier social and political history, and that it did not appear suddenly without predecessors of any type.[71] Before large manorial villas could exist in the Minoan countryside, imitating palatial architecture and creating microcosms of peaceful and genteel, even luxurious, living, the agrarian base of the rural countryside had to develop over a period of many years. The LM I farmhouse at Chrysokamino may represent a late survival of the type of small, semi-independent farmstead that helped develop the farming practices and small-scale craftwork traditions that would lead to the development of larger villages, towns, and splendid villas. That Chrysokamino was founded in the Final Neolithic and continued to exist in LM I and LM III is a testament to the stability of the system of land management it used, but it is also proof of the limits of its immediate landholdings, which were never able to support anything more grand. Whether the development of complex political systems in Minoan Crete is regarded as evolution or revolution,[72] the Chrysokamino domestic unit did not participate much in these dynamic developments, except as a recipient of outside influences and directions. The residents developed a stable pattern early in the region's history, but the farmhouse did not evolve as many other sites did.

Chrysokamino is not unique. During MM and LM I, the island of Crete had a number of small, isolated domestic buildings that occupied the landscape between the larger villages and towns.[73] Excavated examples in eastern Crete include Chalinomouri,[74] Chamaizi,[75] Karoumes A,[76] Cheiromandres,[77] and Kokkino Phroudi.[78] Unexcavated farmhouses are also known from other parts of Crete; six are found in the region of Hagia Photia.[79] Similar situations seem to have existed elsewhere in prehistoric Greece, including the mainland, where both small, independent farms and larger estates occupied the areas between urban complexes.[80]

Chalinomouri, the nearest of these excavated farmhouses, affords an especially close parallel for Chrysokamino.[81] Like Chrysokamino, it was engaged in craftwork as well as farming, operating in the countryside as a semi-independent domestic complex that was less complex than a village. Like Chrysokamino, it survived into LM III.

71. Niemeier 1997.
72. Branigan 1970a, p. 204; Cherry 1983.
73. See Bevan 2002, p. 224, fig. 5.
74. Soles and Davaras 1996, p. 209, fig. 17; *Mochlos* IA.
75. Davaras 1973.
76. Chryssoulaki 1999, pl. 7:β.
77. Chryssoulaki 1999, pl. 7:α.
78. Chryssoulaki 1999, pl. 7:γ.
79. Tsipopoulou and Papacostopoulou 1997, p. 206.
80. Palmer 2001, pp. 52–65.
81. *Mochlos* IA, pp. 103–132.

Isolated sites such as Chrysokamino and Chalinomouri do not fit perfectly into some of the land use systems proposed for Minoan Crete. MacGillivray has suggested that sites with isolated buildings such as Chalinomouri belong to the same category as villages such as Tylissos and Vasiliki,[82] but the variability in the scale of these sites implies serious cultural differences. The Minoan settlement pattern was complex and probably multitiered.[83] It encompassed many different types of habitation sites, including isolated tiny houses, larger farms, villas, hamlets, villages, towns, and cities ruled by palaces.

Several of the Cretan farmsteads seem to have had defensive capabilities in addition to their agrarian functions. Pyrgos Chrysokaminou is an unexcavated small fort at the top of the highest cliff east of Chrysokamino, with a commanding view of the sea (Fig. 1.5). The small, isolated domestic buildings that survived into LM I were all units that combined farming with a strategic location. Chiromandres, Kokkino Phroudi, and Karoumes A seem to have been defensible farmhouses set on the Minoan road system; they have been regarded as guard stations for the roads.[84] Chaimaizi and Pyrgos Chrysokaminou are hilltop forts that dominate their surroundings. Chalinomouri is set so that it controls a small landing place for ships at the eastern end of the Mochlos Plain. The Chrysokamino domestic unit is also set at a strategic location overlooking the upper end of the ravine that controls access to the Kambos from the sea. Its situation is better suited to observation and control of the region than the nearby tiny sites that were present in EM III to MM II, and this commanding position could be one of the reasons it survived into LM I and LM III.

As is the case with the previous period, political allegiances are difficult to reconstruct for this era. Some writers have suggested the region was part of a Malia state.[85] Others have supposed a local palace at Gournia[86] or that decisions were ultimately coming from Knossos.[87] Political alliances are impossible to prove for the local situation at Chrysokamino, but the basic changes that occur in MM III/LM I indicate outside political forces are more likely than simple reactions to the local topography or some other internal factor, because the consolidation that took place at Chrysokamino is similar to the nucleation visible elsewhere. At Pseira, for example, the town expanded to the adjoining hill after MM IIB, and it almost doubled its geographic size in a single period of construction, implying the arrival of many new residents who must have come from beyond the island.[88] Radical population changes, taking place at the same time but in different locations, suggest someone was making decisions regionally in regard to the size and location of settlements.

As elsewhere in this part of Crete, the period between LM IB–early LM II and mature LM II ended with a local destruction. The habitation site was unoccupied for a while, and the later architecture had a different plan of larger scale. New architectural construction techniques were employed after the destruction, but the economy of the new settlement was still based on agriculture and animal husbandry.

82. MacGillivray 1997, p. 22.
83. Driessen 2001.
84. Tzedakis et al. 1989; Tzedakis et al. 1990; Chryssoulaki 1999.
85. Cherry 1986, p. 21, no. 2, fig. 2; Cadogan 1995.
86. Soles 1991.
87. Betancourt 2004c.
88. *Pseira* IV, areas C and D.

LATE MINOAN III

The LM III period was the high point for the economic exploitation of the Chrysokamino territory. The new building was much larger than its LM I predecessor. It had an entrance on the east that allowed access to an enclosed courtyard. Rooms for living and working were constructed of large stones, some of them megalithic. Objects found in the excavation included everything expected from a semi-independent establishment: pottery, including vessels for cooking, serving, and storage; stone tools of many types, with a large number of querns of the type used for grinding grain; animal bones of several species, including sheep/goats, cattle, and swine; marine shells; and loomweights and other objects used in the domestic economy. The site managed its surrounding farmland and used it successfully for crops and animal husbandry.

Behind all of these situations lies the Minoan agrarian economy. One of the most important conclusions resulting from this study is a better understanding of Minoan land management. The evidence survives best from the Late Minoan III period, but it is surely applicable earlier; it suggests that the system in use at Chrysokamino divided the land into specific parts, which were used and maintained differently. The farmhouse was larger than in LM I, and it surely had more people living in it. The mature agricultural management system must be the basis for the productivity that allowed the residents to support a larger population with diversified agriculture and animal husbandry involving several species.

The system of dividing the land into different agricultural zones described in Chapter 21 was fully mature by LM III. This agricultural system would have required land for fields, gardens, and pasture. Analogies with later Greek and Roman farming methods show that ancient practices treated vegetables and field crops very differently. Because of the effects of the local topography and climate (see discussion in Chap. 2), the local conditions would have been very dry, with greater variations in temperature, humidity, wind conditions, and amount of precipitation compared to some of the other parts of Crete. Plants grown in gardens would have needed special care, including extra watering and the addition of organic matter to the soil for nutrients and moisture retention. This extra care left traces in the archaeological record in the form of organic matter in the soil, the chemistry from the manure, and sherds that were carried out to the land along with the household refuse used as fertilizer (Chap. 21).

In the Chrysokamino territory, a location south of the settlement has all of the characteristics that suggest it was reserved for gardens (Fig. 21.1). Here, crops that had to be tended individually and whose products were harvested frequently for fresh produce could be managed easily. This land is recognizable in the archaeological record because it is not separated from the settlement by any topographic barriers, because it is relatively flat, and because it has a heavy scatter of sherds on its surface. The sherds are different from sherds found in settlements; they are small, worn, and isolated from other fragments from the same vessel. They occur within the soil at a depth that suggests tilling. The crops would have included grapes, vegetables, herbs, and other garden plants. Pulses, probably a Minoan staple,[89] would also have been tended here as well.

---

89. Sarpaki 1992.

Low hills with phyllitic soil constitute a separate class of land (Fig. 21.3). They would have been used for dry farming, as they still are in this part of Crete. The phyllitic soil retains moisture better than the red soils, it has more nutrients, and it is well suited to the growing of olives and grain. At Chrysokamino, the phyllitic soils do not have the heavy concentration of small, worn sherds found on the land identified as garden plots, indicating that they were never used for crops like vegetables requiring repeated deposits of manure that were tilled into the soil.

The excavation of the habitation site yielded a large corpus of animal bones, indicating that meat played a substantial role in the residents' diet. Sheep and/or goats were the dominant species. Bones were already present in large numbers in LM I, and they continued to be deposited in LM III, indicating that the occupants enjoyed a steady supply of meat. They would also have had access to the hides, leather, and other skin products that are so useful to early societies,[90] as well as to wool for weaving. They clearly kept flocks and herds, and the topography suggests that the eroded hill of Chalepa is, in fact, suitable for grazing but not for growing crops. It lacks the evidence, present elsewhere, for terraces, sherds on the surface, and deep soil useful for farming. It is likely that some of the flocks and herds were moved seasonally to upland meadows, particularly during the summer when the low land covered with terra rossa loses much of its moisture.

Transhumance was a normal practice in Classical Greece.[91] Analogy with other prehistoric pastoral economies suggests it was probably a regular part of ancient Cretan animal husbandry,[92] and Watrous has made a convincing case for the seasonal movement of even some complete settlements during the Minoan period.[93] It is unlikely, however, that the entire population of Chrysokamino would have moved. At least in the LM I and LM III periods, when the evidence is most complete, the diversity of the domestic economy, with craftwork as well as varied agricultural land use, indicates permanent habitation by at least part of the community.

The evidence presented here supports the view of Halstead that outlying Minoan agricultural establishments engaged in highly diversified agriculture and animal husbandry, in contrast to what may have been more specialized palatial productions.[94] Animals included sheep and/or goats, cattle, and pigs. Crops required gardens as well as fields. This situation had many advantages for an isolated farmstead. Different animals could subsist on different types of pastures, and diseases in one species would not necessarily have spread to others. Crop failures in a single harvest would not have been disastrous if additional harvests could be realized at other times of the year. In addition, different crops required labor at different times of the year, so the work force could be maximized, without long periods of inactivity. The residents of the territory achieved a satisfactory system that resulted in a long-term, stable economy.

90. Driel-Murray 2000.
91. Whittaker 1988.
92. Cherry 1988.
93. Watrous 1977.
94. Halstead 1992.

THE POST-MINOAN ERA

Like other coastal settlements in this part of Crete, the habitation site at Chrysokamino was abandoned in LM IIIB. This was part of a general movement throughout much of the Aegean.[95] This shift in settlement patterns resulted in more inland sites and fewer undefended coastal installations. Most scholars would attribute the change to a disruption of the relatively stable and peaceful situation that prevailed during the height of the Mycenaean economic empire. The majority view is that pirates and marauders made the exposed sites too dangerous, and most people preferred to live elsewhere. Mochlos and Gournia were abandoned or nearly abandoned at about the same time.

New foundations and increased population at existing sites can be seen on the high hills south of modern Kavousi.[96] The residents used Minoan pottery and other artifacts, but the population appears too small to account for all of the coastal residents. It has also been proposed that some people emigrated to Cyprus.[97]

During the 1st millennium B.C., the residents eventually moved back down from the peaks south of modern Kavousi. They resettled the region between the modern village and the sea, although they would never again live on the coastal hillside within the Chrysokamino territory itself. The largest and most important village near the territory was at Kephalolimnos, on the inland side of the coastal hills.[98] The site is not excavated, but it is covered with pottery, and a Roman or Byzantine millstone is a prominent local landmark (Figs. 19.5, 19.6).

The coastal territory of the Bronze Age farmstead must not have been used much during the Classical and Roman periods. Pottery from these periods is absent from the surface, a situation paralleled on the bare and equally dry land of nearby Pseira.[99] The inland slopes of the hills, however, were certainly exploited.

Small amounts of pottery attest to increased use of the region in the Byzantine era. Several sherds from a large Byzantine amphora were found on the surface of the habitation site (App. J, **J-1**). They suggest that some of the ruins may have stood high enough to provide a little shelter; unfortunately, the walls were neither photographed nor described before the construction of a modern mandra in the mid-20th century when many of the stones were robbed. Although no Byzantine architecture has been identified at Chrysokamino, this part of Crete was so heavily settled at this time that people must have visited the territory regularly, whether they farmed there or not.

The slopes of the low hills were farmed during the Venetian period, and the height of the post-Bronze Age activity was reached during the Ottoman occupation. Many sherds from these periods are present in the territory. Probably most of the surviving terrace systems are from the 15th to 19th centuries, as suggested by the excavation of **AF 22b** and its adjoining small cave (Apps. G, K). In recent centuries, the town has been at Kavousi,[100] a settlement whose roots, like those of Chrysokamino, are in the Neolithic period.[101]

95. For bibliography, see Desborough 1964, 1972; Drews 1993.

96. Gesell 1990; Coulson 1990; Haggis 1993b; for more recent bibliography, see Day and Snyder 2004; Eliopoulos 2004; Klein 2004; Glowacki 2004; Nowicki 2004.

97. Karageorghis 1992.

98. Boyd 1901, p. 156; Pendlebury 1939, p. 375.

99. Seager 1910.

100. Faure 1975, p. 32; 1989, p. 343; Spanakis 1991, pp. 322–323.

101. Haggis 2005, p. 38.

The use of the land during these later periods was very different from the pattern in the Bronze Age. The changes have been documented through the identification and description of manmade features in the landscape (App. G). The transformation in land management is best understood by examining the tiny sites that occur in spatially related clusters.[102] They reveal a new settlement pattern for the local territory in the post-Minoan era. The Minoans lived on the coastal side of the hill, and they must have walked out to their gardens and fields and returned home with their harvest and tools. Because the later residents lived farther away, they needed field houses for storage of tools and harvests, threshing floors near the fields to avoid carrying bulky harvests back home, and wells in the fields to provide water.

Terracing

The Chrysokamino territory has several areas with agricultural terraces supported by stone walls (Fig. 16.3). The walls are all made of locally available stones, gathered from the vicinity and not modified in any way. The locations are recorded in Appendix G. Surface pottery shows that most of the terraces were used in Ottoman times, as is to be expected for the period of maximum usage at the end of a long tradition.

The excavation of terrace **AF 22b** provides detailed evidence for one of the large terrace systems (App. K). To build this terrace, the land was cut back at the uphill side, and a one-faced wall of small stones was constructed without mortar on the downhill side. The space behind the wall was then filled in with soil. If this type of terrace construction began at the bottom of the hill, soil from each successively higher terrace could have been shoveled onto the adjoining lower level when the adjacent higher slope was cut back. At the top of the system called AF 22, the highest terrace (**AF 22a**) is rather small, and it would not have yielded enough soil to fill both itself and **AF 22b**, so soil from the excavation of the small cave (Apps. G and K, **AF 9**) was used for part of the fill. The system as a whole was not constructed evenly, but it followed the lay of the land, combining features of the stepped terrace and the braided terrace as defined by Rackham and Moody.[103]

Pottery from the excavation comes from the Ottoman period, and this date is probably applicable for the present configuration of the entire system. It is possible, however, that earlier terraces may have been modified when the final system was built. A small amount of Venetian pottery comes from within the soil of terrace **AF 22**, and it must attest to earlier use of the hillside. An occasional Minoan sherd also occurs on the surface of this terrace system. The Ottoman pottery from the lower level of terrace **AF 22a**, most likely providing the date of the wall, is from the 18th century.

Land Use

This project clearly demonstrates that Minoan land use at Chrysokamino was fundamentally different from the local land use in later periods. The Minoan period provided no evidence for boundary walls or any other indications of private ownership. All of the territory seems to have been farmed by people who knew each other so well that physical markers identifying boundaries were unnecessary. From more recent times, the remains of

102. On the application of this methodology, see Russell 1985, p. 30.
103. Rackham and Moody 1992.

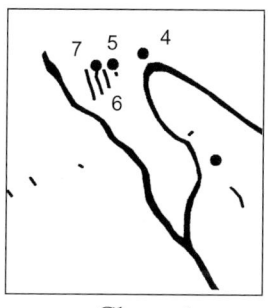 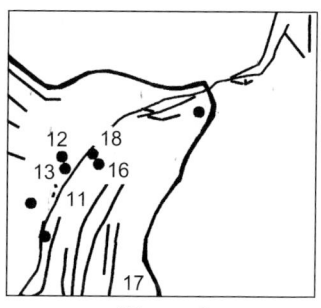 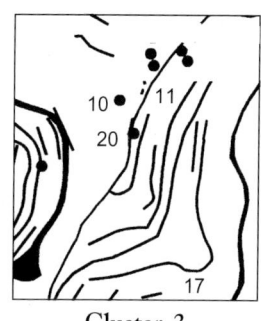

Figure 22.1. Map of Clusters 1–3, with closely related anthropogenic features

eight field houses showed that the territory was divided between several households, each of which required its own permanent fixtures for storage or other functions. Short boundary walls also existed, giving a clear sign that the limits between fields could be the subject of disputes if they were not marked in a permanent way. A basic change in land use is thus apparent.

Several of the anthropogenic features from the post-Minoan periods occur in groups that suggest individual ownership/usage (App. G). Clusters often consist of a group of agricultural terraces, a small building, and a threshing floor. Sometimes a well occurs nearby.

A brief discussion of three clusters shows the pattern (Figs. 16.2, 22.1):

Cluster 1 (Fig. 22.1, left)
    AF 5 and **AF 7**. Field houses (consecutive?)
    AF 4. Threshing floor
    AF 6. Agricultural terrace walls

Cluster 1 is a group of closely spaced and obviously related features. The presence of a threshing floor (App. G, AF 4) indicates a use with grains or some other crop requiring threshing, and the latest phase (mid-20th century) of dry farming on some of the other slopes of Chomatas suggests barley as a likely candidate. Field houses used for storage and for shade in this part of Crete are usually single-roomed structures with a rectangular shape. Because of their different degrees of preservation, the two examples here are more likely to be successive rather than contemporary.

Cluster 2 (Fig. 22.1, center)
    AF 13 and **AF 16**. Field houses (consecutive?)
    AF 18. Threshing floor
    AF 12. Well
    AF 11 and **AF 17**. Agricultural terrace walls

Cluster 2 has features that are similar to those of Cluster 1. As in Cluster 1, the two field houses were probably not used at the same time. The addition of a well, one of three in the territory under examination, shows that the labor expended in digging was preferred to the difficulty of transporting water from a distant source.

Cluster 3 (Fig. 22.1, right)
    AF 10. Mound of stones, probably a field house
    AF 20. Threshing floor
    AF 11 and **AF 17**. Agricultural terrace walls

The mound of stones in this cluster is almost certainly a collapsed building, apparently one of rather small dimensions. It probably served as the field house for threshing floor **AF 20**. The threshing floor is close to agricultural terrace walls (App. G, **AF 11** and **AF 17**), where it would have been convenient at harvest time. These terrace walls extend to the vicinity of Cluster 2, and a boundary between the two parts of the landscape may have existed, but it is not visible archaeologically.

In addition to these clearly associated groups of features, other groups also exist (Fig. 16.2). The small cave (App. G, **AF 9**) is near a large group of agricultural terraces (App. G, **AF 22**), and it was probably excavated for use as a field house. This practice is not uncommon in cases where marginal land is being cultivated.[104] A well (**AF 2**) is associated with a nearby boundary wall (App. G, **AF 3**).

The territory also has features associated with animal husbandry. The oval pen south of the habitation site (App. G, **AF 32**) and its successor (App G, **AF 28**) were intended to hold flocks and herds. Some of the field houses could have been used as shelters for shepherds.

These clusters and the more isolated features show that the later system of land use was completely different than the situation during the Bronze Age. The territory was divided between many users, with individual sections that accommodated dry farming on agricultural terraces by nonresident farmers who traveled out to their tiny plots of land from Kavousi and its predecessors like Kephalolimnos. The eight field houses suggest that the land was subdivided into about six separate plots (assuming that the nearby pairs are successive rather than contemporary).

Most of the post-Minoan features are associated with small-scale agriculture and animal husbandry. They furnish evidence for small individual holdings, not large agrarian estates. The Late Minoan farm was quite large in comparison. The later features probably also represent multiple periods of use, and often only the most recent period is clearly visible. It seems likely that the coastal strip was farmed and grazed during many periods after the abandonment of the Minoan settlement.

The Situation at the Beginning of the 21st Century

Since the end of the 20th century, the Chrysokamino territory has been used mainly for recreation, especially fishing, and for grazing animals. At the end of its long period of use, the coastal strip between Chordakia and the sea has been extensively overgrazed. Even phrygana, an assortment of small spiny shrubs and thistles, is restricted in distribution. The coastal strip is so bare of vegetation that a contrast with the inland area is visible from the air. The satellite photograph published as Figure 1.3 was taken in 1988 before the construction of the road from the earlier track to the excavated habitation site. The boundary between wasteland and olive groves is a prominent feature.

104. Halstead 2000, p. 118.

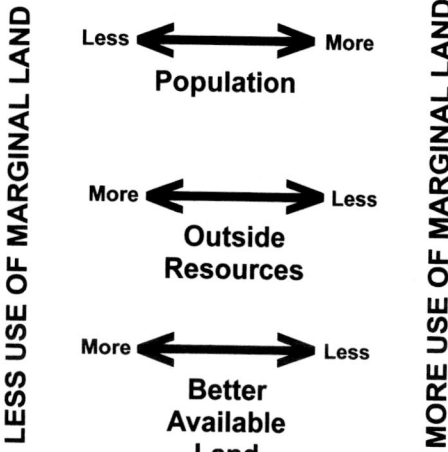

Figure 22.2. Pressures influencing greater or lesser use of marginal land for agriculture

Comments

A fundamental change in land use occurred when the Chrysokamino farm was abandoned in LM IIIB, and most of the residents moved to higher locations away from the coast. The coastal territory at this location would never again have a permanent settlement. It would, instead, alternate between periods of almost complete abandonment, as in the present day, and periods of use by farmers who traveled out to their fields from a village located away from the coast.

Like the more recent abandonment of the coastal strip, the decision not to farm the seacoast after the Bronze Age must have been related to several factors, including relative prosperity. A three-way relationship between size of population, amount of available land, and other components of the region's economy has profoundly affected decisions about farming marginal land. Land that is difficult to farm successfully, yielding only a small return for substantial investment in time and effort, will not be used in every period. The principles are illustrated in Figure 22.2. More population, fewer outside resources, and a decreased availability of better land exert pressure to increase the exploitation of marginal land. If these factors are reversed, the pressure to use marginal land decreases.

In the late 20th century, external contributions to the economy (outside resources) and the availability of better land have caused the coastal strip to be abandoned. Being a waiter in a restaurant, for example, or raising olives in the Kambos is financially and socially more attractive than spending long hours trying to raise crops or animals in a desolate landscape.

During the early 1st millennium B.C., the decrease in population and the concomitant availability of more land was probably responsible for the move away from the coast. With migrations from Crete to Cyprus and other places in the East, the people who remained chose better farmland than the exposed and dangerous coastal strip with its dry and marginal soil. Climatic change that caused the land to become drier must have intensified the process of abandonment.

It was probably an increase in population that prompted reuse of the area in the Venetian and Ottoman periods. During all the late periods when the land was farmed, the farmers traveled out to the fields from homes located inland. These new farming practices continued to be used until the early 20th century. As Lloyd observed in the case of the Roman Empire, "There is little doubt that the peasant commuting to the fields from town, village, or hamlet was a common sight."[105]

In conjunction with this new strategy, the farmers needed to construct field buildings and threshing floors *(alonia)* in the countryside. All of the permanent installations recognized in the present study were built after the Minoan period (App. G), and the excavation of one of them, along with the closest part of its nearby terrace wall system, shows that the final era of use postdated the Venetian period (App. K).

The survey provides several examples of the small field installations used as agricultural processing facilities in the former territory of Chrysokamino (App. G). They typically consisted of a small building or shallow cave, a threshing floor, and a series of terrace walls. Elsewhere in Greece, *alonia* were used for processing a variety of crops, including figs and grapes that needed to be dried.[106] In this part of Crete, however, the flat areas were used primarily for threshing grain, and all of the examples noted at Chrysokamino have vertical stones at their edges to keep the grain contained during the threshing process.

Walls made of rough stones were used as boundary markers before the modern fences were built (App. G). They divided the public land at the coast from the inland private land, but they were so flimsy that they only survive in a few places. These early boundary markers show that the governmental regulations that currently operate and provide for a balance between open land and private land has a long history in the region, but no evidence survives to suggest it operated during the Bronze Age. Like the agricultural-processing facilities, the boundary markers seem to be a post-Minoan phenomenon.

The use of the coastal hills for growing grains continued into the first half of the 20th century at Chrysokamino. In 1947, the United States started the Marshall Plan, which provided funding to help rebuild Greece after the destruction of its economy during World War II. A part of the plan provided money for deep wells in the Kambos to permit irrigation.[107] As a result, the land was planted in olives, a crop that was gaining in favor throughout Crete because of its nutritional value and other aspects of its production.[108] After the transformation of the land to support olive culture, the coastal side of the hills, where the Bronze Age workshop and its associated settlement and burial cave were located, was increasingly neglected because the irrigation system did not reach this far. Although the coastal strip was no longer used at the close of the 20th century, aside from recreation and animal husbandry (by only one shepherd), the inland slopes of Chomatas, called Chordakia, were completely planted in olives and a few other trees. Chordakia and the Kambos, divided into small individually owned plots, supported olives and garden vegetables that were tended by residents of Kavousi, who traveled out to their fields from their homes in the village. By this time, the land management system that favored the location of settlements in the midst of the gardens and fields had ceased to exist for many centuries.

105. Lloyd 1991, p. 235.
106. Lee 2001, p. 61.
107. Allbaugh 1953, pp. 258–263.
108. Riley 2002.

APPENDIXES

APPENDIX A

Petrography and X-Ray Diffraction Analyses of Slags and Furnace Chimneys

by George H. Myer and Philip P. Betancourt

Samples of the slags and furnace chimneys from Chrysokamino were analyzed by optical thin section and by X-ray powder diffraction in the Mineralogy Laboratory, Department of Geology, Temple University. The D. M. Organist Laboratory (Newark, Delaware) prepared the thin sections. The methodology followed standard practices for optical thin section analysis.[1] The presentation of the results follows the format used by Myer, McIntosh, and Betancourt.[2]

The analyses by X-ray diffraction used a Rigaku XRD powder diffractometer model DMAX/B with a copper X-ray tube source, horizontal goniometer (radius 185 mm) with total accuracy within 0.02 degrees, and curved crystal graphite monochronometer analysis system. The continuous scan mode, which counts integrated intensity per specified sample interval, produced diffraction scans run at 10 degrees per minute, with a sampling interval of 0.05. All data were recorded on magnetic disc, analyzed by Rigaku software, and printed and plotted for hard copy use.

PETROGRAPHY OF THE SLAGS

The slags at Chrysokamino are dark masses with a glassy appearance (see Chap. 10). Because many pieces that appear similar when seen without magnification have very different compositions, the class as a whole can be described as heterogeneous in detail but similar in overall appearance. The differences are sufficient to suggest variations in the original materials as well as different conditions in the furnaces during the pyrotechnological process.

Two different classes of slag can be distinguished. One class consists of condensation on the inside of chimney walls, and the other class consists of solid pieces of dark vitreous material that formed from the melt. Analyses of four samples are presented here.

1. Bambauer, Taborsky, and Trochim 1979.

2. Myer, McIntosh, and Betancourt 1995.

Figure A.1. Slag sample CHR 97. Petrographic thin section showing black dendritic crystals of magnetite in a glassy matrix; plane-polarized light, width of field 2.5 mm.

1. CHR 97 (vitreous coating in the interior of a chimney fragment, Fig. A.1), a slag containing opaque dendritic inclusions (magnetite) occurring as branching dendritic crystallizations, with no fayalite or pyroxene.
2. CHR 93 (vitreous coating in the interior of a chimney fragment), a slag containing radiating crystallizations of magnetite, with minor olivine.
3. CHR 87 (slag fragment, Fig. A.2), a slag containing opaque dendritic inclusions (magnetite) occurring as branching dendritic crystallizations, along with green crystal points and masses of pyroxene and masses of cuprite and fayalite.
4. CHR 99 (slag fragment, Figs. A.3, A.4), a slag fragment consisting of two dissimilar bands, with a considerable percentage of clinopyroxene in the glassy matrix along with inclusions of an opaque mineral, fayalite, cuprite, and unidentified phases.

Slag with Black Dendritic Inclusions

Number of samples examined: 1 (CHR 97, slag deposited on the interior of a chimney fragment)

Matrix (Glass Groundmass)

A. Optical properties and color
   1. Under plane-polarized light: mottled, with color variation from tan to yellow brown to dark brown; some of the color variation is zonal around the opaque inclusions.
   2. Under cross-polarized light: isotropic.

B. Overall grain size and modality of inclusions and voids: inclusions vary widely, from under 62 micrometers to over 2 mm; voids vary widely, from under 62 micrometers to over 1 mm; ca. 50% inclusions; ca. 10% voids.

C. Overall preferred orientation of inclusions and voids: random.

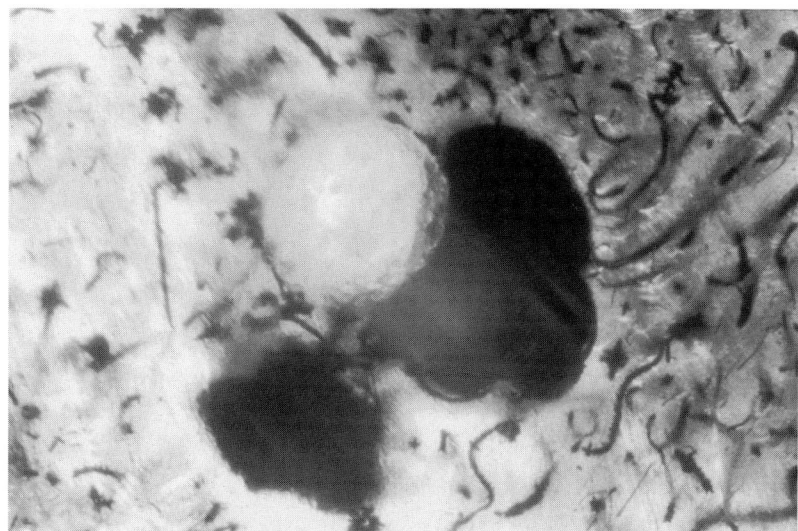

Figure A.2. Slag sample CHR 87. Petrographic thin section showing magnetite (black, thin curved crystals), fayalite (pale, circular, at upper center), and pyroxene (dark rounded masses) in a glassy matrix; plane-polarized light, width of field 800 micrometers.

Figure A.3. Slag sample CHR 99. Petrographic thin section showing pyroxene crystals (radiating), magnetite (black, at top left of center), and cuprite (circular, upper center) in a glassy matrix; cross-polarized light, width of field 800 micrometers.

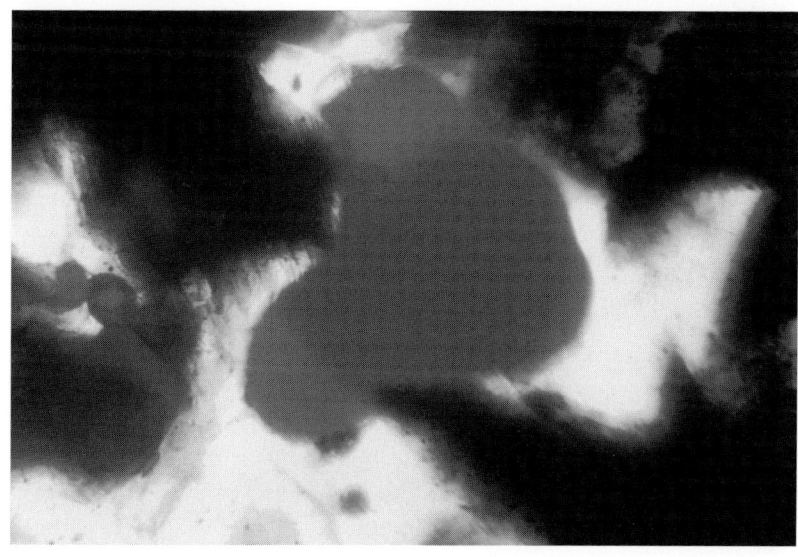

Figure A.4. Slag sample CHR 99. Petrographic thin section showing pyroxene crystals (dark) and cuprite (black triangular mass with rounded corners at center) in a glassy matrix; plane-polarized light, width of field 2.5 mm.

D. Inclusions
  1. Magnetite dendrites.
      Size range: from under 1 micrometer up to ca. 10 × 250 micrometers.
      Shape: multiple orthogonal branching growth patterns with the crystallization initiated along both primary and secondary branches; usually like a two-sided comb, with isolated cross-sections visible as well.
      Comments: opaque dendritic quench crystals with many branches; isometric; identified as magnetite by X-ray diffraction; in some locations the rows of long dendrites are at right angles to one another, creating a gridlike pattern; sometimes iron alteration products are visible as red brown colorations at the margins of the opaque crystals.
  2. Quartz.
      Size range: from ca. 10 micrometers up to ca. 2 mm.
      Shape: subrounded to rounded.
      Comments: single grains; undulose extinction; some grains have acicular inclusions (rutile?); some grains are rimmed with the opaque dendritic crystals (i.e., the quartz acted as a nucleation surface on which the crystals grew).
  3. Chert.
      Size range: from ca. 10 micrometers up to ca. 1 mm.
      Shape: subrounded to subangular.
      Comments: microgranular texture.

Voids

A. Size range: up to over 500 micrometers.

B. Shape: mostly oval.

C. Comments: voids from entrapped gas.

Modality

Matrix (glass groundmass): ca. 50%. Opaque dendrites: ca. 38%–40%. Voids: ca. 10%. Quartz: under 1%. Chert: under 1%.

Discussion

Sample CHR 97 consists of a glassy deposit on the interior of a chimney fragment. The matrix, a mottled glass, includes abundant dendritic crystallizations, which may be identified as magnetite that crystallized when the slag cooled quickly (see App. B). The abundant magnetite indicates the slag formed under reducing conditions. Quartz grains, which are unaltered relics from the original charge, have also been noted in slags of Chalcolithic date from Israel.[3]

The X-ray diffraction analysis of this sample shows magnetite (the dendrites) and quartz (the quartz grains and the chert grains). Other peaks are too ambiguous for certain identification. No fayalite or other members of the olivine group can be detected.

3. Bachman 1980, pp. 108–110.

Slag with Radiating Magnetite Inclusions

Number of samples examined: 1 (CHR 93, vitreous coating deposited on the interior of a chimney fragment).

Matrix (Glass Groundmass)

　A. Optical properties and color
　　1. Under plane-polarized light: mottled, with color variation from tan to yellow brown to dark brown.
　　2. Under cross-polarized light: crystallized (devitrified), consisting of interlocking blades with some pleochroism (most likely a postburial reaction, not an original characteristic of the slag).

　B. Overall grain size and modality of inclusions and voids: inclusions vary widely, from under 62 micrometers to over 2 mm; voids vary widely, from under 62 micrometers to over 1 mm; ca. 15% inclusions; ca. 10% voids.

　C. Overall preferred orientation of inclusions and voids: random.

　D. Inclusions
　　1. Magnetite dendrites.
　　　Size range: from under 1 micrometer up to ca. 60 micrometers.
　　　Shape: small radiating groups of crystals.
　　　Comments: opaque dendritic quench crystals consisting mostly of small radiating groups; identified as magnetite by X-ray diffraction; sometimes Fe alteration products are visible as red brown colorations at the margins.
　　2. Chert.
　　　Size range: ca. 30 micrometers.
　　　Shape: subrounded.
　　　Comments: microgranular texture.
　　3. Carbonate.
　　　Size range: up to ca. 100 micrometers.
　　　Shape: films.
　　　Comments: amorphous; postburial depositions.
　　4. Olivine group.
　　　Size range: under 10 micrometers.
　　　Shape: spherical masses.
　　　Comments: alteration of the glass (devitrification); it is possible that the devitrification is a postburial alteration.

Voids

　A. Size range: up to over 500 micrometers.

　B. Shape: mostly oval.

　C. Comments: voids from entrapped gas.

Modality

Matrix (glass groundmass): ca. 85%. Voids: ca. 10%. Magnetite: ca. 4%–5%. Chert: under 1%. Carbonate: under 1%. Olivine: under 1%.

Discussion

This sample (CHR 93) consists of a glassy deposit on the interior of a chimney fragment. The matrix, a mottled glass, contains only a few inclusions, which may be identified as magnetite that crystallized when the slag cooled quickly. The glassy matrix is extensively altered (devitrified) from weathering.

The X-ray diffraction analysis of this sample shows magnetite (the opaque crystals), quartz (the chert grains), hedenbergite (in the devitrified glass), and a hydrated copper chloride (not present in the thin section prepared from this sample). No olivine could be detected in the X-ray analysis, although the optical examination shows that a tiny amount is present (apparently too little to show up in a bulk analysis).

Slag with Black Dendritic Inclusions and Pyroxene

Number of samples examined: 1 (CHR 87, fragment of loose slag).

Matrix (Glass Groundmass)

A. Optical properties and color
   1. Under plane-polarized light: pale green to brown.
   2. Under cross-polarized light: isotropic where pale in color and anisotropic where brown.

B. Overall grain size and modality of inclusions and voids: inclusions up to ca. 20–30 micrometers; voids vary widely; isotropic glass ca. 5%; anisotropic glass ca. 80%; inclusions ca. 10%–15%; voids under 2%.

C. Overall preferred orientation of inclusions and voids: random.

D. Inclusions
   1. Magnetite dendrites.
      Size range: mostly ca. 5 x 10–20 micrometers.
      Shape: branching dendrites, usually like a two-sided comb, with isolated cross-sections visible as well.
      Comments: opaque dendritic quench crystals with many branches; isometric; identified as magnetite by X-ray diffraction.
   2. Pyroxene.
      Size range: ca. 10 micrometers.
      Shape: crystal points and masses enveloping the dendritic needles; sometimes the pyroxene is nucleated in the magnetite.
      Comments: green.
   3. Fayalite.
      Size range: ca. 10 micrometers up to ca. 50 micrometers.
      Shape: spherical masses.
      Comments: yellow to yellow red masses with internal radiating crystalline structures.

4. Cuprite.
      Size range: under 10 micrometers.
      Shape: Spherical masses.
      Comments: red in color.

Voids

A. Size range: up to over 100 micrometers.

B. Shape: mostly oval.

C. Comments: voids from entrapped gas.

Modality

Anisotropic matrix (glass groundmass): ca. 80%. Magnetite dendrites: ca. 10%. Isotropic matrix (glass groundmass): ca. 5%. Voids: under 2%. Pyroxene: under 5%. Fayalite: under 5%. Cuprite: under 1%.

Discussion

This sample consists of a glassy matrix with inclusions of magnetite dendrites, pyroxene, cuprite, and fayalite. The magnetite forms curved acicular crystals within the glass matrix. The sample was consumed in the preparation of the thin section, so that it was not available for examination by X-ray diffraction.

Slag with Pyroxene, Cuprite, and Fayalite Inclusions

Number of samples examined: 1 (CHR 99, fragment of loose slag).

Matrix (Glass Groundmass)

A. Optical properties and color
   1. Under plane-polarized light: isotropic glass with flow banding (with great differentiation between the bands).
   2. Under cross-polarized light: black.

B. Overall grain size and modality of inclusions and voids: inclusions vary widely, from under 62 micrometers to over 1 cm; voids vary widely, from under 62 micrometers to over 1 mm; ca. 60%–70% inclusions; under 1% voids.

C. Overall preferred orientation of inclusions and voids: usually parallel to the banding

D. Inclusions
Band 1 (pale colored glass)
   1. Pyroxene.
      Size range: average ca. 100–200 micrometers in diameter.

Shape: euhedral twinned crystals in radiating clusters.

Comments: green to dark green to black or brown; pleochroic green to brown (weak to moderate pleochroism) disseminated in the glass; quench crystals from rapid cooling; differentially zoned, with different interference colors at the crystal margins.

2. Fayalite.
    Size range: average ca. 20–40 micrometers in diameter.
    Shape: spherical.
    Comments: yellow to yellow red masses with internal radiating crystalline structure.
3. Cuprite.
    Size range: average ca. 20–100 micrometers in diameter.
    Shape: spherical to cubic.
    Comments: red; interstitial to the pyroxene.
4. Opaque mineral.
    Size range: average ca. 1–12 micrometers in diameter.
    Shape: nearly spherical.
    Comments: opaque; possibly elemental iron (or tenorite?).
5. Rock grains.
    Size range: 1 mm × 250 micrometers to 1 mm × 1.25 mm.
    Shape: rounded.
    Comments: isolated grains composed of interlocking crystals; anisotropic; not identified.

Band 2 (brown glass, strongly flow-banded; dense w/ microscopic inclusions)
1. Opaque mineral.
    Size range: average ca. 1–12 micrometers in diameter.
    Shape: nearly spherical.
    Comments: similar to band 1 but much more abundant.
2. Pyroxene.
    Size range: average ca. 100–200 micrometers in diameter.
    Shape: euhedral twinned crystals in radiating clusters.
    Comments: similar to band 1 but less abundant.
3. Fayalite.
    Size range: average ca. 20–40 micrometers in diameter.
    Shape: spherical.
    Comments: similar to band 1 but less abundant.
4. Cuprite.
    Size range: average ca. 20–100 micrometers in diameter.
    Shape: spherical to cubic.
    Comments: similar to band 1 but less abundant.

Voids

A. Size range: up to over 100 micrometers.

B. Shape: mostly oval.

C. Comments: voids from entrapped gas.

Modality

Band 1

Pyroxene: ca. 70%. Matrix (glass groundmass): ca. 25%. Voids: under 2%. Opaque mineral: under 1%. Cuprite: under 1%. Fayalite: under 1%. Rock grains: under 1%.

Band 2

Matrix (glass groundmass): ca. 70%. Pyroxene: ca. 15%. Opaque mineral: ca. 5%. Voids: under 2%. Cuprite: under 1%. Fayalite: under 1%.

Discussion

Sample CHR 99 consists of a slag with two dissimilar bands. The matrix contains pyroxene, fayalite, cuprite, and other unidentified inclusions.

X-ray diffraction analysis of a bulk sample from this slag indicates that four minerals are sufficiently crystallized to yield patterns: a clinopyroxene, probably augite (the pyroxene crystals), cuprite (the red cubic crystals), fayalite (brown radiating spheres), and elemental iron (probably the opaque inclusions).

Two types of glassy bands can be recognized in this slag, one with abundant clinopyroxene crystals and the other with fewer crystals but with more abundant, tiny spherical inclusions of an opaque mineral, possibly elemental iron. The slag, which was produced under strongly reducing conditions, includes tiny inclusions of fayalite (the Fe end member of the olivine group). Red crystals of cuprite (20–100 micrometers across) are also present, but whether they formed as a part of the original process or in postburial oxidation is uncertain.

The relative amounts of fayalite and pyroxene provide an indication of the degree of reduction during the smelting process. The presence of large amounts of fayalite is a characteristic of slags formed under highly reducing conditions, as opposed to the occurrence of pyroxene, which would be typical of more oxidizing conditions.[4] When substantial amounts of pyroxene and the oxide cuprite are visible in slags from Chrysokamino, as in this slag, less magnetite and fayalite are present, suggesting formation under more oxidizing conditions than the slag recorded as no. 3 above (CHR 87).

CHRYSOKAMINO FURNACE CHIMNEY FABRIC

Number of samples examined: 6 (CHR 91, CHR 92, CHR 93, CHR 94, CHR 97, CHR 98)

Matrix (Groundmass)

    A. Optical properties and color
        1. Under plane-polarized light: mottled.
        2. Under cross-polarized light: anisotropic.

    B. Overall grain size and modality of inclusions and voids: inclusions vary widely, from under 62 micrometers to over 5 mm; voids vary

---

4. Hauptmann, Bachmann, and Maddin 1994, p. 6.

widely, from under 62 micrometers to over 1 cm; ca. 50%–60% inclusions; ca. 10%–20% voids.

    C. Overall preferred orientation of inclusions and voids: usually parallel to the surface of the furnace.

    D. Silt-sized inclusions (under 62 micrometers)
        1. Quartz, type 1.
            Size range: up to 62 micrometers.
            Shape: subangular to angular.
            Comments: single grains; the smaller size range of the larger type 1 quartz grains.
        2. Quartz, type 2.
            Size range: up to 62 micrometers.
            Shape: subangular to angular.
            Comments: multiple grains; the smaller size range of the larger type 2 quartz grains; derived from the phyllite.
        3. Mn Oxide(s)(?).
            Size range: up to 62 micrometers.
            Shape: irregular.
            Comments: opaque films and borders on grains and voids.
        4. Chlorite.
            Size range: ca. 5–6 × 2 micrometers in section.
            Shape: flakes.
            Comments: colorless under plane-polarized light; first-order gray interference colors under cross-polarized light.
        5. Carbonate.
            Size range: ca. 10 micrometers.
            Shape: spherical.
            Comments: very few grains.

Voids

Type 1
    A. Size range: up to over 2 cm.
    B. Shape: varies widely: irregular; cylindrical; various plant forms.
    C. Comments: voids from burned out organic matter, including chaff (Fig. A.5).

Type 2
    A. Size range: up to over 1 cm.
    B. Shape: elongated.
    C. Comments: shrinkage cracks.

Inclusions above Silt-Size (above 62 Micrometers)

        1. Carbonate.
            Size range: up to 3 mm in size.
            Shape: irregular.
            Comments: biogenic carbonate rock; altered by heat; may

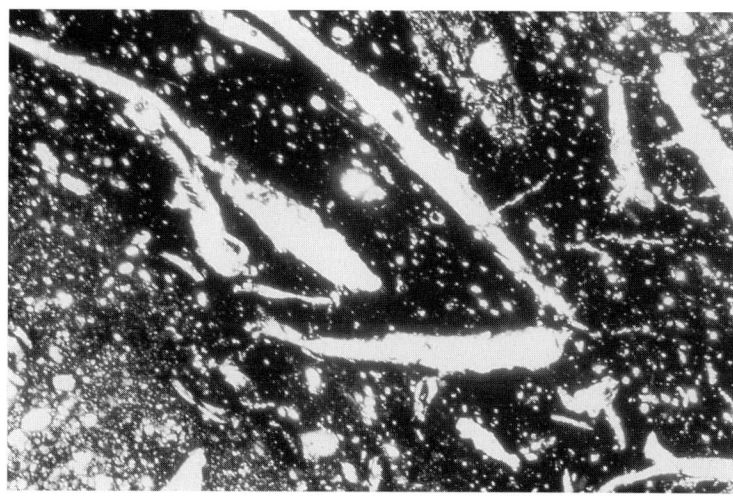

Figure A.5. Furnace chimney sample CHR 97. Petrographic thin section showing clay fabric with voids from burned out chaff; cross-polarized light, width of field 6 mm.

have oolitic structure with spherical to ovoid to irregular oolites; contains foraminifera; one grain in CHR 98 is attached to 1.5 mm quartz grain consisting of low-grade metamorphic quartz with fluid inclusions.

2. Shale.
    Size range: up to 5 mm in size.
    Shape: subrounded.
    Comments: contains grains of quartz; some fragments highly altered by heat.
3. Quartz, type 1.
    Size range: up to 0.5 mm in size.
    Shape: angular.
    Comments: single grains; clear; uniform extinction; precise boundaries.
4. Quartz, type 2.
    Size range: up to 1 mm in size.
    Shape: angular.
    Comments: multiple grains; derived from the phyllite.
5. Chert.
    Size range: up to 500 micrometers in size.
    Shape: rounded, subangular, and angular.
    Comments: granular texture; CHR 94 contains a subrounded chert grain with included Fe-rich rhombic crystals (siderite?).
6. Ferruginous siltstone.
    Size range: up to 5 mm in size.
    Shape: subangular to rounded.
    Comments: sedimentary rock; consists of quartz grains cemented by hematite; may contain small fragments of shale.
7. Phyllite.
    Size range: up to 600 micrometers in size.
    Shape: subrounded.
    Comments: consists of quartz, feldspar, mica, and chlorite; quartz has undulose extinction and multiple grains; CHR 89 has a grain of phyllite containing a subhedral crystal of epidote with light brownish green pleochroism in plain-polarized light and typical

"stained glass window" appearance of second-order interference colors in cross-polarized light.

Modality

Matrix (groundmass): 20%–30%. Voids: 10%–20%.
Silt-sized inclusions: 1. Quartz, type 1: 10%. 2. Quartz, type 2: under 5%. 3. Mn Oxide(s)(?): under 1%.
Larger inclusions: 1. Carbonate: 15%. 2. Shale: 15%. 3. Quartz, type 1: under 5%. 4. Quartz, type 2: under 5%. 5. Chert: under 5%. 6. Siltstone: under 5%. 7. Phyllite: under 5%.

Discussion

The fabric is consistent with a local source in the region of the east side of the Gulf of Mirabello. The quartz and phyllite inclusions are petrographically similar to those found in this part of Crete. Manganese oxide inclusions are also present in clay deposits from the region, including deposits near the site of Chrysokamino. All of the constituents are common, and no distinctive elements have been recognized by this methodology that would allow a precise source to be identified.

ANALYSIS OF SLAGS BY X-RAY DIFFRACTION

Analyses by X-ray powder diffraction was performed on four samples of slag, two of which were angular broken pieces of glassy slag containing microscopic inclusions and two of which were vitrious coatings on the interior surfaces of chimney fragments. The following mineralogical phases were sufficiently well crystallized to provide X-ray patterns in bulk samples:

CHR 96 (glassy slag)
    Akermanite
    Diopside
    Delafossite
    Dolomite (?)

CHR 99 (glassy slag)
    Augite
    Cuprite
    Fayalite
    Elemental iron(?)

CHR 93 (coating on the interior of a chimney fragment)
    Magnetite
    Quartz
    Hedenbergite
    Unnamed hydrated copper chloride

CHR 97 (coating on the interior of a chimney fragment)
    Magnetite
    Quartz

SEM/EDAX Analysis

by Yannis Bassiakos

This study presents analyses of metallic prills and slag by SEM/EDAX. The analyses were made at the Laboratory of Archaeometry, Institute of Materials Science, N.C.S.R. "Demokritos," Hagia Paraskevi, Attikis, Greece.

Table B.1 presents the analyses of copper prills arranged by increasing arsenic content. This set of measurements differs from the PIXE results shown in Appendix D in that these measurements were made on clean, uncorroded metal. The arsenic contents are substantially higher in this set of measurements than they are in the values obtained from PIXE analysis (App. D); they range from 0% to 26%, with an average of 5.51%.

Samples of the slag were sawed, mounted, and analyzed with SEM/EDAX. The results are shown in Table B.2.

TABLE B.1. SEM/EDAX ANALYSES OF COPPER PRILLS ARRANGED BY INCREASING ARSENIC CONTENT

SEM/EDAX Analysis No.	Arsenic Content in Weight Percentage
24	0.00
9	0.20
26	0.50
34	1.50
22	1.10
5	4.10
11	5.30
12	5.50
45	7.37
16	10.00
48	26.00

TABLE B.2. SEM/EDAX ANALYSES OF METALLURGICAL REMAINS

Analysis No.	Mg	Al	Si	P	S	K	Ca	Ti	Cr	Mn	Fe	Ni
1	nd	nd	3.4	nd	nd	0.1	1	nd	nd	nd	4.6	nd
2	1.6	7.4	41	0.8	0	1.6	12	0.5	0.1	0.2	33	0.02
3	0.9	7.5	42.1	0.9	nd	1.9	12	0.4	0.1	0.1	34	nd
4	nd	nd	0.2	0.7	22	0.1	0.2	0.1	0.1	0.06	2.4	0.2
5	nd	0.5	0.1	nd	0.7	nd	0.1	nd	0	0.04	2.7	0.2
6	0.3	2.8	2.6	0.1	0.2	0.2	0.7	1.3	0.1	0.2	88	0.06
7	0.6	24.5	47.6	1	0	0.4	3	0.1	nd	nd	18	0.1
8	0.6	5.7	29.5	0.15	0.1	1.3	5	0.7	0.1	0.05	56	0.07
9	0.3	nd	0.2	0.06	0	0	0.1	0.1	0	nd	2.7	0.06
10	0.5	5.9	37	0.7	0.1	1	26	0.6	0	0.06	26	nd
11	2.5	nd	0.3	nd	0.4	nd	0.4	0	nd	nd	1.9	0.07
12	2.4	nd	0.2	nd	0.1	nd	0.1	0	nd	nd	0.2	0.2
13	0.4	1.1	1.3	0.05	0.2	0	0.9	0.2	0	0.3	95	0.08
14	0.5	6.8	37.7	0.8	0.1	1.4	24	0.8	0.1	0.2	27	nd
15	1.3	5.8	37.5	0.8	0.1	1.2	26	0.6	0.1	0.2	25	nd
16	nd	0.4	0.4	nd	0.6	nd	0.5	0.1	nd	nd	2.3	0.3
17	nd	0.1	1.4	0.7	24	0.1	1	0	nd	nd	6	0.14
18	10	8.8	36.5	1.2	0.1	2.1	2.1	1.1	0	0.2	28	nd
19	17.1	1.5	58.4	0.03	0.1	1	6.4	0.5	nd	0.1	14	nd
20	0.8	7.7	36.7	1	0.1	1.8	21	0.9	0	0.06	29	nd
21	0.5	6.9	34.5	0.6	0.2	1	20	0.7	nd	0.4	34	0.06
22	0.3	0.2	1.3	0.1	0.4	nd	0.8	nd	nd	0.02	4	1
23	0.4	2.7	4.7	0.08	0	0.7	4.6	1	0.2	0.4	85	0.2
24	nd	nd	0.2	0.1	0.1	0.1	0.1	0.1	0.1	0.05	1.9	0.02
25	0.2	0.4	3.4	0.5	24	nd	4.8	0.1	nd	<0.1	11	0.05
26	nd	0.1	0.6	0.02	0.2	nd	0.4	0.1	nd	0.06	4.2	0.3
27	0.2	4.3	39.2	0.6	nd	0.7	14	0.4	nd	0.2	37	0.03
28	0.2	2	1.4	nd	nd	nd	0.6	0.7	0.1	0.03	95	0.1
29	0.5	6.2	33.7	0.7	0.2	1.2	30	0.6	nd	0.4	24	0.04
30	nd	nd	0.2	0.6	24	nd	0.6	nd	nd	nd	5.7	0.4
31	nd	0.2	1.8	1.3	32	0.2	2.9	0.1	nd	0.06	8.7	0.1
32	0.8	6.2	37.9	nd	<0.1	0.7	18	0.8	<0.1	0.3	33	<0.1
33	3.3	4.8	0.4	nd	0.1	nd	0.4	1.6	0.3	0.2	89	0.08
34	nd	0.1	1.4	<0.1	<0.1	nd	0.8	<0.1	nd	<0.1	4.9	1.4
35	1.1	19.7	54	<0.1	0.1	2.1	13	1.2	nd	0.2	8.1	nd
36	1.2	28.9	47	0.3	0.1	0.6	18	0.3	0.1	<0.1	3.5	nd
37	1.7	15.1	51	0.3	0.1	2.3	18	1	0.1	0.3	8.8	nd
38	1.2	2.7	2	0.07	0.2	0.1	0.7	23	0.2	0.3	69	0.06
39	1.5	20.4	60.7	nd	0.2	3	2.3	1.3	9.1	0.3	9.2	0.03
40	1.1	19.5	64.5	nd	nd	2.8	1.6	1.2	0	0.2	7.9	0.02
41	2.0	15.3	53.5	nd	nd	2.34	12.8	0.97	0.06	0.27	11.4	0.01
42	6.37	22.8	53	nd	nd	5.05	0.71	1.03	0.14	0.04	10.6	nd
43	1.78	20.4	63.2	0.03	nd	4.00	1.54	0.84	0.04	0.15	7.5	nd
44	3.5	14.5	22	nd	nd	0.9	1.48	1.1	0.65	0.33	55	0.27
45	nd	nd	0.32	0.17	nd	0.11	0.15	0.05	nd	0.05	1.25	0.95
46	0.86	4.5	42	0.86	0.31	0.94	32.6	0.45	0.23	0.26	14.4	nd
47	nd	0.08	1.93	nd	25.4	nd	1.53	0.07	0.01	0.01	4.18	nd
48	nd	nd	3.06	nd	3.28	nd	2.19	0.08	nd	nd	2.01	0.08
49	nd	nd	0.32	nd	24.5	nd	0.34	0.04	nd	0.02	1.93	0.16

TABLE B.2 (cont.)

Analysis No.	Cu	Zn	As	Pb	Au	Sb	V	Cl	Na	I	Ba
1	0.8	nd	nd	0.6	90	nd	nd	nd	nd	nd	nd
2	1.6	0.2	0.2	nd	nd	nd	nd	nd	nd	nd	nd
3	0.3	0.1	0.1	nd	nd	nd	nd	nd	nd	nd	nd
4	73	0.3	1.4	nd	nd	nd	nd	nd	nd	nd	nd
5	91	0.3	4.1	nd	nd	0.5	nd	nd	nd	nd	nd
6	3.3	0.2	nd	nd	nd	nd	0.2	nd	nd	nd	nd
7	2.3	0.2	1.6	nd	nd	nd	nd	0.3	nd	nd	nd
8	0	0.1	nd	nd	nd	nd	nd	nd	nd	nd	nd
9	96	0.6	0.2	nd	nd	nd	0	nd	nd	nd	nd
10	1.6	0.2	0.3	nd	nd	nd	nd	0.1	nd	nd	nd
11	89	0.2	5.3	nd	nd	nd	nd	nd	nd	nd	nd
12	91	0.5	5.5	nd	nd	nd	nd	nd	nd	nd	nd
13	0.1	0.4	nd	nd	nd	nd	nd	0	nd	nd	nd
14	0.8	0.2	0.2	nd	nd	nd	nd	0	nd	nd	nd
15	1	nd	0.2	nd	nd	nd	nd	0	0.2	nd	nd
16	84	nd	10	nd	nd	nd	nd	1.3	nd	nd	nd
17	63	nd	3.7	nd	0.2	nd	nd	0	nd	nd	nd
18	0.1	nd	0.1	nd	nd	nd	nd	0.1	nd	nd	nd
19	1.1	nd	0.1	nd	nd	nd	nd	0.1	nd	nd	nd
20	0.5	nd	0.1	nd	nd	nd	nd	0	nd	nd	nd
21	0.6	nd	0.1	nd	nd	nd	nd	0.1	0.5	nd	nd
22	88	nd	1.1	nd	nd	nd	nd	nd	nd	3	nd
23	0.2	nd	nd	nd	nd	nd	nd	0.1	nd	nd	nd
24	97	0.2	nd	nd	0.1?	nd	nd	0.5	nd	nd	nd
25	52	0.3	nd	nd	nd	nd	nd	4.2	nd	nd	nd
26	88	nd	0.5	nd	0.5	nd	nd	5.3	nd	nd	nd
27	2.5	0.5	0.4	nd	nd	nd	nd	0.1	nd	nd	nd
28	0	0.3	0.1	nd	nd	nd	nd	nd	nd	nd	nd
29	0.4	nd	nd	nd	nd	nd	nd	nd	0.3	nd	0.9
30	68	nd	0.8	nd	nd	nd	nd	nd	nd	nd	nd
31	53	nd	nd	nd	nd	nd	nd	nd	nd	nd	nd
32	1.3	nd	nd	nd	nd	nd	nd	0.2	0.3	nd	nd
33	0.1	nd	nd	nd	nd	nd	nd	<0.1	nd	nd	nd
34	91	nd	0.5	nd	nd	nd	nd	<0.1	nd	nd	nd
35	0.6	nd	nd	nd	nd	nd	nd	nd	nd	nd	nd
36	0.1	nd	nd	nd	nd	nd	nd	nd	0.2	nd	nd
37	1.5	nd	0.3	nd	nd	nd	nd	nd	nd	nd	nd
38	0.1	nd	nd	nd	nd	nd	nd	nd	nd	nd	nd
39	0.3	nd	nd	nd	nd	nd	nd	0.6	0.2	nd	nd
40	0.3	nd	nd	nd	nd	nd	nd	0.7	0.1	nd	nd
41	1.2	nd	0.26	nd	nd	nd	nd	nd	nd	nd	nd
42	0.03	nd	nd	nd	nd	nd	nd	0.24	nd	nd	nd
43	0.44	nd	nd	nd	nd	nd	nd	0.2	nd	nd	nd
44	0.36	nd	nd	nd	nd	nd	nd	nd	nd	nd	nd
45	89.5	nd	7.37	nd	nd	nd	nd	0.06	nd	nd	nd
46	0.4	0.04	nd	nd	nd	nd	nd	0.15	nd	nd	1.94
47	62.7	0.27	3.73	nd	nd	nd	nd	nd	nd	nd	0.11
48	63.7	0.18	26	nd	nd	nd	nd	nd	nd	nd	nd
49	70.5	0.26	1.93	nd	nd	nd	nd	nd	nd	nd	nd

Notes: Concentrations of less than 0.1% are not reliable, and they are given only as indications. The long calcareous-silicate phases (Ca>Si) are fayalite-like. The long-dark phases consist predominantly of Si. Minute skeletal crystals of magnetite contain mainly Fe but also Si, Ca, and Ti. Cu prills coexist with mattes. Limited idiomorph phases exist in which Ti predominates, and Si, S, Ca, Ba, and Fe are also included. Single grains of barite ($BaSO_4$) are apparent. The abbreviation "nd" means not determined.

The 49 analyses listed in Table B.2 were performed on 10 samples.

Sample M 18-3-(B). Slag containing prills of copper and one prill of gold
1. Gold grain by elements
2. Bulk analysis by oxides
3. Glassy matrix by oxides
4. Sulfide rim surrounding copper prill (by elements)
5. Copper prill by elements
6. Magnetite by oxides
7. Glassy matrix by oxides
8. Fayalite by oxides
9. Copper prill by elements

Sample M 18-3-(A). Slag containing copper prills
10. Bulk by oxides
11. Copper prill by elements
12. Copper prill surrounded by matte rim, denser, "clearer" Cu prill by elements
13. Magnetite crystal by oxides
14. Glassy matrix by oxides

Sample M 18-3-(C). Slag containing copper prills
15. Bulk by oxides
16. Copper prill by elements
17. Copper matte
18. Glassy matrix by oxides
19. Fayalite by oxides
20. Fayalite by oxides containing high Ca

Sample M 19-1-(B)
21. Bulk analysis by oxides
22. Copper prill by elements
23. Magnetite by oxides
24. Copper prill by elements
25. Matte by elements

Sample M 19-1-(C)
26. Copper prill by elements
27. Bulk by oxides
28. Magnetite by oxides

Sample M 19-1-(A)
29. Bulk analysis by oxides
30. Matte by elements
31. Matte by elements

Sample M 19-1-(D)
32. Bulk analysis by oxides
33. Magnetite crystal by oxides
34. Copper prill by elements

Sample N-17-2-(A)
- 35. Area of slag composed of glassy matrix and "fayalites," bulk analysis by oxides
- 36. "Idiomorph" fayalite by oxides, elongated crystals
- 37. Glassy matrix, bulk analysis by oxides
- 38. Magnetite crystal by oxides
- 39. Ceramic part by oxides
- 40. Ceramic part by oxides

Sample N-19-2-(D)
- 41. Glassy matrix, bulk analysis by oxides
- 42. Ceramic area, bulk analysis by oxides
- 43. Vitrified area, bulk analysis by oxides
- 44. Skeletal magnetite, spot analysis by oxides
- 45. Copper prill by elements

Sample N-19-2-(C)
- 46. Bulk analysis by oxides
- 47. Matte A, analysis by elements
- 48. Copper prill, analysis by elements
- 49. Matte B, analysis by elements

APPENDIX C

Lead Isotope and Chemical Analyses of Slags from Chrysokamino

by Zofia Stos and Noel Gale

Although slag heaps of various sizes are not uncommon in the Aegean, they are generally found in the proximity of relevant mineral deposits. The mode of occurrence of copper slag at Chrysokamino thus seems unusual. Perhaps the most frequent appearance of copper slags has been reported from northern Greece[1] where relatively rich copper ores are also found. Many undated (but probably Archaic) copper slag heaps are scattered in the region of the Othrys Mountains in Phtiotis, where copper was mined not so long ago in Limogardion and Sfaki. On the Cycladic islands of Kythnos and Seriphos, both ancient copper slag heaps and copper ores can still be found.[2] However, scatters of copper slag are present on the island of Kea both at Kephala[3] and below the church of Hagios Simeone,[4] and copper ores on this island are rare to nonexistent, with some iron ores containing less than 5% copper. The copper extraction site at Chrysokamino seems, contrary to Mosso's claim,[5] to have no nearby occurrences of copper minerals,[6] as we confirmed when we first visited the site in 1983. The general position of the site, which is set on an elevated coastal cliff, is quite similar to the locations of the slag heaps on Kythnos and Seriphos, but its location seems not to have been selected for its proximity to the source of copper ore.

THE NATURE AND CHEMICAL ANALYSIS OF THE SLAGS AT CHRYSOKAMINO

The slag heap at Chrysokamino appears first to have been described by Hawes[7] and independently by Mosso.[8] Mosso, however, does not seem to have visited it, basing his account on firsthand observations by Hazzidakis. Later, Mosso repeated his earlier account, almost word-for-word, in his book on Mediterranean archaeology.[9] Davies had little to report about ore

1. Papastamataki 1986.
2. Gale et al. 1985; Gale and Stos-Gale 1989; Gale 1989; Stos-Gale 1989.
3. *Keos* I.
4. Caskey et al. 1988.
5. Mosso 1910, pp. 289–291.
6. Branigan 1968, pp. 50–51; Betancourt et al. 1999, p. 352; see also the geological report in this volume.
7. Hawes et al. 1908, p. 33.
8. Mosso 1908, pp. 518–521.
9. Mosso 1910, pp. 289–291.

deposits on Crete[10] and presented hardly any information of substance about Chrysokamino in his book,[11] though he did also mention it in passing in an earlier article,[12] in which he accepted the cave as a worked-out copper mine. Although Fiedler described the Skouries site on Kythnos,[13] he did not include Crete in his admirable and comprehensive survey of matters geological and botanical in Greece, since Crete was then under Egyptian rule and did not become part of Greece until 1913.

We first visited Chrysokamino in 1983 as part of our long-standing work on the archaeometallurgy of Minoan Crete. We were guided to it by an elderly inhabitant of the nearby village of Kavousi. The site, described at length elsewhere in this report, is located on a cliff top exposed to the wind, much like the Bronze Age copper smelting site of Skouries on Kythnos.[14] The site contains a considerable amount of copper slag with small copper prills. In the laboratory in Oxford, we ascertained through electron microprobe analysis that some of the copper prills contain up to several percent of arsenic, a feature not inconsistent with the arsenical copper used in the Early Bronze Age and directly comparable to the analyses reported for some of the copper prills in the Early Bronze Age copper slags at Skouries on Kythnos.[15] The arsenic content in the copper prills of slags from Chrysokamino is confirmed by the independent analyses reported in this volume by Bassiakos (App. B). A distinguishing feature of the Chrysokamino slag heap, first published by Mosso in 1908,[16] is the omnipresence of thick coarse pottery sherds perforated with holes 1.5–2 cm in diameter, often fused and slagged on the interior surfaces, which are clearly the broken remains of furnace chimneys, not crucibles as Mosso thought.

Mosso implied that he believed the slag heap at Chrysokamino dated to Minoan times.[17] Schachermeyr[18] reported finding a sherd of EM II "mottled ware"[19] at this site. However, Faure believed the site postdated the Middle Ages and was probably relatively recent, perhaps related to the work of gypsies.[20] He based his opinion on the state of vitrification of the slags, the use of a siliceous/calcareous flux, and the assumed use of chalcopyrite ore (for which he had no evidence).

Branigan, who was perhaps influenced by Faure, visited the site in 1967 and briefly described both the cave and the slag site, wrongly accepting the coarse ware sherds as crucible fragments.[21] His discussion veered between characterizing the operations at Chrysokamino as the simple melting of copper and the smelting of copper ores. He found no decorated sherds at the slag site and was unaware of the EM II Vasiliki Ware sherd found by Schachermeyr, and so he had no conventional archaeological means of dating the metallurgical operations. Chemical analyses of the slag from Chrysokamino, made by Sargent for Branigan, suggested a high recovery of copper metal, revealed the presence of calcium, and suggested to Sargent that a temperature of about 1150°F (621°C) or more was attained "during the smelting operations."[22] All these features suggested to them a relatively modern date for the slag, "not earlier than about the 12th century AD." Branigan was led to conclude that Chrysokamino was a site used by "itinerant bronze-workers probably using ready-smelted copper which they carried with them," and that they may have belonged to a minority group, the Chalkiades, "operating in the 14th century AD or thereabouts." Since

10. Davies 1935, pp. 266–268.
11. Davies 1935, pp. 7, 113, 264, 270 (n. 1 gives an analysis of the slag by Desch, which misses the calcium content).
12. Davies 1932, p. 987.
13. Fiedler 1841, p. 97.
14. Fiedler 1841; Gale et al. 1985; Gale and Stos-Gale 1989; Stos-Gale 1989, 1998.
15. Gale et al. 1985, p. 90.
16. Mosso 1908, p. 519, fig. 24; 1910, p. 291, fig. 164.
17. Mosso 1908, 1910.
18. Schachermeyr 1938, p. 473.
19. Vasiliki Ware; see Betancourt 1979.
20. Faure 1966, pp. 47–48.
21. Branigan 1968.
22. Branigan 1968, p. 50.

1150°F is well below the melting point of copper, we suspect that 1150°C was intended, and that it was judged too high a temperature to be attained in Bronze Age furnaces.

At the time when Faure and Branigan wrote, the study of ancient copper smelting and slag was in its infancy. The dating to the Early Bronze Age of the Skouries slag heap on Kythnos,[23] however, along with the proof that slags from Skouries had melting points between about 1050°C and 1200°C[24] and contained calcium,[25] shows that evidence based on the physical and chemical properties of the slag from Chrysokamino did not rule out a Minoan date for it. Our examination of the Chrysokamino slag heap produced a few samples of copper ores lying intermingled with the slags. These ores were all oxidized ores, mostly malachite, and suggest that this slag heap bears witness to the smelting of oxidized copper ores. This suggestion is strengthened by the petrological microscopic examination of these slags by Bassiakos (Apps. B and F), who found a number of malachite inclusions in the thin sections examined. A sample of ore found by Betancourt on the beach at Agriomandra near Chrysokamino proved, on examination in Oxford, to be of malachite, and may suggest that copper ores arrived at Chrysokamino by sea.

SOURCES OF COPPER ORES ON CRETE IN THE ARCHAEOLOGICAL LITERATURE

Branigan has shown that the island of Crete led the way in European copper metallurgy in the 3rd millennium B.C.,[26] but the sources of this metal used for the numerous daggers and axes dating to the EM period are not immediately obvious. Mosso made a definite but ill-founded claim that a copper mine existed on the island of Gavdos and in a cave near Chrysokamino.[27] Although he never visited either site, he wrote that ore containing "protoxide of copper" was found somewhere there; at the same time he admitted that none of the rock samples sent to him by Hazzidakis from the walls of the cave contained any copper ore. Faure, from his personal observations, concluded that the cave was natural, rather than being a mine, and that it never contained copper minerals.[28]

About 30 years ago, several papers appeared implying that the Minoans were mining copper ore on Crete. Faure listed 22 sites on Crete where he asserted that copper minerals were found.[29] Typical descriptions by Faure of such occurrences were similar to the one that he gave of a "copper mineralisation" near Kritsa (Mirabello Bay):

> At one and a half hours' walk to the WSW of Kritsa one finds a great variety of eruptive rocks, bordering the high plateau of Katharo.... In such a context the geographer Raulin is the first to have described in 1845 grains of pyrites disseminated in a Cretaceous contact deposit ... I have seen traces of nickel sulphide. For

23. Stos-Gale 1998, pp. 719–720.
24. Gale et al. 1985, pp. 86–89.
25. Gale et al. 1985, p. 87, table 2.
26. Branigan 1968, 1974; cf. Pendlebury 1939, p. 280, and Caskey 1964.
27. Mosso 1908, pp. 518–521; 1910, pp. 289–293.
28. Faure 1966, pp. 47–48.
29. Faure 1966, 1980.

many years local prospectors have mentioned the presence of carbonates of copper to the south of Katharo. For 80 years the peasants of Selakano and of Kritsa have sold to foreigners or sent to the Archaeological Service bronze weapons and votive axes coming from an unknown site, probably to the NE of Selakono. One can hardly doubt the existence of one or many centres for the treatment of metal between Selakano and the Katharo whilst the mountain is rich in toponyms . . . describing furnaces, generally of charcoal; but everyone knows that at all times one has roasted and reduced impure minerals with this fuel.[30]

All this sounds highly speculative and does not describe any real features of ancient mineral exploitation such as galleries, spoil heaps, or real, relatively large, outcrops of copper minerals, or indeed of any ore minerals observed by Faure other than those of iron and nickel.

Branigan reassessed the hypothesis of the existence of Cretan copper ore sources in the early seventies when he conducted several seasons of archaeological survey in south central Crete and described several other occurrences of copper ore deposits.[31] Apart from the occurrence at Chrysostomos near Laseia, which indeed contains malachite and azurite (as does the occurrence at Sklavopoula in west Crete),[32] other mineralizations are mostly described as consisting of "copper minerals in schist." In claiming the discovery of an Early Bronze Age metal source in Crete, Branigan describes some "copper-bearing pebbles" found at the site of Fournou Korifi. Apparently, chemical analyses of these pebbles were made and "the analysis of two of the pebbles produced 1% and 2.2% of copper."[33] On visiting this site in 1983, we were unable to find any copper-bearing pebbles at the location described, nor did we see any in the conglomerate cap of the nearby hill, though a heap of pebbles otherwise answering to Branigan's description was found. No crucibles, furnace fragments, or pieces of slag were found. It seems probable that, if a copper source ever existed here, it was extremely small and of no importance except for strictly local needs.

On the basis of their surveys, both Branigan and Faure concluded that there are enough copper ore deposits on Crete to have satisfied Minoan demands for this metal in the Early Bronze Age. Muhly and Rapp conducted a survey on Crete in 1974, which discounted many of the alleged copper deposits reported by Faure; they stated that they found that "a number of the copper deposits discussed in the literature are represented only by green stains on rocks and some are serpentine rather than copper."[34] They concluded that Crete probably should not be considered a major producer of copper in the Bronze Age. Becker, in the course of extensive fieldwork on Crete conducted primarily to establish soft stone sources, also demonstrated that many alleged copper sources on Crete have been incorrectly identified.[35] Finally, according to a Cretan commercial mining engineer with whom we have corresponded: "The occurrences cited by our mutual friend Professor Faure refer mostly to small samples found as float. In many cases the alleged site has produced no further evidence."[36]

30. Faure 1966, p. 49 (translation by the authors).
31. Branigan 1971; Blackman and Branigan 1975, 1977.
32. This is confirmed in both cases by our own observations in 1983 and later.
33. Branigan 1971, p. 12.
34. Wheeler, Maddin, and Muhly 1975, p. 32.
35. Becker 1976.
36. This information comes from a letter in our archives from Michael Diallinas, who is well known to many Aegean archaeologists, in reply to Noel Gale in October 1977.

THE GEOLOGISTS' VIEW OF COPPER DEPOSITS ON CRETE

The standard handbook of southeastern European mineral deposits does not mention any copper ores at all on Crete,[37] while the Greek section of the later UNESCO metallogenic map of Europe mentions only the copper deposit at Sklavopoula.[38] The only mineral deposits of note on Crete mentioned in the current geological literature are gypsum and iron. The Greek economic geologists from IGME working on Crete in the 1990s denied the existence of all but traces of copper ores on the island.

However, it must be admitted that there is often a certain lack of understanding among modern economic geologists concerning the relatively limited requirements of the metallurgists of five thousand years ago. The modern interest in the nonferrous metallic mineral deposits of the world tends to be focused on the deposits of Australia, Siberia, America, Africa, and the Far East. Over fifty years ago, Bateman wrote:

> From early times until 1800 copper was widely produced in small quantities. From 1801 to 1810 the annual world production was only 18,200 tons, equivalent to less than one month's production of some present-day mines.... The tremendous growth in the use of copper is indicated by the fact that of the total world copper produced in the last 100 years about 80 percent was mined in the last 25 years.... Annual production normally ranges from about 2 to 2.5 million tons of metallic copper.[39]

The copper deposits of the whole of the Mediterranean are quite meaningless when one thinks on such a scale. Consequently, modern economic geological handbooks and maps may in some cases be somewhat misleading when it comes to the search for ancient metal sources.

With the help of Michael Diallinas, Kostas Zervantonakis, and many most helpful geologists from the IGME office in Chania, we conducted a relatively systematic survey of the mineral occurrences on Crete in 1979, 1983, and 1987. We were especially interested in the occurrences described by Faure and Branigan, and thank both for their generous correspondence with us. Our main aim was the collection of ore samples for lead isotope analyses. Copper minerals on Crete are definitely present on two sites known to the archaeologists: one is that of Chrysostomos, southwest of Andiskari on the central south coast, and the other is Sklavopoula in west Crete. The copper carbonate mineralization in metamorphic rocks near the ruined chapel of St. John Chrysostomos is quite interesting. Faure mentions that in 1952 a Greek firm attempted to mine copper ore there, and Branigan confirms the report.[40] Malachite and azurite, as well as limonite, are still visible in the walls of the large open pit that remains from this venture of the 1950s. It is difficult to assess how much copper mineralization might have been present on this site some four thousand years ago. It is very unlikely that the mineralization was ever very extensive, but some oxidized copper ores might have been visible here and might have been used to a small extent in antiquity. The occurrence is surrounded by Minoan

37. Dunning, Mykura, and Slater 1982, with report by Marinos 1982, p. 237.
38. Anastopoulos and Koukouzas 1984.
39. Bateman 1946, p. 479.
40. Faure 1966, p. 52; Branigan 1968, p. 51.

and later sites; on one occasion, J. A. MacGillivray, walking with us around Chrysostomos, noticed several Early Minoan sherds and the remains of a tholos tomb on a field just above the pit. However, no traces of copper smelting, slag, or ancient galleries were found. The copper occurrence in west Crete at Sklavopoula (Selinou), south of Rethymno, is high in the mountains, just below a military radar station. A comparatively recent mining activity here in 1962[41] left spoil heaps of mainly iron/manganese ores with some dispersed oxidized copper minerals. The site has no evidence of any ancient exploitation of these minerals, and the geochemistry of the ores suggests that copper here is only a relatively minor component of iron ores. No slags were found anywhere in the region. Iron deposits have also been exploited to a small extent southwest of Chania. In some of the mines, copper also occurs as a minor component of the ore, and we were able to collect some samples of copper minerals at Kambanos.[42] Some other very minor occurrences of copper minerals also occur relatively near the mine of Chrysostomos, south of Hagia Triada and Phaistos on the south coast, at Lasaia, Lebena, and Miamou.

The results of our survey, combined with all reliable geological information about mineral deposits on Crete, indicate that a significant metal extraction (copper, lead, and silver) from the local ores in Bronze Age times is very unlikely. To widen our survey for possible metal extraction slags on Crete, we enlisted the help of the archaeologist Krzysztof Nowicki, who surveyed large areas of the mountains in Crete. He showed us the site near Kato Zakros, in the Valley of Death,[43] but no copper slags were present. Moreover, no copper slag heaps have been located other than that at Chrysokamino, though of course slag layers have occasionally been found in the excavations of Minoan sites such as Malia.[44] It may be worth pointing out here that the structures at such places as Kato Zakros, which some have sought to identify as Bronze Age copper furnaces, are clearly not designed for this purpose.

CHEMICAL ANALYSES AND DATING OF SLAGS FROM CHRYSOKAMINO

Chrysokamino is the only site found on Crete so far where copper smelting, the extraction of copper from ores rather than the making of artifacts from copper metal, was practiced. In 1994, we were able to obtain two thermoluminescence (TL) dates on two different examples of the characteristic perforated furnace sherds that we had collected in 1983 and in the 1990s from Chrysokamino.[45] Both dates fell into the 3rd millennium B.C.: 2420 ± 345 B.C. and 2710 ± 365 B.C.[46] These dates substantiated the earlier indication of the antiquity of Chrysokamino, which came from the sherd of Vasiliki Ware found there by Schachermeyr.[47]

Petrographic sections of many of the Chrysokamino slags revealed the presence of numerous copper metal prills.[48] Chemical analyses of these prills made in Oxford using an electron microprobe fitted with wavelength dispersive XRF showed that many of the prills contained arsenic at the several percent level, similar to copper prills in the Bronze Age slags from

41. The mining was for iron; see Faure 1966, p. 57.
42. Kambanu in Davies 1935, p. 268.
43. Faure 1966, p. 47, n. 1.
44. Poursat 1985.
45. The dates are courtesy of M. Tite, RLAHA, Oxford.
46. Stos-Gale 1998, p. 721.
47. Schachermeyr 1938, p. 473.
48. We first had a sample of the slag from Michael Diallinas when we visited him in his offices in Herakleion in 1978, and we subsequently collected more samples when we visited the site in 1983.

Skouries on Kythnos.[49] This finding has now been confirmed and extended by Bassiakos in this volume (Apps. B and F).

The pioneering work of Bachmann showed that useful information about the copper smelting techniques used at a particular site can be obtained from bulk chemical analyses of the slags and their interpretation by thermodynamic phase diagrams, though such inferences should, if possible, be aided by mineral identification based on petrography of the slags.[50] Experimental work on silicate melts of various chemical compositions also allows inferences to be made about slag melting points and furnace temperatures, as well as about slag viscosities and densities, in turn permitting estimates of the efficiency of separation of metal from the slag phase in the furnace.[51] Several slag samples collected from the Chrysokamino heap were, therefore, powdered and chemically analyzed for 22 elements, using the technique of inductively coupled plasma atomic emission analysis (ICP-AES). This work was undertaken at the Royal Holloway College, London. Analyses for arsenic content by wavelength dispersive X-ray fluorescence were also undertaken by the analytical laboratories of the British Geological Survey, Keyworth. All analyses were checked against concurrent analyses of international geochemical standards. The chemical analyses are presented in Table C.1. It is evident that, as noted by Faure and Branigan,[52] these slags are quite rich in calcium, although the EBA slags from Skouries on Kythnos also contain calcium at medium high levels.[53]

The bulk chemical compositions, especially the relatively high calcium contents, and the ternary diagrams into which they classify, seem to be in accord with the petrographic studies (Apps. A and B), which reveal pyroxene (including augite), fayalite, and magnetite to be present in the samples of loose slag from Chrysokamino.

Bachmann[54] has shown how such chemical analyses of slags may be interpreted in terms of the quaternary phase system CaO—FeO—Al_2O_3—SiO_2, which, to include other elements, he expanded to:

$$CaO\{+BaO+Na_2O+K_2O\}—FeO\{+MnO+MgO\}—Al_2O_3—SiO_2$$

The basic system plots in space as a tetrahedron, which projects into four ternary systems, but Bachmann and Zacharias, in order to include wollastonite, found it more useful to work in terms of five ternary systems, which include the minerals anorthite, gehlenite, wollastonite, Al_2O_3, SiO_2, and FeO, along with others.[55] They specify the computations that are necessary to interpret chemical analyses of slags in terms of these five ternary diagrams. We have programmed these computations using MathCad, including the calculation of the basicity number, and have further used MathCad to program the computations necessary to use Shaw's method to compute slag viscosities as a function of temperature.[56]

---

49. Gale et al. 1985, p. 90, table 3.
50. Bachmann 1980.
51. Shaw 1972; Bottinga and Weill 1972; Bottinga, Weill, and Richet 1982; Bottinga, Richet, and Weill 1983.
52. Faure 1966, p. 48; Branigan 1968, p. 50.

53. Gale et al. 1985, p. 87, table 2.
54. Bachmann 1980.
55. Bachmann 1980, pp. 120–131.
56. Shaw 1972. Subsequently we found that Ford (1992) has published on the Internet a convenient program (CHEMCAST), which computes a number of basic chemical/petrological parameters and includes viscosities using Shaw's method and densities following Bottinga, Weill, and Richet 1982 and Bottinga, Richet, and Weill 1983.

TABLE C.1. CHEMICAL ANALYSES OF CHRYSOKAMINO SLAG

Sample	SiO_2	Al_2O_3	Fe_2O_3	MgO	CaO	Na_2O	K_2O	TiO_2	MnO	Ba
CHR2	33.91	5.37	19.68	1.33	30.53	0.87	0.85	0.26	0.19	5960
CHR3	42.37	5.51	34.64	0.70	11.88	0.72	0.53	0.23	0.21	920
CHRYA	38.06	5.38	23.08	1.22	25.70	1.19	0.89	0.25	0.22	6580
CHRYB	32.84	7.45	34.40	1.42	17.75	1.52	0.89	0.36	0.34	6460
CHRYC	34.05	7.06	34.16	1.53	17.30	1.29	0.96	0.33	0.32	4660
CHRYD	27.59	5.50	40.08	1.15	17.93	1.41	0.63	0.27	0.34	8180
CHRYE	41.56	6.14	19.53	1.09	23.44	0.83	0.94	0.28	0.12	1521
CHRYF	28.44	3.76	37.08	0.81	10.65	0.56	0.62	0.16	0.08	>20000
CHRYG	47.53	6.22	15.64	1.31	22.32	0.95	0.95	0.30	0.25	14240
CHRYH	45.48	6.49	15.46	1.55	23.44	0.99	0.97	0.31	0.27	14280
CHRYJ	35.22	7.49	30.16	1.36	16.24	1.57	0.90	0.35	0.85	4570
CHRYK	31.59	5.47	42.04	1.19	14.12	0.33	0.80	0.26	0.14	532
CHRYL	46.62	6.32	15.33	1.35	22.70	0.95	0.94	0.30	0.25	13630
CHRYM	35.93	5.04	31.56	1.23	18.71	0.48	0.78	0.26	0.16	6020

One of the most useful parameters that can be obtained from assigning slags, via their chemical compositions, to the appropriate ternary phase diagram is the liquidus temperature, as a reasonable estimate of the lower limit of the furnace temperature attained in the production of a suite of ancient slags. It has been shown that the application of the crude SiO_2—FeO—CaO ternary diagram, much used in the past for this purpose,[57] predicts far too high liquidus temperatures, sometimes by many hundreds of degrees, and that Bachmann's approach produces liquidus temperatures very much nearer the truth.[58] Another method for estimating the liquidus temperatures of slag silicate melts is one developed by Nathan and Van Kirk as an initial part of their attempt to devise a computer model for fractional crystallization in geological magmas.[59] They derived empirical equations, in terms of the fractional cation composition of the magma as derived from its chemical analysis, for the estimation of the liquidus temperature of a number of minerals in a multicomponent silicate liquid at atmospheric pressure. In comparing the predictions of the Nathan and Van Kirk equations with an analyzed set of 975 glasses produced at known temperatures and oxygen fugacities, Ford found that the equations predicted the measured liquidus temperatures to better than 40°C, which is perhaps adequate for archaeometallurgical purposes.[60] However, one should bear in mind that the empirical Nathan and Van Kirk equations are based on data sets directly applicable chiefly to geological melts, and they really stand in need of recalibration for the more iron-rich compositions of typical archaeometallurgical slags. Moreover, the minerals treated by Nathan and Van Kirk do not include some that are more applicable to those found in slags.[61]

Table C.2 shows the results of applying the Bachmann computations to the analyzed slags from Chrysokamino. The table shows the

57. E.g., Conophagos 1980; Milton et al. 1976.
58. Gale et al. 1985.
59. Nathan and Van Kirk 1978.
60. Ford 1981.
61. Bachmann 1982.

TABLE C.1 (cont.)

Sample	Co	Cr	Cu	Li	Ni	Sc	Sr	V	Y	Zn	Zr	Pb	As
CHR2	16	90	5980	28	47	7	465	104	30	127	63	18	260
CHR3	21	127	12380	17	21	12	231	540	26	1915	58	721	810
CHRYA	87	101	5820	38	60	7	596	191	29	194	62	50	900
CHRYB	46	137	19080	30	523	12	304	170	25	329	82	66	1060
CHRYC	35	120	11660	28	258	12	289	110	26	142	81	36	850
CHRYD	31	130	14150	21	353	11	270	127	20	246	63	61	540
CHRYE	24	109	23230	30	83	8	250	215	24	589	62	92	1940
CHRYF	40	60	16470	19	43	6	1225	101	17	861	42	56	570
CHRYG	23	107	16940	30	51	8	497	214	30	675	72	47	2550
CHRYH	18	114	7870	31	32	9	550	230	32	582	74	35	1820
CHRYJ	55	152	22850	28	602	15	206	154	28	335	78	45	510
CHRYK	13	83	7480	26	36	8	233	282	39	267	67	40	170
CHRYL	20	110	10670	30	40	9	502	218	30	603	75	37	1860
CHRYM	29	87	15630	47	169	7	365	141	39	700	70	96	600

Values for oxides are percentages; values for elements are ppm.

calculated selection quotients Q_1, Q_2, Q_3, the phase diagrams selected as appropriate for each slag, the ages of the appropriate principal mineral phases, the basicity numbers, the densities using the approach of Bottinga and colleagues,[62] the viscosities computed using Shaw's method, and the liquidus temperatures computed by the method devised by Nathan and Van Kirk. The Bachmann approach shows that the Chrysokamino slags classify into two ternary phase diagrams, with the majority falling into the Anorthite—Wollastonite—FeO diagram and three falling into the Anorthite—SiO_2—FeO diagram. Figure C.1 shows the 12 slags that plot in the Anorthite region or the Hercynite region of the AN—WO—FeO ternary diagram, and it shows that the slag liquidus temperatures probably fall in the 1200–1250°C region. The three slags that fall in the AN—SiO_2—FeO diagram (Figure C.2) suggest liquidus temperatures of about 1200°C. On the other hand, the calculations of liquidus temperatures using the approach of Nathan and Van Kirk, reported in Table C.2, suggest slightly lower liquidus temperatures of about 1120–1170°C, though these estimates are based on the minerals augite and olivine. Because it is probable that the operating temperatures of ancient smelting furnaces were typically at least 100°C above the melting temperature of the slags,[63] it would seem that at Chrysokamino the furnace temperatures were in the vicinity of 1250–1350°C; this agrees with experimental reconstructions of copper smelting in simple shaft furnaces.[64]

Even assuming a minimum furnace temperature of 1200°C, the slag viscosities are of the order of 20 poise, slightly higher than the Rio Tinto lead slags,[65] but still not very high. Combining the calculations of slag density in Table C.2 and the figure of 7.8 g cm^{-3} for the density of molten copper at 1200°C,[66] one can use Stokes' law to compute the rate

62. Bottinga, Weill, and Richet 1982; Bottinga, Richet, and Weill 1983.
63. Freestone 1988.
64. Merkel 1983, 1990; Bamberger and Wincierz 1990.
65. See Craddock et al. 1985.
66. Lang 1978.

TABLE C.2. RESULTS OF APPLYING THE BACHMANN COMPUTATIONS TO THE ANALYZED SLAGS FROM CHRYSOKAMINO

Sample	Q_1	Q_2	Q_3	Phase Diagram	Basicity No.	AN	SiO_2	FeO	AN	WO	FeO	Density* at 1200°C	Ln(Viscosity)** at 1200°C	T°C[‡] Olivine	T°C[‡] Augite
CHR2	0.093	0.092	1.012	3?	1.383				55.02	23.93	21.05	2.98	2.2	1148	1178
CHR3	0.077	0.235	0.326	2	0.9	38.39	27.18	34.43				3.01	3.91	1091	1132
CHRYA	0.083	0.107	0.775	3	1.185				50.32	26.08	23.60	2.96	2.94	1131	1167
CHRYB	0.134	0.206	0.649	3	1.332				42.56	22.57	34.87	3.12	1.7	1140	1149
CHRYC	0.122	0.202	0.604	3	1.278				41.81	23.42	34.77	3.10	1.96	1138	1152
CHRYD	0.117	0.152	0.772	3	1.733				39.48	19.45	41.07	3.28	0.47	1133	1140
CHRYE	0.087	0.136	0.639	3	0.951				50.38	29.17	20.45	2.87	4.16	1123	1162
CHRYF	0.078	0.174	0.447	2	1.366	39.72	17.05	43.23				3.24	1.64	1108	1129
CHRYG	0.077	0.141	0.547	3	0.78				51.31	32.24	16.45	2.77	5.51	1111	1158
CHRYH	0.084	0.14	0.599	3	0.84				52.35	31.01	16.64	2.79	5.05	1123	1164
CHRYJ	0.125	0.224	0.558	3	1.138				42.94	24.89	32.17	3.03	2.62	1132	1149
CHRYK	0.102	0.202	0.506	3? 2?	1.421	43.67	13.54	42.79	34.97	22.25	42.78	3.25	1.37	1122	1136
CHRYL	0.08	0.141	0.565	3	0.797				51.80	31.87	16.33	2.77	5.38	1115	1159
CHRYM	0.083	0.14	0.591	3	1.207				42.02	25.36	32.62	3.07	2.57	1123	1156

*Density computed by the methods given by Bottinga, Weill, and Richet 1982; Bottinga, Richet, and Weill 1983.
**Viscosity quoted as Log_e(Viscosity) as computed by the methods given by Shaw 1972.
[‡]Liquidus temperatures computed from the formulae given by Nathan and Van Kirk 1978.

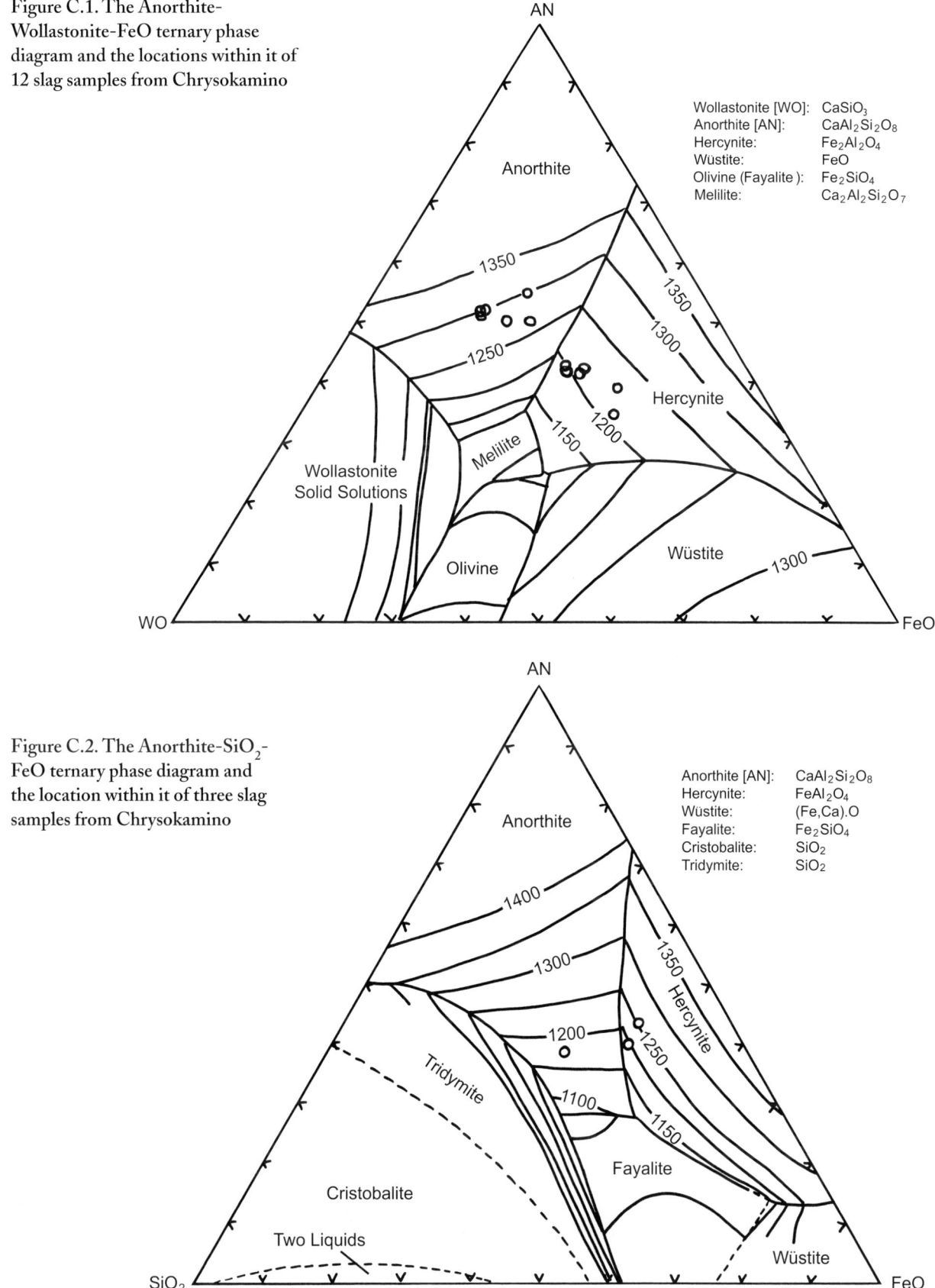

Figure C.1. The Anorthite-Wollastonite-FeO ternary phase diagram and the locations within it of 12 slag samples from Chrysokamino

Figure C.2. The Anorthite-SiO$_2$-FeO ternary phase diagram and the location within it of three slag samples from Chrysokamino

of separation of molten copper metal from molten slag within the furnace. For a spherical copper droplet of 1 mm in diameter, Stokes' law gives a rate of settling through the molten slag of 8 cm per minute, which for a copper droplet of 0.1 mm in diameter drops to 0.08 cm per minute. Given the relatively small size of the Chrysokamino furnaces (Chaps. 7 and 14) and the fact that ancient copper smelting furnaces were operated for several hours,[67] copper droplets of 1 mm in diameter and above could sink through the slag many times over, while very small copper droplets of the order of 0.1 mm in diameter or less would be trapped in the slag, as observed (see Apps. B and F). Such very small copper prills could not easily be broken from the slag by mechanical means, nor would this be necessary, since separation of copper droplets larger than 1 mm in diameter from the molten slag by gravitational settling during operation of the furnace would have been very efficient.

LEAD ISOTOPE ANALYSES OF COPPER AND LEAD ORES FROM CRETE

Lead isotope analyses of copper and lead ores collected during our surveys in Crete are listed in Table C.3, while lead isotope analyses of the Chrysokamino slags appear in Table C.4. The comparatively small number of analyses of minerals from Crete, 28 in total, reflects directly the scarcity of copper ores in the few occurrences. At Chrysostomos and Sklavopoulou, copper is visible as green oxidation on the surface of the host rock, though in both localities richer ores bearing malachite and azurite are present on the ore dumps from 20th-century exploitation. At Sklavopoulou, the rocks are impregnated with iron and manganese minerals, with some copper carbonates. At Chrysostomos, the metamorphic host rock contains some limonite and dispersed copper carbonates. The lead ore occurrence at Ano Varsamonero is relatively small, with exposed veins of fine-grained galena. We analyzed two samples of this galena using instrumental neutron activation: the silver content in the samples was 150 ppm and 240 ppm, the gold content in both around 0.1 ppm. The silver content at this level would have been too low for extraction in the Bronze Age,[68] so that the deposit at Ano Varsamonero cannot have been a Minoan silver source. In some of the iron mines in western Crete, it is possible to find relatively rich copper minerals, and a few such copper-rich minerals were collected from Kambanos as well as Sklavopoulou.

Unfortunately, we were not able to find equally good copper minerals in any other mines, though weak copper mineralization was found at Lasaia and Lebena near the village of Lenta on the central southern coast of Crete. The currently available lead isotope data for Cretan ores are plotted on two mirror diagrams in Figure C.3.

Lead isotope analyses of these ores show that the copper ores in south central Crete and those from Sklavopoulou and western Crete have very

67. Bamberger and Wincierz 1990.
68. Gale and Stos-Gale 1981a, 1981b; Gale, Stos-Gale, and Davis 1984; Pernicka et al. 1983.

TABLE C.3. LEAD ISOTOPE ANALYSES OF COPPER ORES AND GALENAS FROM CRETE

Sample	Occurrence	Description	$^{208}Pb/^{206}Pb$	$^{207}Pb/^{206}Pb$	$^{206}Pb/^{204}Pb$
CST 6A	Chrysostomos	Malachite	2.08110	0.83980	18.618
CST A1	Chrysostomos	Malachite	2.08187	0.83880	18.652
CST1A	Chrysostomos	Malachite	2.08014	0.83992	18.616
CST3	Chrysostomos	Malachite	2.08280	0.83828	18.715
CST4	Chrysostomos	Malachite	2.07765	0.83746	18.682
CHR4	Chrysostomos	Malachite	2.07895	0.84035	18.580
CHR4A	Chrysostomos	Malachite	2.08103	0.84077	18.599
CHR3	Chrysostomos	Malachite	2.07834	0.83913	18.573
CST2	Chrysostomos	Malachite	2.08253	0.84080	18.568
CST1	Chrysostomos	Malachite	2.08180	0.84320	18.483
CST5	Chrysostomos	Malachite	2.08381	0.83862	18.692
Kam200	Kambanos	Copper ore	2.07179	0.83242	18.923
Kam201	Kambanos	Copper ore	2.06447	0.83253	18.950
Kam21g	Kambanos	Copper ore	2.06773	0.83455	18.879
Kamd1	Kambanos	Copper ore	2.06751	0.83312	18.947
LAS1	Lasaia	Copper ore	2.08554	0.84286	18.544
LEB1	Lebena	Copper ore	2.08067	0.84110	18.540
MIAM 1	Miamou	Galena	2.07745	0.84033	18.614
LAQ	Miamou	Galena	2.07676	0.84019	18.591
MLB	Miamou	Galena	2.07275	0.83998	18.579
skav1	Sklavopoulou	Copper ore	2.06482	0.83356	18.935
skav2	Sklavopoulou	Copper ore	2.06982	0.83479	18.897
skav3	Sklavopoulou	Copper ore	2.06450	0.83396	18.981
skav4	Sklavopoulou	Copper ore	2.07143	0.83510	18.833
VARS 1	Ano Varsamonero	Galena	2.09771	0.8511	18.399
VARS 2	Ano Varsamonero	Galena	2.09393	0.85058	18.436
AN1	Ano Varsamonero	Galena	2.09589	0.85083	18.405
ANV100	Ano Varsamonero	Galena	2.09519	0.85033	18.394

different lead isotope geochemistry. The galena from Ano Varsamonero is isotopically different from both of these groups. Our earlier lead isotope analyses of Minoan copper oxhide ingots from Hagia Triada and Kato Zakros[69] show clearly that all of these ingots have lead isotope analyses completely different from those of any copper ore from Crete yet analyzed, so that present evidence shows that these copper oxhide ingots were not made from copper extracted from any known Cretan copper ores, nor from known Cypriot copper ores.[70]

69. Gale and Stos-Gale 1986; Gale 1999.

70. See Gale 1991, 1999; Stos-Gale 1993.

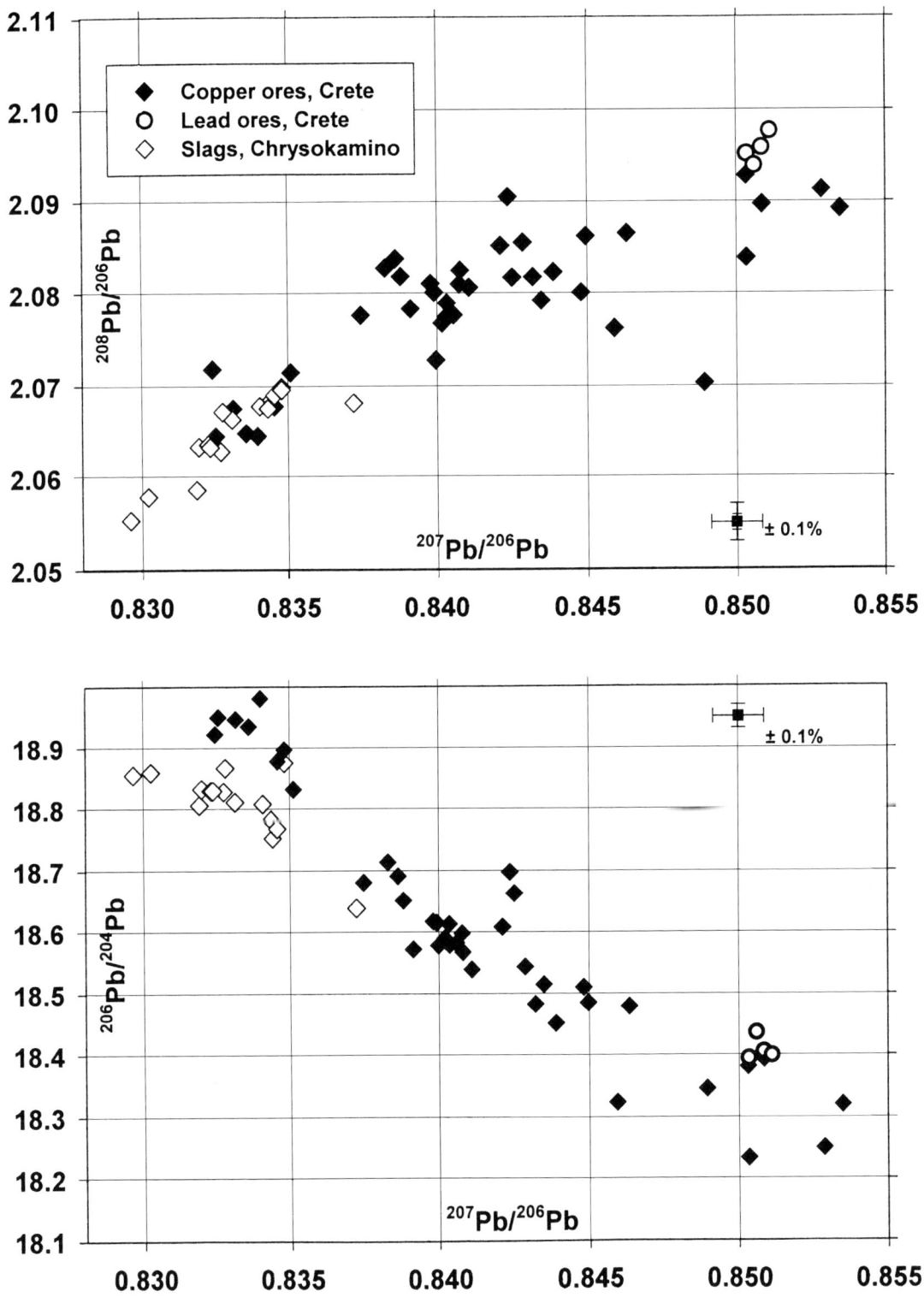

Figure C.3. Lead isotope compositions of ores from Crete compared with those for slags from Chrysokamino

TABLE C.4. LEAD ISOTOPE RATIOS OF PIECES OF CHRYSOKAMINO SLAG

Sample	$^{208}Pb/^{206}Pb$	$^{207}Pb/^{206}Pb$	$^{206}Pb/^{204}Pb$
CHR2*	2.05868	0.83189	18.807
CHRY E	2.05793	0.83024	18.860
CHRY A	2.06281	0.8327	18.829
CHRY C	2.05539	0.82963	18.856
CHR3*	2.06808	0.83722	18.640
CHRY D1	2.06829	0.83438	18.754
CHRY B	2.06776	0.83403	18.809
CHRY I	2.06632	0.83309	18.812
CHRY H	2.06751	0.83432	18.784
CHRY M	2.06715	0.83276	18.868
CHRY J	2.06332	0.83196	18.833
CHRY G	2.06358	0.83228	18.830
CHRY D2	2.06896	0.83454	18.769
CHRY L	2.06953	0.83479	18.876
CHRY K	2.06326	0.83234	18.830

Pieces were collected in August 1983. The numbers are Isotrace Laboratory numbers. Samples marked with an asterisk were analyzed in 1985 with lesser accuracy than the remaining samples, which were analyzed on the new mass spectrometer in 1995.

LEAD ISOTOPE ANALYSES OF SLAGS FROM CHRYSOKAMINO

In 1983 we collected some slags from the surface of the site at Chrysokamino for chemical analysis and for lead isotope analyses to be compared with the Cretan ores. The isotopic data obtained for some of these slags is listed in Table C.4.

Two of these analyses were made in 1985 and published by Stos-Gale.[71] As Figure C.3 shows, the lead isotope analyses of these 14 slag pieces form quite a compact group, clearly different isotopically from the analyzed Cretan ores. The superficial similarity of the isotopic composition of ores from western Crete with the slags from Chrysokamino, apparent on the $^{208}Pb/^{206}Pb$ diagram, can be quite definitely rejected in view of the different $^{206}Pb/^{204}Pb$ ratios of these two groups of ores. The copper ores from Chrysostomos and the occurrences near Lasaia and Lenta are very different in all three lead isotope ratios.

On the presently available evidence, some of the slags from Chrysokamino show identical lead isotope ratios with copper ores from Lavrion,[72] while other Chrysokamino slags have lead isotope compositions matching copper slags and ores from the EC II smelting site of Skouries on the Cycladic island of Kythnos[73] and copper ores or slags from Seriphos (see Table C.5). The lead isotope plot for the ores and slags from these sites is

71. Stos-Gale 1998, p. 724, Chrysokamino 1 and 2.
72. For the data for these ores, see Stos-Gale, Gale, and Annetts 1996.
73. Gale et al. 1985; Gale and Stos-Gale 1989; Stos-Gale 1998.

TABLE C.5. CYCLADIC ORES AND COPPER SLAGS ISOTOPICALLY CONSISTENT WITH SLAGS FROM CHRYSOKAMINO

Sample	Island	Region	Type of Sample	$^{208}Pb/^{206}Pb$	$^{207}Pb/^{206}Pb$	$^{206}Pb/^{204}Pb$
Orkos 3	Kea	Orkos	Copper mineral in a rock	2.06922	0.8338	18.815
Orkos 1	Kea	Orkos	Copper mineral in a rock	2.07010	0.83384	18.818
Orkos 2	Kea	Orkos	Copper mineral in a rock	2.06919	0.83400	18.790
KYT 10 publ	Kythnos	Skouries-H. Ioannis	Slag heap, slag	2.06796	0.83364	18.798
KYT 52b	Kythnos	Skouries-H. Ioannis	Slag heap, slag	2.06998	0.83367	18.827
KYT 3 publ	Kythnos	Skouries-H. Ioannis	Slag heap, slag	2.06896	0.83373	18.837
KYTO1 publ	Kythnos	Skouries-H. Ioannis	Slag heap, oxidized Cu ore	2.06892	0.834	18.789
KYTO2 publ	Kythnos	Skouries-H. Ioannis	Slag heap, oxidized Cu ore	2.06949	0.8345	18.783
KYT 20	Kythnos	Skouries-H. Ioannis	Slag heap, slag	2.0697	0.83461	18.822
Zogo2 publ	Kythnos	Zoghaki	Mine, oxidized Cu ore	2.06972	0.83397	18.819
KYT 17	Kythnos	Skouries-H. Ioannis	Slag heap, slag	2.07011	0.83421	18.816
KYT T30	Kythnos	Skouries-H. Ioannis	Slag heap, slag	2.07012	0.83404	18.815
KYT 19	Kythnos	Skouries-H. Ioannis	Slag heap, slag	2.07015	0.83420	18.815
Zogo3 publ	Kythnos	Zoghaki	Mine, oxidized Cu ore	2.07016	0.83426	18.813
KYTO3 publ	Kythnos	Skouries-H. Ioannis	Slag heap, Cu ore	2.07024	0.83415	18.811
SerAvy prill	Seriphos	Avyssalos	Slag heap, Cu prill from slag	2.06836	0.8345	18.813
KEF 3	Seriphos	Kefala	Slag heap, slag	2.06893	0.83459	18.782
KONDO 1	Seriphos	Kondouro	Mine, oxidized Cu ore	2.07016	0.83408	18.81
ASPYR3	Siphnos	Aspros Pyrgos	Fe-Cu ore	2.06841	0.83499	18.819

compared with the data obtained on samples of slags from Chrysokamino in Figure C.4. This result seems very surprising at first, but it needs to be examined in terms of the paucity of copper ores in Crete and the known copper mineralizations in the Eastern Mediterranean. The Aegean islands, the Peloponnese, and Attica are the lands with copper ore deposits that are closest to the island of Crete.

Starting from the northwest, it has been suggested that the Minoan settlement of Kastri on Kythera was used as a base for exploitation of marble and other decorative stone from Laconia and perhaps also for murex shells.[74] Laconia has also been mentioned as a possible source of copper, lead, and silver. Together with geologists from the Institute of Geology and Mineralogical Exploration, we surveyed Laconia for copper and lead ores in 1987. Some copper mineralization is, indeed, present in several locations. However, the amount of the minerals in all cases is quite small; there are no mines, but only isolated outcrops in the mountains, with small, modern trial/survey trenches. No slag heaps that would provide evidence for smelting of metals were found during the survey or reported by the geologists. Samples of ores from the copper outcrops and from the large lead-zinc deposit in Molai have been analyzed for their lead isotope composition, and they do not match any of the Bronze Age copper or lead artifacts. The largest copper deposit is in the northeast part of the

74. Coldstream and Huxley 1984.

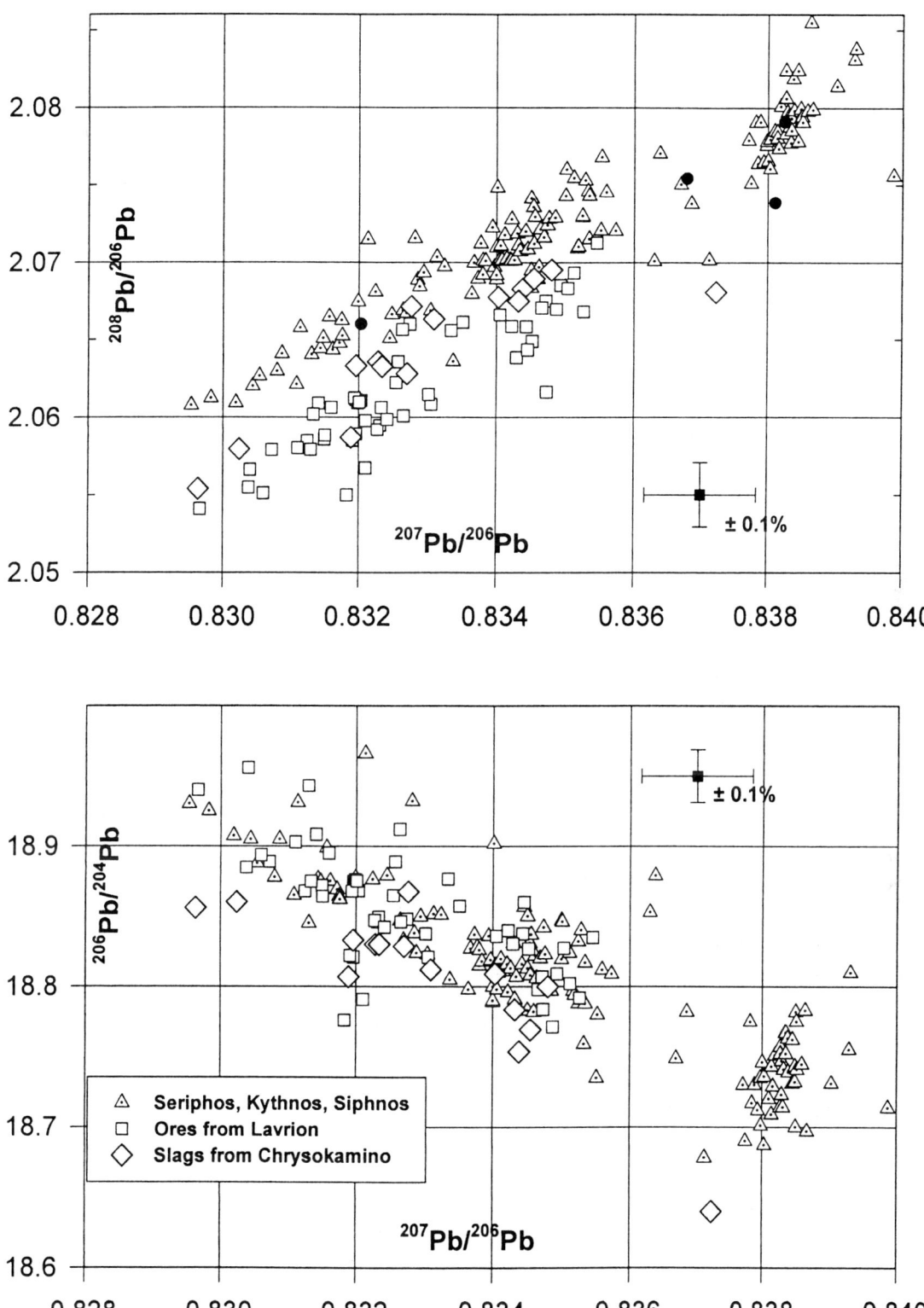

Figure C.4. Lead isotope compositions of slags from Chrysokamino and ores from Lavrion and the Cyclades

Peloponnese at Ermioni. The main mineralization is deep and accessible only by modern mining techniques, but a number of surface expressions of the ores would, in principle, have been accessible to Bronze Age miners. However, the lead isotope analyses of ores from the Ermioni deposits do not match those for Bronze Age copper-based artifacts, so that these ores were not used for production of Bronze Age copper objects.

Northeast from Ermioni, on the Attic coast, is the polymetallic ore deposit near the port of Lavrion (Laurium, modern Lavrio). South from Lavrion, toward Crete, are the Cycladic islands where, on Kythnos and Seriphos, there are copper ores and ancient slags. Silver, lead, and possibly also copper were mined on the Cycladic island of Siphnos. No copper ores are known from the Dodecanese islands or the Aegean coast of Turkey. The sea distance from Crete to Cyprus going toward the island of Rhodes is approximately twice as long as the sea route from Knossos to Lavrion. Therefore, indeed, the Cycladic islands and Lavrion were for the Minoans the closest sources of copper and lead-silver ores.

The amount of slag at Chrysokamino is quite small, and the lead isotope compositions of the analyzed pieces show compositions that would result from using the Lavrion and Cycladic ores separately, or mixed in one smelting charge. It is possible that the whole smelting operation was a result of only a few expeditions in the search for ore on the shores nearest to the northern coast of Crete.

COMPARISON OF LEAD ISOTOPE ANALYSES OF SLAGS FROM CHRYSOKAMINO WITH EM–MM COPPER ARTIFACTS

We have lead isotope and elemental analyses of 118 Early and Middle Minoan copper-based artifacts. On the basis of lead isotope data, Lavrion and the Cyclades are the dominant sources of copper and lead in EM–MM II Crete.

The lead isotope ratios closest to the analyzed pieces of slag from Chrysokamino are found among EM–MM artifacts from several sites. However, since very similar lead isotope ratios appear also among the LM artifacts, copper from Lavrion clearly was used on Crete over a long period of time and must have been extracted mostly somewhere else, not at Chrysokamino (perhaps not in Crete at all), since it would have been possible to transport copper metal to Crete. Many of these objects are made of tin bronze. If they were cast in palace or settlement workshops, the tin would have been added to copper there. Artifacts showing the best lead isotope match with the slags are plotted on Figure C.5 and listed in Table C.6.

Figure C.5 *(opposite)*. Lead isotope analyses of some Prepalatial and Protopalatial copper-based alloy artifacts from Crete compared with slags from Chrysokamino, copper ores from Lavrion, and ores and slags from the Cyclades

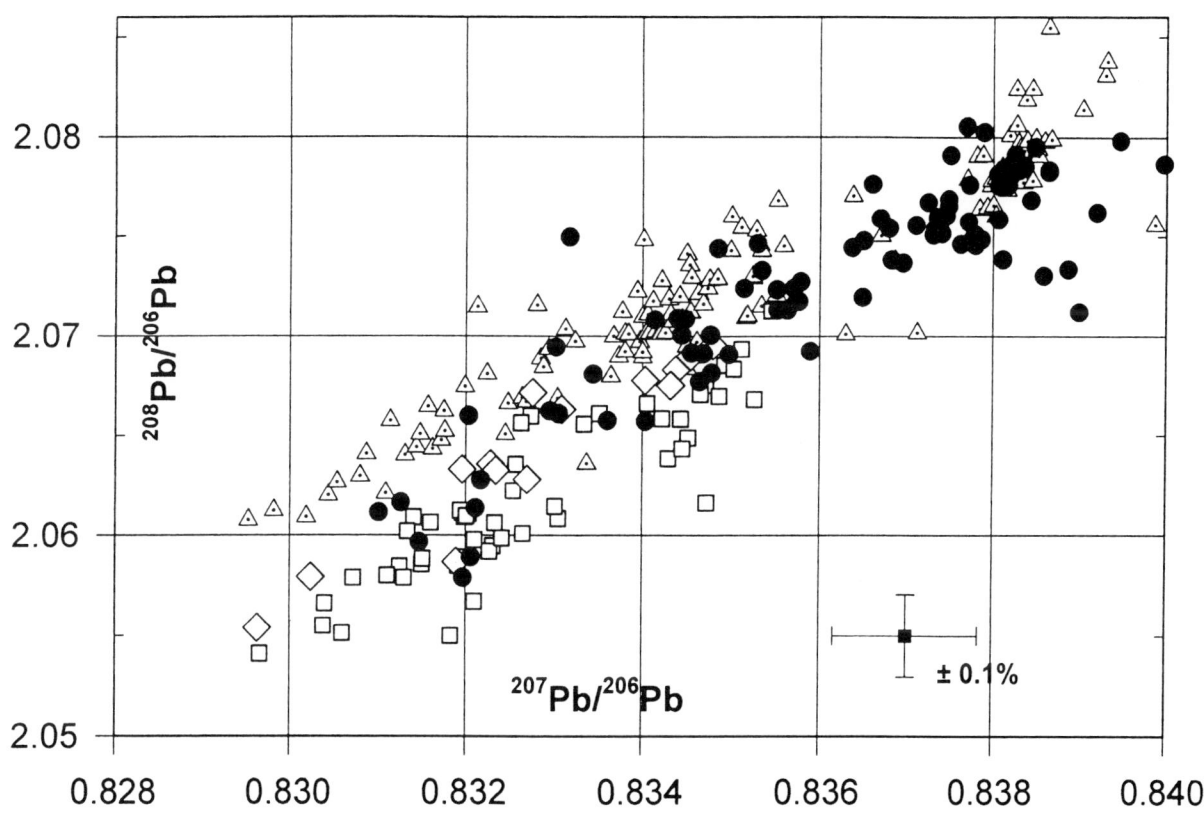

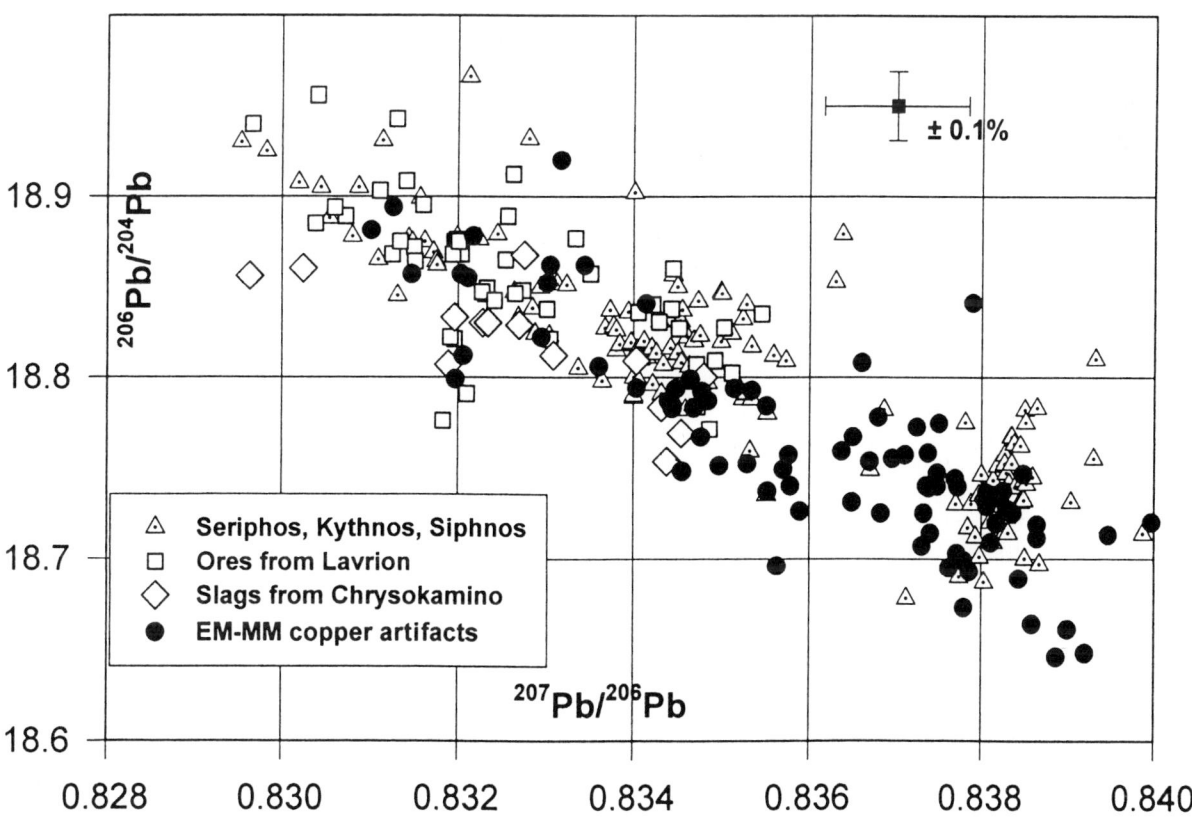

TABLE C.6. EARLY MINOAN COPPER-BASED ARTIFACTS ISOTOPICALLY CONSISTENT WITH THE SLAGS FROM CHRYSOKAMINO

Artifact No.	Site	Description	Chronology	$^{208}Pb/^{206}Pb$	$^{206}Pb/^{207}Pb$	$^{206}Pb/^{204}Pb$
HM 1557/St 9376	Mochlos	dagger, triangular	EM–MM	2.05966	0.83147	18.857
HM 1556/St 9375	Mochlos	dagger	LM?	2.05789	0.83197	18.799
HM 1265	Hagia Triada	dagger, triangular	EM IIA–MM IA	2.066	0.83203	18.857
HM 1930/St 9383	Platanos	dagger, tongue	EM?	2.06137	0.83211	18.855
ANM 4660	Hagia Photia	chisel	EM I–IIA	2.06278	0.83217	18.878
ANM 4667	Hagia Photia	chisel	EM I–IIA	2.06624	0.83295	18.822
HM 2041/St 9368	Pyrgos	dagger, long	EM I–II	2.06944	0.83302	18.852
St 9447	Krasi	dagger with 3 rivets	EM IIA–MM IA	2.0661	0.83305	18.862
ANM 4671	Hagia Photia	dagger	EM I–IIA	2.06807	0.83344	18.862
HM 1195/St 9425	Koumasa	dagger, long	EM IIA–MM IA	2.06576	0.8336	18.806
HM 2005/St 9432	Marathokephalo	dagger, triangular	EM IA–MM IA	2.06572	0.83403	18.794
ANM 4669	Hagia Photia	spearhead	LM I (?)	2.07083	0.83414	18.841
HM 2047/St 9366	Pyrgos	punch/awl	EM I–II	2.0709	0.8344	18.787
ANM 4656	Hagia Photia	chisel	EM I–IIA	2.07006	0.83444	18.783
HM 2040/St 9369	Pyrgos	dagger, long	EM I–II	2.07085	0.83448	18.794
ANM 4662d	Hagia Photia	fishhook	EM I–IIA	2.06915	0.83455	18.748
ANM 4670	Hagia Photia	dagger, long	EM I–IIA	2.06771	0.83464	18.799
HM 1852/St 9399	Platanos	dagger, long	EM I–IIB	2.06912	0.83468	18.783
HM 1182/St 9423	Koumasa	dagger with midrib	EM IIA–MM IA	2.07006	0.83476	18.767
HM 1167/St 9427	Koumasa	dagger, long	EM I–MM II	2.06814	0.83477	18.792
ANM 4674	Hagia Photia	saw	EM I–IIA	2.07439	0.83484	18.787
ANM 4655	Hagia Photia	awl (?)	EM I–IIA	2.06907	0.83497	18.751
ANM 4663	Hagia Photia	chisel	EM I–IIA	2.07241	0.83514	18.794
HM 1434/St 9429	Porti	dagger, long	EM–MM	2.07464	0.83529	18.752
HM 318/St 9364	Palaikastro	dagger, triangular	EM III–MM I	2.07330	0.83534	18.793
MP/70/146	Myrtos, Pyrgos	pin	EM III–MM I	2.07234	0.83551	18.784
HM 1498/St 9443	Kalathiana	dagger, fragment	EM IIA–MM IA	2.07133	0.83552	18.737
HM 1932/St 9381	Platanos	dagger, long	EM IIA–MM IA	2.07132	0.83563	18.696
ANM 4675	Hagia Photia	chisel	EM I–IIA	2.07242	0.8357	18.749
HM 1552/St 9374	Mochlos	dagger, long	EM IIA–MM IA	2.07177	0.83576	18.757
HM1294/St 9411	Hagia Triada	dagger, long	EM IIA–MM IA	2.07275	0.83578	18.74
HM 1188/St 9428	Koumasa	dagger, long	EM IIA–MM IA	2.06923	0.83589	18.726
C-70.22; C1009B	Kea	Hagia Eireini, copper lump	EC	2.06169	0.83108	18.89
16159	Naxos	Unknown	EC	2.05964	0.83172	18.841
KPH.9	Kea	Kephala	EC	2.06519	0.83239	18.842
5807	Euboia	Manika, chisel	EH	2.06353	0.8325	18.847
AE 234 (1893.64)	Amorgos	Amorgos, dagger	EC 2	2.06344	0.8326	18.815
C-9.100; C775-C	Kea	Hagia Eireini, Cu slag	EC–MC	2.06635	0.83261	18.844

TABLE C.6 (cont.)

Artifact No.	Site	Description	Chronology	$^{208}Pb/^{206}Pb$	$^{206}Pb/^{207}Pb$	$^{206}Pb/^{204}Pb$
1927.1358	Amorgos	Amorgos, dagger	EBA–MBA	2.06536	0.83365	18.823
7206a	Lemnos	Poliochni, chisel	EH?	2.07097	0.83371	18.856
16125	Syros	Chalandriani, punch	EC 2	2.06494	0.83372	18.805
AE 236	Amorgos	Amorgos, flat axe	EC 2	2.06563	0.83394	18.794
	Euboia	Manika, chisel	EH	2.06982	0.83416	18.827
	Cyclades, Syros	Chalandriani, tweezers	EC 2	2.0667	0.8344	18.811
1969/12.31.5	Kythnos	Kythnos, flat axe	EC	2.07005	0.83455	18.8
	Euboia	Manika, spatula	EH?	2.06595	0.83462	18.793

ACKNOWLEDGMENTS

The authors would like to thank very warmly the Director and members of the Institute of Geological and Mineral Exploration in Athens and Chania for their great help in surveying Greek copper and lead/silver deposits. On Crete, we were particularly assisted by Michael Diallinas and Kostas Zervantounakis. Research in the Isotrace Laboratory was financed by the Science and Engineering Research Council, the Natural Environment Research Council, the British Academy, the Leverhulme Trust, and the University of Oxford. We are very grateful to INSTAP for funding our work between 1995 and 2001. For chemical analyses, we thank the Department of Earth Sciences at the Royal Holloway College and the British Geological Survey.

Arsenic Content of Copper Prills: A Study Applying PIXE

by Susan C. Ferrence and Charles P. Swann

This study presents analyses by PIXE of copper prills included in slag from the Final Neolithic to Early Bronze Age workshop at Chrysokamino, focusing on the arsenic content as measured from corroded surfaces in comparison with the values obtained from the more pure prill interiors. The study is of interest because many artifacts survive only as corrosion products, and the degree to which these corrosion products reflect the original arsenic content is important in the estimation of the original composition (i.e., is the surface arsenic enhanced because of the removal of other elements, does it remain stable, or is it depleted?). The study addresses the magnitude of the arsenic content of the prill surfaces, the degree of heterogeneity both within individual prills and between prills, and the differences between corroded prill surfaces and the more pure prill interiors.

INTRODUCTION

The many analyses of metal objects from the Aegean Early Bronze Age show that the metallurgists of this period made extensive use of arsenical copper.[1] Some scholars have suggested that arsenical copper has advantages over unalloyed copper and that early Aegean metallurgists, who were aware of these advantages, deliberately produced alloys.[2] Others have challenged both the concept that arsenical copper offered advantages and the assumption that all cases of copper with high arsenic content were produced intentionally.[3] Uncertainties exist in regard to whether the arsenic content was intentional or accidental, whether it was present in the copper ores or was added later at the smelting or casting stage, and whether similar situations existed in all Aegean workshops. The debate has been hampered by several factors, including the minimal number of analyses from anything other than finished products. In particular, only a few analyses have been published from Aegean copper smelting sites.

1. Renfrew 1967; Charles 1967; Gale and Stos-Gale 1989.
2. Charles 1967.
3. Budd and Ottaway 1991; Budd 1991.

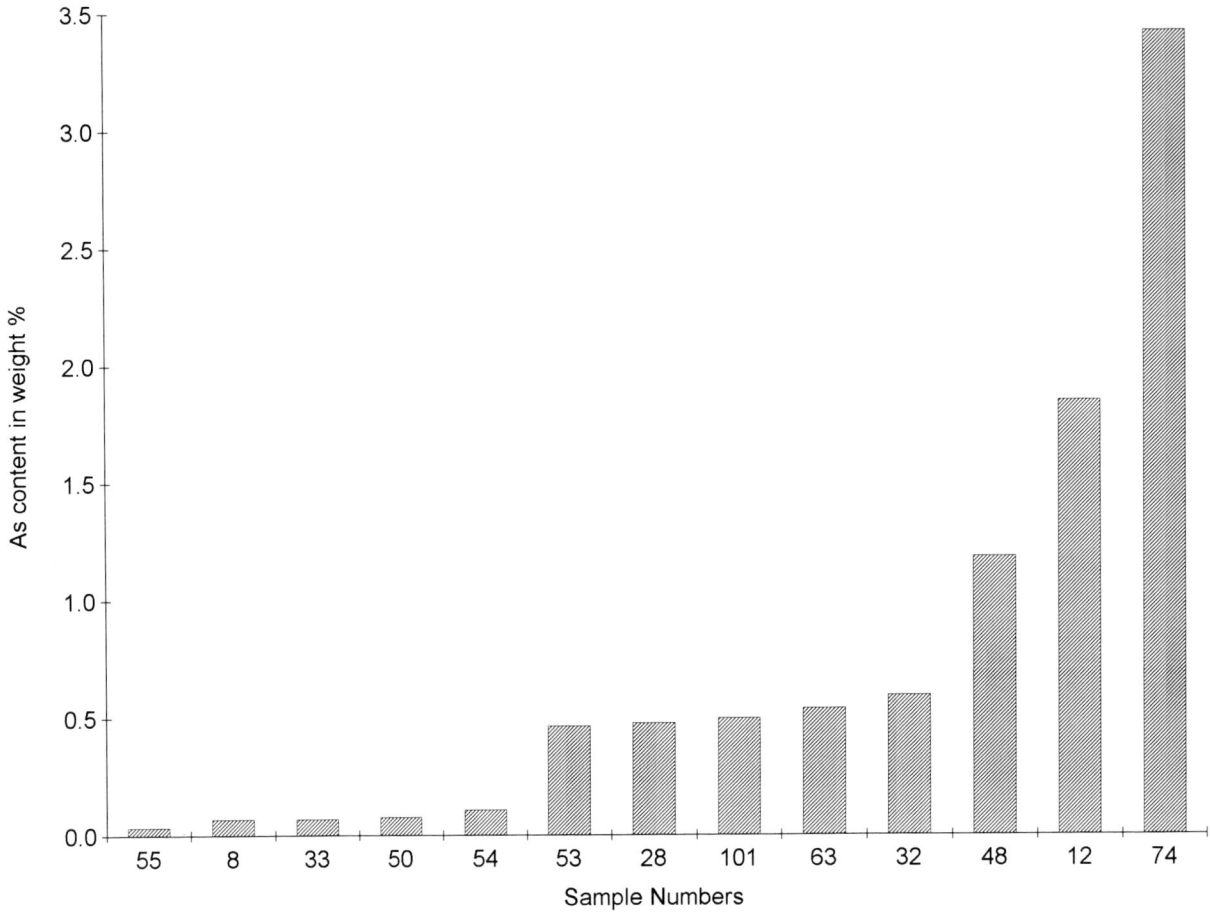

Figure D.1. PIXE analyses of copper prills arranged by increasing arsenic content

PIXE MEASUREMENTS

Samples of prills from the Chrysokamino slag heap were analyzed at the Bartol Research Institute at the University of Delaware. The analysis used air (helium) proton-induced X-ray emission spectroscopy,[4] an analytical tool for the study of archaeological artifacts that has been employed successfully since about 1975.[5] With the additional use of appropriate X-ray filters, the observation of all the elements from sodium through antimony and the heavy elements such as lead is possible with good detection limits. The great advantages of such systems are their basically nondestructive character, their multielemental nature, the good detection limits for most conditions, and the rapidity of data collection; the fact that PIXE is a technique for surface analysis can be both a positive and negative characteristic depending on the intent of the study.

RESULTS AND DISCUSSION

Figure D.1 presents the analyses of copper prills by PIXE, arranged by increasing arsenic content of the samples. The PIXE measurements were conducted on the corroded surfaces of tiny masses and lenses of copper

4. For comments on the methodology, see Swann, Ferrence, and Betancourt 2000.
5. Swann and Fleming 1988, 1990; Fleming and Swann 1992.

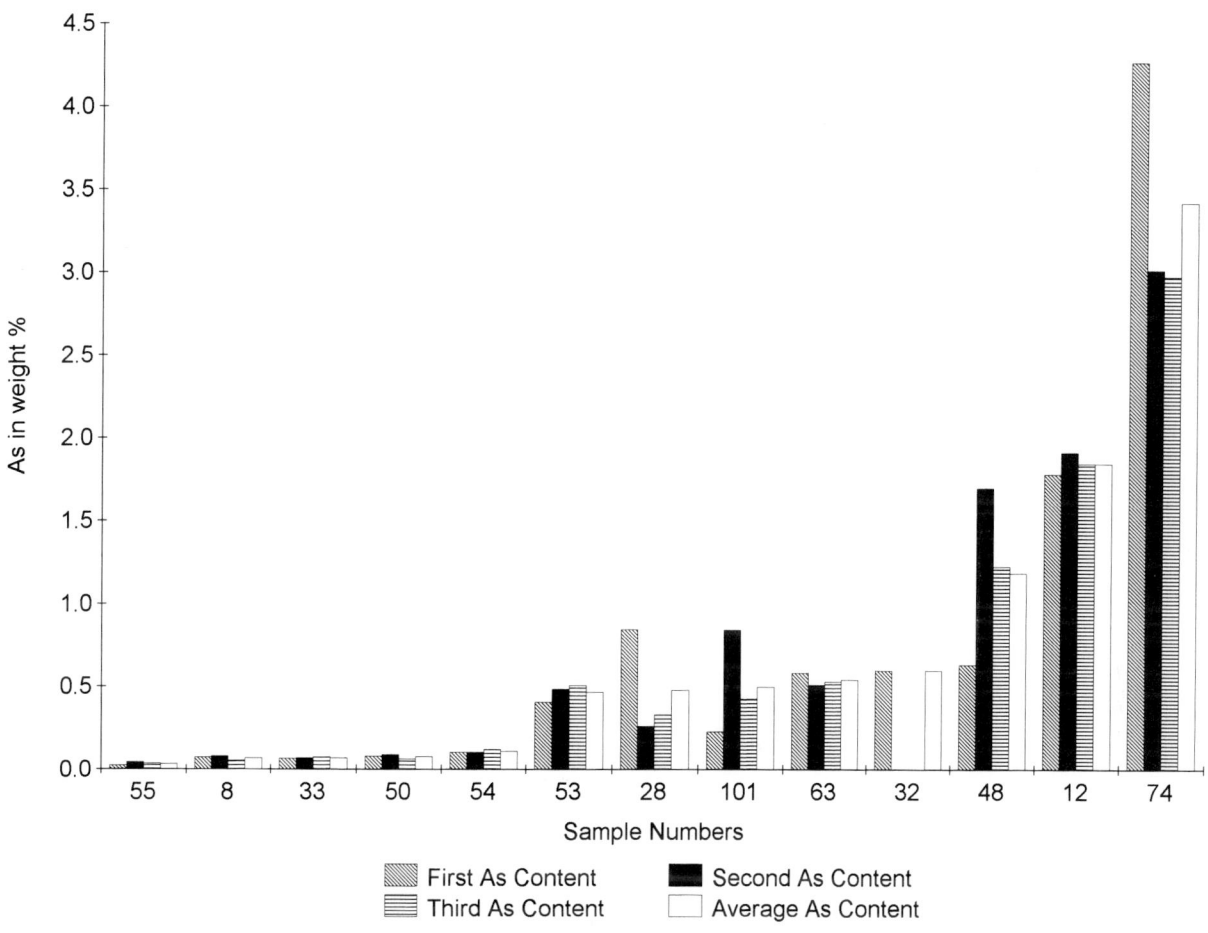

Figure D.2. PIXE measurements showing the variation within individual prills

included within glassy masses of slag. Where the size of the prill permitted, three PIXE measurements were made on each prill, with the measurements 100 micrometers apart. Arsenic contents vary from <0.03% to 3.42%, with an average of 0.72%.

Figure D.2 illustrates the variations within individual prills as shown by the PIXE measurements given in Figure D.1. The three arsenic values for each prill are shown as the first three bars in each group, with the average for each prill presented as the last bar in each group. Variation can be as little as 0.01% or as great as 1.29%.

The measurements presented here show that some of the copper prills obtained from Final Neolithic to Early Bronze Age smelting operations at Chrysokamino contain arsenic in variable amounts. Percentages of arsenic between 2% and 6% have been regarded as significant amounts because of arsenic's antioxidant qualities during casting and its ability to improve hardness.[6] The amounts recorded from Chrysokamino in tiny prills at the microscopic level cannot necessarily be related to the gross amount in larger masses of copper because of the high variability detected.

This amount of arsenic is compatible with the methodology reconstructed for Chrysokamino, where the copper was formed as tiny prills disseminated through a glassy slag. Experiments have shown that incorporation of arsenic as an impurity within smelted copper occurs during the

6. Budd and Ottaway 1991.

smelting process if arsenic minerals are present in the original charge and if the temperature is not high enough to result in a molten metal that would be removed by tapping the furnace.[7] During some conditions, including those suggested for this site, some of the prills would be trapped within the slag even if some of the copper collected at the base of the furnace. As indicated elsewhere, under the conditions present at Chrysokamino only the larger prills would migrate to the bottom of the furnace and collect as molten metal, leaving the smaller ones within the slag.

The lower arsenic values obtained from corroded surfaces compared with clean surfaces (see the results obtained by Bassiakos in App. B) shows that the corrosion tends to deplete the arsenic content contained within the prills. This condition is presumably a postburial action that occurred during the processes of corrosion. In terms of methodology, it is important to recognize this result because it means that values obtained from corroded copper surfaces are useful as minimum values and can be used to document the occurrence of arsenical copper.

7. Pollard et al. 1991.

Slag Analysis by Wavelength Dispersive Spectrometry

by Christine M. Thompson

The following research was carried out at the Center for Materials Research in Archaeology and Ethnology and the Center for Geochemical Analysis at the Massachusetts Institute of Technology. The samples were analyzed by Wavelength Dispersive Spectrometry on a JEOL JXA-733 Superprobe with a beam current of 10nÅ, probe diameter of 1 micrometer, and counting times of 40–80 seconds per element.

Five pieces of metallurgical slag from the smelting workshop at Chrysokamino were analyzed (numbered CHR 55, 48, 50, 8, and 74). They were all dark, vitreous specimens from the Final Neolithic to Early Minoan III–Middle Minoan IA slag pile. All the samples contained small, visible prills of copper.

In the following charts, "nd" means "not detected" and "na" means "not analyzed."

SAMPLE CHR 55

Analyses were made at two opposing points of the sample's matrix, and results indicate a nearly homogeneous composition, which is a silicate high in iron and barium (Points 1 and 2). Both the light and dark "needles" are comprised of the same elements as the matrix. The dark crystals within the matrix (Point 5, shown in Fig. E.1) are an Fe-oxide. The dark crystals in the backscattered electron image (Fig. E.1) appear light in the photomicrograph (Fig. E.2).

TABLE E.1. CHR 55, POINTS 1–5 (Figs. E.1, E.2)

Point	SiO_2	Al_2O_3	Fe_2O_3	CaO	BaO	Cu	Total	Notes
1	35.75	2.44	32.39	3.24	23.83	trace	97.65	matrix
2	37.05	3.04	33.42	3.69	19.60	trace	96.8	note 1
3	28.83	nd	73.57	1.10	0.32	trace	103.81	note 2
4	29.80	2.62	35.15	1.08	31.16	trace	99.81	note 3
5	0.06	0.44	101.15	0.03	0.12	trace	101.80	note 4

Notes: 1. Same matrix opposite Point 1; 2. Dark "needles"; 3. Light "needles"; 4. Dark crystals.

TABLE E.2. CHR 55, POINTS 6, 7 (Figs. E.1, E.2)

Point	Pb	Fe	Cu	As	Sb	Cl	S	Total	Notes
6	nd	3.1	70.48	nd	0.01	nd	25.75	99.34	1
7	nd	0.54	99.13	0.82	0.23	nd	0.30	101.02	2

Notes: 1. Circular inclusion; 2. Prill.

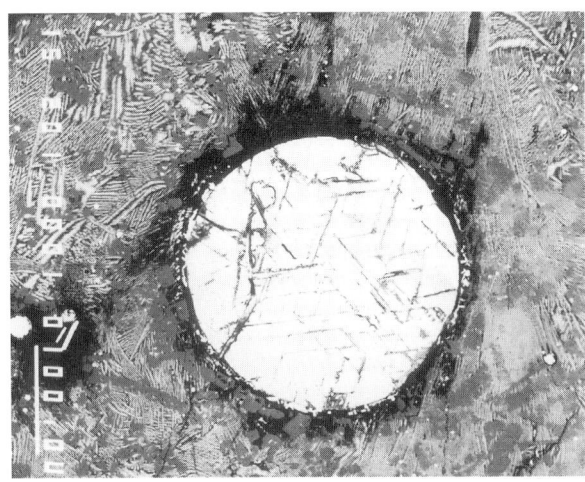

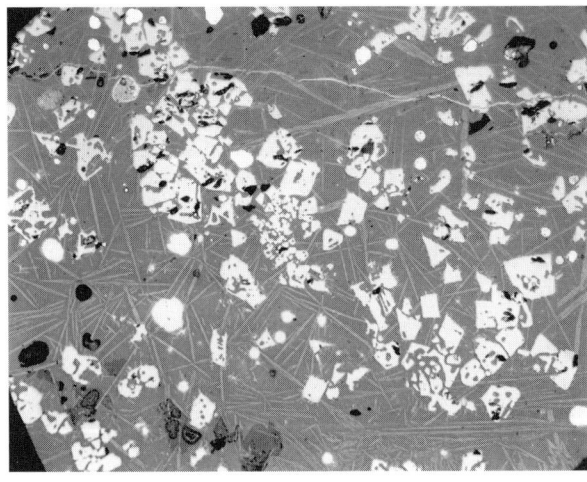

Figure E.1 *(left)*. Backscattered electron image of CHR 55 (200×) with large copper sulfide inclusion, "needles," dark crystals, and prills.
Photograph by C. Thompson

Figure E.2 *(right)*. White light photomicrograph of CHR 55 (200×) with "needles" and small prills.
Photograph by V. Pigott

SAMPLE CHR 48

Sample CHR 48 is comprised of three concentric regions. The innermost region contains Cu-As-oxides (Points 1 and 2). The surrounding region is comprised of Cu-oxides with high levels of Cl (Point 4). Point 3 represents the sample's dominant, exterior region, which is primarily Cu sulfide (Cu$_2$S).

TABLE E.3. CHR 48, POINTS 1–4 (Fig. E.3)

Point	Pb	Fe	Cu	As	Sb	Cl	S	O	Ca	Si	Total
1	nd	0.10	44.91	18.82	0.37	0.73	0.04	26.67	0.88	0.13	92.65
2	nd	0.20	42.28	18.48	0.33	1.12	0.04	26.88	0.88	0.13	90.34
3	nd	0.01	78.98	nd	0.04	nd	19.68	na	Na	na	98.71
4	nd	nd	58.73	nd	0.04	14.91	0.01	na	Na	na	90.63

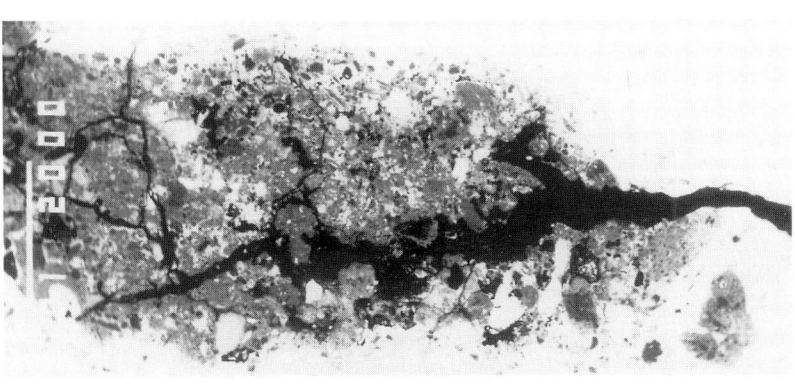

Figure E.3. Backscattered electron image showing the core of CHR 48 (40×). Photograph by C. Thompson

SAMPLE CHR 50

The matrix of CHR 50 is primarily an Fe-Ca-Al-Silicate (Point 1). Ba-sulfate ($BaSO_4$) is associated with Cu-chloride (CuCl) in the remainder of the matrix. The dendrites contain iron oxide with small amounts of aluminum (Point 2). Analyses of the prills reveal high levels of copper, small amounts of iron, and trace amounts of arsenic and antimony (Point 3).

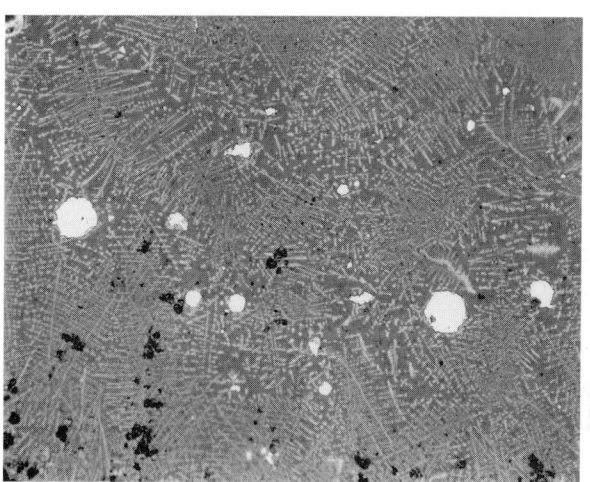

Figure E.4. Photomicrograph of CHR 50 (400×) with dendrites and prill field. Photograph by V. Pigott

TABLE E.4. CHR 50, POINTS 1, 2 (Fig. E.4)

Point	SiO_2	Al_2O_3	Fe_2O_3	CaO	BaO	Cu	Total	Notes
1	44.77	6.39	32.65	14.86	0.18	nd	98.85	matrix
2	0.94	3.93	96.42	0.42	nd	trace	101.71	dendrite

TABLE E.5. CHR 50, POINT 3 (Fig. E.4)

Point	Pb	Fe	Cu	As	Sb	Cl	S	Total	Notes
3	nd	3.52	97.53	0.57	0.18	0.11	0.06	101.97	prill

SAMPLE CHR 8

Fe-Ba-Ca silicates comprise the matrix and needles of CHR 8. Cu-sulfide was detected at Point 4 and is representative of the circular inclusion shown in Figure E.5. Another Cu-sulfide inclusion (not shown) contained an inner region of metallic copper with arsenic (Point 1).

TABLE E.6. CHR 8, POINTS 1–3 (Fig. E.5)

Point	SiO_2	Al_2O_3	Fe_2O_3	CaO	BaO	Cu	Total	Notes
1	34.17	2.64	36.99	8.06	15.73	trace	97.59	matrix
2	29.56	1.17	64.02	4.60	2.82	trace	102.17	needles
3	0.20	0.36	98.20	0.09	0.39	trace	99.24	dendrite

TABLE E.7. CHR 8, POINTS 4, 5 (Fig. E.5)

Point	Pb	Fe	Cu	As	Sb	Cl	S	Total	Notes
4	nd	0.56	80.41	0.21	nd	nd	20.06	101.24	1
5	nd	1.27	97.87	2.97	nd	nd	nd	102.39	2

Notes: 1. Circular inclusion; 2. Circular inclusion.

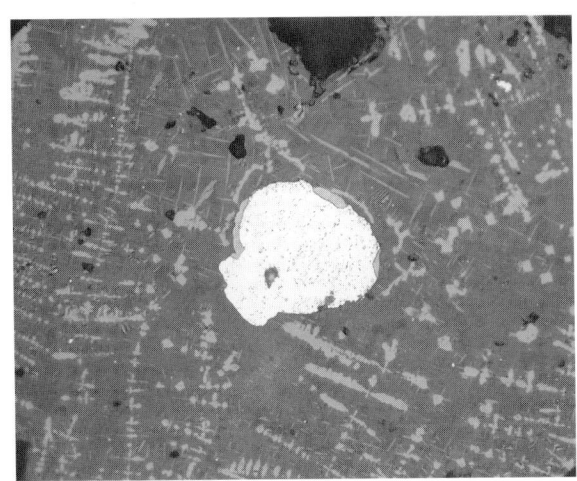

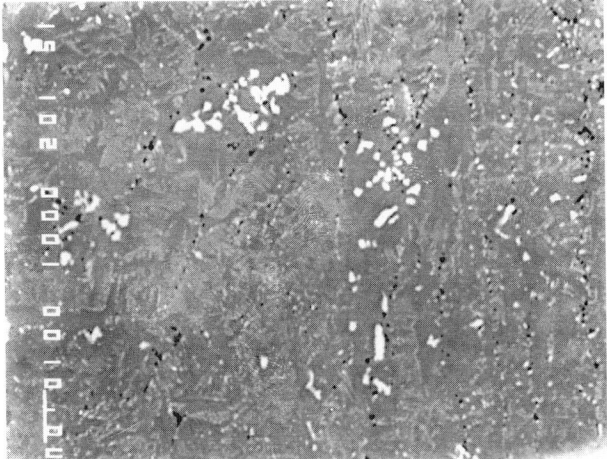

Figure E.5 *(left)*. Photomicrograph of CHR 8 (400×) with "needles," dendrites, and circular inclusion of copper sulfide at center. Photograph by V. Pigott

Figure E.6 *(right)*. Backscattered electron image of CHR 74 (1000×) with matrix and dendrites. Photograph by C. Thompson

SAMPLE CHR 74

The matrix of CHR 74 was a Ca-Fe-Al-silicate, and the dendrites contained iron oxide with measurable amounts of aluminum. The prills were comprised of copper with minor amounts of arsenic and antimony.

TABLE E.8. CHR 74, POINTS 1, 2 (Fig. E.6)

Point	SiO_2	Al_2O_3	FeO	CaO	BaO	Cu	Total	Notes
1	37.26	8.77	15.54	35.48	0.07	trace	97.12	matrix
2	1.19	5.67	93.41	1.74	0.13	trace	102.14	dendrite

TABLE E.9. CHR 74, POINT 3 (Fig. E.6)

Point	Pb	Fe	Cu	As	Sb	Cl	S	Total	Notes
3	nd	0.14	102.3	0.93	0.30	na	na	103.67	prill

Reconstruction of the Copper Smelting Process Based on the Analysis of Ore and Slag Samples

by Yannis Bassiakos and Mihalis Catapotis

The application of scientific techniques in the study of Early Bronze Age metal artifacts and metallurgical finds has a very long history in the Aegean.[1] However, although evidence for early metal production has been available since the middle 1980s,[2] our understanding of copper smelting technologies during the Early Bronze Age is still limited. This is due to the lack of systematic excavations of metallurgical sites and to the preoccupation of archaeometallurgical studies in the Aegean with questions of provenance at the expense of technological considerations. Therefore, as the only systematically excavated smelting site in the Aegean, Chrysokamino provides a unique opportunity for a detailed technological study of a copper smelting process.

This report presents the results of analyses of ore and slag samples from Chrysokamino conducted at the Laboratory of Archaeometry, N.C.S.R. "Demokritos." By integrating archaeological and analytical data, we propose a provisional reconstruction of the copper smelting process that was taking place at the site.

MORPHOLOGY AND GEOLOGICAL CONTEXT OF THE SITE

The metallurgical site of Chrysokamino is located on a promontory that is part of a small isolated peninsula[3] consisting of terraced hillsides on the eastern edge of the Gulf of Mirabello.

Most of the typical geological formations of eastern Crete are present in this area (for the geology, see Farrand in Chap. 2). The stratigraphically lower autochthonous unit, constituting approximately half of the peninsula, is a carbonate formation known as Plattenkalk, present also in Epirus, in the Peloponnese, and in other Cretan areas. It consists of a solid, relatively thinly bedded or slabby gray to bluish semimetamorphic limestone, interlaced with conspicuous white to brown calcitic veins. Sporadically visible in the same formation are some folds, either coarsely crystalline or with intercalations of marly and green limestones. The first overthrusting

1. E.g., Lamb 1936; *Keos* I; Craddock 1976; Gale and Stos-Gale 1981a; Mangou and Ioannou 1999; Pernicka et al. 1990; Stos-Gale 1993.

2. E.g., Gale et al. 1985; Spitaels 1984; Wagner and Weisgerber 1985.

3. This is a submerging tectonic horst; Flemming, Czartoryska, and Hunter 1973; Fortuin 1978.

sheet is the unit of the Phyllite-Quartzite nappe, a formation exhibiting low-grade metamorphism, which is more intense in its stratigraphically lower parts. Most constituents of this phyllitic unit are powdery (silty) and mineralogically altered as a result of in situ weathering of the calcite-rich, fine-grained rock. Much less extended in the site is the dolomite of the second successive overthrusting sheet, which belongs geologically to the Tripolitza nappe (Figs. 2.1, 2.3) and is predominantly gray to dark gray and massively bedded. It occurs in the area of the habitation site and between it and the smelting location.

An interesting geological feature of the area is a weathering-resistant carbonate ledge traversing the local phyllites with a strike of N40-55E and a dip of 40–50S.[4] In our visits to the area, we noted that the selective weathering of the phyllites has often resulted in the exposure of the ledge, forming localized protuberances. In those cases, iron (hydr)oxides can be seen on the lower contact between the carbonate ledge and the phyllites. Similar iron (hydr)oxide crusts can be seen frequently on rock fragments among the tumble covering the slope on the east of the promontory. Those thin crusts appear to develop locally into a weak iron mineralization, mainly hosted in the Plattenkalk/phyllite contact. The existence of weak calcite-accompanied or quartz-accompanied mineralizations is, in our experience, not uncommon in the Aegean HP/LT polymetamorphic formations (e.g., in Attica, in the Cycladic complex, and elsewhere) where the repeated metamorphic action causes mobilization and condensation of formerly disseminated metallic components, relocated by means of ledge formation. The possible role of the local iron mineralization as a source of fluxing material in the copper smelting process at Chrysokamino is addressed later in this report.

In contrast, no copper mineralization has been identified in the vicinity of the metallurgical site, suggesting that copper ores were brought to Chrysokamino from elsewhere. The original interpretation of the nearby cave of Theriospelio as a copper mine[5] has been rejected by Diallinas and others who have examined the cave but found no evidence of mineralization.[6]

METALLURGICAL REMAINS AND SAMPLES

Excavations at the site revealed many tons of dark-colored slag present, mostly in small pieces (under 2 cm in size) or completely pulverized (see Chap. 10). Among the finds were tens of thousands of perforated ceramic fragments identified as chimneys for furnaces (Chap. 7), remains of pot bellows (Chap. 8), tuyeres (Chap. 9), and a small number of stone tools (Chap. 6). A few minute fragments of copper ore (malachite, azurite, and chrysocolla) were revealed during the excavation. Our study concentrated on the examination of ores and slags from the site; Myer and Betancourt studied the furnace chimney fragments (App. A).

Three surface finds of copper ore were collected by Bassiakos after the completion of the excavations (August 2000) and were used for laboratory analysis in order to provide information on the raw materials used

4. Betancourt et al. 1999, p. 352.
5. Mosso 1910, p. 219.
6. Branigan 1968, p. 50; see Chap. 18 in this volume.

in the smelting process. The fragments were of very small size (less than 2 cm). Macroscopically, all three appeared to consist of green copper ores adhering to quartz. The green copper mineral was identified in the field as malachite, $CuCO_3 \cdot Cu(OH)_2$. Iron hydroxides were visible on two of the ore fragments (ORE-1 and ORE-2), both as individual particles and as stains on the surface of quartz.

Forty-six slag samples were selected for this analytical project, mostly deriving from passes 3 (sample series B) and 4 (sample series A) in trench N 20. One sample (sample series C) derived from the surface material. The samples were small, irregularly shaped fragments rarely exceeding 2–3 cm in size. Porosity was limited and of small size (<1 mm), whereas vitrification varied both among and within individual slag fragments. Carbonate crusts derived from the calcareous environment thickly coated most fragments (with the crusts up to ca. 1 mm thick). This epigenetic material had often penetrated to their interior through open porosity and cracks. Despite the fragmented state of the slag and the presence of epigenetic material, both of which prevented a detailed macroscopic examination, many of the largest slag fragments still had visible textural features that could be associated with slag tapping, i.e., the removal of liquid slag from the furnace through a hole. Minute copper prills (<1 mm) and small silica inclusions (<2 mm) could be seen occasionally with a magnifying lens.

An iron ore fragment was inadvertently included in the slag samples collected for laboratory analysis. The size of the fragment was about 3 to 4 cm, and macroscopically it appeared to consist of yellow orange iron hydroxides. This was considered to be an important find as it could have been associated with the use of fluxing agents in the process.

ANALYTICAL OBJECTIVES, METHODOLOGY, AND INSTRUMENTATION

The aim of this study was to understand the technology employed for the production of copper at Chrysokamino. Breaking down the smelting process into its constituent parts, we set the following questions as guidelines for the analytical work:

1. What types of copper ores were processed, and what were the associated gangue minerals?
2. Is there evidence for the use of fluxing agents or other additions in the charge?
3. What was the nature of the smelting process (e.g., reduction smelting, matte smelting, etc.)?
4. What were the operational conditions (temperature and redox conditions) during smelting?
5. Was the slag tapped or was it left to solidify inside the furnace?
6. What were the copper losses in the slag?
7. What was the product of the smelting process (i.e., what was the chemical composition of the smelted copper)?

As the fragmented state of the material did not allow detailed macroscopic examination of the samples, more emphasis was placed on laboratory techniques. Mineralogical analysis was conducted using optical microscopic methods. Polished specimens were examined on a LEICA DM polarizing microscope using a range of magnifications from ×25 to ×1000. Quantitative chemical analysis of ore and slag samples was conducted using a Philips 500 Scanning Electron Microscope (SEM) coupled with an Energy Dispersive Spectroscopy (EDS) detector (see App. B). In most cases, the chemical composition was obtained by multiple area scans on pelletized specimens impregnated into resin. In cases where the size of the sample was small, chemical composition was obtained by multiple area scans on the polished specimens used for microscopic analysis. SEM-EDS was also used for the chemical analysis of individual phases within samples. Finally, the Fe^{2+}/Fe^{3+} ratio of five slag samples was obtained by Mössbauer spectroscopy.[7]

ANALYTICAL RESULTS: THE ORES

Most macroscopic observations concerning the copper ores were generally confirmed microscopically, as the predominant copper minerals were malachite with some azurite, $Cu_3(CO_3)_2(OH)_2$, while the gangue minerals were predominantly quartz with small amounts of iron hydroxides and siliceous phases. Examination of sample ORE-1 under the polarizing microscope also revealed small quantities of sulfidic minerals inside the oxidized matrix. The identification of those minerals with the aid of SEM-EDS point analysis showed the predominant sulfide to be pyrite (FeS_2), occasionally containing small quantities of barium, covellite (CuS), chalcopyrite ($CuFeS_2$), and some chalcocite (Cu_2S) (Fig. F.1). The mineralogy and texture of those inclusions indicates that they constituted residual hypogene sulfides, which had survived the weathering process inside the oxidation zone of a copper (vein?) deposit. No sulfidic inclusions were found in the other two ore samples.

The bulk chemical composition of the three samples was obtained by EDS analysis (multiple area scans) of the polished sections. Sample ORE-3 consisted of a very fine flake of malachite adhering to a quartz fragment, so two separate sets of analyses were conducted, one for the ore and one for the gangue mineral. As shown in Table F.1, the EDS data support the phase identifications based on microscopic analysis. In the first place, the generally low sulfur contents reflect the predominance of oxidized minerals in the ore samples. In the case of ORE-1, however, the residual sulfides identified microscopically slightly raise the sulfur content. Secondly, quartz is shown to be the predominant gangue mineral, followed by iron-hydroxides and aluminum, while the levels of lime are extremely low, never exceeding 0.3%. The average content of arsenic and nickel in all ore samples is well below 0.2%, although occasionally the percentages reached slightly higher levels (0.5%) in individual area scan analyses.

Turning to the iron ore fragment, microscopic examination confirmed that the dominant mineral in the ore is goethite, -FeO(OH), showing

7. We are grateful to Dr. A. Simopoulos from N.C.S.R. "Demokritos" for conducting this analysis and processing the analytical results.

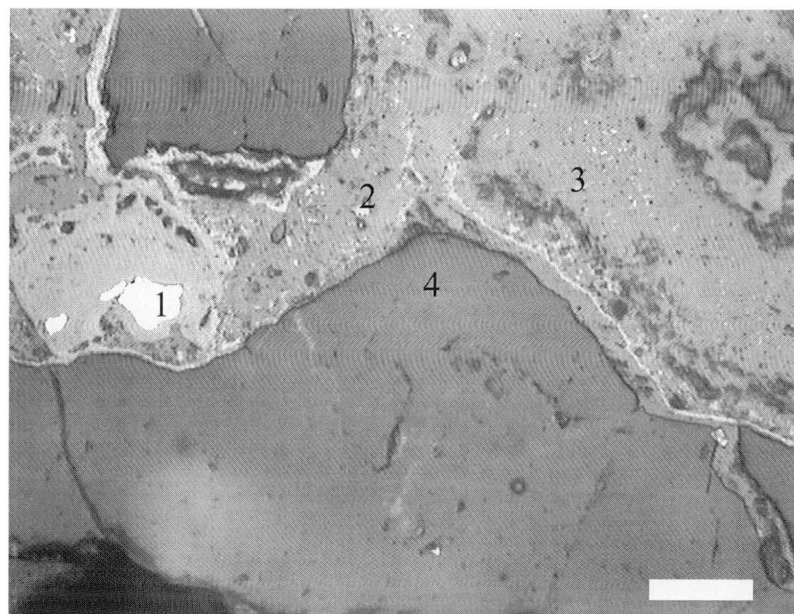

Figure F.1. Copper ore (ORE 1) from Chrysokamino showing residual sulfides (1: pyrite; 2: covellite) in a matrix of oxidized copper/iron minerals (3) and grains of quartz (4); reflected light. Scale: 30 micrometers

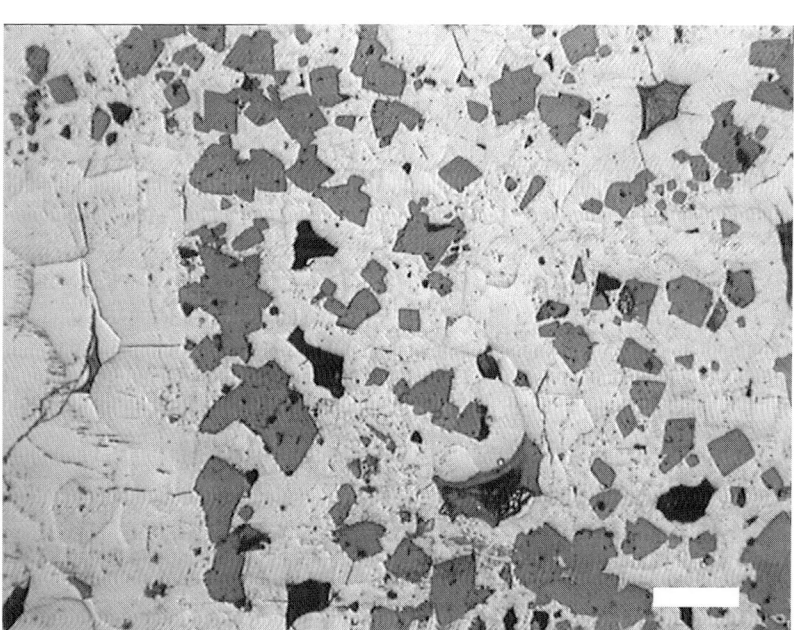

Figure F.2. Iron ore from Chrysokamino showing rhomboid crystals of calcite (dark gray) embedded in a matrix of iron (hydr)oxides (light gray); reflected light. Scale: 0.1 mm

characteristic botryoidal texture in cross-polarized light, locally altered to "limonite" ($Fe_2O_3 \cdot nH_2O$) (Fig. F.2). Small rhomboid crystals of calcite (ca. 100–300 micrometers) were dispersed throughout the ore, raising the CaO content to 25% (Table F.1). The silica, on the other hand, was just 5%. Mineralogically, therefore, the iron ore is different from the copper ores found at the site. Considering also that the iron ore contains no traces of copper, it follows that the iron ore fragment does not represent gangue mineral discarded during some process of in situ beneficiation but rather a material that was separately introduced to the metallurgical assemblage.

TABLE F.1. EDS ANALYSIS (AVERAGE COMPOSITION) OF COPPER ORE AND IRON ORE SAMPLES

Sample	SiO_2	FeO	CaO	Al_2O_3	MgO	MnO	Na_2O	K_2O	CuO	SO_3	Cl_2O	As_2O_3	NiO	TiO_2	P_2O_5	BaO	ZnO
ORE 1	57.60	6.53	0.14	1.51	0.92	0.01	0.76	0.01	29.93	1.52	0.15	0.05	0.08	0.04	0.46	0.14	0.12
ORE 2	54.14	1.41	0.25	1.52	0.52	1.09	0.54	0.10	39.17	0.17	0.18	0.08	0.01	0.12	0.59	0.04	0.06
ORE 3 (ore)	1.53	0.29	0.15	0.92	1.27	0.08	3.37	0.00	91.31	0.03	0.67	0.00	0.10	0.02	0.19	0.07	0.00
ORE 3 (gangue)	97.21	0.09	0.00	0.47	0.28	0.00	0.66	0.07	0.04	0.00	0.04	0.00	0.00	0.04	1.10	0.00	0.02
IRON ORE	4.86	67.36	23.25	0.47	1.29	0.20	1.32	0.12	0.04	0.09	0.23	0.06	0.16	0.00	0.29	0.21	0.08

Results normalized.

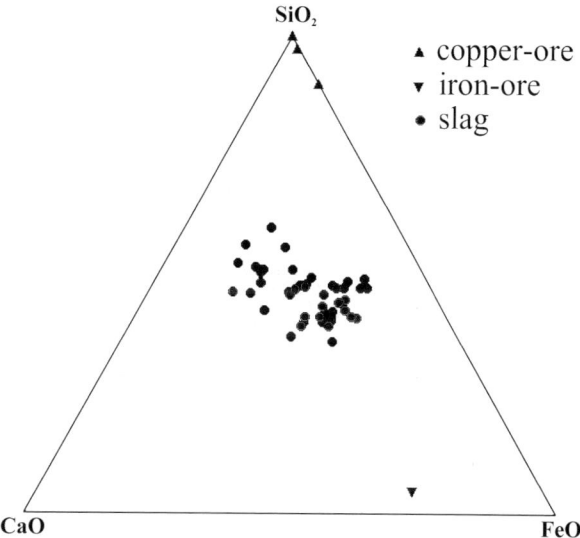

Figure F.3. The position of ores and slags from Chrysokamino in the ternary system CaO-FeO-SiO$_2$

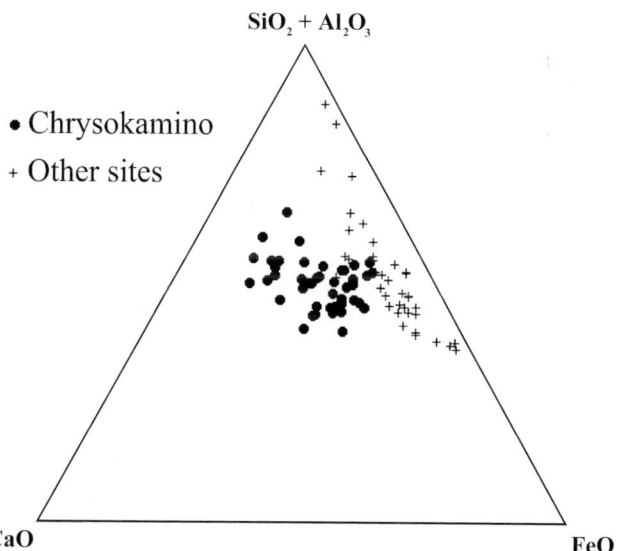

Figure F.4. The position of slags from Chrysokamino and other EBA smelting sites from the southern Aegean in the ternary system CaO-FeO-SiO$_2$ (+Al$_2$O$_3$)

ANALYTICAL RESULTS: THE SLAG

According to the results of the EDS analysis of pelletized samples, the slag from Chrysokamino belongs to the system CaO-FeO$_x$-SiO$_2$, and it also contains aluminum and smaller quantities of MgO, MnO, and alkalis (Table F.2). Most samples fall in the area of olivines and pyroxenes, corresponding to the lowest liquidus temperatures in this slag system (Fig. F.3).

It should be noted that the high levels of lime characterizing the copper smelting slags from Chrysokamino are uncommon in prehistory. Known examples of copper slag with high lime levels come from only a small number of Near Eastern smelting sites, including the Chalcolithic sites F2 and 39 at Timna,[8] the EBA site of Shahr-i-Sokhta in Iran,[9] and the Roman site of Beer Ora at Timna.[10] They are certainly unique in the Aegean Early Bronze Age (Fig. F.4).

8. Merkel and Rothenberg 1999.
9. Hauptman, Rehren, and Schmitt-Strecker 2003.
10. Bachmann 1980.

TABLE F.2. EDS ANALYSIS OF SLAG PELLETS

Sample	CaO	SiO_2	FeO	Al_2O_3	MgO	MnO	Na_2O	K_2O	CuO	SO_3	Cl_2O	As_2O_3	NiO	TiO_2	P_2O_5	BaO	ZnO
A01	28.46	36.34	20.49	5.93	1.59	0.27	1.73	0.81	0.41	0.38	0.16	0.15	0.01	0.28	0.95	1.96	0.10
A02	28.98	45.12	12.04	5.45	1.81	0.33	1.76	0.73	0.78	0.18	0.09	0.24	0.04	0.12	0.97	1.29	0.06
A03	11.00	39.45	33.04	7.60	1.53	0.66	2.72	0.76	1.29	0.21	0.29	0.07	0.03	0.36	0.65	0.29	0.06
A04	20.76	31.29	34.58	5.45	1.61	0.30	2.35	1.10	0.52	0.49	0.10	0.05	0.00	0.21	0.62	0.44	0.14
A05	18.64	33.78	31.55	6.55	1.70	0.10	2.14	1.20	1.51	0.20	0.20	0.18	0.06	0.22	0.80	1.12	0.06
A06	27.98	37.89	16.03	7.02	3.38	0.86	2.66	0.96	0.67	0.30	0.14	0.08	0.02	0.22	0.84	0.93	0.02
A07	19.39	36.00	30.78	5.30	1.16	0.20	2.90	0.78	0.55	0.17	0.13	0.09	0.00	0.28	0.81	1.27	0.19
A08	19.36	35.53	32.97	5.33	1.54	0.21	1.21	0.98	0.92	0.19	0.10	0.07	0.03	0.19	0.75	0.61	0.02
A09	20.50	34.39	31.01	5.53	1.52	0.18	2.15	1.04	1.35	0.25	0.22	0.20	0.06	0.24	0.65	0.66	0.07
A10	24.83	43.28	16.49	6.79	1.51	0.27	2.61	1.13	0.63	0.21	0.12	0.16	0.03	0.46	0.89	0.56	0.01
A11	17.94	45.81	12.67	10.76	1.59	0.20	4.41	1.10	1.15	0.12	0.17	0.13	0.00	0.41	0.50	2.98	0.06
A12	25.11	46.77	11.31	7.24	2.43	0.40	1.99	1.11	0.98	0.16	0.20	0.37	0.01	0.38	1.05	0.46	0.01
A13	19.71	47.89	18.23	5.95	0.95	0.35	2.25	0.70	1.14	0.17	0.16	0.17	0.04	0.25	1.01	0.86	0.18
A14*	22.47	38.43	22.01	7.68	2.38	0.40	2.40	1.00	1.11	0.02	0.09	0.00	0.13	0.21	0.50	1.18	0.00
A15	11.92	39.38	29.52	8.85	1.58	0.34	3.26	0.73	2.30	0.20	0.32	0.07	0.09	0.41	0.80	0.20	0.04
A16*	13.70	40.24	31.22	5.93	1.42	0.21	2.46	0.57	1.70	0.19	0.06	0.66	0.00	0.19	0.89	0.10	0.48
A17*	18.33	34.85	31.84	7.35	1.76	0.34	1.79	0.91	0.92	0.33	0.03	0.13	0.05	0.08	0.48	0.82	0.00
A18*	9.85	38.97	33.82	8.62	1.75	0.30	3.71	0.93	0.92	0.02	0.07	0.06	0.04	0.53	0.37	0.04	0.00
A19*	14.66	38.99	28.03	8.58	2.34	1.15	2.68	1.22	0.78	0.07	0.05	0.00	0.06	0.36	0.73	0.30	0.00
A20	17.38	40.36	24.03	7.22	1.52	0.39	1.97	0.68	2.57	0.13	0.18	0.63	0.03	0.29	0.80	1.45	0.37
A21	27.42	32.45	27.68	4.85	1.31	0.24	2.07	0.66	0.89	0.22	0.11	0.24	0.05	0.16	0.59	1.01	0.05
A23	9.25	40.12	32.02	9.17	1.92	0.52	2.87	1.29	0.97	0.30	0.13	0.00	0.04	0.46	0.70	0.10	0.14
A25	20.35	42.98	20.99	6.40	1.43	0.15	1.99	0.91	1.75	0.14	0.24	0.21	0.01	0.27	0.97	0.93	0.29
A26	21.41	42.20	24.63	3.90	1.23	0.16	1.77	0.64	2.02	0.11	0.20	0.07	0.06	0.21	0.77	0.47	0.16
A28	19.76	41.95	25.31	5.69	1.33	0.15	1.65	0.66	0.98	0.19	0.22	0.05	0.01	0.25	0.71	0.94	0.15
A29	18.83	36.88	29.11	6.99	1.60	0.29	2.76	0.85	0.75	0.22	0.09	0.07	0.00	0.32	0.65	0.55	0.05

TABLE F.2 (cont.)

Sample	CaO	SiO$_2$	FeO	Al$_2$O$_3$	MgO	MnO	Na$_2$O	K$_2$O	CuO	SO$_3$	Cl$_2$O	As$_2$O$_3$	NiO	TiO$_2$	P$_2$O$_5$	BaO	ZnO
B01	24.05	35.31	28.43	5.09	1.49	0.23	1.45	1.11	0.44	0.30	0.10	0.12	0.04	0.24	0.92	0.65	0.06
B02	22.62	38.57	22.63	4.92	1.34	0.20	1.21	0.64	2.23	0.15	0.12	0.44	0.01	0.01	0.84	3.84	0.23
B03	25.88	43.88	15.14	5.53	1.46	0.31	1.96	0.80	1.58	0.03	0.13	0.34	0.00	0.17	0.96	1.73	0.09
B05	22.29	34.65	27.10	5.71	2.70	0.32	2.58	0.81	1.09	0.14	0.17	0.36	0.07	0.20	0.87	0.86	0.08
B06	15.34	35.76	33.02	6.29	2.06	0.16	3.04	1.06	0.97	0.39	0.16	0.03	0.03	0.28	0.65	0.69	0.07
B07	14.56	34.86	35.85	6.13	1.56	0.17	2.93	0.85	1.06	0.34	0.24	0.03	0.08	0.16	0.68	0.48	0.03
B09	31.05	38.34	13.25	5.73	5.86	0.29	1.61	0.81	0.67	0.19	0.14	0.21	0.04	0.35	0.83	0.54	0.10
B11	26.87	41.40	17.42	5.70	1.48	0.25	2.29	0.74	1.11	0.07	0.17	0.35	0.07	0.19	0.83	0.95	0.12
B12	14.34	38.69	28.75	9.21	2.47	0.38	1.89	1.02	1.20	0.21	0.21	0.04	0.01	0.41	0.68	0.46	0.04
B13	20.71	35.86	30.49	5.49	1.64	0.13	1.55	0.89	0.97	0.24	0.13	0.08	0.03	0.25	1.02	0.53	0.01
B14	14.97	34.73	34.53	7.02	1.62	0.28	3.00	0.83	1.51	0.30	0.13	0.03	0.00	0.28	0.60	0.14	0.05
B15	13.98	36.72	31.11	8.02	1.61	0.52	2.72	0.64	2.85	0.19	0.25	0.17	0.04	0.24	0.63	0.28	0.03
B17*	19.97	40.53	24.95	6.37	1.25	0.48	2.77	0.80	0.64	0.27	0.03	0.18	0.05	0.33	0.74	0.65	0.00
B18	17.34	38.36	28.01	7.45	1.58	0.37	2.25	0.96	1.01	0.31	0.18	0.18	0.04	0.35	0.88	0.66	0.06
B20	23.97	33.46	27.56	5.64	1.82	0.26	2.89	0.90	0.85	0.25	0.14	0.05	0.00	0.18	0.76	1.25	0.05
B22	25.98	43.43	16.66	5.06	1.14	0.21	2.29	0.58	1.31	0.09	0.14	0.47	0.00	0.27	0.83	1.44	0.11
B23	15.36	36.31	30.38	8.00	1.85	1.39	3.25	1.05	0.44	0.50	0.16	0.11	0.04	0.31	0.71	0.15	0.01
B24*	19.46	33.37	32.09	7.76	1.41	0.56	1.96	0.96	0.68	0.42	0.03	0.20	0.00	0.15	0.33	0.52	0.10
B26	22.38	40.90	23.87	5.24	1.36	0.14	1.86	1.06	0.71	0.26	0.14	0.04	0.06	0.15	0.98	0.65	0.20
C01	17.58	35.77	31.31	7.30	1.48	0.30	3.06	0.81	1.00	0.09	0.01	0.00	0.05	0.35	0.54	0.30	0.06

Analyses marked with an asterisk have been obtained by multiple area scans (see text). Results normalized.

The average copper content of the slag is ca. 1%. This low amount is indicative of a very efficient smelting process with low copper losses. Arsenic and nickel, which are generally concentrated in the metallic phase rather than the slag,[11] are also found, therefore, at very low levels (<0.5% on average).

Examination of polished sections of slag samples under the polarizing microscope showed a wide range of mineralogical phases, textures, and intergrowths, heterogeneity being notable both within and between individual samples. A brief description of the major phases identified is given below.

SILICATES AND GLASS

Although slag textures range from fully crystallized to fully vitrified, the glassy component is usually dominant. Silicate crystalline formations are present in most slag samples, giving them a fine dendritic texture indicative of rapid precipitation from a liquid solution. Their chemical composition corresponds to the pyroxenes and olivines, most samples approximating the composition of the compounds hedenbergite and $CaO.FeO.SiO_2$ olivine (Table F.3). The silicate crystals are embedded in a glassy matrix that often shows red internal reflections due to the presence of bands of finely dispersed copper and (possibly) cuprite inclusions.[12] We should note that a considerable part of the identified "glass" matrix might be in fact cryptocrystalline, as indicated by the chemical similarity between the glassy and siliceous phases in the slag (Table F.3).

IRON OXIDES

Wustite (FeO) and magnetite (Fe_3O_4) are frequent in most slag samples, the latter usually being the dominant oxide. Wustite is always present in dendritic form, whereas magnetite crystals are generally present as large skeletons near the center of the slag fragments, changing to numerous aligned dendrites with excessive growth along a single direction near the surface. This difference is clearly associated with differences in the cooling rate of various areas in the slag cake: closer to the surface, the drop in temperature is steeper, with the rapid crystal growth producing fine-armed dendrites whose shape and size is primarily determined by kinetic factors. According to the results of Mössbauer analysis of five slag samples, the predominant valency state of iron is Fe^{2+}, with the Fe^{2+}/Fe^{3+} ratio ranging from 80:20 to 90:10 (Table F.4).

A notable feature of many slag samples is the presence of thin bands of magnetite present at the interior of the fragment (Fig. F.5). In many cases those bands run uninterruptedly through the sample, but they are more commonly discontinuous. The microstructure of the slag, however, is always different on the two sides of the band. A similar textural feature has been identified by Okafor[13] in early iron smelting slags from Nigeria and has been shown to reflect successive tap cycles. Whether the presence of those bands indicates a similar process at Chrysokamino is questionable,

11. Yazawa 1980, p. 381.
12. Cf. Hauptman, Rehren, and Schmitt-Strecker 2003.
13. Okafor 1993, p. 446.

TABLE F.3. EDS ANALYSIS OF MAIN COMPONENTS OF SILICATE AND GLASS PHASES IN SLAG SAMPLES

Sample	Phase	CaO	SiO_2	FeO	Al_2O_3	MgO	Na_2O	K_2O
A04	silicate	27.63	33.93	34.71	1.01	1.99	0.23	0.42
A04	silicate	25.66	36.33	28.94	5.69	1.26	0.85	1.07
A05	silicate	22.44	42.19	19.80	8.88	1.01	2.32	2.86
A05	silicate	28.36	34.18	33.27	0.47	2.76	0.57	0.18
A13	silicate	20.85	47.32	21.21	5.29	2.41	0.42	0.55
A14	silicate	32.72	40.78	12.15	6.63	3.07	3.92	0.34
A14	silicate	33.14	40.15	12.00	6.63	3.77	3.39	0.31
A14	silicate	33.12	40.56	12.44	7.08	3.09	3.12	0.24
A14	silicate	23.14	40.43	21.64	7.97	2.64	1.50	1.21
A15	silicate	20.69	43.04	25.36	7.84	2.12	0.64	0.00
A18	silicate	10.53	32.97	46.34	7.33	1.09	0.35	0.80
A19	silicate	13.61	36.75	35.58	8.97	2.94	1.02	0.89
A19	silicate	13.39	36.58	33.90	8.61	3.43	1.22	1.00
B01	silicate	25.76	34.02	32.29	3.13	1.88	1.24	0.73
B01	silicate	26.08	34.12	31.94	2.85	1.91	1.24	0.74
B01	silicate	26.44	33.97	32.65	2.28	2.58	1.03	0.57
B01	silicate	34.57	41.01	13.87	4.38	3.46	1.87	0.37
B01	silicate	34.35	40.65	14.08	4.85	2.78	2.22	0.66
B01	silicate	34.76	40.77	13.07	4.18	4.22	2.13	0.41
B12	silicate	12.30	30.95	39.97	11.08	3.76	0.48	0.09
B13	silicate	33.94	40.64	15.91	4.26	2.46	2.02	0.44
B13	silicate	33.14	40.84	15.93	4.33	2.40	2.10	0.60
B17	silicate	20.77	42.52	27.21	6.28	0.89	0.96	0.63
B20	silicate	28.45	36.24	25.33	5.71	1.43	1.40	0.91
A05	glass	22.28	39.97	28.48	6.27	1.82	0.86	0.05
A13	glass	27.05	49.48	12.88	6.02	1.11	1.41	0.58
A14	glass	21.96	39.91	26.15	8.26	2.61	0.34	0.12
A14	glass	21.46	41.76	24.52	8.31	2.07	0.70	0.25
A14	glass	33.49	40.22	12.38	7.23	2.85	2.99	0.29
A14	glass	23.07	39.20	23.15	7.69	2.37	2.21	1.07
A17	glass	20.96	39.47	24.77	7.25	1.93	3.35	1.22
A18	glass	10.33	41.58	33.17	8.75	1.63	3.06	1.09
A19	glass	15.33	39.92	28.84	8.79	2.03	3.13	1.37
A19	glass	15.40	39.00	28.83	8.21	2.64	2.81	1.39
B05	glass	24.70	38.07	25.31	6.71	2.20	2.09	0.69
B12	glass	15.53	38.92	27.81	8.95	1.53	4.20	1.71
B13	glass	15.31	41.02	30.11	6.69	0.92	1.45	2.28
B13	glass	15.00	41.46	30.95	7.78	0.55	0.55	1.57
B17	glass	21.09	40.68	26.76	5.92	1.47	2.31	1.19
B24	glass	22.36	37.54	24.83	7.91	1.91	3.26	1.02

Results not normalized.

TABLE F.4. RESULTS OF MÖSSBAUER SPECTROSCOPY AND CALCULATED FE^{+2}/FE^{+3} RATIOS OF SLAG SAMPLES

Sample	Solidified phases				Valencies of iron		
	Hedenbergite	Ferrous (unidentified)	Intermediate (unidentified)	Magnetite	Fe^{+2}	Fe^{+3}	Fe^{+2}/Fe^{+3}
A17	40	35	0	25	83.3	16.7	5.0
A18	80	0	20	0	80.0	20.0	4.0
A19	40	50	10	0	90.0	10.0	9.0
B17	80	0	20	0	80.0	20.0	4.0
B24	40	40	0	20	86.7	13.3	6.5

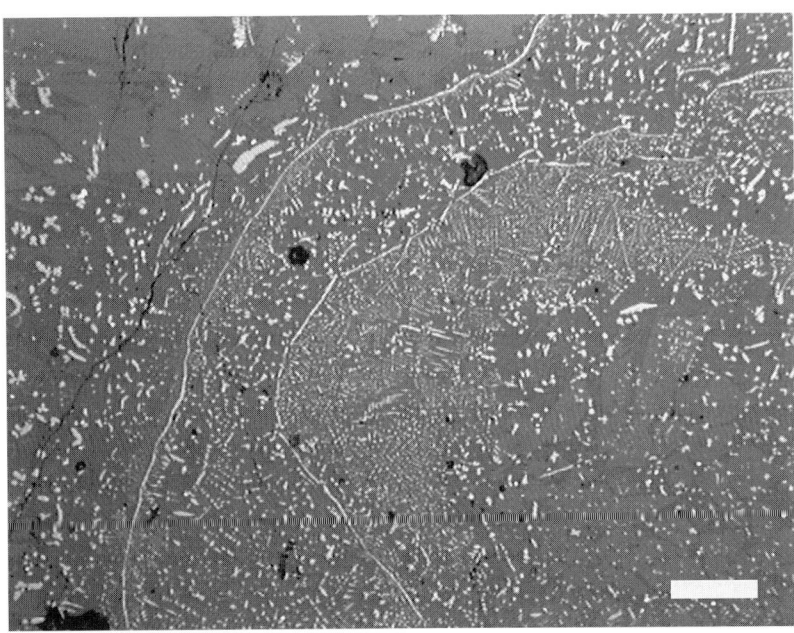

Figure F.5. Sample A04: magnetite "bands" forming three slag "layers." **Reflected light.** Scale: 50 micrometers

as the thickness of individual "slag-layers" rarely exceeds 2–3 mm. However, as the presence of magnetite bands is a strong indication of solidification in contact with the air, it seems very plausible to argue that the bands formed by the "folding" of a relatively viscous slag as it was tapped out of the furnace. The possibility of slag tapping at Chrysokamino is considered in more detail in the discussion of the analytical results.

Copper Prills

Copper prills are common in most slag samples. Their diameter ranges from a few micrometers to 1 mm in some rare cases. According to the results of EDS analysis, their chemical composition corresponds to the quaternary alloy Cu-Fe-As-Ni (Table F.5). Arsenic levels are highly fluctuating, often reaching or even exceeding the level of solid solubility in copper, which is ca. 8%.[14] This results in the formation of the arsenic-rich Cu_3As phase present in many prills in the slag (Fig. F.6).[15] The nickel content of copper

14. Northover 1989.
15. Cf. Lechtman and Klein 1999, p. 521, fig. 24.

TABLE F.5. EDS ANALYSIS OF COPPER PRILLS IN SLAG SAMPLES

Sample	Cu	Fe	As	Ni	S
A05	93.66	2.81	2.65	0.88	0.00
A05	90.36	1.30	6.99	1.25	0.10
A05	94.74	2.34	2.57	0.35	0.00
A05	94.82	1.36	2.85	0.74	0.23
A05	95.67	0.71	2.90	0.72	0.00
A09	96.62	2.00	0.83	0.37	0.17
A09	96.05	1.01	2.46	0.32	0.16
A09	95.06	0.65	4.04	0.09	0.15
A09	96.43	0.43	2.75	0.24	0.15
A11	89.26	1.65	7.78	0.41	0.90
A14	87.94	2.20	6.59	0.77	2.50
A14	90.13	1.28	7.78	0.78	0.03
A15	97.52	2.29	0.00	0.02	0.16
A15	97.30	2.31	0.09	0.00	0.30
A15	95.91	2.48	0.93	0.45	0.23
A15	95.33	3.33	0.36	0.90	0.08
A15	89.52	1.23	8.44	0.81	0.00
A15	88.83	0.30	9.83	0.97	0.07
A15	93.34	1.17	3.65	1.72	0.12
A15	90.78	1.84	4.00	0.50	2.89
A16	96.15	3.54	0.22	0.04	0.05
A16	96.97	3.03	0.00	0.00	0.00
A16	98.24	1.66	0.10	0.00	0.00
A16	98.57	1.34	0.00	0.08	0.00
A17	90.69	2.21	4.94	1.97	0.19
A18	90.20	2.47	5.24	1.99	0.10
A18	92.42	3.08	3.19	1.22	0.10
A18	91.23	2.46	4.86	1.25	0.20
A19	61.71	0.45	23.11	14.70	0.03

TABLE F.5 (cont.)

Sample	C	Fe	As	Ni	S
B05	88.96	2.30	7.79	0.89	0.07
B05	86.70	1.95	10.47	0.81	0.07
B05	72.49	1.79	22.84	2.76	0.12
B05	91.36	1.23	6.82	0.53	0.06
B05	67.11	1.31	31.00	0.38	0.19
B17	64.82	1.78	32.61	0.67	0.10
B17	74.15	0.74	24.75	0.27	0.09
B17	67.07	3.85	26.26	2.72	0.10
B20	88.62	2.78	7.47	1.12	0.01
B20	88.76	3.64	6.62	0.95	0.04
B20	85.99	4.89	8.19	0.88	0.05
B20	88.00	2.86	8.09	0.99	0.06
C01	95.81	2.07	0.96	1.05	0.12
C01	94.53	3.09	1.22	1.01	0.14
C01	95.20	2.69	0.94	1.07	0.09
C01	94.23	3.37	1.55	0.79	0.07
C01	93.96	3.28	1.89	0.48	0.39
C01	95.43	2.39	1.40	0.67	0.11
C01	93.26	4.71	1.07	0.89	0.07
C01	95.71	1.77	1.45	0.92	0.15
C01	94.75	2.11	2.13	0.84	0.17
C01	95.11	2.35	1.44	0.90	0.20
C01	94.97	3.39	0.45	1.10	0.09
C01	95.94	0.35	2.18	1.29	0.23
C01	94.89	2.77	1.55	0.71	0.08
C01	93.99	1.48	3.32	1.03	0.18

Results normalized.

prills is also variable, but most values are between 0.5% and 2%. Iron content has a mean value of 2.2% but reaches 5% in some cases. No -iron dendrites were identified microscopically, possibly due to their extremely small size.[16] The only exception is a large prill (Diam. = ca. 1 mm) found in sample A18 containing massive -iron dendrites and sulfidic inclusions embedded in a copper matrix. The composition of this prill was found by EDS analysis to be 69.7% Fe, 23.7% Cu, 3.9% As, and 0.8% S. Finally, small matte particles are often embedded in copper prills, but this phase is usually present as "rims" surrounding the metallic prills (Fig. F.6).

16. Roeder, Sculac, and Notis 1984.

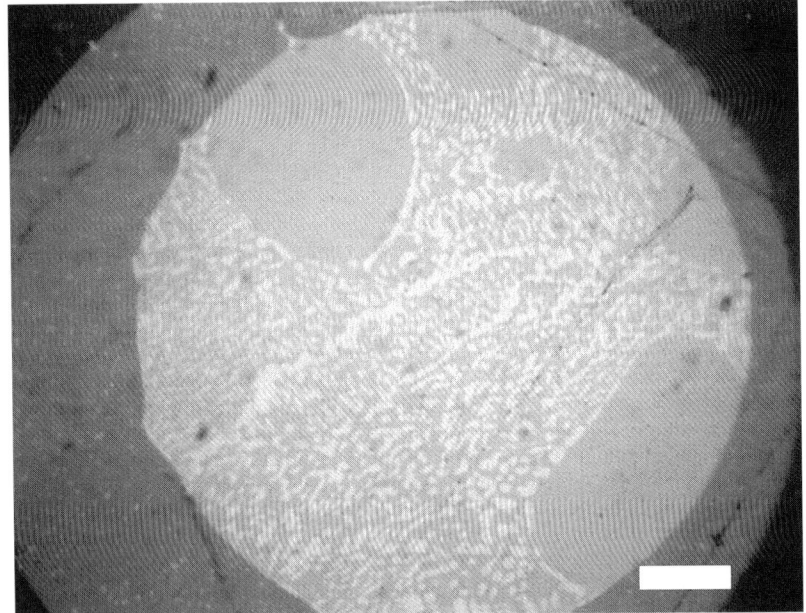

Figure F.6. Sample B18: Copper prills showing massive precipitation of a second arsenic-rich phase (Cu_3As). The prill is surrounded by matte. Reflected light.
Scale: 10 micrometers

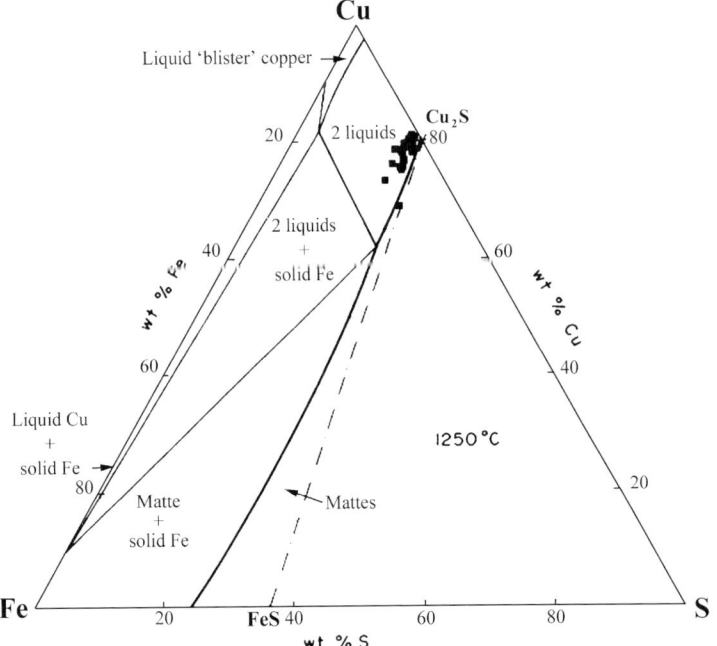

Figure F.7. The position of matte inclusions present in slags from Chrysokamino in the ternary system Cu-Fe-S

Matte Inclusions

Although matte inclusions are common, they are significantly less common than the copper prills with which they are often associated (see above). Most inclusions have characteristic Widmastätten structure resulting from solid-state precipitations from the sulfide solution.[17] As shown by the results of EDS analysis (Table F.6), the average iron content is 4%, while nickel and arsenic are usually below 1%. The matte compositions deviate from equilibrium compositions having a slightly higher copper content (Fig. F.7). This is due to the fact that matte compositions were acquired using area

17. Cf. Doonan 1996, p. 57.

TABLE F.6. EDS ANALYSIS OF MATTE INCLUSIONS IN SLAG SAMPLES

Sample	Cu	Fe	As	Ni	S
A04	77.42	2.69	0.60	na	19.29
A05	79.18	1.83	0.30	0.17	18.52
A05	78.68	2.20	0.40	0.21	18.51
A05	79.37	2.57	0.24	0.00	17.82
A09	79.08	2.48	0.17	0.00	18.27
A17	72.43	6.33	0.40	0.30	20.54
A18	74.26	5.98	0.34	0.25	19.17
A18	68.58	9.37	0.05	0.09	21.90
A18	75.26	5.44	0.08	0.00	19.23
A19	79.39	1.65	0.70	0.53	17.72
A19	79.64	1.40	0.30	0.28	18.39
A19	78.17	1.01	1.66	1.16	18.01
A19	77.85	5.18	0.28	0.30	16.38
B01	75.55	4.60	1.02	0.07	18.76
B01	79.51	1.53	0.96	0.24	17.77
B01	78.70	1.30	2.10	0.33	17.57
B05	76.85	4.32	1.63	na	17.20
B05	77.90	3.14	0.90	na	18.05
B05	77.11	3.11	2.23	na	17.55
B05	76.78	4.27	1.27	na	17.68
B05	78.94	2.08	1.35	na	17.63
B12	80.39	0.95	0.11	0.03	18.53
B13	75.66	4.53	0.07	0.04	19.70
B17	75.43	1.87	4.44	0.25	18.02
B17	78.13	2.38	1.25	0.42	17.82
B17	79.53	1.46	0.60	0.53	17.88
B17	75.46	4.84	0.94	0.19	18.57
B20	79.02	3.20	0.17	0.10	17.51
B20	75.72	6.78	0.00	0.01	17.50
B20	72.38	9.17	0.77	0.14	17.54
B20	75.00	5.88	0.29	0.06	18.77
B24	68.53	9.24	0.30	0.46	21.46
B24	72.47	6.61	0.68	0.69	19.56
B24	73.25	6.43	0.44	0.45	19.42

Results normalized ("na" = not analysed).

scan analyses, therefore incorporating precipitated and mechanically entrapped phases. Such inclusions in the matte are visible microscopically and include globules of metallic copper and a light blue copper-arsenic phase.[18] If those phases were not included in the analysis, the copper content of matte inclusions would range from ca. 75% to almost 80%.

Undissolved Materials

Very few inclusions in the slag represent raw materials that were not dissolved in the slag matrix during smelting. Small silica grains (0.5–2 mm) with signs of severe fragmentation are occasionally present, and their subangular shape suggests that they were not in the process of dissolution. Only a single silica grain with signs of ongoing dissolution was identified. In rare cases, fine dispersions of green and red copper minerals, probably reflecting residual copper ore (malachite and cuprite?), were associated with the quartz grains. Remains of iron ore are also present, either as minute, heavily altered ore fragments or, more often, as agglomerations of iron oxides, which have clearly resulted from the decomposition of larger inclusions.[19]

Other Inclusions

Cuprite (Cu_2O), a mineral associated with slightly reducing conditions in the smelting furnace, is uncommon in the slag from Chrysokamino. When present, it is usually found in clearly corroded areas of the samples, suggesting that it is a corrosion product rather than a phase formed during the smelting process. Delafossite ($Cu_2O.Fe_2O_3$), another mineral associated with slightly reducing atmospheres, is uncommon in slags from Chrysokamino, being present in only two samples. Small fragments of ceramic material are present in four slag samples, suggesting that some of the material from the furnace walls was dissolved in the slag. Minute gold grains (2–5 mm) are occasionally identified, usually as dispersions in the glass matrix aligned with copper prills. EDS analysis of a gold grain gave 90% Au, 4.6% Fe, 3.4% Si, 1.0% Ca, 0.8% Cu and 0.6% Pb.

It was not possible to formulate a coherent system for the classification of the slag samples on the basis of the microstructural features described above, mainly due to the mineralogical heterogeneity of most samples and the absence of obvious, distinct "groupings." Nonetheless, to facilitate the interpretation of the analytical results, six typical microstructures were selected to represent the range of mineralogical phases and textures identified (Fig. F.8).

Although a correlation between chemical composition and microstructure exists, it should not necessarily be taken to indicate the presence of different types of slag in the assemblage. Instead, as the chemical composition of the slag samples falls in an area of the $CaO-FeO_x-SiO_2$ system with three precipitation areas (pyroxenes, olivines, and iron oxides), even the slightest chemical differences could have resulted in the formation of different phases (or different relative abundances of the various phases) during solidification. To be specific, higher levels of iron in the molten slag favor (a) the formation of olivines $2(Ca, Fe)O.SiO_2$ rather than pyroxenes $(Ca, Fe)O.SiO_2$ and

18. Cf. Lechtman and Klein 1999, p. 515, fig. 14.
19. Cf. Hauptman, Rehren, and Schmitt-Strecker 2003, p. 204, fig. 6.

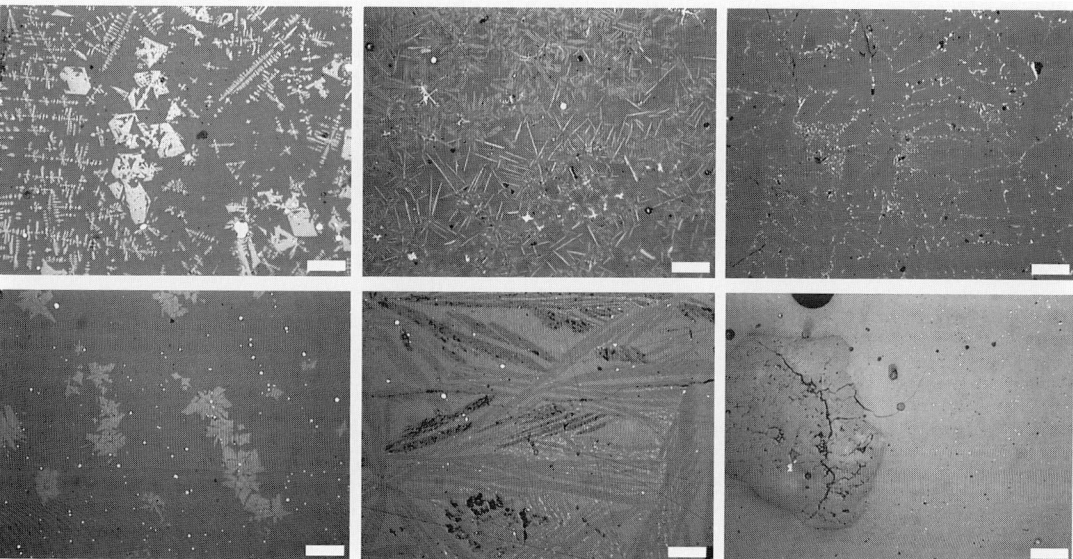

Figure F.8. Typical microstructures found in slags from Chrysokamino. Reflected light. Top left: A04, A07, A08, A09, A17, A20, A21, A25, A29, B01, B05, B06, B07, B12, B13, B14, B15, B18, B20, B24, C01. Top center: A03, A15, A16, A18, A23. Top right: A01, A06, B09. Bottom left: A14, A19, A28, B17, B26. Bottom center: A02, B03, B22. Bottom right: A10, A11, A12, A13, B02, B11, B23. Scale: 50 micrometers

(b) the precipitation of magnetite and/or wustite. Conversely, higher levels of silica and/or lime promote the formation of pyroxenes, the suppression of the formation of free iron oxides, and the presence of unfused silica grains due to the supersaturation of the liquid slag.

DISCUSSION OF RESULTS: MASS BALANCE

The investigation of the mass balance of a smelting process is a powerful means to visualize the nature and relative quantities of materials going in and out of the furnace. The approach followed here is based on the simple premise that all elements present at the output of the process (i.e., in the slag and metallic phases) should be accounted for by materials present at the input (raw materials). Therefore, through the comparison of copper ores, slag, and metallic inclusions, it is possible either to confirm their chemical similarity or to postulate the use of additional materials that do not appear in the archaeological record.

COPPER ORES AND SLAG

A striking difference between the chemical composition of the gangue minerals in the copper ores and the slags from Chrysokamino is that the latter contain significant levels of iron oxides and lime (Fig. F.3). Unless it is assumed that all three copper ore samples included in this study are entirely different from those actually smelted at Chrysokamino, the consistently high levels of iron and lime in the slag must be accounted for by another material present in the furnace during the process. The role of fuel ash as a source of lime in early copper smelting slags has been emphasized in the literature but can only account for some 6%–8% of CaO in the bulk composition.[20] Erosion of the calcareous clay from the furnace wall during smelting might have been an additional source of lime in the slag, but only a minor one.

20. Merkel 1990; Merkel and Rothenberg 1999; Shugar 2003.

A plausible interpretation for the chemical composition of the slag samples derives from a consideration of the function of the fragment of iron ore found at Chrysokamino. As already mentioned, the presence of iron ore at a copper smelting site could reflect the use of such ores as fluxes, especially since the predominant gangue mineral in the copper ore is silica. According to the results of EDS analysis, the iron ore sample contains 67% FeO and 23% CaO, so its use as a fluxing agent in the smelting process (together with the fuel ash and perhaps some material from the erosion of the furnace wall) would explain the increased iron and lime levels in the slag.

Another interesting hypothesis is that the fragment of goethite has derived from the local weak iron mineralization hosted in the Plattenkalk/phyllite contact (see above). Detailed comparative analyses using geological samples are necessary in order to examine this possibility.

Metallic Phase and Copper Ores

A comparison between the chemical composition of the metallic prills found in the slag and the composition of copper ores reveals yet another discrepancy. All copper prills (except for the prills in sample A16) contain high levels of arsenic and ca. 1% nickel. Yet those elements were not present at significant levels in the three copper ores analyzed. Although the limitations of the EDS technique for the analysis of elements at levels near or below 0.5% should not be ignored, the Cu/As and Cu/Ni ratios in the ores are far too low to have resulted in the percentages of As and Ni found in the copper prills. More conclusive results would derive from the analysis of more copper ores and the use of techniques like AAS, ICP, NAA, or WDS for chemical analysis. On present evidence, however, we suggest that the arsenic and nickel found in the copper prills did not come from the copper ores but from another material that was added to the charge. This suggestion is also supported by the results of lead isotope analysis of slag samples from Chrysokamino (App. C), which show significant variability, probably reflecting the smelting of ores from various sources. Unless it is assumed that the copper ores from the various sources were all characterized by similar arsenic/nickel concentrations, these elements were probably introduced into the smelting furnace separately. It should be noted that the suggested addition of arsenical minerals and iron fluxes could have influenced the isotopic composition of the slag (and the smelted copper), rendering any comparison with ore deposits misleading. However, although identification of the specific ore source(s) may be problematic, the isotopic variability characterizing the slags from Chrysokamino is a strong indication for the smelting of ores from multiple deposits.

At present, no evidence from the site indicates the type of arsenical minerals that might have been used at Chrysokamino. Experimental studies on the production of arsenical copper have shown that various types of arsenic-bearing minerals (including arsenides, arsenates, and sulfarsenides) can be co-smelted with copper ores to produce copper-arsenic alloys.[21] In fact, that early Aegean metallurgists recognized and exploited such arsenical minerals has been established recently by Doonan and colleagues,[22]

21. Lechtman and Klein 1999; Shimada and Merkel 1991; Rostoker and Dvorak 1991.

22. Doonan et al. forthcoming.

who presented evidence from the EM I–II site of Poros Katsambas near Knossos suggesting production of arsenical copper by addition of roasted löllingite (an iron arsenide) in melted copper.

The fluctuation in the arsenic content of prills results from the uncontrollable formation of the volatile compound As_2O_3 during the process and does not, therefore, reflect variable arsenic in the raw materials. Nickel, on the other hand, is usually collected in the metallic phase during smelting, so its consistently low level in the prills suggests a similarly low level in the raw materials.

THERMODYNAMIC PROFILE AND OPERATIONAL PARAMETERS

Investigation of the complex physicochemical phenomena taking place during the smelting process at Chrysokamino is a necessary step for the reconstruction of the technology employed for the extraction of copper from its ores. For this, it is necessary to make accurate estimations of the prevalent conditions during the process and to explain how they would promote both the transformation of cupriferous ores into a copper-rich product and its separation from the slag phase.

Temperature

The standard method for the estimation of the operating temperature of ancient smelting processes involves the calculation of the liquidus temperature of the slag. The liquidus temperature is, in turn, obtained from the phase diagram corresponding to the slag under investigation. For this purpose, Bachmann[23] has proposed a procedure that allows the selection of the most appropriate subsystem in the Al_2O_3-CaO-FeO-SiO_2 system based on the relative abundances of the various oxides present in the slag.

Although this is a useful approach, it should not be employed uncritically in the study of early copper smelting processes. First of all, as Bachmann himself argues,[24] it is necessary to examine whether unfused materials (very common in prehistoric slags) are present in the slag samples because their inclusion in the calculation of the liquidus temperature will produce overestimated smelting temperatures. Secondly, in the moderately reducing conditions characterizing early copper smelting practices, the first phase to precipitate from the liquid slag (thus determining the liquidus temperature) is very often magnetite. This compound, however, is not always included in the ternary diagrams employed in archaeometallurgical studies. This is because modern phase diagrams usually describe slag systems in equilibrium with metallic iron and therefore under strongly reducing conditions ($pO_2 = 10^{-11}$–10^{-12} atm). Under such conditions, however, iron is present in the slag in its ferrous state, and no magnetite is precipitated. Moreover, a recent study of the liquidus temperature of Al_2O_3-MgO-CaO-FeO-Fe_2O_3-SiO_2 slags by Kongoli and Yazawa[25] has shown that the effects of common oxides (e.g., Al_2O_3, MgO, or CaO) on the liquidus temperature of the slag may be different under reducing and oxidizing conditions.

23. Bachmann 1980.
24. Bachmann 1980, p. 110.
25. Kongoli and Yazawa 2001.

The use of phase diagrams for the determination of the liquidus temperature is valid in the case of the slags from Chrysokamino because of the limited presence of unfused materials. Given the abundance of magnetite in the slag samples, we employed a phase diagram[26] that corresponds to moderately reducing conditions ($pO_2 = 10^{-8}$ atm). The solid phases predicted by this diagram are in good agreement with the mineralogy of the slag samples, therefore confirming our selection. From the phase diagram, the liquidus temperature of most slag samples is found to lie between 1150 and 1230°C. In silica-rich samples, the liquidus exceeds 1300°C, but this value is probably an overestimation due to the increased presence of unfused silica grains in them. Overall, raising the temperature to ca. 1300°C would have produced a slag of sufficiently low viscosity.[27] Such temperatures were attainable in the Early Bronze Age.[28]

Redox Conditions

The study of the redox conditions (i.e., the reducing or oxidizing capacity of the furnace atmosphere) during smelting is usually based on the investigation of elements exhibiting multiple valency states in typical copper smelting conditions, the most important being copper and iron.[29] Copper phases associated with moderately reducing conditions (pO_2 = ca. 10^{-5} atm at 1300°C), such as cuprite (Cu_2O) and delafossite ($Cu_2O \cdot Fe_2O_3$), are rare in slags from Chrysokamino (see above), suggesting that conditions were more reducing.

At more reducing conditions, estimations of the partial pressure of oxygen at a given temperature can be deduced from the relative abundance of ferrous and ferric cations, taking into consideration the effects on their activities by other oxides present in the slag.[30] Using the Fe^{2+}/Fe^{3+} ratio provided by Mössbauer spectroscopy and extrapolating from the data available in the literature,[31] a range of pO_2 values from $10^{-7.5}$ to $10^{-9.5}$ atm at 1300°C is derived.

Similar redox conditions can be deduced from a consideration of the copper content in the matte, which ranges from 75% to almost 80%, excluding the precipitated phases (see section on matte analysis). Such matte compositions under metal saturation (i.e., when co-present with metallic copper) correspond to pO_2 values from ca. 10^{-9} to 10^{-7} atm.[32] Finally, the iron content in copper prills, with an average value of 2.2%, is also compatible with such intermediate redox conditions.[33]

Thermodynamic Profile of the Smelting Process

The pO_2 values calculated above are compatible with the phases identified in the slag, so they probably describe accurately the local redox conditions at the area in the bottom of the furnace where the slag was collected. However, the numerous perforations on the ceramic shaft, on the one hand, and the combined action of the pot bellows and the strong winds, on the other, would have created extremely complex and variable conditions inside the furnace during the smelting process.

26. Kongoli and Yazawa 2001, p. 587, fig. 11.
27. Cf. Kresten 1986, p. 44, fig. 1.
28. Merkel 1990.
29. Hauptmann, Weisgerber, and Bachmann 1989.
30. Turkdogan 1983, p. 237.
31. Timucin and Morris 1970, p. 3196, figs. 11, 12.
32. Sridhar, Togurl, and Simeonov 1997, p. 195, fig. 5; Yazawa 1980, p. 379, fig. 2.
33. E.g., cf. Merkel 1990 and Shugar 2003.

Although the complete pO_2/temperature profile in the furnaces at Chrysokamino can only be investigated through experimental work, it is safe to assume that in contrast to a typical shaft furnace, the presence of perforations subjected the descending charge to more oxidizing conditions that resembled typical conditions of ore-roasting processes. In fact, similar perforated cylindrical furnaces have been used for the roasting of copper ores.[34] Such conditions would have had no significant effect on the copper carbonates or the iron hydroxides (apart from their calcination due to the increasing temperature), but they would have burned off much of the sulfur present in the charge and removed much of the arsenic as As_2O_3.[35]

Near the bottom of the furnace, the atmosphere was probably more reducing and the temperature much higher, enabling the formation of slag and the reduction of copper oxides to metallic copper. The reducing conditions and the presence of a cover of liquid slag then allowed most of the remaining arsenic to be collected in the metallic copper. Nickel was almost entirely concentrated in the liquid copper. A small amount of iron, determined mainly by the redox conditions, also entered the metallic phase. Finally, since iron sulfides readily react with copper oxides producing metallic copper and/or copper sulfides,[36] the formed matte phase would gradually become copper-rich.

According to this model, the output of the smelting process probably consisted of a slag phase, a Cu-As-Fe-Ni alloy phase, and a high-grade matte. Those are the phases that were identified in the smelting slags from Chrysokamino.

Slag-Metal Separation

Strong indications suggest that liquid slag was tapped from the furnace during the smelting process. Firstly, many slag fragments have characteristic flow features that are typically associated with tapped slag. Secondly, the magnetite bands inside the slag samples reflect "foldings" occurring during the viscous flow of molten slag in contact with the air. Finally, the texture of most mineralogical phases present in the slag samples is indicative of rapid crystallization, a fact that could be associated with solidification outside the furnace. It should be noted, however, that a significant amount of furnace slag (up to 50% of the total amount)[37] probably remained inside the furnace after the completion of the smelting process.

Although much of the copper produced during the smelting process was probably collected at the bottom of the furnace where it formed an ingot, a significant part remained trapped in the slag (both tapped and furnace slag), requiring crushing to achieve further separation. This is indicated by the highly fragmented state of the slag found at the site, slag cakes being absent and slag pieces rarely exceeding 2–3 cm in size.[38]

The very low copper content found in all analyzed slag samples suggests that the entrapped metal was not finely dispersed in the slag matrix but was instead concentrated in sizable nodules that could be easily collected by crushing the slag into small pieces (1–2 cm) and handpicking. However, the evidence shows that this might not have always been the case. The

34. Craddock 1995, p. 169; we thank Oli Pryce for drawing our attention to the similarity between the Chrysokamino furnaces and the furnaces reported from Yeke, Katanga.

35. Cf. Lechtman and Klein 1999, p. 510.

36. Biswas and Davenport 1976, p. 81; Rostoker, Pigott, and Dvorak 1989, p. 73.

37. Merkel 1990.

38. See Chap. 10.

amount of pulverized slag found at Chrysokamino is probably too large to have derived from the simple crushing of the slag, and it could suggest that smelters often needed to grind the slag to a very fine size to expose minute copper inclusions. Such fine metal particles would have been impossible to collect by hand; instead, a "secondary" beneficiation process (probably washing) would have been required to separate the metal from the slag.

It is not possible, on present evidence, to explain the different degrees of comminution of the solidified slag for the recovery of entrapped copper. Perhaps the laborious grinding/washing process was used for the more viscous furnace slag or was undertaken after a less successful smelting operation. Examination of the pulverized slag from the site (despite the obvious analytical difficulties) may help to determine whether it is chemically and/or mineralogically different from the larger slag fragments.

There are various types of evidence from Chrysokamino that could be associated with the processes described above. They include not only the small number of stone tools found at the site, but also the two shallow, cup-shaped depressions carved on the bedrock that might have served as "mortars" for this purpose.[39] These finds, however, could equally have served other purposes including the preparation of the raw materials or even (in the case of stone tools) the destruction of the furnace after the completion of the smelting process.[40]

To conclude, it should be emphasized that the overall copper losses during the metallurgical process at Chrysokamino were extremely low. Based on the average copper content in the slag samples (ca. 1%), the total amount of copper entrapped in ca. one ton of slag would be approximately 10 kg.

The Product

The main product of the smelting process was a quaternary copper-arsenic-iron-nickel alloy. The arsenic content was probably highly variable due to the uncontrollable formation of the volatile As_2O_3. This argument is supported by the results of the EDS analysis of copper prills, which show considerable fluctuation in the arsenic content even within a single slag fragment. The nickel content was relatively constant because this element is generally concentrated in the metallic copper, while iron probably showed minor fluctuations due to the variable redox conditions. As high iron contents make copper very brittle,[41] an additional refining stage was necessary, although there is no evidence, at present, to indicate that this process was conducted at the site. Some arsenic was also removed in the fumes during the refining process, although if a reducing atmosphere was retained the loss would have been insignificant. The chemical composition of the copper prills from Chrysokamino is not unknown in the Aegean. Co-presence of arsenic above 2% and nickel at about 1% in copper has also been found in the copper prills of a slag fragment from the EBA smelting site of Skouries on Kythnos.[42] A similar chemical pattern is found in many Early Minoan artifacts.[43]

Whether this reflects the use of mineralogically similar raw materials or the exploitation of the very same sources is impossible to say on present

39. Betancourt et al. 1999, pp. 354–366.
40. Betancourt et al. 1999, p. 366.
41. Papadimitriou 2001.
42. Gale et al. 1985, p. 90, table 3.
43. Branigan 1968, p. 48, table 15.

evidence. In any case, this similarity suggests that technologies comparable to that employed at Chrysokamino might have also been used in other areas in the Aegean for the production of arsenical copper during the Early Bronze Age.

A secondary product of the smelting process at Chrysokamino was a high-grade matte, although, on present evidence, we cannot estimate the amount produced in each smelting operation. If sulfur was mainly introduced into the charge by the residual sulfides present in the copper ores, then based on the analysis of the ore samples, the amount of matte produced would have probably been minimal. If, however, the arsenical mineral added to the charge was a sulfarsenide/sulfarsenate, then more matte would have been produced. Further work is required to investigate the possible presence of matte in the output of the smelting product and its apparent absence in the metallurgical remains.

CONCLUSIONS

Chrysokamino was the locus of highly sophisticated smelting activities characterized by careful selection and mixing of raw materials and manipulation of the conditions in the furnace to enable the production of arsenical copper and its separation from the gangue materials with minimal copper losses in the slag.

The copper ores used in the smelting operations were copper carbonates, occasionally containing some residual sulfides, associated with quartz and small amounts of iron (hydr)oxides. Calcite bearing iron ore, similar to a fragment found at the site, was probably used as a flux. It is possible that this material was collected from the weak iron mineralization hosted in the nearby Plattenkalk/phyllite contact, which we located during a geological reconnaissance of the Chrysokamino peninsula. Finally, an arsenical mineral containing low levels of nickel was probably added to the smelting charge. Unfortunately, the nature of this mineral cannot be established on present evidence. However, the use of iron arsenides for the production of arsenical copper at the EM site of Poros Katsambas[44] could suggest that a similar mineral was employed at Chrysokamino.

The combined effects of air penetrating from the perforations on the shaft, the air blasts from the bellows, and the fuel supply created a very complex pO_2/temperature profile inside the smelting furnaces at Chrysokamino. The maximum temperature attained (near the combustion zone) was approximately 1300°C as deduced from the chemical composition, mineralogy, and texture of the slag. The redox conditions were probably highly variable; at the bottom of the hearth, the pO_2 was ca. 10^{-8} atm, but more oxidizing conditions probably prevailed at higher levels in the furnace. A significant amount of arsenic and sulfur was, therefore, removed as the charge descended in the shaft. The output of the smelting process consisted of a Cu-As-Fe-Ni alloy, a lime-rich ferrosilicate slag, and a seemingly small amount of high-grade matte. The slag was periodically tapped to enable a continuous smelting process. Most metal was concentrated at the bottom of the hearth forming an ingot, but some remained entrapped in the slag,

44. Doonan et al. forthcoming.

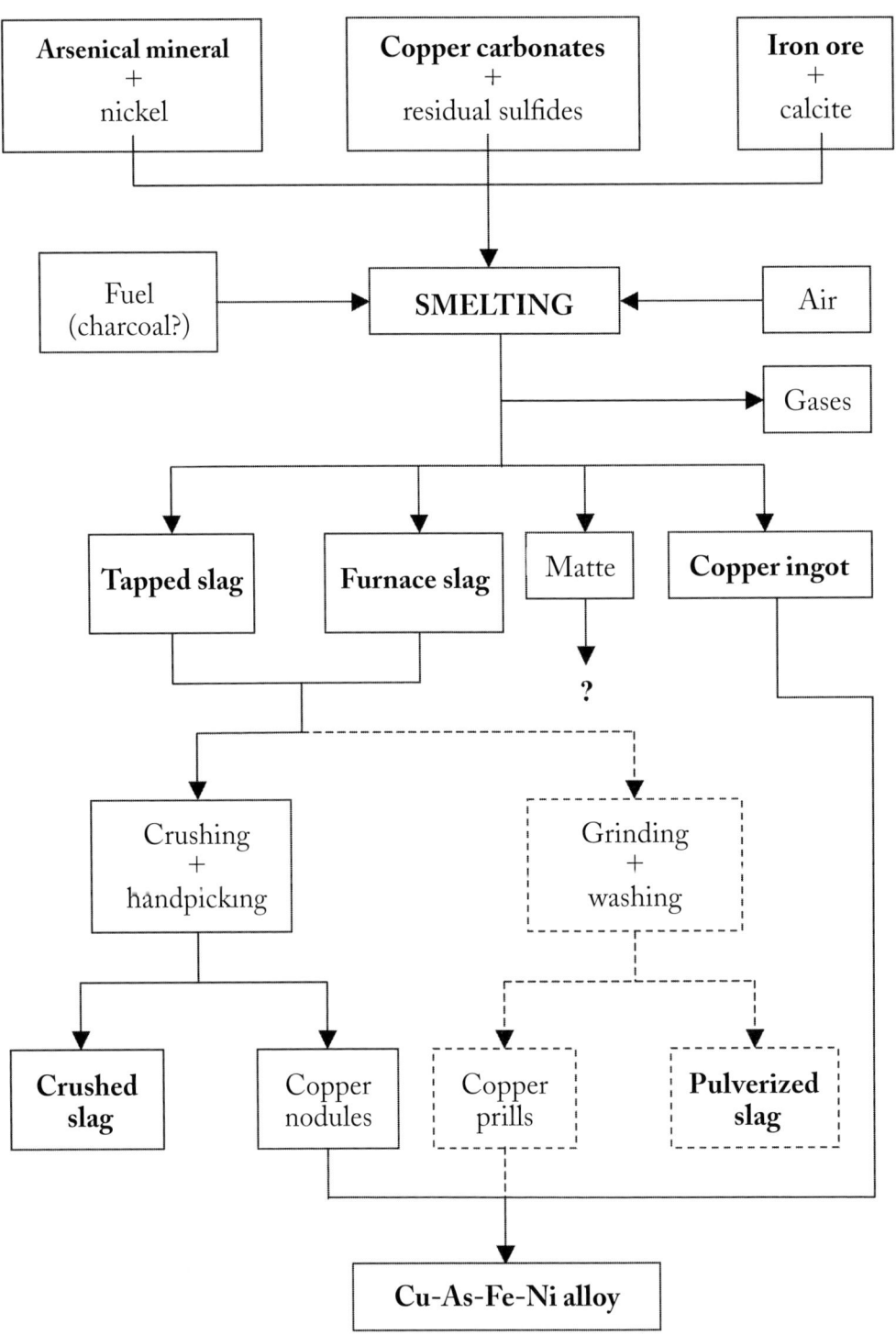

Figure F.9. Schematic representation of the proposed copper smelting process at Chrysokamino

which then needed to be crushed and/or pulverized to enable further separation. A simple remelting in a relatively reducing atmosphere would have been necessary to remove the iron present in the smelted copper without causing significant arsenic losses. However, no evidence of this process has been found at the site.

Although the technological questions set at the beginning of this report have been explored in varying degrees of detail, it is possible to propose a provisional reconstruction of the smelting process at Chrysokamino that could serve as a working hypothesis for future research (Fig. F.9). More archaeological, analytical, and experimental work is essential for an enhanced understanding of the metallurgical processes taking place at the site.

Register of Anthropogenic Features

by Philip P. Betancourt, Lada Onyshkevych, and William B. Hafford

A careful examination of the Chrysokamino territory was conducted between 1995 and 1999. It included an instrument survey as well as a recording of anthropogenic features. An anthropogenic feature (AF) was defined as a location with visible architecture (including terrace walls or groups of terrace walls), a sherd scatter, or some other human alteration of the natural landscape.

The vegetation within the territory that included the metallurgy workshop consisted primarily of low bushes (Fig. G.1). Visibility was generally good, and features were clearly visible. A total of 41 Anthropogenic Features were recorded within the Chrysokamino territory. Many of them were at locations previously noted by the Kavousi-Thriphti Survey.[1] The close view of the landscape presented here may be regarded as an extension of the previous survey, focusing on a small part of the originally surveyed territory in greater detail. It differs from the earlier survey in that it makes note of features from all periods, not just from the Bronze Age. The following features were recorded (for their locations, see Fig. 16.2).

AF 1. Terrace walls
AF 2. Well[2]
AF 3. Wall segments, perhaps a boundary wall[3]
AF 4. Threshing floor
AF 5. Field house (?)
AF 6. Terrace walls
AF 7. Field house
AF 8. Sherd scatter[4]
AF 9. Venetian/Ottoman to modern cave (see App. K)
AF 10. Mound of stones, probably a field house[5]
AF 11. Terrace walls
AF 12. Well
AF 13. Field house
AF 14. Terrace walls
AF 15. Field house
AF 16. Field house
AF 17. Terrace walls (Chordakia)

---

1. Mook and Haggis 1990; Haggis 1992, 1993b, 1995, 1996b, 2005.
2. Haggis 2005, locus 49.
3. Haggis 2005, locus 49.
4. Haggis 2005, locus 52.
5. Haggis 2005, locus 49.

Figure G.1. The metallurgical workshop and its nearby territory, as seen from the habitation site (AF 29)

AF 18. Threshing floor
AF 19. Wall, perhaps a boundary wall
AF 20. Threshing floor
AF 21. Boundary walls
AF 22. Terrace walls
AF 23. Metallurgy workshop[6]
AF 24. Field house
AF 25. Terrace walls
AF 26. Boundary wall
AF 27. Boundary wall
AF 28. Sheepfold *(Mandra)*[7]
AF 29. Chrysokamino habitation site[8]
AF 30. Terrace walls
AF 31. Terrace walls
AF 32. Oval enclosure[9]
AF 33. Field house
AF 34. Theriospelio cave[10]
AF 35. Terrace walls
AF 36. Terrace walls
AF 37. Terrace walls
AF 38. Terrace wall
AF 39. Well
AF 40. Road and house walls
AF 41. Metal fence

The methodology devised for the recording of these features is discussed in Chapter 15. The special circumstances for this highly specialized survey were dictated by the project's goals, the topography, and other factors, such as the need to determine if any tiny features associated with mining or beneficiation of ore were present in the vicinity of the smelting workshop. Among the main considerations that determined the special methodology were the small size of the territory to be examined and the existence of a prior, traditional, regional surface survey. The previous survey had already

6. Haggis 2005, locus 88.
7. Haggis 2005, locus 50.
8. Haggis 2005, locus 50.
9. Haggis 2005, locus 52.
10. Haggis 2005, locus 55.

collected pottery, and few artifacts remained on the surface. Consequently, pottery collection for this project was limited to a few specific features, a sampling strategy that could be described as selective rather than comprehensive. This selectivity decreases the probability that the pottery is representative.[11] On the other hand, since all pottery collecting has value,[12] some of the evidence permits different conclusions from those reached by the survey of Haggis, and the new collection allows these conclusions to be illustrated with specific ceramic evidence.

The recording system was designed to meet the specialized needs of this survey. Locations were recorded by electronic means with a Total Station using a grid established for the region in 1995. Each manmade feature was given a number, and groups of terrace walls were lettered sequentially from the highest to the lowest on the slope of the hill. Locations often defined as "sites" in traditional surveys were divided into several components at the preliminary stage of investigation.[13] Their designations were later joined together when study warranted it. For several reasons (including ease in map references and in digital recording), all manmade features were given feature numbers (including modern ones such as the metal fence, **AF 41**). Dates for the Minoan and post-Minoan pottery were assigned by Betancourt and Poulou-Papadimitriou respectively.

REGISTER OF ANTHROPOGENIC FEATURES

AF 1   Terrace walls   Fig. G.2

Grid coordinates: E5459 N4851 (at center of area).
H.: 120 masl (at center of area).
Visibility: good.
Dim.: ca. 70 m north–south × 108 m east–west; ca. 7,500 m².
Date: unknown.
Pottery: none collected.
Comments: Four parallel agricultural terrace walls are west of the modern gate to the Chrysokamino area, uphill from the road and downhill from the modern fence. All the walls here are constructed of irregularly shaped blocks of carbonate, ranging in size from 0.06 × 0.06 × 0.06 to 0.3 × 0.3 × 0.3 m, with the largest stones positioned at the bottom of the walls. Soil has eroded over the walls, and parts of the walls have washed away.

Wall **AF 1a**, the highest wall on the slope, was made with smaller stones than the lower three walls. It is the least well preserved of the four walls, and it uses bedrock in its construction.

Wall **AF 1b** is 27 m downhill (north) of **AF 1a**. It is preserved to a maximum height of 1.1 m, and it is up to 0.2 m thick. It is founded on bedrock.

Wall **AF 1c** is 16.2 m north of **AF 1b**. Its greatest preserved height is 0.6 m (0.2 m thick). Bedrock is used only in small parts of the wall.

Wall **AF 1d** is 16 m north of **AF 1c**. It is the same thickness as the others, with a preserved height of up to 0.7 m.

AF 2   Well   Fig. G.3

Grid coordinates: E5363.5 N4896.6.
H.: 124 masl.
Visibility: good.

---

11. This has been noted by many authors, e.g., Binford 1968a, p. 13; Hope Simpson 1983; Cherry, Gamble, and Shennan 1978.

12. Whallon 1983.

13. On the difficulties of site definition, see the comments by Wells 1996, pp. 17–21.

Figure G.2. Location of anthropogenic features **AF 1**, **AF 2**, and **AF 3**, uphill from the road at the right of the photograph. View looking north from the intersection marked **AF 40** on Figure 16.2.

Figure G.3. Well, AF 2

Diam.: 1.45 m interior, 2.2 m exterior; H. 0.45–0.80 m.
Date: Byzantine or later.
Pottery: none collected.
Comments: The southwestern slope of Chomatas has two peaks. A well is located on the saddle between them, uphill from the habitation site (**AF 29**). The wellhead is built of unworked fieldstones without mortar, forming a circular feature. It includes both blocky and flat stones of various sizes, up to ca. 0.3 × 0.4 × 0.15 m.

AF 3 Wall segments   Fig. G.4

Grid coordinates: E5362.4 N4890.7 (at center of wall).
H.: 124.4 masl (at center of wall).
Visibility: good.
Dim.: ca. 82.2 m north–south (length of reconstructed wall).
Date: Byzantine or later.
Pottery: none collected.
Comments: Three segments of wall are located near the well recorded as **AF 2**. They cannot be terrace walls because they are on a flat area at the ridge of the hill. The stones are very irregular, and the wall can never have stood very high. Wall **AF 3a** meanders along the saddle for 23 m. Its height ranges from 0.07 to 0.17 m, and

Figure G.4. Segment of a wall, possibly a boundary marker (AF 3)

its thickness is 0.1 to 0.3 m. It is made of unworked fieldstones laid without mortar. Wall **AF 3b** is 4.2 m south of **AF 3a**. It consists of seven stones in a line. Wall **AF 3c**, 2.7 m long and 0.2–0.4 m thick, is northwest of **AF 3b**. It is mostly buried.

The most likely explanation for these sections of wall is that they form a boundary marker, dividing the public land on the coastal side of the ridge from the private land inland. The well (**AF 2**) is just barely inside the wall (land side). The boundary today, marked by a modern fence (**AF 41**), makes all of the hilltop public land, including this feature.

AF 4   Threshing floor   Fig. G.5

Grid coordinates: E5296.9 N4993.2.
H.: 121.8 masl.
Visibility: good.
Diam.: ca. 7.9 m.
Date: probably Venetian or Ottoman.
Pottery: none collected.
Comments: This oval or circular threshing floor near **AF 5** has a floor of hard-packed soil and a raised soil perimeter with vertically placed phyllite slabs, 0.25–0.6 m in length. Slabs remain standing in place at the north and west. They are all unworked.

Figure G.5. Destroyed threshing floor with a few standing stones at the perimeter (AF 4)

Figure G.6. Collapsed field house (AF 5)

AF 5   Field house   Fig. G.6

Grid coordinates: E5269.3 N4983.6.
H.: 117 masl.
Visibility: good.
Dim.: 9.1 × 9.7 m.
Date: probably Venetian or Ottoman.
Pottery: none collected.

Comments: The remains of a one-roomed structure built of drywall masonry, primarily of flat phyllite slabs, lies between **AF 4** and **AF 7**. Stones are small in the north and west walls (average size 0.15 × 0.10 × 0.05 m) and larger in the eastern portion of the north wall (average size 0.9 × 0.25 × 0.4 m). Although it is not certain that the preserved north wall meets the natural bedrock outcrop on the east, it is likely that this outcrop was used as part of the east wall. This bedrock outcrop jogs in to the west, but it maintains a primarily north–south line for a distance of 5.9 m. It is 1.2 m high at its highest point, but no evidence of a wall on the bedrock survives. Although the north wall is the best preserved one, even this wall is not more than two courses high in most places. It is 9.1 m long (height between 0.29 and 0.70 m). The poorly preserved west wall, which is 9.7 m long and 0.7 m thick, is laid on bedrock. The other walls have collapsed.

The field house is near a threshing floor (**AF 4**) and terraced fields (**AF 6**), suggesting that this group of features forms a unit.

AF 6   Terrace walls   Fig. G.7

Grid coordinates: E5257.0 N4968.1 (at center of area).
H.: 109 masl (at center of area).
Visibility: good.
Dim.: ca. 22 m north–south × 33.3 m east–west; ca. 600 m^2.
Date: probably Venetian or Ottoman.
Pottery: none collected.

Comments: West of **AF 5** is an area of agricultural terrace walls. The walls, built of unworked fieldstones, are about 8–10 m apart. They are mostly carbonate blocks, up to 0.3 × 0.3 × 0.2 m in size, and they follow the contours of the hill, occasionally incorporating bedrock into their construction. This system of terrace walls, in which parallel walls are placed on the slope of a hill, is called a check-dam. The highest wall (**AF 6a**) stands 6–7 courses high in some places (H. 0.8 m).

Figure G.7. One of the terrace walls at AF 6

AF 7   Field house   Figs. G.8, G.9
   Grid coordinates: E5253.8 N4982.2 (at center of area).
   H.: 111.2 masl (at center of area).
   Visibility: good.
   Dim.: ca. 4.4 m north–south × 3.6 m east–west.
   Date: probably Venetian or Ottoman.
   Pottery: none collected.
   Comments: The lower parts of the walls of a small building with one room are downhill from **AF 5**. The structure is built of drywall masonry using flat phyllite slabs. It is oriented northwest–southeast. The southeast wall, 4.15 m long, contains a doorway. The interior is filled with stones fallen from the walls, obscuring the floor. The northwest wall is 4.16 m long, and the northeast wall is 3.45 m long. Wall thickness varies from 0.51 to 0.69 m. Only the lower parts of the walls are preserved. The preserved wall height is up to 0.7–0.9 m.
   It is possible that **AF 5** succeeded this poorly preserved structure.

AF 8   Sherd scatter   Fig. G.10
   Grid coordinates: E5390 N4530 (at center of area).
   H.: 112 masl (at center of area).
   Visibility: good.
   Dim.: ca. 85 m north–south × 170 m east–west.
   Date: FN–LM III.
   Pottery: recorded with **AF 32**.
   Comments: A scatter of EM–LM sherds (recorded as **AF 8**) occurs on the red soil south of the habitation site (**AF 29**). Except for **AF 32**, an oval structure near the southern side of this scatter, no architecture or other feature is visible. The sherds are present across the entire area from the habitation site to south of the oval structure. No specific concentration seems to be present, as the sherds form a fairly even scatter across the entire area. Their fabrics range in date from FN to LM III. The discussion in Chapter 21 suggests that this area is the garden for the habitation site.

AF 9   Cave
   Grid coordinates: E5411 N4702 (at center of interior).
   H.: 111.5 masl.

Figure G.8. Collapsed field house (AF 7) on the slope uphill from the metallurgy site. View looking west toward the metallurgy workshop

Figure G.9. Collapsed field house (AF 7). View looking south

Figure G.10. Looking downhill (south) from the habitation site (AF 29) to the location of a scatter of sherds (AF 8) and an oval enclosure (AF 32)

Visibility: good.
Dim.: ca. 9.2 m north–south × 6.6 m east–west.
Date: Venetian to Ottoman to modern.
Pottery: none collected from the surface.
Comments: A small manmade cave cut into the phyllitic soil has a semicircular entrance 1.5 m high, facing west toward the sea. The interior is oval. It is 1.8 m high near the entrance, 9.2 m wide, and 6.6 m deep. The floor is loose soil with a few scattered angular phyllite stones.

The cave was excavated in 1997 under the supervision of Brigit Crowell (App. K). The excavation discovered that it was carved out in stages, and the proximity to terrace **AF 22b** suggests the cave is contemporary with this feature (Venetian to Ottoman).

AF 10   Mound of stones

Grid coordinates: E5483.5 N4760.7.
H.: 139.8 masl.
Visibility: good.
Diam.: ca. 6.7 m.
Date: Ottoman: see Appendix J, **J-6** and **J-12**.
Comments: An artificial mound of stones 1.2 m high is found on the western ridge of Chomatas, uphill from the habitation site (**AF 29**). The mound seems to be a collapsed field house, and a portion of one wall (0.8 m long) is visible at the south/southwest. Pottery from near the feature is Ottoman.

AF 11   Terrace walls

Grid coordinates: E5504 N4766 (at center of area).
H.: 137 masl (at center of area).
Visibility: good.
Dim.: L. 1.3 m (wall 11a); 2.3 m (wall 11b).
Date: unknown.
Pottery: none collected.
Comments: Agricultural terrace walls in the public land on Chomatas (outside the modern fence, **AF 41**) show that the boundary has not always been where it is now. Two fragments of terrace walls are preserved at this location. Wall **AF 11a** is 0.25–0.3 m high and 0.15 m thick; wall **AF 11b** is 0.3–0.4 m high (2–3 courses high) and 0.2–0.35 m thick. The walls are not completely preserved.

AF 12   Well

Grid coordinates: E5513.5 N4800.1.
H.: 128.5 masl.
Visibility: good.
Diam.: ca. 1.0 m.
Date: probably Venetian or Ottoman.
Pottery: none.
Comments: A well is 5.1 m north of the field house called **AF 13**. The southern part of the wellhead is partly preserved, with a height of 1.05 m, enclosing an area 1.0 m in diameter. The walls are 0.55 m thick, built of fieldstones in a drywall technique. The stones measure 0.1 × 0.1 × 0.05 m to 0.6 × 0.4 × 0.18 m.

AF 13   Field house

Grid coordinates: E5515.8 N4793.2.
H.: 130.5 masl.
Visibility: good.

Dim.: 4.2 × 5.1 m.
Date: probably Venetian or Ottoman.
Pottery: none collected.
Comments: The ruins of a small, one-roomed building are 5.1 m south of **AF 12**. The building is oriented from north to south, with an opening 0.9 m wide in the east wall. The rubble walls are 0.55 m thick, preserved up to three stones high (0.55 m). Stones are up to 0.3 × 0.3 × 0.2 m in size. The interior of the room is filled with stones from the collapsed walls. An east–west wall 6.2 m long, probably part of a terrace wall, is near the building.

AF 14   Terrace walls

Grid coordinates: E5665 N5047 (at center of wall).
H.: 84.7 masl (at center of wall).
Visibility: good.
Dim.: ca. 91 m northwest–southeast × 15 m northeast–southwest; 1,200 m².
Date: unknown.
Pottery: Minoan pottery seen; none collected.
Comments: A group of nine terraces forms a system of check-dams in the ravine downhill from the modern gate to the territory of Chrysokamino. Walls range from 4.5 to 11.4 m in length, and they are placed at intervals of 3.2 to 9.15 m in horizontal distance and 0.6 to 2.6 m in vertical distance. The lowest few walls curve slightly southwest, following the direction of the ravine's slope. The walls range in height from 0.5 to 1.5 m. Most of the stones are small, ca. 0.2 × 0.2 × 0.15 m in size. All sherds seen on the surface in this area are Minoan. No field houses or other features are nearby.

AF 15   Field house   Fig. G.11

Grid coordinates: E5636.8 N4843.0.
H.: 113 masl.
Visibility: good.
Dim.: ca. 3 × 4 m.
Date: Ottoman or later.
Pottery: Ottoman pottery seen; none collected.
Comments: A small Ottoman or later field house with a single room provides information for comparison with the less well-preserved small buildings in the Chrysokamino territory. It is built on bedrock, using the drywall technique. The orientation is northeast–southwest, with the entrance in the northeast wall. Stone sizes vary from 0.1 × 0.1 × 0.1 m to 0.4 × 0.7 × 0.5 m. The floor measures 3 × 4 m. Rubble walls 0.5 m thick support a flat roof built of wooden crossbeams layered on top with sticks, brush, plastic sheet, and gravel. The interior is mostly covered with mud and plaster. The floor is soil and bedrock with a bedrock bench at the back. Wooden beehives are sometimes placed nearby.

AF 16   Field house

Grid coordinates: E5545.9 N4796.2.
H.: 126.7 masl.
Visibility: moderate.
Dim.: ca. 3.9 m × 4.0 m.
Date: Ottoman.
Pottery: see Appendix J, **J-13**.
Comments: A poorly preserved building with only the north and west walls still partly in place is built of unworked stones in the drywall technique. Stones range from 0.1 × 0.1 × 0.1 m to 0.45 × 0.5 × 0.15 m. A sherd from the feature is

Figure G.11. Standing field house (AF 15)

Ottoman. Terrace walls on Chordakia (**AF 17**) and a threshing floor (**AF 18**) are nearby.

AF 17 Terrace walls (Chordakia)

Grid coordinates: E5508 N4620 (at center of area).
H.: 110 masl (at center of area).
Visibility: poor to moderate to good.
Dim.: ca. 296 m north–south × 141 m east–west; 42,000 m².
Date: Minoan to modern.
Pottery: Appendix H, **AF 17.1** to **AF 17.3** and Appendix J, **J-3**, **J-9**, and **J-14** (EM III to LM IIIB, Venetian, Ottoman).

Comments: The east slope of Chomatas is covered with terrace walls, which show many stages of rebuilding and reuse. Most of the terraces are about 5 to 15 m apart. The name Chordakia is used for this terraced slope. A light scattering of sherds on the terraces indicates that the region has been used for agriculture since the Minoan period (or earlier?). Sherds published in this volume illustrate the range.

The area is planted in olive trees. The olives here and in the adjacent plain called the Kambos were planted after the Marshall Plan provided funds in 1947 for deep wells for irrigation.[14] Before then, the terraces were used for grains.

Wall **AF 17a** is closest to the modern fence (**AF 41**). It is preserved to a height of ca. 0.5 to 1 m, with an average thickness of ca. 0.4 m. It is mostly of dolomite blocks (from ca. 0.15 × 0.15 × 0.1 to 0.8 × 0.8 × 0.2 m). Bedrock is incorporated into the wall in some places.

Wall **AF 17b** is ca. 12–13 m downhill from **AF 17a**. The preserved part of the built portion of this wall ranges from ca. 0.6–0.8 m. It abuts bedrock cliffs, some of which drop as much as 5 m. Dolomite blocks are used in the construction, with stones up to ca. 0.3 × 0.25 × 0.15 m.

Wall **AF 17c** is ca. 6 m down the hill from **AF 17b**. It is ca. 0.6–0.9 m high and up to ca. 0.3 m thick. Dolomite blocks up to ca. 0.3 × 0.3 × 0.25 m are used in the construction.

14. Allbaugh 1953, pp. 258–263.

Wall **AF 17d** is from 0.7 to 3 m high, with a thickness of 0.3–0.4 m. Dolomite blocks up to 0.7 × 0.25 × 0.2 m are used in its construction, but most stones are smaller than this. The southern end uses some phyllite and bedrock. The northern part is washed away.

Walls **AF 17e** and **AF 17f** consist of several separate segments, up to 1 m or more in height. Materials and techniques are similar to those used in the other walls of this region.

AF 18   Threshing floor

Grid coordinates: E5540.9 N4803.2 (at surviving walls).
H.: 125.9 masl.
Visibility: moderate.
Dim.: 4.9 m (length of surviving walls).
Date: probably Ottoman.
Pottery: none collected.
Comments: A threshing floor is 4.7 m north of the field house recorded as **AF 16**. The shape is unknown (not preserved). It has a soil floor. The perimeter, which survives only in two sections, consists of thick, vertically set carbonate slabs with an average height of 0.5 m and a width of 0.35 m. The slabs are not worked.

AF 19   Wall

Grid coordinates: E5724.9 N4903.3 (at center of wall).
H.: 129.4 masl.
Visibility: good.
Dim.: ca. 28.5 m (length of wall).
Date: Byzantine or later.
Pottery: none collected.
Comments: A long wall, 28.5 m in length, is constructed with many vertically set stones, a technique used in Byzantine and later times in this part of Crete for walls that were not parts of buildings. The date for this technique is provided by several Byzantine walls on Pseira, including a well-dated wall from the 5th to 9th century A.D. that ran east to west across the Minoan plateia.[15] The orientation for **AF 19** is west/northwest to east/southeast. The height is 0.15 to 0.85 m, and the thickness is 0.8 m. The wall is probably a boundary wall.

AF 20   Threshing floor

Grid coordinates: E5495.5 N4736.1.
H.: 136 masl.
Visibility: poor.
Dim.: not measurable.
Date: probably Venetian or Ottoman.
Pottery: none collected.
Comments: A threshing floor, mostly concealed under a pile of loose fieldstones, is not sufficiently exposed to be measured. Its perimeter stones are vertically set phyllite slabs.

AF 21   Boundary walls   Figs. G.12–14

Grid coordinates: E5601 N4840 (at center of area).
H.: 119 masl.
Visibility: good.
Dim.: ca. 39 m east–west.

15. *Pseira* IV, p. 139.

Figure G.12. Sections of wall (AF 21), probably a boundary wall

Figure G.13. View of AF 21 from above

Date: Byzantine or later.
Pottery: Appendix H, **AF 21.1** and **AF 21.2**.
Comments: Two walls are west of the field house called **AF 15**. They are roughly parallel with the modern fence, and they are on the crest of the hill, so they cannot be terrace walls. They have two faces, and the preserved parts are clearly constructed as double-faced walls with rubble fill between the two courses of stones. The wall sections appear to be a predecessor of the modern fence.

One section of wall has an oblique corner. The part that is east of the corner is 15.8 m long, 0.1–0.35 m high, and up to 0.3 m thick. It is constructed of carbonate fieldstones up to 0.25 × 0.3 × 0.2 m in size. An additional 15.9 m of wall is west-northwest of the corner. The thickness in this section reaches 0.55 m, with double stones; its height is up to 0.25 m.

Figure G.14. Section of wall (AF 21)

The second section of wall is oriented north/northeast–south/southwest. It is 11.4 m long (as preserved). Stones range from very small (0.05 × 0.05 × 0.10 m) at the south/southwest end to very large (0.7 × 0.8 × 0.3 m) at the north/northeast end. Its thickness is 0.3 m, and it has a height of 0.35 to 0.8 m.

AF 22   Terrace walls   Fig. 2.16

Grid coordinates: E5366 N4667 (at center of area).
H.: 105 masl (at center of area).
Visibility: moderate.
Dim.: ca. 172 m north–south × 165 m east–west; 21,000 m².
Date: Venetian to Ottoman.
Pottery: Appendix J, **J-4**, **J-5**, and **J-7** (from the excavation of **AF 22b**); a very few Minoan and Ottoman sherds were seen elsewhere in the area but were not collected.

Comments. A series of 15 terrace walls for a check-dam system is in a ravine between the cave called **AF 9** and the sea. They vary considerably in preserved length, height, and spacing. While some of the walls form continuous curves extending to both sides of the ravine, many of them extend only 30–40 m along one side of the ravine and run into other terrace walls in this same system. The spacing between walls can be as much as 20 m. The walls are generally over 2 m high, although some of them rise much less (only a single course). Thickness varies from 0.2 to 0.8 m. One terrace wall (**AF 22o**) has between 15 and 20 courses preserved and stands 1.5 m high. Where preserved, the stones are mostly carbonate blocks that range in size from 0.1 × 0.1 × 0.1 to 0.5 × 0.3 × 0.3 m.

An excavation was made in terrace **AF 22b** in 1997 under the supervision of Crowell (see App. K). The excavation discovered that the terrace was built in the Venetian or Ottoman period, and at least some of the wall was definitely Ottoman. No evidence for earlier activity was found. Samples of the soil from this terrace are discussed in Appendix L.

AF 23   Metallurgy workshop   Fig. 1.7

Grid coordinates: E5015 N5001 (at center of area).
H.: 38 masl.
Visibility: good.
Dim.: 34 × 10 to 15 m.
Date: FN to EM III–MM IA.
Pottery: see Chapter 5.
Comments: see Part II in this volume.

AF 24 Field house

Grid coordinates: E5286 N4544.
H.: 118.5 masl.
Visibility: good.
Dim.: ca. 2.1 × 2.1 m.
Date: Ottoman or modern.
Pottery: none collected.

Comments: West of the habitation site (**AF 29**), two rough stone walls are built up against the bedrock to form a small shelter. The northern wall has two small windows ca. 0.5 × 0.3 × 0.2 m in size. The two walls are about 1 m high. They are built in drywall technique using carbonate fieldstones ranging in size from 0.1 × 0.1 × 0.1 m to 0.3 × 0.3 × 0.3 m. The shelter is from the Ottoman period or later, and people tending crops or animals probably used it as a field house.

AF 25 Terrace walls

Grid coordinates: E5747 N4918 (at center of area).
H.: 129 masl (at center of area).
Visibility: good.
Dim.: ca. 70 m north–south × 35 m east–west; ca. 2,000 m².
Date: unknown.
Pottery: none collected.

Comments: East of the road near **AF 15** is an area with three terrace walls, the highest of which is close to the probable boundary wall called **AF 19**. The highest wall (**AF 25a**), mostly washed away, is preserved only two stones high (0.2 m). The building material, mostly carbonate blocks, varies from 0.1 × 0.1 × 0.1 m to 0.2 × 0.4 × 0.25 m. The thickness is 0.25 m. Wall **AF 25b** is lower on the hill, to the west, and it is also poorly preserved. It is mostly composed of large, yellow, irregular blocks of limestone, with a few smaller stones (range 0.15 × 0.15 × 0.1 m to 0.55 × 0.65 × 0.25 m). Two courses are preserved (height 0.55 m) with a thickness of 0.25 m. Between 15–20 m lower on the hill is wall **AF 25c**, also mostly washed away. It is 2–3 courses high (0.3–0.65 m) with a preserved thickness of up to 0.5 m.

AF 26 Boundary wall?

Grid coordinates: E5662.8 N4838.4 (at center of wall).
H.: 109.2 masl (at center of wall).
Visibility: good.
Dim.: 17.7 m (preserved length of wall)
Date: Ottoman (?).
Pottery: Ottoman pottery seen; none collected.

Comments: A poorly built rubble wall runs perpendicular to the modern road near **AF 25**. It is only one course high (0.15 to 0.40 m), and its thickness incorporates up to two or three rows of stones in some parts (0.35 to 0.70 m thick). The wall is above ground, and although it appears to be fairly recent, part of it has washed away. It is probably a boundary wall.

AF 27 Boundary wall Fig. G.15

Grid coordinates: E5677.4 N4868.3 (at center of walls).
H.: 117 masl.
Visibility: good.
Dim.: ca. 19.5 m east–west (length of preserved walls).
Date: Byzantine or later.
Pottery: none collected.

Comments: Walls constructed of limestone blocks and vertical slabs of phyllite in two rows with a fill of rubble between them (compare **AF 19** and **AF 21**).

Figure G.15. Section of boundary wall (AF 27)

The main section of wall continues the line of **AF 21**, which is west of the modern road. The thickness is 0.6 m. The primary wall runs along an east–west line up and down the hill, and what appears to be a cross-wall lies 5 m from the easternmost end of the wall. The cross-wall is only 1.5 m long. It consists of a single row of large fieldstones (from 0.15 × 0.15 × 0.1 m to 0.3 × 0.3 × 0.25 m). Farther west, the main east–west wall meets another wall running northwest and southeast for a distance of 7.5 m. This wall also uses two stones for its width (up to 0.8 m), and although fewer vertical stones are present, it is probably related to the uphill wall. Two courses are preserved to a height of 0.4 m. Stone sizes vary from 0.2 × 0.1 × 0.1 m to 0.5 × 0.35 × 0.2 m.

AF 28   Sheepfold

Grid coordinates: E5384 N4560 (at center of feature).
H.: 120 masl.
Visibility: moderate.
Dim.: ca. 7.2 m north–south × 20.9 m east–west.
Date: middle of the 20th century.
Comments: About the middle of the 20th century, a shepherd built a sheepfold over the south side of the Chrysokamino habitation site (**AF 29**) using stones from the archaeological site. The sheepfold has been recorded by Haggis.[16] It was dismantled during the course of excavation.

AF 29   Habitation site (Chap. 17)   Figs. G.16–18

Grid coordinates: E5388 N4569.
H.: 123 masl.
Visibility: good.
Dim.: ca. 20 m north/northeast–south/southwest × 26 m east/northeast–west/southwest.

16. Haggis 2005, p. 115, locus 50.

Figure G.16. General view of the habitation site (**AF 29**) before excavation

Figure G.17. The habitation site (**AF 29**) before excavation, showing the east side, looking south

Figure G.18. The main entrance to the habitation site (**AF 29**), looking north from outside the building

Date: FN–LM IIIB.
Pottery: see Chapter 17; see also Appendix J, **J-2** and **J-3** (Venetian).
Comments: see Chapter 17.

AF 30   Terrace walls

Grid coordinates: E5222 N4663.
H.: 50 masl.
Visibility: good.
Dim.: ca. 30 m north–south × 26 m east–west; ca. 700 m².
Date: unknown.
Pottery: Minoan and Ottoman pottery seen; none collected.
Comments: Walls downhill from **AF 22** and **AF 39** are probably an extension of the large terraced area **AF 22**. Two sections of wall are visible. One section is built inside a steep ravine, and it prevents the sides from eroding (compare **AF 31**, which is similar), and the other wall terraces additional land south of it. The two walls are well preserved. They are made of dolomite fieldstones using drywall masonry, and they stand as much as 1 m or more in height.

AF 31   Terrace walls

Grid coordinates: E5463 N5073 (at center of area).
H.: 25 masl (at center of area).
Visibility: moderate.
Dim.: ca. 105 m northwest–southeast (length of system).
Date: unknown.
Pottery: Minoan and Ottoman pottery seen; none collected.
Comments: Fishermen from Kavousi routinely use the ravine at this location to walk down to the sea (for sport fishing). Rubble walls here are built of drywall masonry to create a series of terraces. They are made with carbonate blocks and stand over 1 m high. Some sections of wall are at the edge of the ravine, designed to prevent its walls from eroding (compare **AF 30**). The terracing may be an extension of the terrace system recorded as **AF 36**.

AF 32   Oval enclosure

Grid coordinates: E5397 N4491.
H.: 106 masl.
Visibility: good.
Dim.: ca. 42 m north–south × 69 m east–west; 2,300 m² of interior space in the enclosure.
Date: Minoan activity leaving a sherd scatter; oval structure is Ottoman, 18th century A.D.
Pottery: Appendix J, **J-8** (X 1345, from within the construction of the wall of the oval enclosure); additional sherds form a pottery scatter that extends across the hillside both within the structure and uphill and slightly downhill from it; Appendix H, **AF 32.1** to **AF 32.53** (FN to LM III).
Comments: An oval enclosure with the east side open is located downhill from the settlement (**AF 29**). This entire area has a scatter of Minoan sherds on it, but a sherd found in the stones used as the lowest course and foundation for the wall (Appendix J, **J-8**) provides an 18th-century date for this architectural feature, which is almost completely above ground. The enclosure is almost certainly a pen for sheep and goats. It is built of stones without any mortar (drywall masonry). The sherds that form the pottery scatter over the area including this late feature are not related to it. The area is apparently a part of the garden for the Minoan habitation site (see Chap. 21).

AF 33 Field house

Grid coordinates: E5471.8 N4514.5.

H.: 94.8 masl.

Visibility: good.

Dim.: The east wall is 5.5 m long, the west wall is 4.8 m long, the north wall is 2.9 m long, and the south wall is 3.1 m long (all measurements on the exterior).

Date: Ottoman.

Pottery: Appendix J, **J-33.1** and **J-33.2** (LM I to LM IIIA/B); Ottoman pottery seen (not collected).

Comments: A small one-room building is constructed of uncoursed carbonate fieldstones without mortar (drywall masonry). The upper parts of the walls have collapsed into the interior, covering the floor. The building is oriented north–south with a doorway 0.38 m wide in the east wall. A niche is built into the north end of the west wall, in the interior of the building. It is 0.37 m wide by 0.38 m high by 0.3 m deep. The construction technique uses larger stones in the lower part of the wall and smaller stones higher up. The northeast corner, where the walls are best preserved, is 1.4 m high.

AF 34 Theriospelio cave   Fig. 1.11

Grid coordinates: E5180.7 N5146.9 (at entrance).

H.: 44 masl (at entrance).

Visibility: good.

L.: ca. 60 m.

Date: FN–EM III or later.

Pottery: see Chapter 18.

Comments: see Chapter 18.

AF 35 Terrace walls

Grid coordinates: E5437 N4966 (at center of area).

H.: 80 masl (at center of area).

Visibility: moderate to good.

Dim.: ca. 50 m north–south × 8 m east–west (estimated ca. 400 m^2).

Date: unknown.

Pottery: none collected.

Comments: A terraced area north of the modern road has poorly preserved walls of carbonate fieldstones. The extent of this system cannot be measured accurately because of the poor preservation.

AF 36 Terrace walls

Grid coordinates: E5497 N5085.

H.: 40 masl.

Visibility: moderate.

Dim.: ca. 31 m north–south × 17 m east–west; ca. 400 m^2.

Date: unknown.

Pottery: none collected.

Comments: This area is probably related to **AF 31** and **AF 35**. Walls are built of carbonate fieldstones. Two walls can be seen. **AF 36a** is 31 m in length, and **AF 36b** is approximately the same length.

AF 37 Terrace walls

Grid coordinates: E5204 N4880 (at center of area).

H.: 75 masl (at center of area).

Visibility: good.

Dim.: ca. 12 m north–south × 84 m east–west; ca. 800 m².
Date: unknown.
Pottery: none collected.
Comments: Terrace walls uphill from the metallurgy location are built of carbonate and phyllite blocks, creating a check-dam system. The terraces are all of phyllitic soil. Walls are not well preserved.

AF 38 Terrace wall

Grid coordinates: E5221 N4599 (at center of area).
H.: 75 masl (at center of area).
Visibility: good.
Dim.: ca. 4 m north–south × 5 m E–W; 16 m².
Date: unknown.
Pottery: none collected.
Comments: A terrace wall south of **AF 30**, built of carbonate fieldstones, is probably part of the same system recorded as **AF 30** and **AF 22**.

AF 39 Well   Fig. G.19

Grid coordinates: E5313.8 N4676.5.
H.: 86 masl.
Visibility: good.
Diam. of exterior: ca. 2 m.
Date: Ottoman or later.
Pottery: none collected.
Comments: A circular well consists of a shaft lined with stones. It is dug into the soil covering the bedrock on this hillside, on a terrace above the ravine that runs downhill from near **AF 9**. The interior is ca. 0.80–0.90 m in diameter. The well is partly filled with stones.

AF 40 Road and house walls   Fig. G.20

Grid coordinates: E5336 N4821 (at intersection on west slope of Chomatas).
H.: 115 masl (at intersection on west slope of Chomatas).
Visibility: good.
Dim.: average width ca. 5 m.

Figure G.19. Opening of a well (AF 39)

Figure G.20. The modern road (AF 40) showing the section constructed in May 1996, as seen from the habitation site (AF 29)

Date: road is modern; walls are Minoan.
Pottery: none collected.
Comments: A graded but unpaved road is one of the most easily visible landscape features of the Chrysokamino territory. The portion of the road from the intersection at E5336 N4821 to the habitation site (**AF 29**) was constructed in May of 1996. The other parts of the road are older (graded in the mid-20th century or earlier, following traditional access routes to the area). House walls are visible in the scarp of the new section of the road.

AF 41   Metal fence

Grid coordinates: E5497 N4744 (at the highest point on Chomatas).
H.: 135 masl (at the highest point on Chomatas).
Visibility: good.
Dim.: not applicable.
Date: modern.
Pottery: none collected.
Comments: A metal fence with two construction periods, both after the middle of the 20th century, is important for the interpretation of earlier stone walls that follow the same general line (**AF 3**, **AF 26**, and **AF 21/27**). The fence is also a useful orientation feature for those using the maps prepared by this project because, like the modern road, it is easily visible.

The modern fence demarcates the private land (which is inland) from public land (the coastal strip, owned by the village of Kavousi). It is the latest in a series of boundary markers at approximately the same place, but the earlier walls are not continuous. The later metal fence clearly defines the border between the territory used by individuals and the public land.

The Minoan Pottery from the Survey

by Philip P. Betancourt

The Minoan pottery presented here is used to illustrate both the date of several of the anthropogenic features recorded in Appendix G and the character of the preserved ceramics associated with these features (Figs. H.1 to H.4). It is all collected from the surface. Many of the sherds are very worn, and their identification is often based on shape and fabric rather than decoration.[1]

AF 17

Early Minoan III–Middle Minoan IA

AF 17.1   (X 1669; from **AF 17d** surface)   Fig. H.1

Vessel, body sherd.
Max. dim. 4.2.
Mirabello Fabric (yellowish red, 5YR 5/6).
Comments: EM III–MM IA; the fabric of X 1669 is typical of EM III, but it continues into the next period.

Late Minoan I

AF 17.2   (X 1299; from **AF 17a** surface)   Fig. H.1

Cooking dish, rim sherd.
Diam. of rim not measurable; max. dim. 5.3.
A phyllite fabric (red, 2.5YR 5/8). Thickened rim.
Comments: LM I; the piece is an ordinary cooking dish, a broad, open shape used in cooking from the EM until the LM IIIC periods (for the shape, see Betancourt 1980). The fabric used here is the phyllite fabric that is most common in LM I (Haggis and Mook 1993; for the petrography of this fabric, see Betancourt and Myer 1995).

Late Minoan I–III

AF 17.3   (X 1668; from **AF 17d** surface)   Fig. H.1

Tripod cooking pot (?), body sherd.

---

1. For fabrics in the region, see Haggis and Mook 1993; Betancourt and Myer 1995; Myer, McIntosh, and Betancourt 1995.

Max. dim. 3.9.

A phyllite fabric with white stone inclusions (red, 2.5YR 5/6).

Comments: In this part of Crete, the phyllite fabric with white stone inclusions is usually LM III.

AF 21

Late Minoan I

AF 21.1   (X 1739; from **AF 21** surface)   Fig. H.1

Pithos, body sherd.
Max. dim. 9.0.
A coarse fabric (dark yellowish brown, 10YR 4/4, with a light reddish brown surface, 5YR 6/4).
Ropework on exterior.

Late Minoan IIIA–B

AF 21.2   (X 1725; from **AF 21** surface)   Fig. H.1

Tripod cooking pot, leg sherd.
Max. dim. 5.3.
A phyllite fabric containing other stones (red, 10R 4/8).
Comments: For the use of the cooking pot, see Betancourt 1980. The phyllite fabric with white stone inclusions is usually LM III.

AF 29

Late Minoan IIIA–B

AF 29.1   (X 326; from the surface at E5397.284 N4588.443)   Fig. H.1

Kylix, base sherd.
Diam. of base 6.
Fine fabric (light reddish brown, 2.5YR 6/4). Traces of dark slip on exterior.
Comments: surface worn. For the shape, see Popham 1969.

AF 32 and AF 8

The pottery from **AF 8** comes from the area between the habitation site and a rocky ridge just south of a large oval animal pen (**AF 32**). The terminus post quem for the pen's construction is in the 18th century. The sherd scatter extends over the ridge and a few meters down the other side of the hill, but it does not continue all the way down the hillside to Lakkos Ambeliou. The scatter contains a heterogeneous mixture of small, worn sherds. Pieces do not join with other fragments, and in most cases the corners are rounded, and the paint is missing. The character of this pottery is different from ceramic pieces found in and near houses, and it is interpreted in this volume as a deposit representing casual sherds tilled

THE MINOAN POTTERY

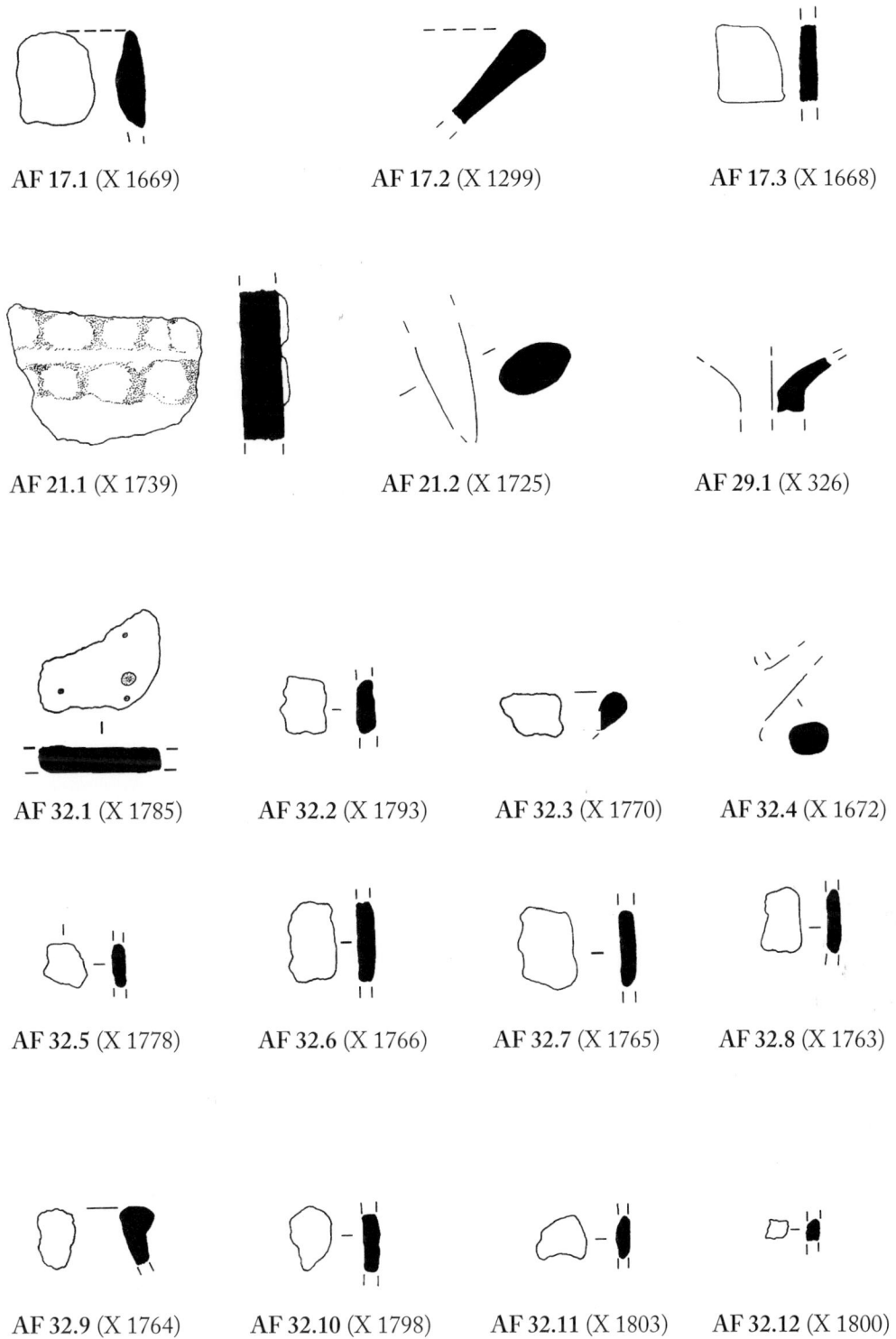

Figure H.1. Minoan pottery from the surface, **AF 17 to AF 32**. Scale 1:3

into the ground along with manure. The interpretation is supported by the results of chemical analysis in related deposits near Lakkos Ambeliou[2] and on Pseira Island.[3]

Final Neolithic

AF 32.1   (X 1785; from **AF 32** surface)   Fig. H.1

Open vessel, body sherd.
Max. dim. 6.2.
A dark-surfaced fabric (mostly black) with phyllite inclusions; burnished.
Open vessel with impressed dots in the interior.
Comments: Probably from the Final Neolithic. The punctations in the interior of the vessel are an unusual element. For the FN in this part of Crete, see the comments in Chapter 5.

Early Minoan III–Middle Minoan II

AF 32.2   (X 1793; from **AF 32** surface)   Fig. H.1

Cooking dish, body sherd.
Max. dim. 3.0.
Mirabello Fabric (dark reddish brown, 5YR 5/4).
Comments: This is a typical cooking dish fragment of the type used in EM III and later times. For the shape, see Betancourt 1980.

AF 32.3   (X 1770; from **AF 32** surface)   Fig. H.1

Cooking dish, rim sherd.
Max. dim. 3.0.
Mirabello Fabric (dark reddish brown, 5YR 3/4).
Comments: Similar to **AF 32.2**.

AF 32.4   (X 1672, from **AF 32** surface)   Fig. H.1

Small vessel, handle sherd.
Max. dim. 3.0.
Mirabello Fabric (brown, 7.5YR 5/4).
Comments: It is possible that this is a handle from a cup, but some small jugs have similar appendages.

Early Minoan III–Late Minoan I

AF 32.5   (X 1778; from **AF 32** surface)   Fig. H.1

Vessel, body sherd.
Max. dim. 2.0.
Mirabello Fabric (brown, 10 YR 5/3).
Comments: Mirabello Fabric is typical of the Middle Bronze Age; its use comes to an end at the beginning of LM I (Haggis and Mook 1993, type 1).

AF 32.6   (X 1766; from **AF 32** surface)   Fig. H.1

Vessel, body sherd.
Max. dim. 3.5.
Mirabello Fabric (brown, 7.5YR 5/4 exterior with very dark gray, 7.5YR N3 interior).

2. Morris 1994, pp. 187, 197.
3. Betancourt and Hope Simpson 1992; Bull, Evershed, and Betancourt 1999.

Middle Minoan–Late Minoan

AF 32.7   (X 1765; from **AF 32** surface)   Fig. H.1

    Vessel, body sherd.
    A coarse fabric (between brown and dark brown, 10YR 4/3).

AF 32.8   (X 1763; from **AF 32** surface)   Fig. H.1

    Closed vessel, body sherd.
    Max. dim. 2.7.
    A coarse fabric (yellowish red, 5YR 5/8 exterior with reddish brown, 5YR 4/3 interior).

AF 32.9   (X 1764; from **AF 32** surface)   Fig. H.1

    Closed vessel, rim sherd.
    Max. dim. 2.6.
    Rolled rim.
    A coarse fabric (red, 2.5 YR 5/6).

AF 32.10   (X 1798; from **AF 32** surface)   Fig. H.1

    Vessel, body sherd.
    Max. dim. 2.5.
    A fine fabric (reddish yellow, 5YR 6/6).

AF 32.11   (X 1803; from **AF 32** surface)   Fig. H.1

    Vessel, body sherd.
    Max. dim. 2.2.
    A phyllite fabric with a few white inclusions (reddish brown, 5YR 5/4).

AF 32.12   (X 1800; from **AF 32** surface)   Fig. H.1

    Vessel, body sherd.
    Max. dim. 1.4.
    A coarse fabric (yellowish red, 5YR 5/6).

AF 32.13   (X 1794; from **AF 32** surface)   Fig. H.2

    Vessel, body sherd.
    Max. dim. 2.7.
    A coarse fabric (red, 2.5YR 5/8).

AF 32.14   (X 1799; from **AF 32** surface)   Fig. H.2

    Vessel, body sherd.
    Max. dim. 1.5.
    A coarse fabric (reddish brown, 5YR 5/4).

AF 32.15   (X 1760; from **AF 32** surface)   Fig. H.2

    Tripod cooking pot, body sherd with leg.
    Max. dim. 4.5.
    A phyllite fabric with many white inclusions (between red, 2.5YR 5/6 and reddish brown, 5YR 4/4). Leg with thick oval section with three vertical slashes on outside of leg.

AF 32.16   (X 1786; from **AF 32** surface)   Fig. H.2

    Bridge-spouted jar, rim sherd.
    A phyllite fabric with many white inclusions (yellowish red, 5YR 4/6). Thickened rim.

Middle Minoan–Late Minoan I

AF 32.17 (X 1780; from **AF 32** surface) — Fig. H.2

  Vessel, handle sherd.
  Max. dim. 3.2.
  A coarse fabric (yellowish red, 5YR 5/6).

AF 32.18 (X 1759; from **AF 32** surface) — Fig. H.2

  Jar, rim sherd.
  Diam. of rim ca. 24–26.
  Mirabello Fabric (light red, 2.5YR 6/6). Outturned rim.

Middle Minoan–Late Minoan III

AF 32.19 (X 1768; from **AF 32** surface) — Fig. H.2

  Vessel, body sherd.
  Max. dim. 3.6.
  A coarse fabric (red, 2.5YR 5/6 exterior with yellowish brown, 10 YR 3/4).

AF 32.20 (X 1769; from **AF 32** surface) — Fig. H.2

  Vessel, body sherd.
  Max. dim. 3.7.
  A coarse fabric (yellowish red, 5YR 5/6 exterior with dark brown, 7.5YR 3/2 core).

AF 32.21 (X 1772; from **AF 32** surface) — Fig. H.2

  Vessel, body sherd.
  Max. dim. 2.7.
  A phyllite fabric with a few white inclusions (red, 2.5YR 4/6).

AF 32.22 (X 1797; from **AF 32** surface) — Fig. H.2

  Vessel, body sherd.
  Max. dim. 3.0.
  A coarse fabric (yellowish red, 5YR 5/6).

AF 32.23 (X 1773; from **AF 32** surface) — Fig. H.2

  Vessel, body sherd.
  Max. dim. 3.0.
  A coarse fabric (red, 2.5 YR 5/6).

AF 32.24 (X 1771; from **AF 32** surface) — Fig. H.2

  Vessel, base sherd.
  Max. dim. 5.4.
  A phyllite fabric with a few white inclusions (reddish brown, 2.5YR 4/4).

AF 32.25 (X 1782; from **AF 32** surface) — Fig. H.3

  Closed vessel, base sherd.
  Diam. of base 32.
  A phyllite fabric with a few white inclusions (reddish brown, 5YR 5/4).

AF 32.26 (X 1783; from **AF 32** surface) — Fig. H.3

  Closed vessel, body sherd.
  Max. dim. 6.8.

THE MINOAN POTTERY

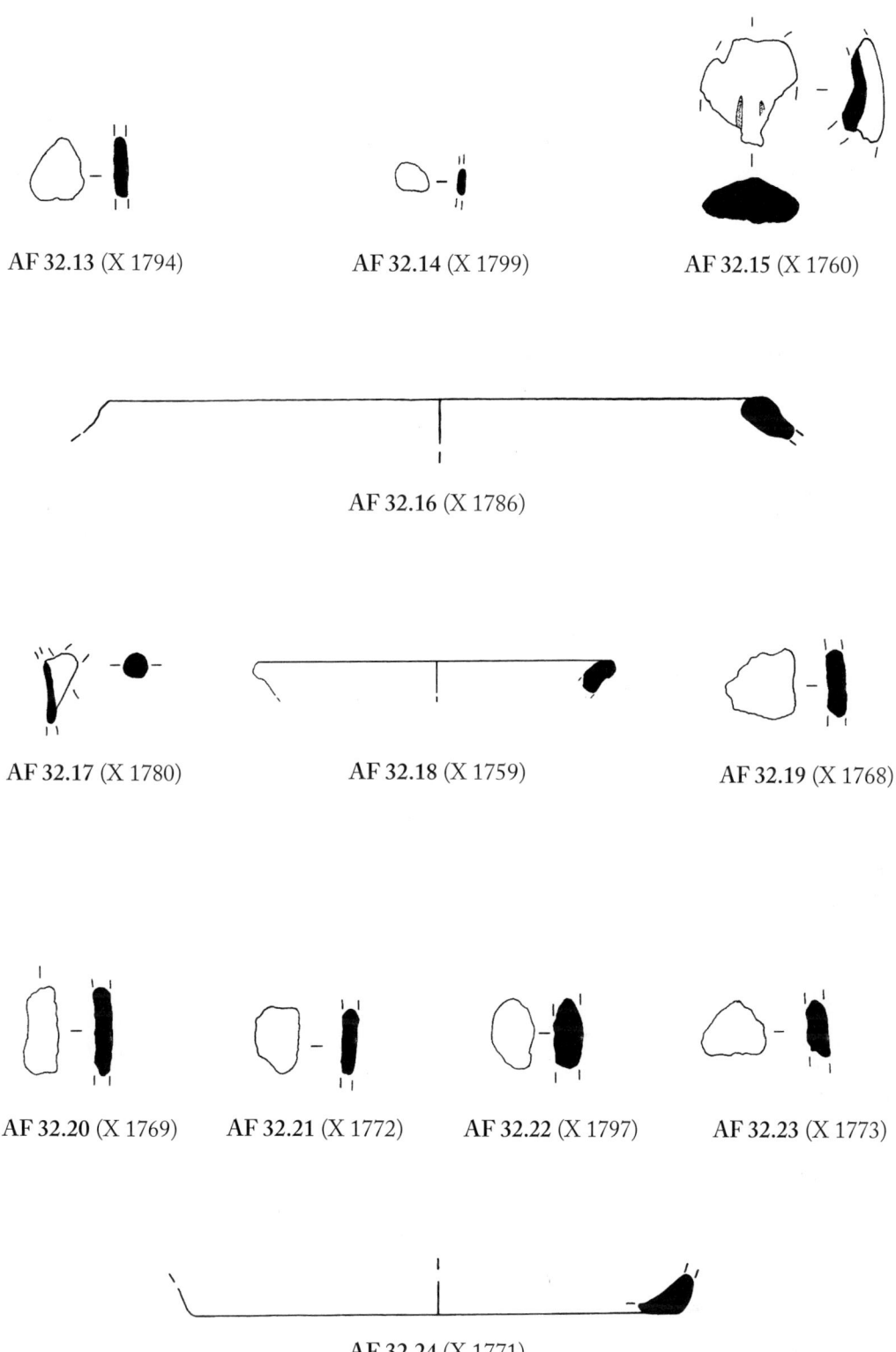

Figure H.2. Minoan pottery from the surface, AF 32. Scale 1:3

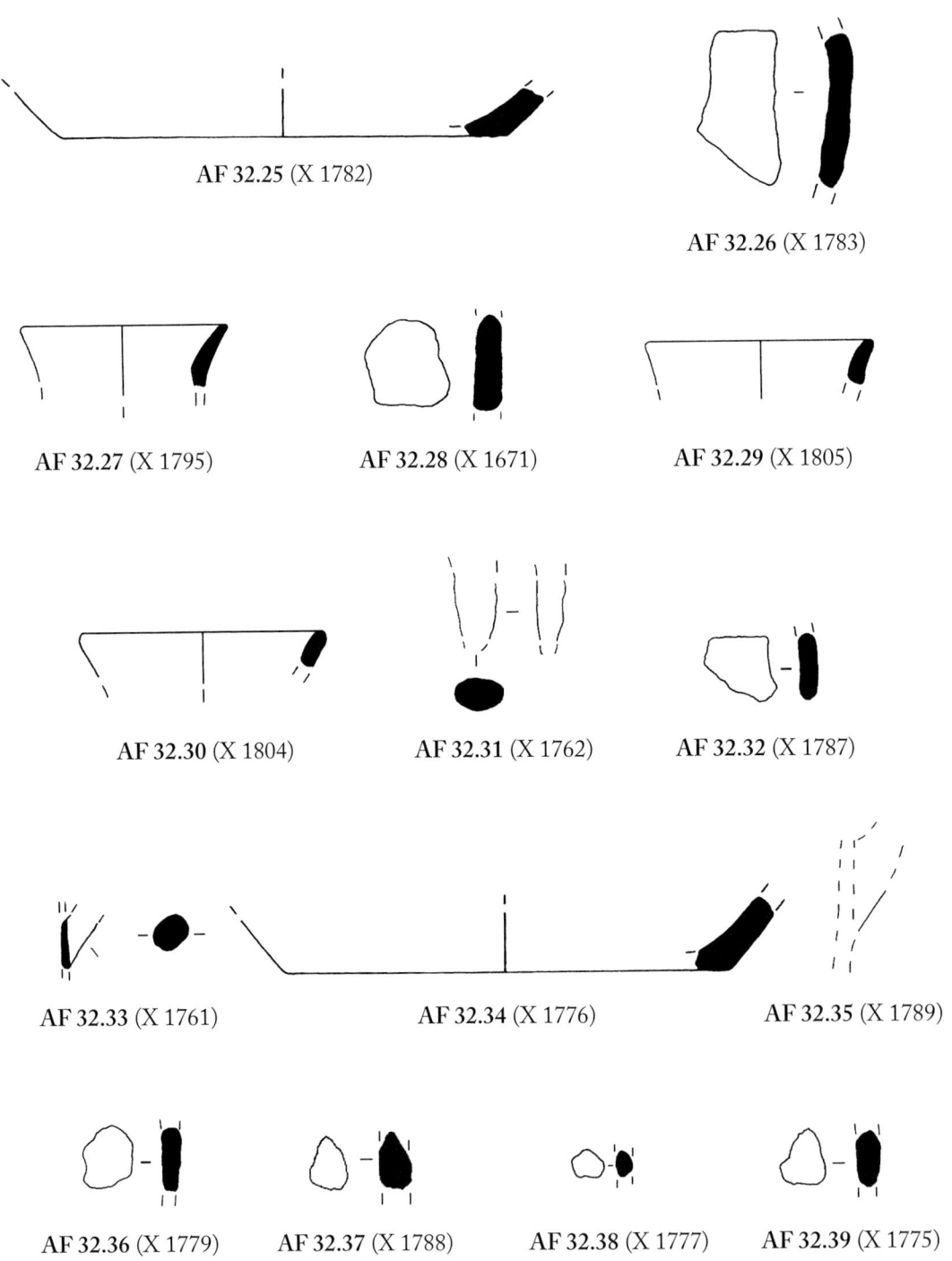

Figure H.3. Minoan pottery from the surface, **AF 32.** Scale 1:3

Convex profile.
A phyllite fabric with many white inclusions (yellowish red, 5YR 5/6).

Middle Minoan III–Late Minoan I

AF 32.27   (X 1795; from **AF 32** surface)   Fig. H.3

Conical cup, rim sherd.
Diam. of rim 7.
A fine phyllite fabric (brown, 7.5YR 5/2).

Late Minoan I

AF 32.28   (X 1671; from south of **AF 32**)   Fig. H.3

Vessel, body sherd.
Max. dim. 4.1.
A phyllite fabric (yellowish red, 5YR 5/6).

AF 32.29   (X 1805; from **AF 32** surface)   Fig. H.3

Cup or tumbler, rim sherd.
Diam. of rim 10.
A fine fabric (yellowish red, 5YR 5/6).

AF 32.30   (X 1804; from **AF 32** surface)

Vessel, rim sherd.
Max. dim. 2.1.
A phyllite fabric with a few white inclusions (yellowish red, 5YR 5/6).

AF 32.31   (X 1762; from **AF 32** surface)   Fig. H.3

Tripod cooking pot, leg sherd.
Max. dim. 3.4.
A phyllite fabric (red, 2.5YR 4/6).
Leg with thick oval section.

Late Minoan I–III

AF 32.32   (X 1787; from **AF 32** surface)   Fig. H.3

Closed vessel, body sherd.
Max. dim. 3.6.
A phyllite fabric (yellowish red, 5YR 5/6).

AF 32.33   (X 1761; from **AF 32** surface)   Fig. H.3

Closed vessel, body sherd with handle.
Max. dim. 2.7.
A phyllite fabric with a few white inclusions (red, 2.5YR 5/6).

AF 32.34   (X 1776; from **AF 32** surface)   Fig. H.3

Closed vessel, base sherd.
Max. dim. 6.3.
A phyllite fabric with a few white inclusions (red, 2.5YR 5/6).

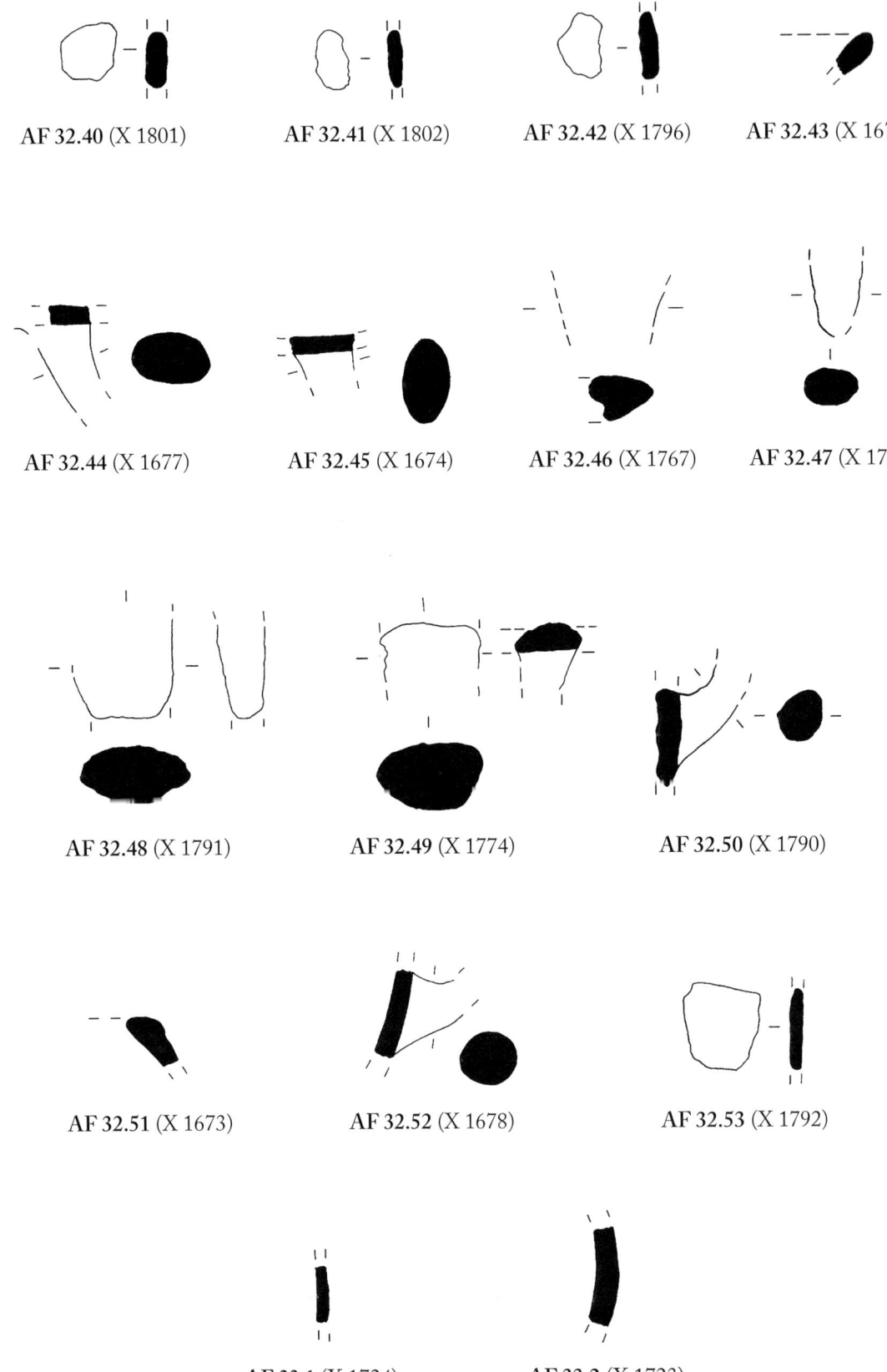

THE MINOAN POTTERY

Figure H.4 *(opposite)*. Minoan pottery from the surface, AF 32, AF 33. Scale 1:3.

AF 32.35 (X 1789; from **AF 32** surface)   Fig. H.3
  Closed vessel, handle sherd.
  Max. dim. 3.7.
  A phyllite fabric with many white inclusions (brown to dark brown, 10YR 4/3).

AF 32.36 (X 1779; from **AF 32** surface)   Fig. H.3
  Vessel, body sherd.
  Max. dim. 2.8.
  A phyllite fabric with a few white inclusions (red, 2.5YR 4/6).

AF 32.37 (X 1788; from **AF 32** surface)   Fig. H.3
  Vessel, body sherd.
  Max. dim. 2.2.
  A phyllite fabric (red, 2.5YR 5/6).

AF 32.38 (X 1777; from **AF 32** surface)   Fig. H.3
  Vessel, body sherd.
  Max. dim. 1.4.
  A phyllite fabric (yellowish red, 5YR 5/6).

AF 32.39 (X 1775; from **AF 32** surface)   Fig. H.3
  Vessel, body sherd.
  Max. dim. 2.2.
  A phyllite fabric (red, 2.5YR 4/6).

AF 32.40 (X 1801; from **AF 32** surface)   Fig. H.4
  Vessel, body sherd.
  Max. dim. 3.1.
  A phyllite fabric with a few white inclusions (yellowish red, 5YR 5/8).

AF 32.41 (X 1802; from **AF 32** surface)   Fig. H.4
  Vessel, body sherd.
  Max. dim. 2.6.
  A phyllite fabric (red, 2.5YR 5/6).

AF 32.42 (X 1796; from **AF 32** surface)   Fig. H.4
  Vessel, body sherd.
  Max. dim. 2.9.
  A coarse fabric (yellowish red, 5YR 5/6).

AF 32.43 (X 1676; from **AF 32** surface)   Fig. H.4
  Cooking dish, rim sherd.
  A phyllite fabric with white stone inclusions (yellowish red, 5YR 4/6).
  Comments: This is often a Late Minoan III fabric in this part of Crete.

AF 32.44 (X 1677; from **AF 32** surface)   Fig. H.4
  Tripod cooking pot, leg sherd.
  Max. dim. 5.4.
  A phyllite fabric with white stone inclusions (red, 10R 4/8).
  Comments: See **AF 32.43**.

AF 32.45   (X 1674; from **AF 32** surface)   Fig. H.4

   Tripod cooking pot, leg sherd.
   Max. dim. 4.7.
   A phyllite fabric with white stone inclusions (yellowish red, 5YR 4-5/6).
   Leg with thick oval section.
   Comments: See **AF 32.43**.

AF 32.46   (X 1767; from **AF 32** surface)   Fig. H.4

   Tripod cooking pot, leg sherd.
   Max. dim. 3.4.
   A phyllite fabric with a few white inclusions (red, 2.5YR 5/6).
   Leg with thick oval section.
   Comments: See **AF 32.43**.

AF 32.47   (X 1784; from **AF 32** surface)   Fig. H.4

   Tripod cooking pot, leg sherd.
   Max. dim. 3.8.
   A phyllite fabric with a few white inclusions (yellowish red, 5YR 4/6).
   Leg with thick oval section.
   Comments: See **AF 32.43**.

AF 32.48   (X 1791; from **AF 32** surface)   Fig. H.4

   Tripod cooking pot, leg sherd.
   Max. dim. 6.0.
   A phyllite fabric (between red, 2.5YR 5/6 and dark reddish brown, 5YR 3/4).
   Leg with thick oval section.

AF 32.49   (X 1774; from **AF 32** surface)   Fig. H.4

   Tripod cooking pot, leg sherd.
   Max. dim. 5.2.
   A phyllite fabric with a few white inclusions (yellowish red, 5YR 5/6).
   Leg with thick oval section.
   Comments: See **AF 32.43**.

AF 32.50   (X 1790; from **AF 32** surface)   Fig. H.4

   Jar, handle sherd.
   Max. dim. 6.5.
   A phyllite fabric with many white inclusions (red, 2.5YR 4/6).
   Comments: See **AF 32.43**.

AF 32.51   (X 1673; from **AF 32** surface)   Fig. H.4

   Jar, rim sherd.
   Diam. of rim ca. 40.
   A phyllite fabric with white stone inclusions (reddish yellow, 5YR 6/8).
   Comments: See **AF 32.43**.

AF 32.52   (X 1678, from **AF 32** surface)   Fig. H.4

   Jar, body sherd with part of handle.
   Max. dim. 5.2.
   A phyllite fabric with white stone inclusions (red, 2.5YR 4-5/8).
   Comments: See **AF 32.43**.

Late Minoan III

AF 32.53 (X 1792; from **AF 32** surface) Fig. H.4
 Closed vessel, body sherd.
 Max. dim. 4.3.
 A fine fabric (reddish brown, 5YR 5/6).
 Comments: Dated on the basis of the fabric.

AF 33

Late Minoan I

AF 33.1 (X 1724; from **AF 33** surface) Fig. H.4
 Vessel, body sherd.
 Max. dim. 2.6.
 A phyllite fabric (yellowish red, 5YR 5/6).

Late Minoan IIIA–B

AF 33.2 (X 1723; from **AF 33** surface) Fig. H.4
 Vessel, body sherd.
 Max. dim. 5.2.
 A phyllite fabric containing other stones (red, 2.5YR 6/6).
 Comments: Possibly a cooking pot? Dated by the fabric.

APPENDIX I

Evidence for Beekeeping

by Susan C. Ferrence and Elizabeth B. Shank

Evidence for possible beekeeping within the geographic territory of the Chrysokamino farmstead was discovered during the summer of 1999. A pottery sherd measuring 5.0 cm long by 4.3 cm wide by 1.4 cm thick was found in a survey of the landscape south of the habitation site (Figs. I.1, I.2). The sherd came from a hilltop that is lower than the elevation of the habitation site. Both hills overlook the Gulf of Mirabello. The sherd was found on the surface among thyme and small pine trees. No sherds from pottery vessels were observed nearby.

The sherd is too small for any reconstruction of its original appearance, but it may belong to a class of vessels resembling large basins with scoring on the interior. The possibility that such vessels might have served as beehives in Crete has been proposed several times,[1] although it has also been suggested that the objects could have been used as graters.[2] The interpretation of the class as beehives has been discussed in detail by Melas, who cites many parallels for incised beehives from Classical Greece and elsewhere.[3] Scoring in the interior would have enabled the honeycombs to adhere to the internal surface of the beehive.

Melas suggests that most beehives were located to the south of, but near to, houses on hillsides covered with "aromatic shrubs and stunted pines."[4] In this context, the sherd discovered at Chrysokamino is significant because of its distant location from the habitation site, indicating that beekeeping was not always limited to the immediate surroundings of a house or farm. The area in which the sherd was found would have been ideally suited for beekeeping and casual pasturage.

The sherd from Chrysokamino is important because most similar fragments have been found in settlements, and this is one of the better arguments against the use of such incised vessels as beehives.[5] Scored vessel fragments were found within the rooms of buildings at several places, including Kommos, Kato Syme Viannou, Pseira, and sites on Karpathos.[6] These occurrences have been interpreted to represent the storage of beehives when they were not in use. Scored vessel fragments have also been found within the habitation site at Chrysokamino.

Melas suggests that Minoan honey production was largely oriented toward domestic use and local trade.[7] Isolated farms needed bees for crop

---

1. See, among others, *Kommos* II, pp. 98, 168, nos. 467, 1483; *Kommos* III, p. 25, no. 439; Floyd 1998, p. 75, no. 263, from Pseira.
2. Floyd 1998, p. 180; Davaras 2001, pp. 81–82.
3. Melas 1999.
4. Melas 1999, p. 487.
5. Davaras 2001, p. 82.
6. For a list, see Melas 1999.
7. Melas 1999, p. 490.

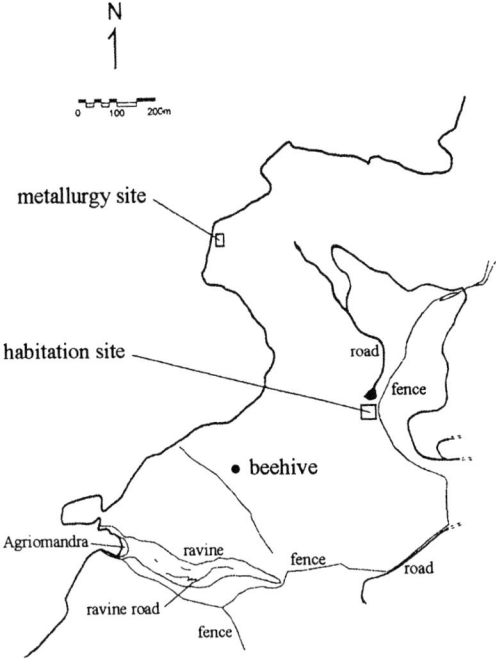

Figure I.1. Location of the beehive fragment found away from any settlement

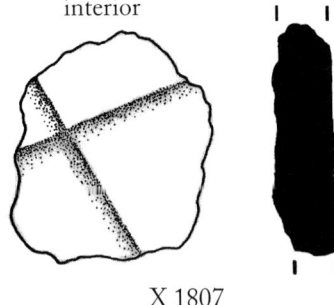

Figure I.2. The beehive fragment

pollination. The honey and its by-products would have been consumed within the household.

Chrysokamino was a self-sufficient farmstead isolated from any village. Raising bees would have been necessary for ensuring the pollination of the crops and running a successful private farm. The farmer could also take advantage of the resulting honey and wax production, making life for his family that much sweeter.

APPENDIX J

The Byzantine to Ottoman Pottery from the Survey

by Natalia Poulou-Papadimitriou

Although only a few pieces of pottery from the Byzantine and later periods are present at Chrysokamino, they are interesting for the information they provide about the region during these periods (Figs. J.1, J.2). No signs of a permanent settlement are present, and no spot has a large concentration of sherds. The pottery fragments occur in isolation, and they probably attest to farming, herding, hunting, or casual travel through the area. Several of the sherds are from closed vessels that would have been useful in carrying water and other liquids, while a few pieces are from dishes and other open shapes.

CATALOGUE

The catalogue follows the format used for the Minoan pottery, with the sherds arranged by periods. Colors are described using the Munsell system.[1]

Byzantine Period

J-1 (X 1485, from the surface of unit XB-42-1) Fig. J.1
  Large amphora, many body sherds.
  Diam. of body (restored) ca. 30.
  A fine fabric (reddish yellow, 5YR 6–7/6).
  Comments: 12th–13th century, probably 13th century. Found on the surface of the Minoan habitation site.
  Venetian Period

J-2 (X 1657, from **AF 29** surface) Fig. J.1
  Open bowl, body sherd.
  Max. dim. 5.9.
  A fine fabric (reddish yellow, 7.5YR 7/6). Creamy off-white glaze on interior and upper part of exterior; greenish yellow drips evenly spaced on exterior.
  Comments: Venetian, 15th–16th century Glazed Painted Ware; a Cretan production.

---

1. Kollmorgen Instruments Corporation 1992.

J-3   (X 1300, from **AF 17** surface)   Fig. J.1
   Amphora, rim sherd.
   Diam. of rim 14.
   A fine fabric (light red, 2.5YR 6/8) with inclusions; without slip.
   Thickened, flat rim; groove under rim.
   Comments: Venetian, 15th–16th century.

Venetian Period or Ottoman Period

J-4   (X 1167, from the excavation of terrace **AF 22b**,   Fig. J.1
      unit Xgamma-1-9)
   Amphora or large jar, body sherd.
   Max. dim. 4.
   A fine fabric with mica inclusions (light red, 2.5YR 6/8).
   Comments: Venetian or Ottoman.

J-5   (X 1235, from the excavation of terrace **AF 22b**,   Fig. J.1
      unit Xgamma-1-9)
   Closed vessel, body sherd.
   Max. dim. 3.4.
   A coarse fabric (red, 2.5YR 5/6, to reddish gray, 5YR 5/2).
   Comments: Venetian or Ottoman.

Ottoman Period

J-6   (X 1656, from **AF 10** surface)   Fig. J.2
   Amphora, body sherd.
   Max. dim. 5.2.
   A coarse fabric (reddish yellow, 5YR 6/6). Almost straight profile.
   Comments: The shape has parallels from Eleftherna from as early as the 14th century, but the hard-fired paste suggests this piece is from the 17th to 18th century.

J-7   (X 1168, from the excavation of terrace **AF 22b**,   Fig. J.2
      unit Xgamma-1-10)
   Plate, base sherd.
   Diam. of base 8.
   A fine fabric (light reddish brown, 5YR 6/4), slipped, with a yellowish glaze over the slip. Ring base. Glazed with "random" drip pattern.
   Comments: Ottoman. Cretan production. A trace of the tripod trivet used in stacking is visible in the glaze in the interior of the vessel.

J-8   (X 1345, from **AF 32** surface)   Fig. J.2
   Flask, body sherd.
   Max. dim. 11.3.
   A fine fabric with inclusions (light red, 2.5YR 6/6).
   Comments: Found within the wall of the oval structure at **AF 32**. 18th century. Undecorated, without slip; from a flask with two handles (a water container).

J-9   (X 1307, from **AF 17a** surface)   Fig. J.2
   Small lekane, rim sherd.
   Diam. of rim 28.

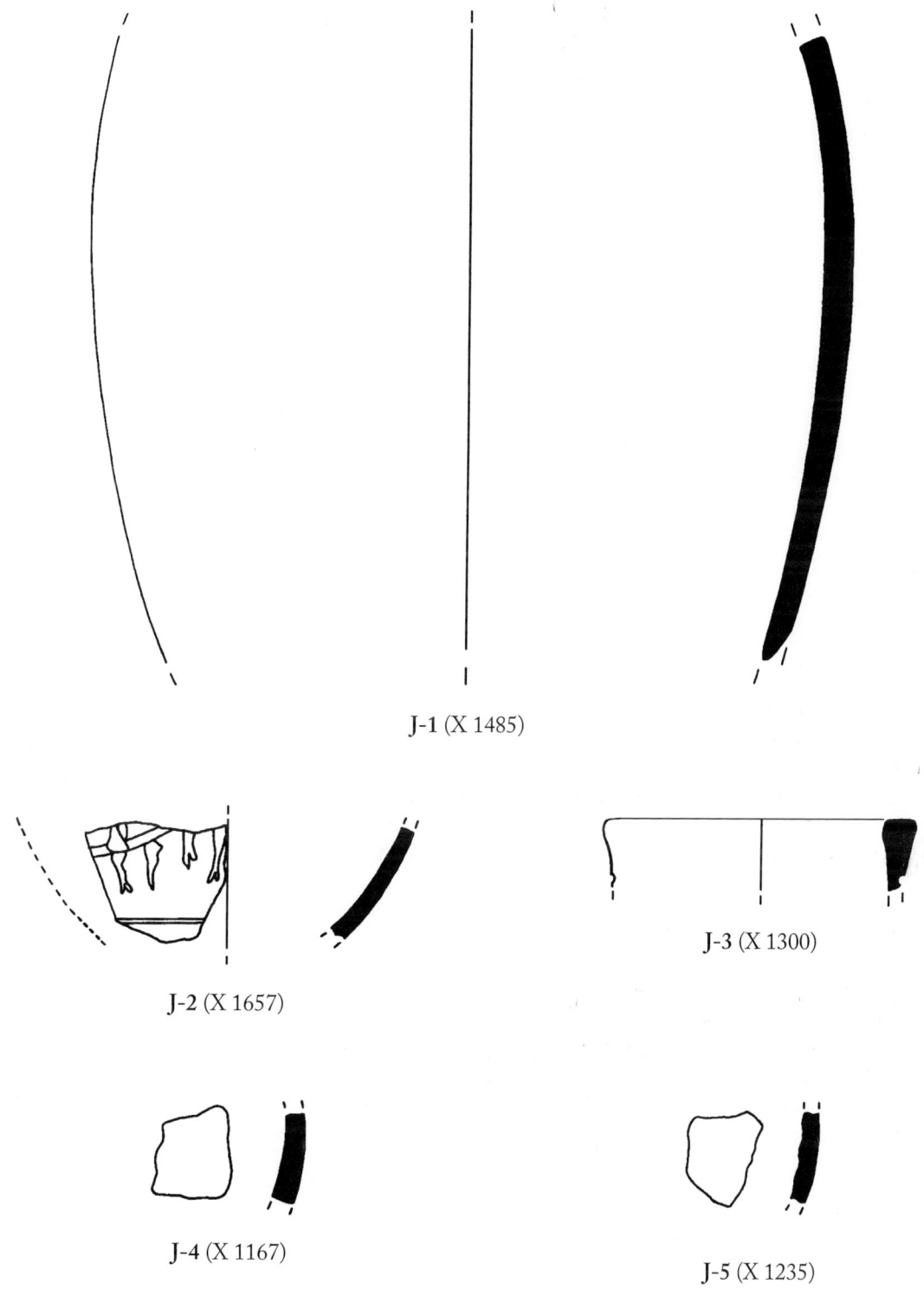

Figure J.1. Byzantine to Ottoman pottery from the surface of the territory (J-1 to J-5). Scale 1:3

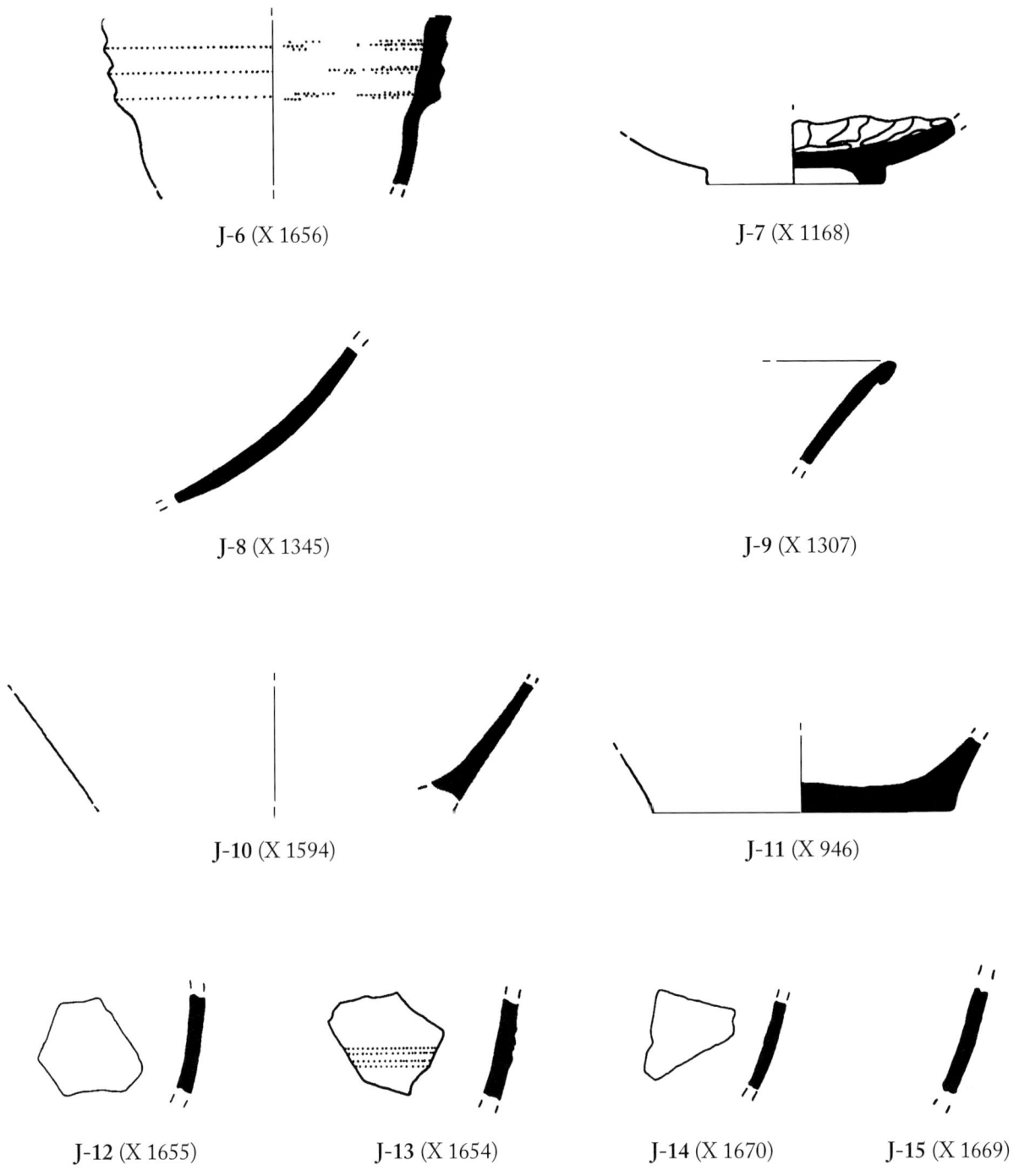

Figure J.2. Ottoman pottery from the surface of the territory (J-6 to J-15). Scale 1:3

A fine fabric (light red, 2.5YR 6/6), with yellow glaze (5Y 8/8) on the interior and exterior of the rim.
Thickened, outturned rim.
Comments: Ottoman, 19th century.

J-10   (X 1594, from the surface of unit XB-33-0)   Fig. J.2

Large lekane, sherd from the beginning of the base.
Max. dim. 9.8.
A fine fabric (dark reddish gray, 5YR 4/2) with inclusions; glaze (dark yellowish brown, 10YR 4/6) on interior.
Comments: Ottoman. Found on the surface of the Minoan habitation site. Glazed Ware (greenish glaze without slip).

J-11   (X 946, from the surface of unit XB-12-1)   Fig. J.2

Large lekane, base sherd.
Diam. of base ca. 10.
A fine fabric (reddish yellow, 5YR 6/6); yellow glaze (2.5Y 7/6) in interior.
Comments: Found on the surface of the Minoan habitation site. Glazed Ware without slip; this very rough ware is found until the 19th century; local Cretan production.

J-12   (X 1655, from **AF 10** surface)   Fig. J.2

Amphora, body sherd.
Max. dim. 5.
A coarse fabric (yellowish red, 5YR 5/8). Straight profile.
Comments: Ottoman.

J-13   (X 1654, from **AF 16** surface)   Fig. J.2

Closed vessel, body sherd.
Max. dim. 4.2.
A coarse fabric (yellowish red, 5YR 5/6). Incised bands.
Comments: Ottoman.

J-14   (X 1670, from **AF 17d** surface)   Fig. J.2

Flask, body sherd.
Max. dim. 5.3.
A fine fabric with inclusions (yellowish red, 5YR 5/6).
Comments: Ottoman.

J-15   (X 1669, from **AF 17d** surface)   Fig. J.2

Flask, body sherd.
Max. dim. 5.1.
A fine fabric with inclusions (red, 2.5YR 5/6).
Comments: Ottoman.

APPENDIX K

THE EXCAVATION OF CAVE AF 9 AND TERRACE AF 22b

by Brigit Crowell and Philip P. Betancourt

A small manmade cave (**AF 9**, Figs. K.1, K.2) and the second highest terrace in a series of step-terraces (**AF 22b**, Figs. K.3, K.4) were excavated in order to better understand the date and character of the agricultural plots near the habitation site. No sherds were present on the surface of either feature. The investigation revealed that both the cave and the terrace were used in Venetian or (more likely) Ottoman times rather than in the Minoan period.

THE EXCAVATION OF CAVE AF 9

A trench was excavated in 1997 in a small cave located near a large group of agricultural terraces (see Fig. 16.2, **AF 9**, and Fig. K.4). The goal of the excavation was to ascertain the function and phases of use for the small cavern and to describe the feature's shape, size, and other characteristics. This cavern is the largest manmade cave within the borders of the Chrysokamino territory, and it is near the Minoan habitation site, so the question of its possible use in the Minoan period was important to the interpretation of the area.

The cave faces north–northwest. It is carved out of soft phyllite, which has almost completely decomposed into brown soil. The entrance is carved into a low cliff (running roughly north–south), and it opens just above terrace **AF 22b**, providing easy passage from the terrace system. The width of the entrance is ca. 3.45 m, and the height is 1.53 m from the modern surface inside the cave to the upper center of the entrance. The sides of the cave mouth are ca. 0.95 m thick at the base, and they rise in an arch above the entrance. Inside the cave, the rock is roughly carved into a dome that is less than 2 m high. Areas of the interior appear to have been cut back to form an irregular oval area of floor, with two areas extending deeper on the north and south sides. The dimensions of the interior are 5.98 m from the entrance to the back wall, with a minimum width of 8.21 m north–south. The floor and the land just outside the entrance slope downhill from east to west.

A trench measuring 1 × 3 m was laid out on an east–west axis, with 2 m extending east into the cave and 1 m extending to the west outside

Figure K.1. Cave AF 9, looking south-southwest (1999)

Figure K.2. The cave's entrance

the entrance. After an initial clearing of modern debris, the trench was excavated in 11 passes of 10 to 15 cm deep levels. The first pass began at an elevation of 111.91 masl, and soil was removed to a depth of 1.79 m.

The soil was fairly consistent and homogeneous in all passes. It was a soft, fine-grained, powdery soil, pale yellow (2.5Y 7/4) when dry and light yellowish brown (2.5Y 6/4) when wet. The only indications of intrusion were found fairly close to the surface. In passes 2 and 3, at 111.80 masl, a lens of darker brown soil was present. It was found only in the eastern part of the trench. At a lower level (111.52 masl), a reddish, burned area with small fragments of charcoal was present. Natural soil was found directly above and below these two deposits. Despite their small size, the deposits are useful because they demonstrate that human or animal activity may be easily recognized within the environment of the cave. The only artifacts recovered were two fragments of thin, corroded iron found near the lower lens of darker soil.

The two lenses of different soil indicate that the cave was carved out in stages. In view of the consistently natural soil found in all low levels, it may be concluded that the cave is fairly recent, having nothing to do with the Minoan period. Its construction was probably associated with one or more of the periods of use of the nearby terrace **AF 22b**.

Figure K.3. Terrace group **AF 22**, looking north-northwest (1999)

Figure K.4. Terrace **AF 22b**, looking north-northwest (1999)

THE EXCAVATION OF AGRICULTURAL TERRACE AF 22B

Terrace **AF 22b** is part of a terrace system near the Minoan habitation site, on the southwest slope of the hill of Chomatas (see Fig. 16.2, **AF 22**). No olive trees grow on any of the terraces in this system, and there is no visible indication of recent agricultural use. It is the second highest manmade terrace in the system, and it is adjacent to the small cave **AF 9** (Figs. K.1, K.2).

A trench was excavated in this terrace to investigate the terrace wall's construction technique and its period of use. It was laid out at a right angle to the terrace wall on an east–west axis, 4 m east–west by 1 m north–south, extending across the level area of the terrace. The upper stones of the wall were missing, so the first few passes excavated only soil. No artifacts were on the surface before excavation.

The trench was excavated in 10 cm passes from the surface to a level close to the base of the wall. The excavation exposed both sides of the terrace wall. It was built with rows of stones of various sizes and irregular shapes, often having flat faces. The material was mostly dolomite, which occurs in large quantities on the surface just south of this location.

The Soil

In the upper levels of the trench, the soil was similar to the soil on the surface of the terrace. It was hard and dry, and it appeared brown when first excavated. It was a compact mixture of soil, phyllite pebbles, and larger stones. Modern snails and roots were found in the upper levels. The soil was pale yellow (2.5Y 7/4) when dry and light yellowish brown (2.5Y 6/4) when wet. It was easily distinguished from the natural soil of the hillside, which lay below it. The description and analysis of the soil is published in Appendix L.

Either the natural soil was cut back to form a depression for the browner soil that the wall supported, or the slope of the hillside was very shallow here before the wall was constructed. The bottom of the fill was higher at the east, where the natural soil was found at 107.75 masl (just over 1 m below the modern surface in the northeast corner). At the west, near the terrace wall, the natural surface was at 107.50 masl.

The Date of the Terrace

The upper courses of the preserved stones of the terrace wall were first encountered in pass 7, at an elevation of 108.10 masl. The excavation ended at 107.50 masl. No artifacts were recovered below a sherd found in association with the upper preserved course of the wall (J-7). Only three sherds were found in the excavation: J-4 (Venetian or Ottoman), J-5 (Venetian or Ottoman), and J-7 (Ottoman). This pottery indicates that the terrace was probably constructed in Venetian or Ottoman times. Ottoman construction of at least part of the wall is attested by the sherd J-7, which rested just behind the face of the wall below its uppermost preserved course.

Conclusions

The excavation helped clarify the use of the terrace behind wall **AF 22b**. It established the extent and depth of fill that the wall supports, as well as the characteristics of the terrace's soil and the underlying natural soil of the hill. The hill may have been cut back to form a depression for the agricultural soil, which was carried in from another location, possibly the nearby cave. The relative absence of intrusive material suggests that little activity occurred at any depth. If the soil had been repeatedly fertilized or redug, it is likely that more sherds or other types of artifacts would have been found in the soil.[1] The presence of an Ottoman sherd (App. J, J-7, X 1168) provides a date for at least the upper part of the terrace wall.

It appears that this system of terraces, which is one of the largest in the Minoan territory, was constructed at least partly in conjunction with Ottoman farming practices. The terrace and the nearby cave (**AF 9**) can probably be associated with the same phase of agricultural activity in the region.

---

1. One can contrast the situation in two Minoan terraces on Pseira, both of which had extensive evidence for activity within the soil; see Betancourt and Hope Simpson 1992.

Soils and Sediments from Natural Deposits at Chrysokamino

by Eleni Nodarou

This study investigates selected samples of soils and sediments from the archaeological site of Chrysokamino.[1] Using low-technology analytical methods, it examines particular physical properties of the sediments, including particle size distribution, organic material content, and magnetic susceptibility. The aim of the analysis is to investigate the properties of the natural deposits and provide comparative data for the archaeological sediments excavated at the Minoan farmhouse. The comparative study offers insights on the provenance of the archaeological soil material and an assessment of anthropogenic factors in soil formation processes. The work presented here forms part of my master's thesis.[2]

MATERIAL AND METHODOLOGY

Material used in the present study consists of 12 samples of ca. 500 g each that were collected in situ from an agricultural terrace (phyllitic sediment), a road exposure (terra rossa), and an alluvial deposit. Field observations were recorded for all locations.

Processing was conducted at the INSTAP Study Center for East Crete where the soil samples were dried, weighed, and sifted through a 2 mm sieve. All gravels were removed and collected, and the fine fraction was subsequently subsampled. The samples were labeled, sealed, and transported to the laboratory of the Department of Archaeology at the University of Sheffield. Before carrying out any analysis, all soil lumps were gently crushed with a mortar and pestle and sifted through a 2 mm sieve.

Analytical Procedure

All samples were analyzed for particle size distribution, organic material content, and magnetic susceptibility (Table L.1).

Particle Size Analysis

The textural properties of a deposit are produced by several factors, including contributions from the parent material, alteration during transport,

---

1. I would like to thank P. Betancourt for permission to carry out this study and for the invitation to publish part of it in this volume. This work has benefited greatly from guidance and advice from C. Frederick. Thanks are owed to F. McCoy for useful comments on an earlier draft of the paper. Sampling permits were kindly provided by the 24th Ephorate of Prehistoric and Classical Antiquities and the Institute of Geology and Mineral Exploration (IGME).

2. Nodarou 1998.

TABLE L.1. ANALYTICAL DATA FROM NATURAL DEPOSITS AT CHRYSOKAMINO

Samples	Mean Depth (cm)	LOI %	Magn. Susc.	Mz	S1	Gravels %	Sand %	Silt %	Clay %	Textural Class
AGRICULTURAL TERRACE										
Zone 1/Sample 1	7.5	3.9	18.32	2.13	4.94	32	25.0	30	13	loam
Zone 1/Sample 2	22.5	4.0	21.79	0.80	5.27	45	19.0	25	11	loam
Zone 2/Sample 3	40.0	3.0	26.44	1.86	5.23	36	24.0	25	15	loam
Zone 2/Sample 4	60.0	2.2	27.74	1.36	5.27	41	23.0	21	15	clay loam
Zone 2/Sample 5	77.5	4.4	28.70	2.20	5.50	38	17.0	27	18	clay loam
Zone 3/Sample 6	97.5	3.7	10.24	6.76	2.49	0.5	7.5	64	28	clay
Zone 3/Sample 7	120.0	2.1	7.71	6.50	2.11	0	5.0	73	22	silt loam
Zone 4/Sample 8	140.0	4.9	13.18	0.96	4.8	43	29.0	15	13	sandy clay
ROAD EXPOSURE										
Zone 1	12.0	9.3	193.16	2.90	5.03	29	28.0	24	19	loam-clay loam
Zone 2a	38.5	4.5	134.72	4.06	4.66	16	38.0	23	23	sandy clay
Zone 2b	38.5	6.4	179.40	2.50	5.66	40	23.0	17	20	clay loam
RED ALLUVIUM		7.9	248.38	8.03	3.02	0	9.0	34	57	clay

aspects of the depositional environment, and postdepositional alteration. Archaeological sediments are culturally deposited and modified, and they can therefore be treated as artifacts. The deciphering of these different inputs helps us to interpret the sedimentological conditions prevailing during the period of human activity and to develop a better understanding of the history of the area under study.[3]

Particle size analysis provides information about the percentage of sand, silt, and clay in sediments. The mean particle size and the sorting (i.e., the homogeneity of grain size distribution) are indicative of the nature of the depositional environment.[4] Particle size distribution reflects formation processes and transport modes, energy or velocity of the transporting medium, and energy or turbulence in the basin of deposition. Based on analyses of particle size distribution at the tell of Sitagroi, Davidson was able to draw conclusions about the tell's rate of buildup and breaks in the continuity of occupation.[5]

The limitations of the method should also be considered:[6]

1. Conclusions must take into account the small size of the sample in relation to the detail of the analysis. Archaeological levels cannot always be fully sampled, and therefore the sample may not be statistically representative of the entire level. This might result in a slightly skewed average.
2. Incomplete disaggregation, organics, and carbonates are factors that must be considered in the results of the analysis.
3. Artifacts and ecofact particles (ceramic sherds, bones, etc.) can often contribute significant weight fractions.

In this research project, particle size analysis has been employed for the description and comparison of the different sediments deriving from different locations and environments. The samples were homogenized through sieving. No ecofacts were discovered in the samples presented here.

3. Gladfelter 1977, p. 522; Aitken 1992.
4. Hassan 1978, p. 206; Aitken 1992.
5. Davidson 1973.
6. Canti 1995, p. 184.

The hydrometer method was used for grains of silt and clay size, and the dry sieve was used for sands and gravels. The entire procedure followed the Standard Method for Particle Size Analysis of Soils (SMPSAS). The mean particle size and the standard deviation of size distribution (sorting) were calculated and plotted against depth in each section.

Organic Material: Loss on Ignition

Organic material in sediments derives mainly from the decay of plant and animal residues. It comprises compounds of carbon and nitrogen, along with carbonates and charcoal. People are sedimentary agents, and human activity changes the physical and chemical traits of a soil. Therefore, in archaeological contexts, high quantities of organic matter may indicate significant contributions from plant and animal sources, horizons of human occupation, or an area of specific human activity. This organic material can be used for diet, climate, and landscape reconstruction. Vegetation, foodstuffs, housing, and roofing material are the most common organic imports.[7]

Among the limitations of the method, it should be mentioned that:[8]

1. Possible confusion can exist between natural organic matter accumulated as part of soil-forming processes and organic matter brought to the site by people during occupation.
2. Organic matter can be lost due to postdepositional alterations.

In the present research, the organic material was treated as a significant factor in the differentiation of the sediments under study and as a criterion for their provenance. The method used for the analysis was loss on ignition, which involved measuring the loss of weight after a 10 g split of the silt and clay fraction of each sample was placed in a muffle furnace for 24 hours at 400°C.

Magnetic Susceptibility

Magnetic susceptibility measurements are widely used in paleoenvironmental studies.[9] Sediments often contain magnetic features resulting from the influx of detrital allochthonous material, which can be used for lithostratigraphic correlations, studies of provenance, and reconstruction of environmental conditions. Human occupation and intervention in the natural environment can also be detected according to the fluctuations of magnetic values.[10]

Determination of magnetic susceptibility was undertaken using the standard MS2 Bartington meter with a Bartington dual-frequency sensor of the MS2B Type and an internal diameter of 36 mm. It was set to low frequency in the SI system, with the range multiplier knob set at 1.0. The samples were passed through a 2 mm sieve, and part of the fine fraction was used to fill the standard cubes for the measurement. Four measurements were taken for each sample, and the average of these values was calculated. The magnetic susceptibility for each sample was derived from the equation Average/(Mass of sample/0.01).

7. Aitken 1992; Stein 1992, p. 195; Ball 1964.
8. Stein 1992.
9. Mullen 1977.
10. Gale and Hoare 1991.

THE AGRICULTURAL TERRACE

Geoarchaeological studies on agricultural terraces have proved significant in providing information on ancient agricultural systems and practices as well as artifact distribution.[11] An agricultural terrace (**AF 22b**, Fig. K.4) was excavated at Chrysokamino in 1997 under the supervision of Crowell (App. K). It is situated at a short distance from the habitation site at Katsoprinos and belongs to the stepped type in the classification of Rackham and Moody.[12] The slope is fairly steep, and the purpose of the extensive terracing was to retain the soil, preserve moisture, and make the cultivation of the area more successful. Terracing was a common practice for the phyllite-quartzite soils of the region under study.

The terrace was investigated in order to record the succession of strata and to discern potential anthropogenic interference. Archaeological finds were scarce (only three sherds), but they provide secure dates for the topographic feature.

The terrace consisted of four zones. Each one was subdivided into two or three subzones and sampled separately according to field observations. The criteria for separating the zones and subzones were soil color (using Munsell color charts), uniformity in color and texture, structure, and boundary differentiation. The samples were sifted through a 2 mm sieve, and the fine fraction was analyzed for organic material (loss on ignition), magnetic properties (magnetic susceptibility), and particle size distribution.

Field observations are reflected on the stratigraphic column (Fig. L.1). Zones 1 and 2 consist of gravelly silty loam of pale yellow color. The individual characteristics and the boundary of Zone 1 are indicative of an Ap horizon, a humic surface with mixed gravels and organic matter. In regard to Zone 2, it is important to investigate whether the abundance of coarse material is a result of colluviation processes or of anthropogenic interference. The analytical data clearly indicate the existence of three different types of sediment:

1. An anthropogenic stratum (0–0.85 m)
2. A natural stratum (0.85–1.3 m)
3. Weathered bedrock (1.3–1.7 m)

Zone 1/Sample 1 bears all the characteristics of a cultivated A horizon. These characteristics include high values for organic content and magnetic susceptibility and the predominance of fine and extremely poorly sorted sand (Mz = 2.13 phi). In comparison with lower levels, particle size distribution shows a greater quantity of gravels, similar quantities of sand and silt, and less clay. Lower values of organic content and magnetic susceptibility, as well as seemingly better sorting, occur deeper than the humic surface of a natural deposit. The analysis for Zone 1/Sample 2 and Zone 2/Samples 3, 4, and 5 show a general decrease in the content of organic matter, a mean particle size of coarse to medium sand, and an extremely poor sorting. At the bottom of Zone 2/Sample 4 (ca. 70 cm below the surface), magnetic properties increase, with a culminating point at the same depth and an increase in the percentage of gravels, while the sorting remains the same. All these properties indicate that the sediment was not the product of in situ weathering; its formation was due to anthropogenic activity.

11. Betancourt and Hope Simpson 1992; Bull, Betancourt, and Evershed 1999, 2001.

12. Rackham and Moody 1996.

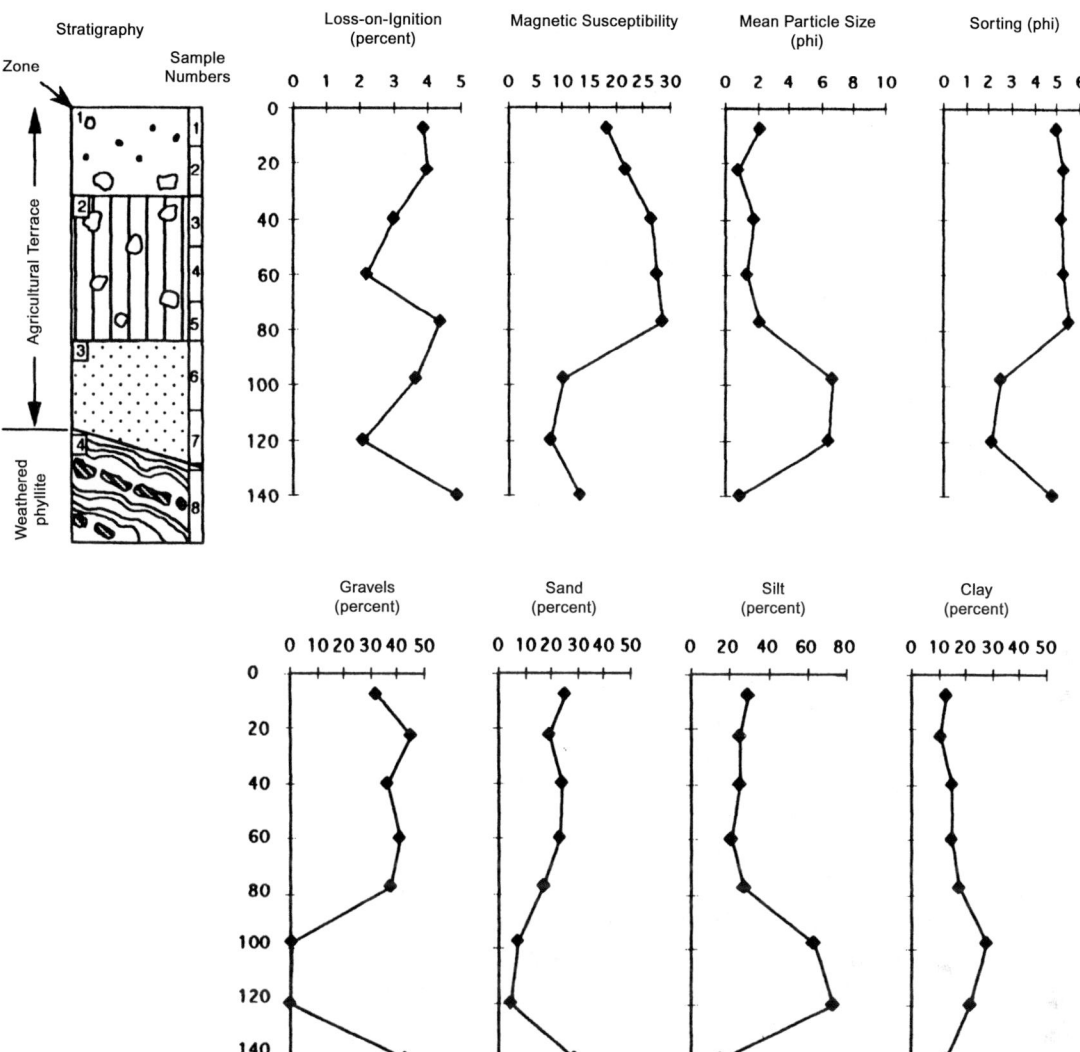

Figure L.1. Analytical data for agricultural terrace AF 22b

The archaeological evidence, though scarce, also favors an anthropogenic interpretation for the sequence. The excavation of the terrace started at the depth of 108.82 masl and was terminated at a depth of 106.86 masl. At the depth of 107.75 masl (excavation unit XD-1-9), which corresponds to Zone 2/Sample 4 of the stratigraphic column, the excavation discovered a body fragment from an unpainted closed vessel (App. J, J-4). It is from the Venetian or Ottoman period. In the next lower stratum (XD-1-10), a base from a plate was found (App. J, J-7). It is Ottoman, dating to the 18th century. Although the possibility of Minoan use cannot be excluded, in the absence of any relevant archaeological evidence, the Venetian to Ottoman finds afford a better date for the construction and use of the terrace. This means that all the inputs and disturbances of the natural soil for the construction of the terrace were Venetian–Ottoman or later.

The boundary between Zone 2/Sample 5 and Zone 3/Samples 6 and 7 marks a considerable change in the nature and properties of the deposit. These strata constitute the soil formed in situ from the weathered bedrock. This level consists of silty and well-sorted phyllitic material. Little organic matter is present, and the values of magnetic susceptibility are low. Little gravel is present, but the amount of clay is higher than in the other strata.

The lowest stratum (Zone 4/Sample 8) consists of partially weathered phyllite and intercalating veins of calcite. It constitutes the base on which anthropogenic activity took place, and it is close to bedrock. This is why there is an increase in the content of gravelly material at the expense of sand and silt. The sorting is extremely poor, and the mean particle size is that of coarse sand (Mz = 0.96 phi).

THE ROAD EXPOSURE

Between Lakkos Ambeliou and the top of the hill where the Minoan farmhouse is located, the landscape is dominated by reddish sediment (terra rossa) over dolomite bedrock (Fig. L.2). The bed is relatively solid in some parts and more weathered in others. An exposure of sediment was sampled to examine the characteristics of this sediment and to provide comparative material for the archaeological sediments from the Minoan farmhouse (Figs. L.3, L.4).

The upper zone (Zone 1, 0–0.24 m below surface) consists of an A horizon that includes remains of modern vegetation (such as roots) and a fair amount of coarse lithic material. As expected, the analyses show high values for organic content and magnetic susceptibility, while the particle size distribution indicates extremely poorly sorted sediment with a mean particle size of very fine to fine sand (Mz = 2.90 phi). Gravels, sand, and silt share almost the same percentage, whereas clays are present in a smaller percentage.

The second zone (Zone 2, ca. 0.24–0.53 m below surface) was divided into two parts that were sampled separately because field observations indicated an area of different soil color. The first sample (Zone 2a) has a dark red color, a low content in organic material, and a low value of magnetic susceptibility. The hydrometer analysis shows that the sample consists of

Figure L.2. The sediment deposit at Lakkos Ambeliou (terra rossa), with the habitation site in the distance

SOILS AND SEDIMENTS FROM NATURAL DEPOSITS 409

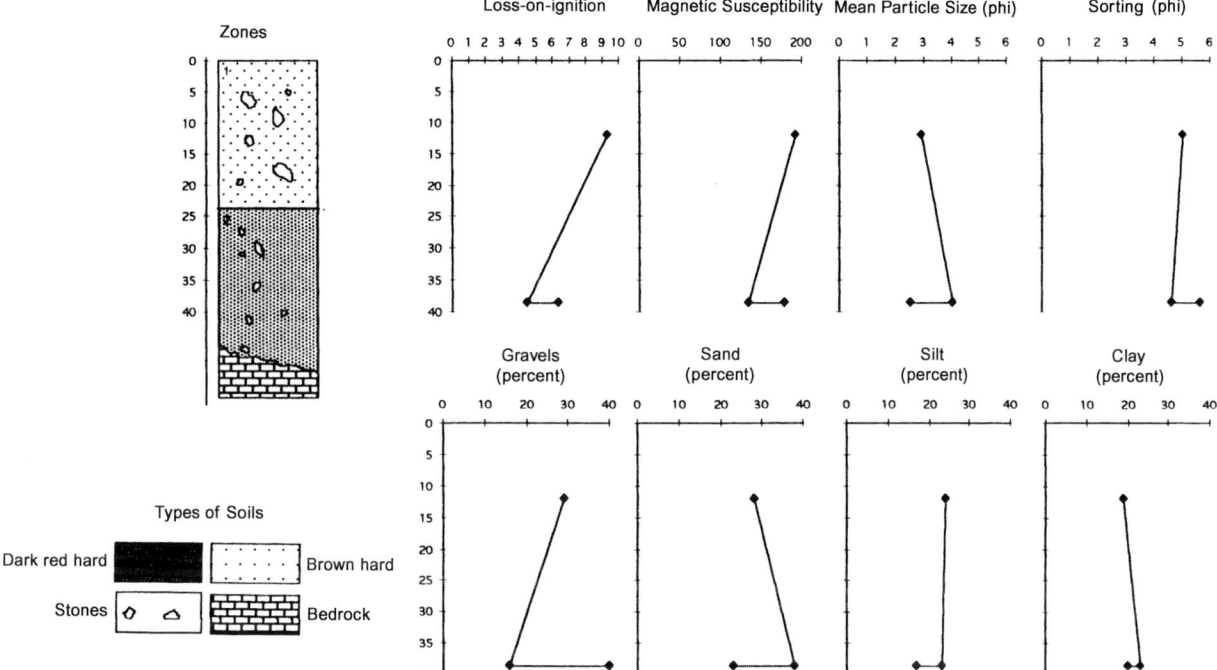

Figure L.3. Analytical data for the road exposure

Figure L.4. Locations of the two sets of analyzed soil samples. Aerial photograph taken at 6:52 a.m. on June 5, 1988. © 1999 Aerial Images, Inc., and 1999 SovInformSputnik

an extremely poorly sorted sediment with a mean particle size of very fine sand to coarse silt (Mz = 4.06 phi). The percentage of gravel-sized particles is lower in this part of the exposure, while sand-sized particles are predominant. Silt-sized and clay-sized particles share similar values.

The differentiated part of the zone (Zone 2b) has a yellowish brown color and higher values of organic content and magnetic susceptibility. The examination of particle size distribution shows that the sample is very poorly sorted, with a mean particle size in the fine sand-sized range. The percentage of gravels varies up to 40%, while sand, silt, and clay fractions vary up to 23%, 17%, and 20% respectively.

Beneath Zone 2 lies the partly weathered bedrock. Sample 2a may contain the products of weathering. Sample 2b is related to the Bt horizon, which explains its higher values.

THE RED ALLUVIUM

The alluvial deposit at Lakkos Ambeliou (Fig. L.2) constitutes a Vertisol, and its properties have been discussed in detail by Morris.[13] The present study was undertaken to compare the properties of this natural deposit with those of a thick deposit of earth excavated in one of the trenches of the Minoan farmhouse (XB-9). The red color and the clayey texture of that deposit contrasted markedly with the white phyllitic sediments found in the other trenches. The aim of the analysis was to elicit information about its provenance and function.

The alluvium at Lakkos Ambeliou was the most obvious source to be tested since it displays the same macroscopic characteristics, the red color, and the fine, clayey texture. Due to the textural homogeneity of the alluvial deposit, a single sample was taken. The analysis shows high values of magnetic susceptibility and organic content, very poor sorting, and a mean particle size near the boundary between very fine silt and coarse clay (Mz = 8.03 phi). The different size classes are dominated by clay-sized particles, which excludes an aeolian source. According to Pye,[14] soils that are not deposited by aeolian action are predominantly silty, containing 10%–25% sand and up to 40% clay, which approximates the case for the deposit at Lakkos Ambeliou. The high water content of the clay particles makes them too heavy to have been carried in large quantity by the wind. The analysis indicates that the properties of the clayey alluvium from Lakkos Ambeliou are similar to those of the archaeological sediment of the Minoan farmhouse. The concentration of this red soil in the archaeological context has been interpreted as a collapsed mudbrick wall.[15]

COMMENTS

The study of the natural deposits from Chrysokamino using a combination of macroscopic observations and low-technology sedimentological analysis offers important information on the properties of various natural deposits. By comparing the values of these samples with those from the

13. Morris 2002.
14. Pye 1992.
15. Nodarou 1998.

archaeological sediments excavated at the Minoan farmhouse, it has been possible to differentiate between natural and anthropogenic deposits and to elucidate the provenance of the latter.

In the case of the agricultural terrace, the analysis permits the distinction between natural deposits and levels affected by human activity. The presence of a pottery fragment of the Venetian/Ottoman period provides a chronological indication for the construction of the terrace. The analysis is informative regarding the provenance of the archaeological material, since the gray white earth excavated from most areas at the Minoan farmhouse shares similar properties with the phyllitic sediments of the agricultural terrace.[16]

The analysis of two red deposits, the red alluvium from the road exposure and the Vertisol at Lakkos Ambeliou, had a twofold aim: to examine the natural properties of those deposits and to compare them with the red earth excavated at the Minoan farmhouse. The former has been shown to be a natural deposit, devoid of human interference and somewhat different from the clayey deposit at Lakkos Ambeliou. Another study has indicated that the latter contains an anthropogenic component,[17] but the present study demonstrates its connection to the Minoan farmhouse. It appears to have provided the raw material for the construction of a mudbrick wall that collapsed and disintegrated when occupation ceased.[18]

The use of basic sedimentological analytical techniques has proved to be an efficient tool in the combined and comparative study of natural and archaeological deposits. This approach provides insights on the provenance of the archaeological sediments, the use of space in archaeological contexts, and the interference of the human agent in natural deposits.

CATALOGUE OF SAMPLES

Samples are listed in the catalogue by zones, with Zone 1 (the upper zone) first. Colors are recorded first as dry, then moist.

Agricultural Terrace

Zone 1

Depth: 0–0.33 m.
Munsell colors: 5Y 7/3 (pale yellow); 5Y 5/3 (olive).
Texture: silt-loam.
Consistency: slightly hard.
Lower boundary: clear smooth.
Structure: medium coarse, subangular blocky, moderate.
Roots: few fine, common large.
Horizon: Ap.
Comments: 35%–40% coarse fragments (<2 mm).

Zone 2

Depth: 0.33–0.85 m.
Munsell colors: 5Y 7/3 (pale yellow); 5Y 5/3 (olive).
Texture: gravelly silt-loam.

16. Nodarou 1998.
17. Morris 2002.
18. Nodarou 1998.

Consistency: slightly hard to hard.
Lower boundary: abrupt.
Structure: medium coarse, subangular blocky, moderate.
Roots: few fine, few very fine.
Comments: 50% coarse fragments (<2 mm).

Zone 3

Depth: 0.85–1.3 m.
Munsell colors: 5Y 8/3 (pale yellow); 5Y 6/3 (pale olive).
Texture: silt-loam.
Consistency: slightly hard.
Lower boundary: abrupt, wavy.
Structure: platy coarse to very coarse.
Roots: common fine.
Comments: <1% coarse fragments (<2 mm); bedding; in situ weathered phyllite, the structure breaking along phyllite formation planes.

Zone 4

Depth: 1.3–1.7 m.
Munsell colors: 5Y 7/3 (pale yellow); 5Y 5/3 (olive).
Lower boundary: approaching bedrock.
Structure: none.
Roots: none.
Comments: Saprolite, integral phyllite weathered product. The phyllite foliation is visible; there are zones/veins of calcite and large aggregates.

Road Cut

Zone 1

Depth: 0–0.24 m.
Munsell colors: 7.5YR 4/4 (brown); 7.5YR 3/4 (dark brown).
Texture: silt to clay.
Consistency: hard.
Lower boundary: clear.
Structure: moderate medium subangular blocky breaking to smaller granular grains.
Roots: common.
Horizon: A.
Comments: 40%–50% coarse fragments.

Zone 2

Depth: 0.24–0.53 m.
Munsell colors: Sample a, 2.5YR 4/6 (dusky to dark red), 2.5YR 3/6 (dark red); Sample b, 5YR 4/6 (yellowish red), 5YR 3/4 (dark reddish brown).
Texture: clay.
Consistency: hard.
Lower boundary: abrupt, wavy.
Structure: very fine, fine subangular blocky, moderate, strong.
Roots: few, fine.
Horizon: Bt.
Comments: 25%–35% coarse fragments; some parts are rocky up to 70%.

Organic Residue Analysis of Pottery Sherds from Chrysokamino

by Ruth F. Beeston, Joe Palatinus, Curt Beck, and Edith C. Stout

We report on the results of organic residue analysis of twelve pottery sherds from the Early Minoan III–Middle Minoan IA site of Chrysokamino. These sherds were excavated from in and near a small structure located in close proximity to copper smelting furnaces where, according to evidence described elsewhere in this volume, arsenical copper was produced. Vessel types represented in this group of sherds include jars, shallow bowls, a rounded cup, and a jar or bellows.

Organic residue analysis is a destructive method in which pottery sherds are ground and samples for analysis are extracted with solvents capable of dissolving a wide range of organic compounds. Organic molecules that were absorbed and protected from degradation by the porous matrix of the ceramic material are thereby released into solution. The technique of gas chromatography-mass spectrometry (GC-MS) is an effective method for the separation, detection, and identification of the minute quantities of often numerous organic compounds present in the sherd extracts. Gas chromatography serves to separate the complex mixture of chemicals by passing it through a heated column in a flowing stream of helium gas. Different compounds require different lengths of time to traverse the column and therefore emerge and are detected at characteristic times (known as retention times). A plot of signal versus time (the total ion chromatogram, or TIC) is generated. As they emerge from the column, molecules are passed into the mass spectrometer, in which they are ionized and broken apart by collisions with high-energy electrons. A collection of fragment ions is produced, and their masses are displayed in a mass spectrum, giving clues about the molecular structure of the compound. The mass spectrum can be compared to spectra of thousands of reference compounds using a computer-searchable database. In some cases, an exact determination of the identity of a compound can be made based on a matching pattern of peaks in the mass spectrum. In other cases, one can readily determine the class of organic molecules to which a compound belongs (fatty acid ester, straight chain alkane, methyl ketone, etc.), and that information, along with the retention time, provides a certain identity. The retention times of known compounds can be measured independently under the same conditions in order to provide confirmation for the identity of a compound. If a peak cannot be assigned to a particular compound, its mass spectrum is still recorded in the hopes that additional information will allow an assignment in the future.

The organic compounds found by GC-MS in pottery sherd extracts can be attributed to three sources. The first source, and the one of primary interest, are those molecules that came in contact with the pottery vessel during its use in ancient times; these compounds were absorbed into the surface and interior of the porous ceramic material where they remained preserved and protected from light and oxygen. With careful interpretation, these compounds provide clues about the original contents and use of the vessel. Unfortunately, human contact during excavation, cleaning, handling, and storage of the sherd can contaminate the sample by introducing a second source of organic compounds. These compounds include plasticizers (from contact with plastic buckets, tools, and bags), chemical sunscreens, cosmetic products, and cleaners, all of which can contribute to the mixture of molecules detected during organic residue analysis. Finally, the laboratory procedures involved in preparation of the sample extract unavoidably introduce an additional source of contaminants. Although in most cases these contaminants are readily recognized and discounted, their presence complicates the analysis, and it could mask the presence of molecules of archaeological significance.

From the list of organic compounds revealed by GC-MS to be present in a sherd, the chemist can attempt to draw conclusions about the nature of the original vessel contents. This difficult step involves examining known botanical (and other) sources of specific compounds and combinations of compounds. Although much is known about the chemical composition of various substances (wine, olive oil, plant extracts, etc.), not all botanical species have been analyzed, and the nature of long-term changes in composition is still poorly understood. Therefore, one must be cautious about overinterpreting the chemical results. Organic residue analysis can provide evidence for the likely use of certain "ingredients" based on the collection of molecules that have been detected. However, the absence of evidence for a specific compound does not rule out its presence; various factors, including extremely low concentrations, can prevent the detection of a compound.

In an effort to assess the reproducibility and limitations of organic residue analysis, and ultimately to enhance the information obtained, three sherds in this study were analyzed in duplicate in separate laboratories equipped with different GC-MS instruments. As discussed in subsequent sections, the collection of molecules confirmed to be present showed some variation.

EXPERIMENTAL PROCEDURES

Sample Preparation

A 3–5 gram sample of each pottery sherd was obtained. Table M.1 provides the sherd number, vessel type, description of ceramic material, and the mass of sample for each of the 12 sherds analyzed in this study. In some cases the entire sherd was used as provided; for sherds weighing >5 g, an

TABLE M.1. DESCRIPTIONS OF SHERDS ANALYZED FOR ORGANIC RESIDUES

EUM No.	Vessel Type	Description of Ceramic Appearance	Wt. of Sherd (if known)
521	bucket jar	brown, grainy	3–5 g
522	bucket jar	thick, light brown with thin light layer on one side	3–5 g
523	jar	reddish and white on one side, black on other	3.08 g
524	shallow bowl	two pieces; brown, grainy with reddish layer one side	3–5 g
525	jar or basin	brown on one side, red on other side	4.11 g
526	jar (or bellows?)	red on outside, brown inner layer; white and black particles	3.10 g
527	jar	red, darker on one side	2.82 g
528	bucket jar	brownish red, darker on one side	3.95 g
547	jar	red with gray sections	3.98 g
548	rounded cup		3–5 g
549	shallow bowl	gray in middle, light red on one side	4.21 g
550	jar, probably a bridge-spouted jar		3–5 g

appropriately sized sample was obtained by breaking the sherd. To prevent contamination, sherds were handled with forceps and were wrapped in paper during breakage.

Each sample was ground for five minutes in a stainless steel mill, and the powder was placed in an unused glass vial equipped with a teflon-lined cap. To the powder, 5 ml of dichloromethane (Aldrich 99.9+%, PRA grade) and 5 ml of diethyl ether (Aldrich anhydrous 99+%, ACS reagent grade) were added. To facilitate extraction, the mixture in the vial was sonicated in an ultrasonic bath for 10–15 minutes at intervals during a two-day period. The temperature of the water in the ultrasonic bath was kept below 40°C. The mixture was transferred to a clean glass centrifuge tube and centrifuged for 3–5 minutes. The clear liquid extract was removed with a glass disposable pipette and placed in a new vial. The volume was reduced to about 1 ml by partly immersing the vial in a 30–35°C water bath and directing a stream of nitrogen gas onto the surface of the extract. The extracts were treated with an ether solution of diazomethane generated from Diazald™ and KOH. Vials containing treated extracts were loosely capped and kept in a ventilated hood for 24 hours. The volume was again reduced under a stream of nitrogen gas until approximately 0.1 ml remained to produce a sufficiently concentrated sample for GC-MS analysis.

Instrumental Analysis

All 12 sherds were analyzed at Davidson College; three of these (EUM 521, 524, and 526) were also analyzed at Vassar College. Table M.2 shows the instrument parameters in both laboratories.

At Davidson College, peaks were identified by comparison of the mass spectrum to those contained in the NIST mass spectral database using the library search capabilities of the SatView software. In addition, authentic samples of normal hydrocarbons, saturated methyl esters, aldehydes, alcohols, ketones, and others were injected under the same conditions for comparison of retention times and mass spectra. These standard compound mass spectra were added to the searchable library. We do not report peaks recognized as contaminants (including plasticizers, modern synthetic chemicals, and impurities introduced in the laboratory). However, some unusual molecules likely to have been introduced through handling are mentioned.

At Vassar College, identification was made using the Wiley Registry of Mass Spectral Data, 6th ed., with the search system Benchtop PBM (Palisade Corporation), as well as an in-house MS DataBase listing standards of authentic samples and spectra from the literature. The reports of the Vassar analyses include some unusual extraneous compounds (but not the ubiquitous contaminants from plastic bags and from laboratory chemicals) whose identification may be helpful to other workers in the field.

Because the instruments at Davidson and Vassar are equipped with different columns and different types of mass analyzers, retention times of individual compounds differ, as do some features of their mass spectra. Some compounds in the three sherds analyzed in both laboratories were detected on both systems, while other compounds were detected and identified on only one of the instruments. It appears that more slowly eluting compounds, such as long chain fatty acid esters and higher alkanes, were detected more readily on the instrument at Vassar. Clearly, if a compound is detected on both systems, that substance is certain to be present in the sherd. But the comparative results between our laboratories indicate that failure to detect a particular compound in one laboratory should not be interpreted as an indication that the sherd did not contain that compound. We have found that different portions of the same sherd, even when analyzed using the same instrument and identical conditions, may differ quite dramatically in the organic residue components that can be detected. We believe, therefore, that all the constituents found in either laboratory are actually present in the three sherds that were analyzed both at Davidson and at Vassar.

RESULTS

Organic residue analysis by gaschromatography-mass spectrometry allows the detection and identification of compounds that are soluble in the nonpolar extraction solvent (ether/dichloromethane) and sufficiently volatile to pass through the gas chromatograph. Polar, water-soluble compounds and nonvolatile, high molecular weight organic molecules (for example, salts, carbohydrates, and other macromolecules) are not detected. In this section we provide a summary of the findings for each of the 12 sherds from Chrysokamino. Data including retention time and mass spectra (molecular ion, base peak, fragment ions) will be included in a later publication.[1]

EUM 521. Washed sherd from a bucket jar, **100** (X 2)

Seven compounds were identified in the extract of sherd EUM 521, including three alkanes, camphor and two related compounds, and triacetin. A large camphor peak is evidence that this sherd contained, or was exposed in modern times, to a medicinal preparation. Well over 100 occurrences

1. Beeston et al. forthcoming.

TABLE M.2. GC-MS INSTRUMENT PARAMETERS

	Davidson College	Vassar College
GC-MS System	Varian Saturn 2000 GC/MS/MS	Hewlett-Packard 6890 GC
	Varian 3800 GC	Hewlett-Packard 5973 MSD
		Hewlett-Packard MSD ChemStation
Mass Spectrometer	ion trap	quadrupole
GC Column	Phenomenex ZB-5	Hewlett Packard HP-1 MS
	30 m × 0.25 mm	15 m × 0.25 mm
Stationary Phase	poly(5% diphenyl; 95% dimethyl-siloxane), 0.25 μ	poly(dimethylsiloxane), 0.25 μ
Split Mode	1:10	1:10 and splitless
Sample Injection	1 μL	1 μL
Injector Temperature	300°C	250°C
Carrier Gas	Helium; flow rate 1.2 ml/min.	Helium; flow rate 1.2 ml/min.
Column Temperature Program	initial 50°C; held for 1 min.; ramped to 250°C at 5°/min.; held for 19 min.	initial 50°C; immediately ramped to 250°C at 5°/min.
Total Run Time	60 min.	40 min.
Ionization	electron impact; 70 eV	electron impact; 70 eV
Filament Delay	2.5 min.	1.5 min.
Masses Recorded	35–500	40–550
Base-Line Correction	yes	yes
Scans Averaged	3 to 5	3 to 5

of camphor are reported by Duke,[2] and this list contains many herbs, including varieties of rosemary, savory, thyme, lavender, basil, hyssop, sage, licorice, and mint. The Merck Index lists several modern industrial uses of camphor, as well as its use as a topical anti-infective and antipruritic. Smaller peaks were identified as the related compounds, campholenal and endo-isocamphonone. Campholenal has been reported in *Thymus funkii* (thyme) and *Helianthus annuus* (sunflower), a native of North America, but no occurrences of endo-isocamphonone are listed[3] or revealed by a *Chemical Abstracts* search.

No conclusions can be based on the presence of the three alkanes, tetradecane (C14), heptadecane (C17), and octadecane (C18). Plant sources of these hydrocarbons are known,[4] but these compounds might also have been introduced by contact with a mineral oil or other petroleum-based lotion or ointment.

The presence of triacetin in this sample is likely to be due to contamination. The only botanical source reported for triacetin is the fruit of *Carica papaya*, which does not occur in the Mediterranean region. It is an antifungal agent (also known as Enzactin and Fungacetin). Therefore, it is a possibility that this constituent in EUM 521 was introduced through handling.

As these results illustrate, a common problem arising in the interpretation of organic residue analysis is the lack of information about a sherd's recent history. While plasticizers and other modern synthetic compounds can be easily recognized and ignored, natural ingredients introduced via lotions, perfumes, insect repellents, and sunscreens cannot always be

2. Duke 2002.
3. Duke 2002.
4. Duke 2002.

distinguished from compounds arising from ancient sources. This points to the need for minimal handling of sherds to be analyzed by archaeologists, conservators, and illustrators and for avoidance of products that might introduce contaminants through contact with sherds.

EUM 522. Sherd from a bucket jar, **101** (X 169)

In this sherd, eight components were identified, including camphor, two alkanes, four aldehydes, and a 2-ketone. No fatty acid methyl esters were detected in the sample analyzed at Davidson. It is possible that incomplete esterification occurred during the methylation procedure, as the Vassar results (described below) revealed that fatty acids were present. As in EUM 521, camphor was detected, but in much lower quantities. Camphor occurs in many medicinal and aromatic herbs; its presence cannot be used to pinpoint a certain plant source. The aldehydes nonanal, decanal, undecanal, and dodecanal can each be attributed to numerous plant sources (28, 23, 5, and 9 occurrences respectively are listed by Duke).[5] Several plants, most notably *Coriandrum sativum* (coriander) and several species of *Citrus* are reported to contain at least three of the four aldehydes observed in EUM 522. Since citrus fruits were not introduced in the Mediterranean until the Middle Ages, these can be ruled out; coriander was known and used as a flavor agent during the Bronze Age.[6] The two hydrocarbons, undecane and dodecane, are common to *Glycyrrhiza glabra* (licorice root), *Pimenta dioica* (allspice, a native only of central America and the West Indies), and *Cymbopogon parkeri* (Skhabar). The ketone, 2-undecanone, has ten known plant sources, including rue and oregano. Although exact botanical sources cannot be pinpointed, these results are consistent with the use of aromatic herbs in the preparation of flavored wine. Pine resin acid esters were not detected at Davidson.

Sherd EUM 522 was also analyzed at Vassar College. Of 24 components, 16 could be identified as ancient constituents, 5 as modern contaminants, and 3 remain unassigned.

The largest group of ancient constituents is a group of 12 fatty acids. They include the nine saturated fatty acids caprylic acid (C8), pelargonic acid (C9), capric acid (C10), undecanoic acid (C11), tridecanoic acid (C13), myristic acid (C14), pentadecanoic acid (C15), palmitic acid (C16), and stearic acid (C18). The shorter, and especially the odd-numbered, members of this series are likely to be degradation products of the larger ones. More significant is the presence of the three unsaturated fatty acids, palmitoleic acid (C16:1), oleic acid (C18:1 cis), and elaidic acid (C18:1 trans). They are a certain indication of plant oils. Olive oil has this composition, but other plant oils cannot be excluded.

The presence of two diterpene resin acid esters, methyl dehydroabietate and its 7-keto-derivative, is a certain indication of pine resin. Its only known use is as an additive to wine to produce retsina.

A trace of borneol, identified with probability but not certainty, is a plant product that is too ubiquitous to assign a more specific origin. Duke lists no fewer than 141 plants containing this compound, including many conifers, but also herbs, e.g., oregano.[7] Biers and colleagues found borneol in two of the Corinthian "plastic" vases they studied.[8]

5. Duke 2002.
6. Mabberley 1997, p. 146.
7. Duke 2002.
8. Biers, Gerhardt, and Braniff 1994, p. 26.

The certain identification of acetanilide poses a puzzle. This compound is not listed as a natural plant product in the compilations of Duke or Karrer.[9] A complete *Chemical Abstracts* search yields only one animal source (the sea urchin *Temnopleuros toreumaticus*) and one plant source *(Peganum harmala)*. The latter is native to the Mediterranean and Asia; it contains hallucinogenic alkaloids[10] and has been used as an intoxicant possibly since the 5th millennium B.C.; it also has a long history in medicine to treat eye diseases, rheumatism, and Parkinson's disease.[11] However, acetanilide has also been reported in numerous analyses of modern waste water, into which it is probably leached by the biodegradation of a number of modern herbicides.[12] Since the presence of *Peganum* alkaloids cannot be confirmed by our method of analysis, it remains uncertain whether acetanilide is an ancient constituent or a contaminant. We have found it previously in a tripod cooking pot from Chania, Crete.

Of five certain and uncommon contaminants, three are degradation products of plasticizers with a 2,2,4-trimethl-1,4-pentanediol structure. 2-Phenoxyethanol is a known ingredient of suntan lotions, and ditolyl ether is certainly a modern synthetic compound, albeit of unknown origin. Together, these constituents suggest a resinated wine with aromatic or medicinal ingredients that was protected from air by a layer of oil (probably olive oil) to prevent its oxidation to vinegar.

EUM 523. Sherd from a jar, 103 (X 211)

The most prominent constituent found in the extract of EUM 523 is camphor. As described previously, this medicinal compound has numerous plant sources, but may also arise from contact with a modern topical preparation. Other organic compounds found in smaller amounts include the alkanes undecane, dodecane, tetradecane, pentadecane, and octadecane. These relatively low molecular weight saturated hydrocarbons provide no diagnostic information. *Glycyrrhiza glabra* (licorice) and *Pimenta dioica* (allspice) are the only plants in which all of these hydrocarbons occur,[13] but the latter possibility is not indigenous to Crete. Other possible origins for the alkanes include contact with plastic or a petroleum product (mineral oil in skin lotion). Two aldehydes and a ketone were also found to be present. Nonanal, decanal, and 2-undecanone (all of which were also detected in EUM 522) each have numerous plant sources, including citrus fruits and fragrant and medicinal herbs. Citrus can be ruled out based on its much later introduction into the region.

EUM 524. Sherd from a shallow bowl, 94 (X 270)

EUM 524 contained a rich collection of organic molecules. A total of 19 compounds, not including known contaminants, were identified. Many of these, including camphor, the alkanes tetradecanae and octadecane, and the aldehydes nonanal and decanal, have been discussed previously. The methyl esters of hexanoic acid (caproic acid, C6), octanoic acid (caprylic acid, C8) and nonanoic acid (pelargonic acid, C9) were identified. These acids are present in a variety of plants (all occur in licorice root, hot pepper, and cinnamon),[14] and the C6 and C8 acids occur in the fat and milk of sheep and goats.[15] Nonanoic acid may arise from the decomposition

9. Duke 2002; Karrer 1958–1981.
10. Harborne, Baxter, and Moss 1999, pp. 204, 213.
11. Mabberley 1997, p. 437.
12. Stampert and Tuorinen 1998.
13. Duke 2002.
14. Duke 2002.
15. Beck et al. 2001; Beck forthcoming.

of oleic acid, an important constituent of olive oil, but this is an unlikely source in this case, since higher molecular weight fatty acids were not detected. Other methyl esters identified in EUM 524 are those of benzoic acid and 2-ethylhexanoic acid. Benzoic acid is sometimes found in our "blank" samples, indicating that this may be a laboratory contaminant in this case. However, it also has a large number of plant sources[16] and has been previously noted in pottery sherds as a marker for Aleppo pine,[17] used to flavor retsina. 2-Ethylhexanoic acid has a number of possible sources, including fruits, herbs, and wine.

Several other compounds of interest were identified in EUM 524. Isophorone, for which the only known botanical source listed by Duke is *Crocus sativus* (saffron),[18] was detected, and its identity was confirmed by analysis of an authentic isophorone sample. In a study of saffron extracts, isophorone was found to be the second most abundant volatile component after safranal.[19] Although it cannot be determined whether this spice was used as a flavor agent, a medicinal agent, or a colorant at Chrysokamino, the presence of saffron in this vessel is very likely. Another compound of interest detected in this sherd is verbenone (2-pinene-4-one), a known constituent of 28 plants, including many varieties of rosemary, hyssop, mint, germander, and sage.[20] It is also present in oil of verbena, from *Verbena triphylla*,[21] and *Verbena officinalis*.[22] The latter variety, also called vervain, was used for a variety of medicinal purposes in ancient Greece and is still used today for treatment of liver ailments.[23]

Two lactones, gamma-octalactone and gamma-nonalactone, were identified with certainty. Four occurrences of each are listed;[24] licorice root is the only plant reported on both lists. Other possible sources are discussed below. Another peak was identified as either 1,3-diacetylbenzene or 1,4-diacetylbenzene, both of which are present in tea leaves.[25] This compound is more likely a modern contaminant.

Finally, three peaks that clearly indicate the presence of pine resin were identified. The methyl ester of dehydroabietic acid ($C_{21}H_{30}O_2$) was detected; two other peaks were identified as $C_{21}H_{34}O_2$ and $C_{21}H_{36}O_2$, compounds with the same skeletal structure as the abietic acid, but with one and no double bond, respectively. These compounds are certain indicators of pine resin. Coupled with the observation of benzoic acid (detected as methyl benzoate) mentioned earlier, the resin is most likely from pine. The collection of molecules identified in this sherd reveal that the resinated wine was flavored with a variety of fragrant and medicinal herbs.

Sherd EUM 524 was also analyzed at Vassar College. Of 29 compounds identified, 22 may be ancient constituents, one is an unassigned mixture, and six are modern contaminants. Six saturated fatty acids are unusual in that they are of low molecular weight: pelargonic acid (C9), capric acid (C10), undecanoic acid (C11), lauric acid (C12), myristic acid (C14), and pentadecanoic acid (C15). The branched anteisopentadecanoic acid is very likely the product of postdepositional bacterial action and hence a contaminant. The absence of palmitic acid (C16) and stearic acid (C18) is striking; these are the most common saturated fatty acids in both animal and plant fats.

16. 49 occurrences reported in Duke 2002.
17. Beck et al. 2001.
18. Duke 2002.
19. Cadwallader, Baek, and Cai 1997, p. 68.
20. Duke 2002.
21. Budavari et al. 1989, p. 1565.
22. Duke 2002.
23. Jashemski 1999, p. 92.
24. Duke 2002.
25. Duke 2002.

Equally curious is the presence of six straight-chain, primary alcohols: 1-nonanol (C9), 1-decanol (C10), 1-dodecanol (C12), 1-tetradecanol (C14), 1-pentadecanol (C15) and 1-octadecanol (C18). The combination of straight-chain fatty acids and straight-chain primary alcohols is usually the result of the hydrolysis of the esters in epicuticular waxes, but the chain length of both the acids and the alcohols is here too short to allow that interpretation. The alcohols found occur in the free state in numerous plants,[26] so that no assignments are possible. Two alkanes, heptadecane (C17) and octadecane (C18) are also too ubiquitous to invite assignment to a single plant source.

Three gamma-lactones with 8, 9, and 10 carbon atoms are known to occur together in a wide range of Old and New World fruits, according to a *Chemical Abstracts* search. They are also reported as degradation products of plant oils (including olive oil) and animal fats, but the absence of palmitic and stearic acids in this sherd rules out these sources.

Acetanilide may be derived from *Peganum harmala*, or it may be the biodegradation product of modern herbicides, as discussed in the case of EUM 522 above.

There are six occurrences of 1,4-diacetylbenzene in the literature (*Chemical Abstracts* search), of which two are found in fruits (peach and blackberry); the other four are found in industrial air and water pollutants. The antiquity of this compound is therefore in question.

Unequivocal evidence that the vessel once contained wine resinated with pine resin is given by the presence of methyl dehydroabietate and 13-ethyl-16-norpodocarpa-8,11,13-triene.

Among the less common contaminants are three degradation products of 2,2,4-trimethyl-1,3-pentanediol plasticizers, 2-phenoxyethanol from suntan lotions, and 2(2-butoxyethoxy)-ethanol.

In summary, the vessel must have contained resinated wine flavored or medicated with a range of plant materials that cannot be identified from the compounds found in the sherd.

EUM 525. Sherd from a jar or basin, 111 (X 525)

Ten compounds were identified in the extract of EUM 525, including the four alkanes dodecane, tetradecane, pentadecane, and octadecane. These may be attributed to any number of plant sources or to contamination with a modern product. As in EUM 524, both isophorone and verbenone were detected and confirmed by comparison with authentic samples. These fragrant compounds are present respectively in saffron and verbena (as well as other aromatic herbs). Camphor, another compound of medicinal value derived from a variety of plants, was detected in EUM 525; its presence in most of the sherds from Chrysokamino suggests either a common ingredient or a common source of contamination.

The aldehydes nonanal and decanal were also identified. Twelve common occurrences of these two compounds are reported,[27] including six varieties of *Citrus* and the spices coriander, dill, ginger, safflower, and lemon verbena. The only ester detected in EUM 525 was the methyl ester of octanoic acid (caprylic acid), 17 occurrences of which are listed by

26. Karrer 1958–1981, pp. 43–45; Duke 2002.
27. Duke 2002.

Duke.[28] The vessel from which this sherd originated most likely contained a medicinal herbal preparation, perhaps a flavored wine.

EUM 526. Sherd from a jar or bellows, **63** (X 685)

The most significant result from the analysis of EUM 526 at Davidson College was the presence of a cluster of at least eight peaks for which the mass spectra revealed molecular ions at 272. The mass spectra of these compounds are consistent with abietadiene, for which numerous isomers are possible. Abietadiene is related to abietic acid, an important constituent of pine resin. When pine resin is heated (or perhaps allowed to decompose over a long time) to produce pine tar, the acid group of abietic acid can be lost through decarboxylation, resulting in compounds with the formula $C_{20}H_{32}$. In addition, two peaks revealed mass spectra with molecular ions at 274 and 270, most likely due to related compounds with one fewer and one additional double bond, respectively. We did not attempt to identify exact structures of these compounds, but their presence is a clear indication of pine resin or pine tar.

Other compounds detected and identified at Davidson include methyl octanoate and methyl decanoate (the methyl esters of caprylic and capric acids) and gamma decanolactone. Several peaks with spectra characteristic of alkanes, but with retention times that do not match the unbranched alkane reference compounds, were also noted.

Sherd EUM 526 was also analyzed at Vassar College. Of 35 compounds identified, 27 are ancient constituents, one is an unassigned mixture, and seven are modern contaminants.

The sherd is rich in straight-chain, saturated fatty acids. The 11 members of this group identified are caproic acid (C6), enanthic acid (C7), caprylic acid (C8), pelargonic acid (C9), capric acid (C10), undecanoic acid (C11), lauric acid (C14), pentadecanoic acid (C15), palmitic acid (C16), and stearic acid (C18). There are no unsaturated fatty acids, suggesting an animal fat rather than a vegetable oil as the source. The presence of anteisopentanoic acid most likely results from postdepositional microbial activity and is hence a contaminant. The branched fatty acid 2-ethylhexanoic acid has been reported in wine;[29] a *Chemical Abstracts* search yielded 135 occurrences, including wine and Old and New World fruit, but it also occurs in the meat and milk of sheep and goats. The latter source is suggestive, since three of the saturated fatty acids (caproic acid, caprylic acid, and capric acid) in this sherd are named after their isolation from the genus *Caper*, or goat. Biers and colleagues identified 2-ethylhexanoic acid in 14 of the 24 Corinthian "plastic" vases they analyzed.[30]

A single alcohol of low molecular weight, 1-nonanol (C9), is too common to be assigned to a particular plant source. Trace amounts of five straight-chain, saturated hydrocarbons (alkanes) with 26, 27, 28, 29, and 30 carbon atoms were detected. These can be attributed either to the plant waxes that coat the leaves, petals, and fruits of virtually all higher plants, or to the insect waxes, of which beeswax is the most familiar. However, neither wax esters nor their degradation products, i.e., long-chain alcohols and fatty acids with more than 20 carbon atoms, have been found in this sherd.

28. Duke 2002.
29. Danzer et al. 1999, pp. 26–34.
30. Biers, Gerhardt, and Braniff 1994, pp. 37–57.

A single lactone, gamma-decanolactone (C10), has more than 100 natural occurrences.[31]

The presence of pine resin is shown by seven diterpenoid compounds: the hydrocarbon 13-ethyl-16-norpodocarpa-8,11,13-triene, along with the resin acid esters methyl dehydroabietate, its 7-keto-derivative, a hydroxy-derivative, and three other only partially identified oxidation products. The identification of methyl benzoate makes it possible to narrow the source of the resin to the Aleppo pine, *Pinus halepensis,* which is indigenous to the eastern Mediterranean.

The modern contaminants include four plasticizers derived from 2,2,4-trimethyl-1,4-pentanediol, 2-phenoxyethanol, and ditolyl ether, i.e., the same contaminants that have been found in sherds EUM 522 and EUM 524.

In summary, the ancient contents of this vessel are identified as a resinated wine with unspecifiable additions of botanicals that may have been flavorings or medicinals.

EUM 527. Sherd from a jar, **110** (X 1658)

Eight compounds were identified in EUM 527, most of which have been discussed previously. The finds included camphor, the methyl ester of nonanoic acid, two lactones, and three alkanes. In addition to these, anisaldehyde was detected. Of 17 occurrences of anisaldehyde reported in Duke, the most likely are *Foeniculum vulgare* (fennel), *Cuminum cyminum* (cumin), and *Pimpinella anisum* (anise). Cultivation of these three plants in the Mediterranean region for flavoring and medicinal purposes dates to Minoan and Mycenaean times.[32] Fennel is still commonly used to treat digestive problems; its medicinal uses in ancient times, as recorded by Dioscorides, include treatments for stomach, eye, and bladder problems, as well as for fever and animal bites.[33]

Possible sources of gamma-octalactone and gamma-nonalactone, also detected in EUM 524, include *Glycyrrhiza glabra* (licorice root) and several types of fruit. Camphor may originate from any of a number of botanical sources.

The presence of methyl nonanoate, an oxidation product of oleic acid, may be indicative of olive oil although no other fatty acid esters were detected. The absence of peaks for other saturated and unsaturated fatty acids may arise from a failure of our methylation procedure. The comparative results reported above for sherds analyzed in both laboratories indicate that the method used at Davidson was not as effective at detecting fatty acids.

EUM 527 was found to contain heptadecane, octadecane, and eicosane, saturated alkanes with 17, 18, and 21 carbons, respectively. As previously mentioned, these are of little diagnostic value since there are a number of known plant sources, including the epicuticular waxes of many fruits, leaves, and flowers, as well as modern products, that could have introduced these as contaminants.

The compound diethyltoluamide (DEET) was also detected in this sherd. Clearly a modern contaminant, DEET is present in insect repellent products.

31. *Chemical Abstracts* search.
32. Mabberley 1997, pp. 159, 230, 454.
33. Jashemski 1999, p. 52.

The composition of the organic residue in EUM 527 suggests the presence of aromatic and/or medicinal plants, although specific botanicals cannot be identified.

EUM 528. Sherd from a bucket jar, **34** (X 1850)

Three interesting compounds were identified in EUM 528. One is anisaldehyde, which was also detected in EUM 527 and is discussed above. This compound most likely originates from the use of fennel, cumin, or anise as either a flavor agent or a medicinal extract. Ethyl salicylate was also found in this sherd. Duke lists only two occurrences for this compound, *Hyacinthus orientalis* (hyacinth),[34] known in the northeastern Mediterranean, and *Filipendula ulmaria* (meadowsweet), which is indigenous to Eurasia.[35] The third compound, identified as isomethyl b-ionone, is not listed by Duke. The mass spectra of isomethyl b-ionone and isomethyl a-ionone are quite similar, however, so either isomer could have been present. The results of a *Chemical Abstracts* search for occurrences of isomethyl b-ionone indicate that this compound is used in the perfume industry and is a minor component of the leaf oil of *Peumus boldus*,[36] a tree native to Chile and therefore not a possible source. Isomethyl a-ionone is also used in perfumes, cosmetics, and hair products. Three occurrences are reported for this compound, including a recent paper listing isomethyl a-ionone as a component of *Aegopodium podagraria*, or ground elder,[37] a plant with a history of medicinal use since at least Roman times.[38] Ground elder was used for the treatment of gout, an inflamation of the joints of the feet and hands. It is possible that the isomethyl ionone found in EUM 528 originated from an ancient plant product, but the possibility of its introduction via a modern cosmetic or perfume cannot be eliminated.

Other compounds identified in sherd EUM 528 include nonanal and two alkanes, dodecane and tetradecane. Due to the widespread occurrence of these compounds, none of these can be considered diagnostic for a particular plant.

EUM 547. Sherd from a jar, **102** (X 144)

This sherd, like EUM 524, was found to contain a rich selection of organic molecules. A total of 22 compounds were identified in EUM 547. Finds include three alkanes (C14, C17, and C18) and four methyl esters (those of 2-ethylhexanoic acid, octanoic acid, nonanoic acid, and decanoic acid). Tetradecane, heptadecane, and octadecane were also detected together in EUM 521. No conclusions can be based on the presence of these compounds, as they might be attributed to a variety of plant sources or to a modern contaminant. The methyl esters of the C8, C9, and C10 acids have a variety of plant sources; all are found in *Glycyrrhiza glabra*, *Cinnamomum aromaticum*, and *Capsicum frutescens*. Of these, only *Glycyrrihiza glabra* (licorice) is indigenous to Crete. Other possible sources of these acids include animal fat (C8, C10) and the degradation of olive oil (C9). The fact that no higher molecular weight fatty acids were detected does not necessarily rule out the presence of oil and fat, as our experimental methods may not have enabled us to detect these.

34. Duke 2002.
35. Mabberley 1997, pp. 227, 283.
36. Bruns and Köhler 1974.
37. Paramonov et al. 2001.
38. Mabberley 1997, p. 11.

The ketones 2-decanone, 2-undecanone, and 2-dodecanone were also detected in EUM 547; their retention times were compared with known samples to confirm the identifications. These compounds are almost certainly attributable to the medicinal plant rue *(Ruta graveolens, Ruta chalepensis)*. Also present (as confirmed by comparison to standard samples) were the corresponding aldehydes, decanal, undecanal, and dodecanal. Several occurrences for each of these are listed by Duke,[39] but only *Coriandrum sativum* and two species of *Citrus* (lemon and petitgrain) are known to contain all three aldehydes. *Citrus* is ruled out based on its absence from Bronze Age Crete; coriander, however, was known very early in the Mediterranean.

Additional compounds detected in EUM 547 include camphor, verbenone, anisaldehyde, gamma octalactone, and gamma nonalactone. Each of these has been identified in other sherds as well. These might have originated from aromatic herbs such as verbena, fennel, cumin, anise, and licorice root, as discussed previously for sherds EUM 524, 525, 527, and 528.

Four additional compounds, 1,2-benzisothiazole, N-methylphthalidimide, 3-bromothieno[3,2-c]pyridine, and diethyltoluamide (DEET) were identified in EUM 547. These are almost certainly contaminants; DEET is present in most insect repellents, but possible sources of the other compounds are unknown.

The variety of compounds identified in EUM 547 is consistent with the presence of various herbs and medicinal plants. The absence of marker compounds for pine resin or fatty acids prevents us from determining whether this was a flavored wine, an oil infusion, or another herbal extract.

EUM 548. Sherd from a rounded cup, **78** (X 210)

In this sherd, nine compounds were detected and identified including camphor, dodecane, octadecane, nonanal, decanal, and the methyl esters of octanoic acid and decanoic acid, all of which have been discussed previously. Due to the number of plant sources known for these compounds, none offer definitive diagnostic information. The remaining two compounds found in EUM 548 were tentatively identified as isomethyl b-ionone and isomethyl a-ionone (one of which was also found in EUM 528). Duke lists four occurrences for isomethyl a-ionone (listed under the synonym a-cetone): *Satureja cuneifolia* (Turkish savory), *Nepeta racemosa* (catmint), *Origanum sipyleum*, and *Micromeria myrtifolia*. Ground elder is another possible source, as discussed above in regard to EUM 528. Both compounds are used in the perfume industry (isomethyl a-ionone is used in soaps with scents of raspberry, cranberry, strawberry, and blueberry, air fresheners, perfumes, and potpourri); therefore, contact with a modern scented product cannot be ruled out as a source of these finds.

EUM 549. Sherd from a shallow bowl, **89** (X 168)

Analysis of this sherd revealed a cluster of peaks, very similar to those found in EUM 526, for which the mass spectra showed a molecular ion at 272. This is consistent with the formula $C_{20}H_{32}$, and the spectra suggest that these compounds are isomers of abietadiene, degradation products of abietic acid derivatives found in pine resin. The abietadienes are consistent

39. Duke 2002.

with the presence of pine tar (produced by heating or decomposition of pine resin) in the pottery vessel from which this sherd was obtained.

In addition to the abietadienes, this sherd was found to contain many of the same compounds that are found in other sherds from Chrysokamino, including five alkanes (C12, C14, C15, C17, and C18), one ester (methyl decanoate), and camphor. These compounds are too widely distributed in nature to allow any definite conclusions regarding the contents of the vessel.

Isophorone and verbenone, which were also found together in EUM 524 and 525, were also identified. They suggest the use of saffron and verbena or another aromatic herb, as verbenone is rather widely distributed. The three gamma-lactones, gamma-octalactone, gamma-nonalactone, and gamma-octalactone, are also present. Occurrences of these include a variety of fruits (including apricot, blueberry, papaya, pineapple, and peach, none of which are native to the Mediterranean), licorice root, and peppermint.

Traces of 2-decanone and 2-undecanone were also noted in this sherd. These occur together only in *Peumus boldus* (boldo tree native to Chile) and *Ruta graveolens* (rue), a medicinal plant known and used in the Bronze Age Mediterranean.

The variety of compounds found in EUM 529 suggests the presence of pine tar (or aged resin) as well as a variety of plant extracts. A resinated wine flavored with aromatic or medicinal botanicals is the most likely source of this collection of molecules.

EUM 550. Sherd from jar, probably a bridge-spouted jar, **82** (X 149)

In the extract from EUM 550, we identified 25 organic compounds, including 8 methyl esters, 3 alkanes, 3 lactones, 2 ketones, and 2 aldehydes. This is the only sample from Chrysokamino in which methyl esters of fatty acids longer than 10 carbons were detected at Davidson. The esters of both myristic acid (C14) and palmitic acid (C16) were identified, along with those of decanoic (C10), nonanoic (C9), octanoic (C8), heptanoic (C7), and hexanoic (C6) acids. For each of these acids, several known botanical occurrences are listed; two plants contain all of these: *Glycyrrhiza glabra* (licorice root) and *Asimina triloba* (pawpaw), a native of eastern North America. Another possible source of these acids, particularly nonanoic, myristic, and palmitic acids, is olive oil. Oleic acid, the most prevalent fatty acid in olive oil, was not detected. However, the nonanoic acid may arise from the decomposition of oleic acid. The other ester found in this sample, methyl 2-ethylhexanoate, has many occurrences, including wine, fruits, herbs, fish, and cheese.

Dodecane (C12), heptadecane (C17), and octadecane (C18) were the only alkanes identified in EUM 550. These are too widespread to permit the assignment of specific sources. The longer alkanes are present in epicuticular waxes on leaves, fruits, and flower petals. However, if such waxes were the source of these hydrocarbons, we would expect to find additional alkanes containing more than 20 carbon atoms.

The lactones gamma-hexalactone, gamma-octalactone, and gamma-nonalactone were all found in this sherd. Licorice root is the most likely source of these three lactones. The combination of the ketones 2-decanone and 2-undecanone found in EUM 550 (and also in EUM 547 and 549), is consistent with the presence of rue, a medicinal herb. Nonanal and decanal each have more than 20 occurrences; no additional diagnostic information may be obtained from the presence of these compounds.

Additional compounds identified in this sherd include camphor, isophorone, 3,3,5-trimethylcyclohexanone (dihydroisophorone), and isomethyl b-ionone (or the a isomer). Camphor is too ubiquitous in nature to provide any certain identification. Isophorone is a major constituent of saffron, as discussed previously. The related compound dihydroisophorone is not listed in Duke, and a *Chemical Abstracts* search does not reveal any likely botanical occurrences for this compound. Isomethyl ionone (the a or b isomer), as noted in the cases of EUM 528 and 548, may be a modern contaminant or a constituent of ground elder, catmint, or Turkish savory.

Two additional compounds, triacetin and N-methylphthalimide, are most likely modern contaminants.

CONCLUSIONS

The sherds from Chysokamino have been found to contain a variety of modern synthetic compounds that can only have been introduced by recent contact with skin care products (medicinal ointments, insect repellents, sunscreens, etc.). This raises the possibility that some of the natural products found could either be ancient residues or modern contaminants. It is highly unlikely, however, that all of the many natural products identified are contaminants.

Thus, the conclusion that emerges from the numerous compounds found in the 12 Chrysokamino samples is that these vessels were not used for the preparation of ordinary food; cooked meals would have left much larger quantities of animal fats and vegetable oils. Rather, these vessels appear to have been used in the making and storing of complex herbal concoctions, several of which included resinated wine, a well-known ancient Greek strategy to make medicines more palatable. The organic residue analysis results from this group of 12 sherds are entirely consistent with the idea that the Chrysokamino hut (or the original location from which the sherds found in its soil originated) served as a workshop for the preparation of herbal extracts from plants such as rue, saffron, coriander, fennel, anise, and licorice root. Since the workers of Chrysokamino were engaged in the decidedly unhealthy activity of smelting arsenical ores to make arsenical copper, it is plausible to infer that these herbal concoctions were intended to alleviate the symptoms of arsenic poisoning, i.e., that they were, in fact, medicinal preparations.

Finally, it is clear from the duplicate analyses of sherds EUM 522, 524, and 526 in two different laboratories that the organic compounds detected in two pieces of the same sherd can vary substantially. Reasons for the contrasting results obtained at Vassar and Davidson include differences in the methylation procedure and in the GC-MS conditions (which resulted in the detection of fewer fatty acid methyl esters at Davidson), the possible introduction of different laboratory contaminants (which could have masked the presence of some marker compounds in each laboratory), and the varying composition of organic compounds in different parts of a sherd. These observations indicate that negative results, the apparent absence of a compound in an organic residue, should not be interpreted as proof that the compound did not exist within the sherd. In order to obtain an exhaustive list of nonpolar, volatile molecules present in a sherd, that sample would have to be analyzed multiple times and under varying experimental conditions, a costly and time-consuming proposition. There is certainly much information to be gained from a single analysis of a pottery sherd; the presence of many compounds can be detected with certainty, especially when both the mass spectrum and the retention time are matched to those of an authentic sample. The list of molecules detected provides evidence for the use of various substances (or types of substances) in the original vessel. However, one must assume that other compounds are also present but undetected. This absence may be due to low concentrations, peak "masking" by other compounds emerging from the column at the same time, or the inability of those compounds to traverse the column and enter the mass spectrometer (as in the case of fatty acids that have not been converted to methyl esters). Continued work in the development of organic residue analysis methodology and sample handling protocols will allow organic residue analysis to make further contributions to an understanding of human behavior in early civilizations.

Petrographic Analysis of Two Final Neolithic Sherds from the Chrysokamino Metallurgy Location

by Eleni Nodarou

Two pottery sherds from the metallurgy location at Chrysokamino were selected for petrographic analysis. They are both from the Final Neolithic period, and they derive from open vessels. The results of the analysis suggest a broadly local provenance and similar manufacturing techniques for both vessels.

Description of Sherds

General Information

Fabric: Semicoarse with metamorphic rock fragments
Samples: sherds **1** (X 410) and **10** (X 611)

Matrix (Groundmass)

A. Optical properties and color
   1. Under plane-polarized light: there is some color differentiation between the core and the margins of the section. The margins are dark brown and the core is orange brown.
   2. Under cross-polarized light: the margins are dark red brown and the core is dark orange brown; the micromass is optically active.

B. Overall grain size and modality of inclusions and voids: coarse inclusions vary from 1.9–0.1 mm long dimension. The fine fraction is <0.1 mm and up to 60 micrometers in its longest dimension. Ratios for inclusions and voids (coarse:fine:voids) are ca. 35:43:8 (X 410) and 15:82:3 (X 611). Voids range in size from 1.14 to 0.07 mm in longest dimension.

C. Overall preferred orientation of inclusions and voids: in sample X 410, the voids display preferred orientation parallel to vessel margins; in sample X 611, they are randomly oriented. In both samples, the nonplastics are randomly oriented.

D. Silt-sized inclusions (under 62 micrometers)
1. Monocrystalline quartz
    Size range: up to ca. 62 micrometers
    Shape: equant, subangular to subrounded
    Comments: straight extinction
2. Phyllite
    Size range: up to ca. 62 micrometers
    Shape: elongate
3. Biotite mica
    Size range: up to ca. 62 micrometers
    Shape: laths
    Comments: Dark mica is present in the matrix, but in some cases it is also a component of the phyllite and quartzite fragments.
4. White mica
    Size range: up to ca. 62 micrometers
    Shape: laths
    Comments: White mica is most likely muscovite. It is present in the matrix, but in some cases it constitutes a component of the phyllite and quartzite.
5. Polycrystalline quartz
    Size range: up to ca. 62 micrometers
    Shape: equant; subrounded
    Comments: In some cases, the quartz grades into quartzite (with evidence of metamorphism).
6. Chert
    Size range: up to ca. 62 micrometers
    Shape: equant; subrounded.
    Comments: fine-grained.

Voids

A. Size range: Voids vary in size from 1.14 to 0.07 mm in their longest dimension.

B. Shape: Vugs are rare; the majority are mesoplanar and macroplanar voids, close to double-spaced. The vugs are irregular in shape; the planar voids are elongate.

Inclusions above Silt-Size

1. Monocrystalline quartz
    Size range: 0.77–0.1 mm in longest dimension
    Shape: equant; angular to subangular
    Comments: straight extinction
2. Chert
    Size range: 1.14–0.2 mm in longest dimension
    Shape: equant or slightly elongate; subrounded
    Comments: fine-grained

3. Metamorphic rock fragments
    Size range: 1.9–0.76 mm in longest dimension (for the quartzite); 0.77–0.4 mm in longest dimension (for the phyllite)
    Shape: the quartzite is equant to slightly elongate; angular-subangular. The phyllite is elongate.
    Comments: The main metamorphic constituent is quartzite. A majority of the grains have undergone alteration or deformation. In some cases, there is evidence of shearing. There are also rare fragments of phyllite, composed of biotite mica, quartz, and white mica. The metamorphic inclusions suggest an environment with low-grade metamorphism. The scarcity and the small size of the metamorphic rock fragments lead us to characterize them as detrital in the fabric.
5. Micrite
    Size range: 0.5–0.2 mm in longest dimension
    Shape: equant; subrounded
    Comments: merging boundaries
6. Sandstone
    Size range: 1.14–0.38 mm in longest dimension
    Shape: equant to slightly elongate; subrounded
    Comments: there are two types; one is composed of unevenly distributed, equant quartz grains with very little clay matrix, occasionally "cemented" with calcite (quartzarenite). The other is composed of small, equant quartz grains and biotite mica laths with more matrix (greywacke).
7. Alkali feldspar
    Size range: ca. 0.12 mm in longest dimension
    Shape: slightly elongate
    Comments: simple twinning
8. Plagioclase feldspar
    Size range: ca. 0.15 mm in longest dimension
    Shape: elongate
    Comments: polysynthetic twinning
9. Mica
    Size range: ca. 0.08 mm in longest dimension
    Shape: laths
    Comments: mainly biotite mica, but there are some laths of white mica, also.
10. Textural concentration features (clay lumps)
    Size range: 0.70 to <0.1 mm in longest dimension
    Shape: equant; subangular to subrounded
    Comments: These features range in color from dark red to very dark brown or black. In some cases, they contain very fine quartz inclusions, while in others they have no inclusions. A few fragments in both samples might be grog; their size ranges from 0.76 to 0.2 mm in longest dimension; their shape is equant, angular to subangular; and the color in cross-polarized light is greenish gray. They have merging boundaries and voids around the margins.

Modality

Matrix (groundmass): ca. 40%–50%
Voids: 3%–8%

Silt-sized inclusions

Monocrystalline quartz: ca. 3%
Phyllite: under 1%
Biotite mica: under 1%
White mica: under 1%
Polycrystalline quartz: under 1%
Chert: under 1%

Larger inclusions

Monocrystalline quartz: 20%–30%
Metamorphic rocks: 5%–10%
Sandstone: 5%
Micrite: under 2%
Chert: 5%
Alkali feldspar: under 2%
Plagioclase feldspar: under 2%
Mica: 3%
Textural concentration features: 3%–7%

Discussion

This is a semicoarse, noncalcareous, red-firing fabric, characterized by the presence of quartz and detrital metamorphic rock fragments. Its composition is not indicative of provenance, but it seems that a red (alluvial?) deposit has been used for the manufacture of these vessels. The detrital quartzite and phyllite fragments point toward a red soil deposit, while the angular shape of most of the grains indicates short distance transportation. These observations, along with the fact that the area around Chrysokamino is dominated by the presence of the Phyllite-Quartzite series, suggest a local origin for this pottery. It is not possible to assign specific provenance, but the red alluvial deposits of the north coast are very likely to have provided the raw material for the manufacture of this pottery.

The two samples examined present textural and compositional similarity. The color of the matrix shows similar firing conditions, consisting of an oxidizing atmosphere and low firing temperature. Moreover, the similarity in the texture of some of the plastic and nonplastic components, such as the chert, the micrite, and the other nonplastic inclusions, lead us to suggest that the two samples share the same provenance in terms of raw material used and technology of manufacture.

REFERENCES

Acheson, P. 1997. "Does the Economic Explanation Work? Settlement, Agriculture, and Erosion in the Territory of Halieis in the Late Classical–Early Hellenistic Period," *JMA* 10, pp. 165–190.

Aitken, J. J. 1992. "Archaeological Sediments as Artifacts. The Dirt of Khok Phanom D, Central Thailand" (diss. Univ. of Otago, Dunedin, New Zealand).

Alexiou, S. 1951. "Πρωτομινωικαί ταφαί παρά το Κανλί-Καστέλλι Ηρακλείου," *CretChron* 5, pp. 275–294.

———. 1979. "Τείχη και ακροπόλεις στη Μινωική Κρήτη (Ο μύθος της μινωικής ειρήνης)," *Κρητολογία* 8, pp. 41–56.

———. 1980. "Προανακτορικές ακροπόλεις της Κρήτης," in *Πεπραγμένα του Δ΄ Διεθνούς Κρητολογικού Συνεδρίου* Α΄, Athens, pp. 9–22.

Alibertis, A. 1994. *Κρήτη: Die Samariaschlucht und ihre Pflanzen*, Herakleion.

Alimov, K., N. Boroffka, M. Bubnova, J. Burjakov, J. Cierny, J. Jakubov, J. Lutz, H. Parzinger, E. Pernicka, V. Radililovskij, V. Ruzanov, T. Širinov, D. Staršinin, and G. Weisgerber. 1998. "Prähistorischer Zinnbergbau in Mittelasien," *Eurasia Antiqua* 4, pp. 137–199.

Allbaugh, L. G. 1953. *Crete: A Case Study of an Underdeveloped Area*, Princeton.

Anastopoulos, J., and C. Koukouzas. 1984. "The Metallogeny of Greece, Sheet 8," in *Mémoire explicatif de la Carte métallogénique de l'Europe et des pays limitrophes*, Paris, pp. 411–427.

Andreopoulou-Mangou, E. 1993. "Χημική ανάλυση χάλκινων αντικειμένων από τη Σκοτεινή Θαρρουνίων," in *Σκοτεινή Θαρρουνίων: Το σπήλαιο, ο οικισμός, και το νεκροταφείο*, ed. A. Sampson, Athens, pp. 435–437.

Andreou, S., M. Fotiadis, and K. Kotsakis. 2001. "The Neolithic and Bronze Age of Northern Greece," in *Aegean Prehistory: A Review*, ed. T. Cullen, Boston, pp. 259–327.

Angélier, J., J.-P. Cadet, G. Delibrias, J. Fourniquet, M. Gigout, M. Guillemin, M.-T. Hogrel, C. Lalou, and G. Perre. 1976. "Les déformations du Quaternaire marin, indicateurs néotectoniques: Quelques exemples méditerranéens," *Revue de géographie physique et de géologie dynamique*, 2nd ser., 18:5, pp. 427–448.

Antonova, I., V. Tolstikov, and M. Treister. 1996. *The Gold of Troy: Searching for Homer's Fabled City*, London.

Atherdon, M. 2000. "Human Impact on the Vegetation of Southern Greece and Problems of Palynological Interpretation: A Case Study from Crete," in Halstead and Frederick 2000, pp. 62–78.

Aurenche, O., and S. K. Kozlowski. 1999. *La naissance du néolithique au Proche Orient*, Paris.

Bachmann, H.-G. 1980. "Early Copper Smelting Techniques in Sinai and in the Negev as Deduced from Slag Investigations," in *Scientific Studies* in *Early Mining and Extractive Metallurgy,* ed. P. T. Craddock, London, pp. 103–134.

———. 1982. *The Identification of Slags from Archaeological Sites* (Institute of Archaeology Occasional Publication 6), London.

Bagolini, B., and F. Lo Schiavo, eds. 1996. *The Copper Age in the Near East and Europe: Thirteenth International Congress of Prehistoric and Protohistoric Sciences, Forlì, Italia 8/14 September 1996,* Forlì.

Ball, D. F. 1964. "Loss-on-ignition as an Estimate of Organic Matter and Organic Carbon in Non-calcareous Soils," *Journal of Soil Science* 15:1, pp. 84–92.

Bambauer, H. U., F. Taborsky, and H. D. Trochim. 1979. *Optical Determination of Rock-Forming Minerals by W. E. Tröger,* Stuttgart.

Bamberger, M. 1985. "The Working Conditions of the Ancient Copper Smelting Process," in Craddock and Hughes 1985, pp. 151–157.

Bamberger, M., and P. Wincierz. 1990. "Ancient Smelting of Oxide Copper Ore," in *The Ancient Metallurgy of Copper,* ed. B. Rothenberg, London, pp. 123–157.

Banou, E. S. 1995a. "Building AA (Seager's House A). The Pottery," in *Pseira* I, pp. 19–21.

———. 1995b. "Building AD Center. The Pottery," in *Pseira* I, pp. 108–124.

———. 1998. "The Pottery, Building AC," in *Pseira* II, pp. 13–26.

Barker, G. 1985. *Prehistoric Farming in Europe,* Cambridge.

Barker, G., and J. Lloyd, eds. 1991. *Roman Landscape,* London.

Bar-Yosef, O. 2001. "From Sedentary Foragers to Village Hierarchies: The Emergence of Social Institutions," in *The Origin of Human Social Institutions* (Proceedings of the British Academy 100), ed. W. G. Runciman, London, pp. 1–38.

———. 2002a. "The Upper Paleolithic Review," *Annual Review of Anthropology* 31, pp. 363–393.

———. 2002b. "The Natufian Culture and the Early Neolithic: Social and Economic Trends in Southwestern Asia," in *Examining the Farming Language Dispersal Hypothesis* (McDonald Institute Monographs), ed. P. Bellwood and C. Renfrew, Cambridge, pp. 113–126.

Bateman, A. M. 1946. *Economic Mineral Deposits,* New York.

Baumann, A., G. Best, W. Gwosdz, and H. Wachendorf. 1976. "The Nappe Pile of Eastern Crete," *Tectonophysics* 30:3/4, pp. 33–40.

Baumann, A., G. Best, and H. Wachendorf. 1977. "Die alpidischen Stockwerke der südlichen Ägäis," *Geologische Rundschau* 66, pp. 492–522.

Baxter, M. 2003. *Statistics in Archaeology,* London.

Beck, C. W. Forthcoming. "Certainty and Doubt in the GC-MS Analysis of Ancient Organic Remains," in *Archaeology Meets Science: Biomolecular and Site Investigations in Bronze Age Greece,* ed. Y. Tzedakis and H. Martlew, Oxford.

Beck, C. W., E. H. Stout, K. M. Woukulich, V. Karageorghis, and E. Aloupi. 2001. "The Uses of Cypriote White-Slip Ware Inferred from Organic Residue Analysis," *Ägypten und Levant* 14, pp. 13–43.

Becker, M. J. 1976. "Soft-Stone Sources on Crete," *JFA* 3, pp. 361–374.

Becker, M. J., and P. P. Betancourt. 1997. *Richard Berry Seager: Pioneer Archaeologist and Proper Gentleman,* Philadelphia.

Beeston, R. F., J. Palatinus, C. W. Beck, and E. C. Stout. Forthcoming. "Organic Residue Analysis: New Finds at Chrysokamino," in *Archaeology Meets Science: Biomolecular and Site Investigations in Bronze Age Greece,* ed. Y. Tzedakis and H. Martlew, Oxford.

Begemann, F., S. Schmitt-Strecker, and E. Pernicka. 1992. "The Metal Finds from Thermi III–V: A Chemical and Lead-Isotope Study," *Studia Troica* 2, pp. 219–239.

Begemann, F., E. Pernicka, and S. Schmitt-Strecker. 1995. "Thermi on Lesbos: A Case Study of Changing Trade Patterns," *OJA* 14, pp. 123–136.

Beile-Bohn, M., C. Gerber, M. Morsch, and K. Schmidt. 1998. "Neolithische Forschungen im Obermesopotamien: Gürcütepe und Göbekli Tepe," *IstMitt* 48, pp. 5–78.

Bernus, S. 1983. "Découvertes hypothèses, reconstitution et preuves: Le cuivre médiéval d'Azelik-Takedda (Niger)," in *Métallurgies africaines*, ed. N. Echard, Paris, pp. 153–171.

Bernus, S., and N. Echard. 1985. "Metal Working in the Agadez Region (Niger): An Ethno-Archaeological Approach," in Craddock and Hughes 1985, pp. 71–80.

Betancourt, P. P. 1977. "Some Chronological Problems in the Middle Minoan Dark-on-Light Pottery of Eastern Crete," *AJA* 81, pp. 341–353.

———. 1979. *Vasilike Ware: An Early Bronze Age Pottery Style in Crete* (*SIMA* 56), Göteborg.

———. 1980. *Cooking Vessels from Minoan Kommos: A Preliminary Report* (*UCLAPap* 7), Los Angeles.

———. 1983. *Minoan Objects Excavated from Vasilike, Pseira, Sphoungaras, Priniatikos Pyrgos, and Other Sites* (University Museum Monograph 47), Philadelphia.

———. 1984. *East Cretan White-on-Dark Ware* (University Museum Monograph 51), Philadelphia.

———. 1985. *The History of Minoan Pottery*, Princeton.

———. 1994–1996. "Μινωικό εμπόριο λίθου: Τα στοιχεία από τη νήσο Ψείρα," *Κρητική Εστία* 5, pp. 47–70.

———. 1995. "Handbook for Trench Supervisors, Chrysokamino Excavations"(INSTAP Study Center for East Crete), Pacheia Ammos.

———. 1997. "The Copper Smelting Site at Chrysokamino, Crete," *Ancient Greek Technology: Proceedings, 1st International Conference, Thessaloniki 1997*, Thessaloniki, pp. 51–54.

———. 1998. "Middle Minoan Objects in the Near East," in *The Aegean and the Orient in the Second Millennium B.C.: Proceedings of the 50th Anniversary Symposium, Cincinnati, 18–20 April 1997*, ed. E. H. Cline and D. Harris-Cline, Liège, pp. 5–12.

———. 1999. "What is Minoan? FN/EM I in the Gulf of Mirabello Region," in *MELETEMATA*, vol. 1, pp. 33–40.

———. 2003a. "Interpretation and Conclusions," in *Pseira* VII, pp. 123–139.

———. 2003b. "Interpreting Ceramic Petrography: The Special Product Model, a New Model for the Distribution of Pottery in Early Minoan Crete," in *METRON: Measuring the Aegean Bronze Age* (*Aegaeum* 24), ed. K. P. Foster and R. Laffineur, Liège, pp. 117–121.

———. 2003c. "The Impact of Cycladic Settlers on Early Minoan Crete," *Mediterranean Archaeology and Archaeometry* 3, pp. 3–12.

———. 2004a. "Coastal Trade: The Eastern Gulf of Mirabello in the Early Twentieth Century," in *Pseira* VIII, pp. 91–97.

———. 2004b. "Traditional Lime Production in the Eastern Gulf of Mirabello Area," in *Pseira* VIII, pp. 99–104.

———. 2004c. "Pseira and Knossos: The Transformation of an East Cretan Seaport," in Day, Mook, and Muhly 2004, pp. 21–28.

Betancourt, P. P., and C. Davaras. 1990. "Αρχαιολογικές έρευνες στην Ψείρα: Περίοδοι 1987–1989," *Αμάλθεια* 82–85, pp. 20–37.

Betancourt, P. P., and C. R. Floyd. 2000–2001. "Ανασκαφή στο Χρυσοκάμινο," *Κρητική Εστία* 8, pp. 9–22.

Betancourt, P. P., C. R. Floyd, and J. D. Muhly. 1997. "Excavations at Chrysokamino, Crete, 1996," *AJA* 101, pp. 374–375.

Betancourt, P. P., and R. Hope Simpson. 1992. "The Agricultural System of Bronze Age Pseira," *Cretan Studies* 3, pp. 47–54.

Betancourt, P. P., and N. Marinatos. 1997. "The Minoan Villa," in Hägg 1997, pp. 15–19.

Betancourt, P. P., J. D. Muhly, W. R. Farrand, C. Stearns, L. Onyshkevych, W. B. Hafford, and D. Evely. 1999. "Research and Excavation at Chrysokamino, Crete, 1995–1998," *Hesperia* 68, pp. 343–370.

Betancourt, P. P., J. D. Muhly, and C. R. Floyd. 1998. "Excavations at Chrysokamino, Crete, in 1997," *AJA* 102, p. 391.

Betancourt, P. P., and G. H. Myer. 1995. "Phyllite Fabrics in Eastern Crete," in *The Ceramics Cultural Heritage*, ed. P. Vincenzini, Faenza, pp. 395–399.

———. 1999. "Copper Smelting Slags from Chrysokamino, Crete," *AJA* 103, p. 273.

Betancourt, P. P., and J. S. Silverman. 1991. *Pottery from Gournia* (University Museum Monograph 72), Philadelphia.

Bevan, A. 2002. "The Rural Landscape of Neopalatial Kythera," *JMA* 15, pp. 217–255.

Biers, W. R., K. O. Gerhardt, and R. A. Braniff. 1994. *Lost Scents: Investigation of Corinthian "Plastic" Vases by Gas Chromatography-Mass Spectrometry* (*MASCAP* 11), Philadelphia.

Binford, L. R. 1968a. "Archaeological Perspectives," in *New Perspectives in Archaeology*, ed. S. R. Binford and L. R. Binford, Chicago, pp. 5–32.

———. 1968b. "Post-Pleistocene Adaptations," in *New Perspectives in Archaeology*, ed. S. R. Binford and L. R. Binford, Chicago, pp. 313–341.

Bintliff, J. L. 1977. *Natural Environment and Human Settlement in Prehistoric Greece* (*BAR-IS* 28), Oxford.

Bintliff, J. L., E. Farinetti, P. Howard, K. Sarri, and K. Sbonias. 2002. "Classical Farms, Hidden Prehistoric Landscapes, and Greek Rural Survey: A Response and Update," *JMA* 15, pp. 259–265.

Bintliff, J. L., P. Howard, and A. Snodgrass. 1999. "The Hidden Landscape of Prehistoric Greece," *JMA* 12, pp. 139–168.

Bintliff, J. L., and A. M. Snodgrass. 1988. "Off-Site Pottery Distributions: A Regional and Interregional Perspective," *CurrAnthr* 29, pp. 506–513.

Bischoff, D. 2002. "Les megaliths de Göbekli Tepe," *Archéologia* 393, pp. 28–37.

Biswas, A. K., and W. G. Davenport. 1976. *Extractive Metallurgy of Copper*, Oxford.

Bittel, K. 1959. "Beitrag zur Kenntnis anatolischer Metallgefässe der zweiten Hälfte des dritten Jahrtausends v. Chr.," *JdI* 74, pp. 1–34.

Blackman, D. J., and K. Branigan. 1975. "An Archaeological Survey on the South Coast of Crete, between the Ayiofarango and Chrysostomos," *BSA* 70, pp. 17–36.
———. 1977. "An Archaeological Survey of the Lower Catchment of the Ayiofarango Valley," *BSA* 72, pp. 13–80.
Blitzer, H. 1993. "Olive Cultivation and Oil Production in Minoan Crete," in *La production du vin et de l'huile en Méditerranée. Oil and Wine Production in the Mediterranean Area*, ed. M.-C. Amouretti and J.-P. Brun, Paris, pp. 163–175.
———. 1995. "Minoan Implements and Industries," in *Kommos* I:1, pp. 403–535.
———. 2004. "Agriculture and Subsistence in the Late Ottoman and Post-Ottoman Mesara," in *The Plain of Phaistos: Cycles of Social Complexity in the Mesara Region of Crete* (Monumenta Archaeologica 23), ed. L. V. Watrous, D. Hadzi-Vallianou, and H. Blitzer, Los Angeles, pp. 111–217.
Bogaard, A., M. Charles, P. Halstead, and G. Jones. 2000. "The Scale and Intensity of Cultivation: Evidence from Weed Ecology," in Halstead and Frederick 2000, pp. 129–134.
Bognár-Kutzián, I. 1976. "On the Origins of Early Copper-Processing in Europe," in *To Illustrate the Monuments: Festschrift S. Piggott*, ed. J. V. S. Megaw, London, pp. 69–76.
Bonacasa, N. 1967–1968. "Patrikiès—Una stazione medio-minoica fra Haghia Triada e Festòs," *ASAtene*, n.s. 29–30, pp. 7–54.
Boroffka, N., J. Cierny, J. Lutz, H. Parzinger, E. Pernicka, and G. Weisgerber. 2002. "Bronze Age Tin from Central Asia: Preliminary Notes," in *Ancient Interactions: East and West in Eurasia* (McDonald Institute Monograph), ed. K. Boyle, C. Renfrew, and M. Levine, Cambridge, pp. 135–159.
Bossert, E.-M. 1967. "Kastri auf Syros: Vorbericht über eine Untersuchung der prähistorischen Siedlung," *ArchDelt* 22, A', pp. 53–76.
Bottema, S. 1980. "Palynological Investigations on Crete," *Review of Paleobotany and Palynology* 31, pp. 193–217.
———. 1994. "The Prehistoric Environment of Greece: A Review of the Palynological Record," in *Beyond the Site: Regional Studies in the Aegean Area*, ed. P. N. Kardulias, Lanham, Md., pp. 45–68.
Bottinga, Y., P. Richet, and D. F. Weill. 1983. "Calculation of the Density and Thermal Expansion Coefficient of Silicate Liquids," *Bulletin minéralogique* 106, pp. 129–138.
Bottinga, Y., and D. F. Weill. 1972. "The Viscosity of Magmatic Silicate Liquids: A Model for Calculation," *American Journal of Science* 272, pp. 438–475.
Bottinga, Y., D. Weill, and P. Richet. 1982. "Density Calculations for Silicate Liquids 1: Revised Method for Aluminosilicate Compositions," *Geochimica et Cosmochimica Acta* 46, pp. 909–919.
Boyd, H. A. 1901. "Excavations at Kavousi, Crete, in 1900," *AJA* 5, pp. 125–157.
———. 1904–1905. "Gournia: Report of the American Exploration Society's Excavations at Gournia, Crete, 1904," *University of Pennsylvania Transactions of the Department of Archaeology, Free Museum of Science and Art* 1, pp. 177–188.
Branigan, K. 1967. "The Early Bronze Age Daggers of Crete," *BSA* 62, pp. 211–239.
———. 1968. *Copper and Bronze Working in Early Bronze Age Crete* (*SIMA* 19), Lund.
———. 1970a. *The Foundations of Palatial Crete*, London.
———. 1970b. *The Tombs of Mesara*, London.
———. 1971. "An Early Bronze Age Metal Source in Crete," *SMEA* 13, pp. 10–14.
———. 1974. *Aegean Metalwork of the Early and Middle Bronze Age*, Oxford.
———. 1984. "Early Minoan Society: The Evidence of the Mesara Tholoi Reviewed," in *Aux origines de l'hellénisme: La Crète et la Grèce. Hommage à Henri van Effenterre*, ed. C. Nicolet, Paris, pp. 29–37.
———. 1988. *Pre-Palatial: The Foundations of Palatial Crete: A Survey of*

Crete in the Early Bronze Age, Amsterdam.

———. 1991. "Mochlos—An Early Aegean 'Gateway Community'?" in *Thalassa, l'Egée préhistorique et la mer: Actes de la troisième Rencontre égéenne internationale de l'Université de Liège, Station de recherches sous-marines et océanographiques (StaReSO), Calvi, Corse, 23–25 avril 1990* (*Aegaeum* 7), ed. R. Laffineur and L. Basch, Liège, pp. 97–105.

———. 1999. "Late Neolithic Colonization of the Uplands of Eastern Crete," in *Neolithic Society in Greece,* ed. P. Halstead, Sheffield, pp. 57–65.

Broodbank, C. 1992. "The Neolithic Labyrinth: Social Change at Knossos before the Bronze Age," *JMA* 5, pp. 39–75.

———. 2000. *An Island Archaeology of the Early Cyclades,* Cambridge.

Brown, A. 1993. *Before Knossos—Arthur Evans's Travels in the Balkans and Crete,* Oxford.

Bruns, K., and M. Köhler. 1974. "Über die Zusammensetzung des Boldoblätterols," *Parfümerie und Kosmetik* 55, pp. 225–227.

Budavari, S., M. J. O'Neil, A. Smith, and P. E. Heckelman, eds. 1989. *The Merck Index: An Encyclopedia of Chemicals, Drugs, and Biologicals,* 11th ed., Rahway.

Budd, P. 1991. "Eneolithic Arsenical Copper: Heat Treatment and the Metallographic Interpretation of Manufacturing Processes," in Pernicka and Wagner 1991, pp. 35–44.

Budd, P., B. Chapman, C. Jackson, R. Janaway, and P. Ottaway, eds. 1991. *Archaeological Sciences 1989: Proceedings of the Conference on the Application of Scientific Techniques to Archaeology, Bradford, September 1989,* Oxford.

Budd, P., and B. Ottaway. 1991. "The Properties of Arsenical Copper Alloys: Implications for the Development of Eneolithic Metallurgy," in Budd et al. 1991, pp. 132–142.

Bull, I. D., R. P. Evershed, and P. P. Betancourt. 1999. "Chemical Evidence for a Structured Agricultural Manuring Regime on the Island of Pseira, Crete during the Minoan Period," in *MELETEMATA,* vol. 1, pp. 69–74.

———. 2001. "An Organic Geochemical Investigation of the Practice of Manuring at a Minoan Site on Pseira Island, Crete," *Geoarchaeology* 16, pp. 223–242.

Bunk, W. G. J., A. Hauptmann, S. Kölschbach, and G. Woelk. 2002. "Wind-Powered Copper Smelting Technology from the 3rd Millennium B.C. at Feinan/Jordan," in *Proceedings of the Fifth International Conference on the Beginnings of the Use of Metals and Alloys: Messages from the History of Metals to the Future Metal Age,* ed. G.-H. Kim, K.-W. Yi, and H.-T. Kang, Gyeongju, Korea, pp. 331–338.

Burford, A. 1993. *Land and Labor in the Greek World,* Baltimore.

Cadogan, G. 1971. "Was There a Minoan Landed Gentry?" *BICS* 18, pp. 145–148.

———. 1990. "Lasithi in the Old Palace Period," *BICS* 37, pp. 172–174.

———. 1995. "Mallia and Lasithi: A Palace-State," in Πεπραγμένα του Ζ΄ Διεθνούς Κρητολογικού Συνεδρίου Α΄, Rethymnon, pp. 97–104.

———. 1997. "The Role of the Pyrgos Country House in Minoan Society," in Hägg 1997, pp. 99–103.

Cadwallader, K. R., H. H. Baek, and M. Cai. 1997. "Characterization of Saffron Flavor by Aroma Extract Dilution Analysis," in *Spices: Flavor Chemistry and Antioxidant Properties* (ACS Symposium Series 660), ed. S. J. Risch and C. Ho, Washington, D.C., pp. 66–79.

Caley, E. R. 1949. "On the Prehistoric Use of Arsenical Copper in the Aegean Region," in *Commemorative Studies in Honor of Theodore Leslie Shear* (*Hesperia* Suppl. 8), pp. 60–63.

Canti, M. 1995. "A Mixed-Method Approach to Geoarchaeological Analysis," in *Archaeological Sediments and Soils: Analysis, Interpretation, and Management,* ed. A. J. Barnham and R. I. Macphail, London, pp. 183–190.

Carothers, J. J. 1992. "The Pylian Kingdom: A Case Study of an Early State" (diss. Univ. of California, Los Angeles).

Carter, T. 1998. "Knowledge Is Power: Craft Specialization and Social Inequality in the Southern Aegean Early Bronze Age," *AJA* 102, pp. 414–415.

Caskey, J. 1964. *Greece, Crete, and the Aegean Islands in the Early Bronze Age,* Cambridge.

Caskey, M., L. Mendoni, A. Papastamataki, and N. Neloyannis. 1988. "Metals in Keos: A First Approach," in *The Engineering Geology of Ancient Works, Monuments, and Historical Sites,* ed. G. P. Marinos and G. C. Koukis, Rotterdam, pp. 1739–1745.

Cauvin, J. 1994. *Naissance des divinités, naissance de l'agriculture: Le révolution des symboles au néolithique,* Paris.

———. 2000. *The Birth of the Gods and the Origins of Agriculture,* trans. T. Watkins, Cambridge.

Cauvin, J., I. Hodder, G. O. Rollefson, O. Bar-Yosef, and T. Watkins. 2001. "The Birth of the Gods and the Origins of Agriculture: Review Feature," *CAJ* 11, pp. 105–121.

Chadwick, J. 1973. *Documents in Mycenaean Greek,* Cambridge.

Chapouthier, F., P. Demargne, and A. Dessene. 1962. *Fouilles exécutées à Mallia. Quatrième rapport: Exploration du palais (1929–1935 et 1946–1960)* (*Étcrét* 12), Paris.

Charles, J. A. 1967. "Early Arsenical Bronzes—A Metallurgical View," *AJA* 71, pp. 21–26.

———. 1980. "The Coming of Copper and Copper-Base Alloys and Iron: A Metallurgical Sequence," in *The Coming of the Age of Iron,* ed. T. A. Wertime and J. D. Muhly, New Haven, pp. 150–181.

———. 1985. "Determinative Mineralogy and the Origins of Metallurgy," in Craddock and Hughes 1985, pp. 21–28.

———. 2000. *Out of the Fiery Furnace: Recollections and Meditations of a Metallurgist,* Cambridge.

Chavakis, I. n.d. Φυτά και βοτάνια της Κρήτης, Athens.

Chernykh, E. N. 1978. *Gornoe delo I metallurgija v drevnejsej Bolgarii,* Sofia.

Cherry, J. F. 1983. "Evolution, Revolution, and the Origins of Complex Society in Minoan Crete," in

Krzyszkowska and Nixon 1983, pp. 33–45.

———. 1986. "Polities and Palaces: Some Problems in Minoan State Formation," in *Peer Polity Interaction and Socio-Political Change,* ed. C. Renfrew and J. F. Cherry, Cambridge, pp. 19–45.

———. 1988. "Pastoralism and the Role of Animals in the Pre- and Protohistoric Economies of the Aegean," in *Pastoral Economies in Classical Antiquity,* ed. C. R. Whittaker, Cambridge, pp. 6–34.

Cherry, J. F., J. L. Davis, A. Demitrack, E. Mantzourani, T. F. Strasser, and L. E. Talalay. 1988. "Archaeological Survey in an Artifact-Rich Landscape: A Middle Neolithic Example from Nemea, Greece," *AJA* 92, pp. 159–176.

Cherry, J. F., J. L. Davis, and E. Mantzourani. 1991. *Landscape Archaeology as Long-Term History: Northern Keos in the Cyclades Islands from Earliest Settlement until Modern Times,* Los Angeles.

Cherry, J. F., J. L. Davis, E. Mantzourani, and T. M. Whitelaw. 1991. "The Survey Methods," in Cherry, Davis, and Mantzourani 1991, pp. 13–35.

Cherry, J. F., C. Gamble, and S. Shennan, eds. 1978. *Sampling in Contemporary British Archaeology,* Oxford.

Chlouveraki, S. 1998. "The Archaeology of Minoan Gypsum: Patterns of Consumption of a Prestigious Stone," *AJA* 102, pp. 391–392.

Chryssoulaki, S. 1999. "Minoan Roads and Guard Houses: War Regained," in *POLEMOS: Le contexte guerrier en Égée à l'âge du Bronze* (*Aegaeum* 19), vol. 1, ed. R. Laffineur, Liège, pp. 75–85.

Coghlan, H. H. 1972. "Some Reflections on the Prehistoric Working of Copper and Bronze," *ArchAustr* 52, pp. 93–104.

Coldstream, J. N., and G. L. Huxley, eds. [1972] 1973. *Kythera: Excavations and Studies Conducted by the University of Pennsylvania Museum and the British School at Athens,* repr. Park Ridge, N.J.

———. 1984. "The Minoans of Kythera," in *The Minoan Thalassocracy: Myth and Reality,* ed. R. Hägg and N. Marinatos, Stockholm, pp. 107–112.

Conophagos, C. E. 1980. *Le Laurium antique,* Athens.

Coulson, W. D. E. 1990. "The Protogeometric Period at Kavousi," in Πεπραγμένα του ΣΤ´ Διεθνούς Κρητολογικού Συνεδρίου Α´, Chania, pp. 185–193.

Courty, M.-A. 1998. "The Soil Record of an Exceptional Event at 4000 B.P. in the Middle East," in *Natural Catastrophes during Bronze Age Civilisations* (*BAR-IS* 728), ed. B. J. Peiser, T. Palmer, and M. E. Bailey, Oxford, pp. 93–108.

Courty, M.-A., P. Goldberg, and R. I. Macphail. 1989. *Soils, Micromorphology and Archaeology,* Cambridge.

Craddock, P. T. 1976. "The Composition of Copper Alloys Used by the Greek, Etruscan, and Roman Civilisations," *JAS* 4, pp. 93–113.

———. 1995. *Early Metal Mining and Production,* Washington, D.C.

———. 2001. "From Hearth to Furnace: Evidence for the Earliest Metal Smelting Technologies in the Eastern Mediterranean," *Paléorient* 26, pp. 151–165.

Craddock, P. T., I. C. Freestone, N. H. Gale, N. D. Meeks, B. Rothenberg, and M. Tite. 1985. "The Investigation of a Small Heap of Silver Smelting Debris from Rio Tinto, Huelva, Spain," in Craddock and Hughes 1985, pp. 199–218.

Craddock, P. T., and M. J. Hughes, eds. 1985. *Furnaces and Smelting Technology in Antiquity* (*BMOP* 48), London.

Craig, J. R., and D. J. Vaughan. 1994. *Ore Microscopy and Ore Petrography,* New York.

Creutzburg, N., et al. 1977. *General Geological Map of Greece. Crete Island. Scale: 1:200,000,* Athens.

Dall'Aglio, P. L., and G. Marchetti. 1991. "Settlement Patterns and Agrarian Structures of the Roman Period in the Territory of Piacenza," in Barker and Lloyd 1991, pp. 160–168.

Danin, A., R. Gerson, and J. Garty. 1983. "Weathering Patterns on Hard Limestone and Dolomite by Endolithic Lichens and Cyanobacteria: Supporting Evidence for Eolian Contribution of Terra Rossa Soil," *Soil Science* 136, pp. 213–217.

Danzer, K., D. de la Calle Garcia, G. Thiel, and M. Reichenbächer. 1999. "Classification of Wine Samples According to Origin and Grape Variety on the Basis of Inorganic and Organic Trace Analysis," *American Laboratory* 31:20, pp. 26–34.

Davaras, C. 1971. "Πρωτομινωικόν νεκροταφείον Αγίας Φωτιάς Σητείας," *AAA* 4, pp. 392–397.

———. 1973. "Νέαι έρευναι εις την ελλειψοειδή οικίαν Χαμαιζίου," in *Πεπραγμένα του Γ΄ Διεθνούς Κρητολογικού Συνεδρίου Α΄*, Athens, pp. 46–53.

———. 1977. "Αρχαιότητες και μνημεία Ανατολικής Κρήτης," *ArchDelt* 27, Β΄ 2 (1972), pp. 645–654.

———. [1981]. *Μουσείον Αγίου Νικολάου*, Athens.

———. 1989a. "Αρχαιότητες και μνημεία Ανατολικής Κρήτης," *ArchDelt* 37, Β΄ 1 (1982), p. 387.

———. 1989b. *Guide to Cretan Antiquities*, 2nd ed., Athens.

———. 2001. "Comments on the Plateia Building," in *Pseira* V, pp. 79–88.

Davaras, C., and P. P. Betancourt. 2004. *The Hagia Photia Cemetery* I: *The Tomb Groups and Architecture*, Philadelphia.

Davey, C. J. 1979. "Some Ancient Near Eastern Pot Bellows," *Levant* 11, pp. 101–111.

Davidson, D. A. 1973. "Particle Size and Phosphate Analysis—Evidence for the Evolution of a Tell," *Archaeometry* 15:1, pp. 143–152.

Davies, N. de G. 1943. *The Tomb of Rekh-mi-re at Thebes*, New York.

Davies, O. 1932. "Bronze Age Mining round the Aegean," *Nature* 130, pp. 985–987.

———. 1935. *Roman Mines in Europe*, Oxford.

Day, L. P., M. S. Mook, and J. D. Muhly, eds. 2004. *Crete beyond the Palaces: Proceedings of the Crete 2000 Conference*, Philadelphia.

Day, L. P., and L. M. Snyder. 2004. "The 'Big House' at Vronda and the 'Great House' at Karphi: Evidence for Social Structure in LM IIIC Crete," in Day, Mook, and Muhly 2004, pp. 63–79.

Day, P. M. 1991. "A Petrographic Approach to the Study of Pottery in Neopalatial East Crete" (diss. Univ. of Cambridge).

———. 1997. "Ceramic Exchange between Towns and Outlying Settlements in Neopalatial East Crete," in Hägg 1997, pp. 219–227.

Day, P. M., L. Joyner, and M. Relaki. 2003. "A Petrographic Analysis of the Neopalatial Pottery," in *Mochlos* IB, pp. 13–32.

Day, P. M., D. E. Wilson, and E. Kiriatzi. 1997. "Reassessing Specialization in Prepalatial Cretan Ceramic Production," in Laffineur and Betancourt 1997, vol. 2, pp. 275–290.

———. 1998. "Pots, Labels, and People: Burying Ethnicity in the Cemetery at Aghia Photia, Siteias," in *Cemetery and Society in the Aegean Bronze Age*, ed. K. Branigan, Sheffield, pp. 133–149.

Demakopoulou, K. 1998. *Κοσμήματα της Ελληνικής Προϊστορίας: Ο Νεολιθικός θησαυρός*, Athens.

Demargne, P. 1945. *Fouilles exécutées à Mallia. Exploration des nécropoles (1921–1933)* (*Étcrét* 7), Paris.

Demoule, J.-P., and C. Perlès. 1993. "The Greek Neolithic: A New Review," *Journal of World Prehistory* 7, pp. 355–416.

Dercksen, J. G. 1996. *The Old Assyrian Copper Trade in Anatolia*, Istanbul.

Dermitzakis, M., V. Karakitsios, and E. Lagios. 1995. "Neotectonic and Recent Deformation of Crete," in *Πεπραγμένα του Ζ΄ Διεθνούς Κρητολογικού Συνεδρίου Α΄*, Rethymnon, pp. 197–212.

De Ryck, I., A. Adriaens, and F. Adams. 2003. "Microanalytical Metal Technology Study of Ancient Near Eastern Bronze from Tell Beydar," *Archaeometry* 45, pp. 579–590.

Desborough, V. 1964. *The Last Mycenaeans and Their Successors*, Oxford.

———. 1972. *The Greek Dark Ages*, New York.

Diamond, J. 1995. "Blueprints, Bloody Ships, and Borrowed Letters," *Natural History* 104:3, pp. 16–21.

Dimopoulou, N. 1997. "Workshops and Craftsmen in the Harbour-Town of Knossos at Poros-Katsambas," in Laffineur and Betancourt 1997, vol. 2, pp. 433–438.

Doonan, R. C. P. 1996. "Old Flames, Slags, and Society: Copper Smelting Technology in the Ramsau Valley, Austria during the Bronze Age" (diss. Univ. of Sheffield).

Doonan, R. C. P., P. M. Day, N. Dimopoulou, and V. Kilikoglou. Forthcoming. "Lame Excuses for Emerging Complexity in Early Bronze Age Crete: The Metallurgical Finds from Poros Katsambas," in *Sheffield Studies in Aegean Archaeology*, ed. P. M. Day, Oxford.

Drews, R. 1993. *The End of the Bronze Age*, Princeton.

Driel-Murray, C. 2000. "Leatherwork and Skin Products," in *Ancient Egyptian Materials and Technology*, ed. P. T. Nicholson and I. Shaw, Cambridge, pp. 299–319.

Driessen, J. 2001. "History and Hierarchy: Preliminary Observations on the Settlement Pattern of Minoan Crete," in *Urbanism in the Bronze Age Aegean*, ed. K. Branigan, Sheffield, pp. 51–71.

Duke, J. A. 2002. *Dr. Duke's Phytochemical and Ethnobotanical Databases*, Department of Agriculture, Agricultural Research Service, Beltsville, Md. http://www.ars-grin.gov/cgi-bin/duke.

Dunning, F. W., W. Mykura, and D. Slater. 1982. *Mineral Deposits of Europe* 2: *Southeast Europe*, London.

Durkin, M. K., and C. J. Lister. 1983. "The Rods of Digenes: An Ancient Marble Quarry in Eastern Crete," *BSA* 78, pp. 69–96.

Easton, D. F. 2002. *Schliemann's Excavations at Troia 1870–1873* (Studia Troica Monographien 2), Mainz.

Efe, T., and A. Ilaslı. 1997. "Pottery Links between the Troad and Inland Northwestern Anatolia during the Trojan Second Settlement," in *Poliochni e l'antica età del Bronzo nell'Egeo settentrionale*, ed. C. G. Doumas and V. La Rosa, Athens, pp. 596–609.

Effenterre, H. van, and M. van Effenterre. 1997. "Towards a Study of

Neopalatial 'Villas': Modern Words for Minoan Things," in Hägg 1997, pp. 9–13.

Efstratiou, N. 1997. "Η προϊστορική πασσαλόπηκτη οικοδομική παράδοση του ελλαδικού χώρου: Μια πειραματική αρχαιολογική προσέγγιση," *Ancient Greek Technology: Proceedings, 1st International Conference, Thessaloniki 1997*, Thessaloniki, pp. 391–400.

Eidt, R. C. 1977. "Detection and Examination of Anthrosols by Phosphate Analysis," *Science* 197, pp. 1327–1333.

Eliopoulos, T. 2004. "Gournia, Vronda Kavousi, Kephala Vasilikis: A Triad of Interrelated Shrines of the Expiring Minoan Age on the Isthmus of Ierapetra," in Day, Mook, and Muhly 2004, pp. 81–90.

Elster, E. S., and C. Renfrew, eds. 2003. *Prehistoric Sitagroi: Excavations in Northeast Greece, 1968–1970* 2: *The Final Report*, Los Angeles.

Eres, Z. 2003. "Traditionelle Dorfarchitektur im Istranca-Gebirge," in *Aşağı Pinar* 1, *Archäologie in Eurasien* 15, ed. N. Karul, Z. Eres, M. Özdoğan, and H. Parzinger, Mainz, pp. 155–173.

Esin, U. 1999. "Copper Objects from the Pre-Pottery Neolithic Site of Aşıklı (Kızılkaya Village), Province of Aksaray, Turkey," in Hauptmann et al. 1999, pp. 23–30.

Evans, A. J. 1921–1935. *The Palace of Minos: A Comparative Account of the Successive Stages of the Early Cretan Civilization as Illustrated by the Discoveries at Knossos*, 4 vols., London.

Evans, J. D. 1964. "Excavation in the Neolithic Settlement at Knossos, 1957–60," *BSA* 59, pp. 132–240.

———. 1968. "Knossos Neolithic, Part II: Summary and Conclusions," *BSA* 63, pp. 267–276.

———. 1971. "Neolithic Knossos: The Growth of a Settlement," *PPS* 37, pp. 5–117.

———. 1994. "The Early Millennia: Continuity and Change in a Farming Settlement," in *Knossos: A Labyrinth of History: Papers Presented in Honour of Sinclair Hood*, ed. D. Evely, H. Hughes-Brock, and N. Momigliano, London, pp. 1–20.

Evely, R. D. G. 2000. *Minoan Crafts: Tools and Techniques* 2 (*SIMA* 92), Jonsered.

Facorellis, G., and I. Maniatis. 2002. "Radiocarbon Dating of the Neolithic Settlement of Ftelia on Mykonos," in Sampson 2002b, pp. 309–315.

Fairbridge, R. W. 1972. "Quaternary Sedimentation in the Mediterranean Region Controlled by Tectonics, Paleoclimates, and Sea Level," in *The Mediterranean Sea: A Natural Sedimentation Laboratory*, ed. D. H. Stanley, Stroudsburg, Pa., pp. 99–113.

Farrand, W. R., and C. H. Stearns. 2004. "The Bedrock Geology of Pseira," in *Pseira* VIII, pp. 13–26.

Fassoulas, C. G. 2001. *Field Guide to the Geology of Crete*, Herakleion.

Faure, P. 1964. *Fonctions des cavernes crétoises*, Paris.

———. 1966. "Les minerais de la Crète antique," *RA* 1966, pp. 45–78.

———. 1975. "Villes et villages de la Crète orientale entre 1577 et 1629 (listes inédites)," *Κρητολογία* 1, pp. 28–36.

———. 1979. "Eglises Crétoises sous roches," *Κρητολογία* 9, pp. 53–83.

———. 1980. "Les mines du roi Minos," in *Πεπραγμένα του Δ´ Διεθνούς Κρητολογικού Συνεδρίου* Α´, Athens, pp. 150–168.

———. 1984. "Hydronymes crétois," *Κρητολογία* 16–19, pp. 30–61.

———. 1989. *Recherches de toponymie crétoise*, Amsterdam.

———. 2000. "Ports et mouillages minoens de la côte nord de la Crète," in *Πεπραγμένα του Η´ Διεθνούς Κρητολογικού Συνεδρίου* Α´, Herakleion, pp. 457–469.

Ferrence, S., C. P. Swann, and P. P. Betancourt. 2002. "Analysis of White Pigments on Pottery from Five Bronze Age Minoan Archaeological Sites," *Aegean Archaeology* 5, pp. 47–54.

Fiedler, K. G. 1841. *Reise durch alle Theile des Königreiches Griechenland in den Jahren 1834 bis 1837* 2, Leipzig.

Fimmen, D. 1921. *Die kretisch-mykenische Kultur*, Leipzig.

First Civilization in Europe = *The First Civilization in Europe and the Oldest*

Gold in the World—Varna, Bulgaria, Tokyo 1982.

Fitton, J. L. 1989. "*Esse Quam Videre*: A Reconsideration of the Kythnos Hoard of Early Cycladic Tools," *AJA* 93, pp. 31–39.

Fleming, S. J., and C. P. Swann. 1992. "Recent Applications of PIXE Spectrometry in Archaeology II: Characterization of Chinese Pottery Exported to the Islamic World," *Nuclear Instruments and Methods in Physics Research B* 64, p. 528.

Flemming, N. C. 1972. "Eustatic and Tectonic Factors in the Relative Vertical Displacement of the Aegean Coast," in *The Mediterranean Sea: A Natural Sedimentation Laboratory,* ed. D. H. Stanley, Stroudsburg, Pa., pp. 189–201.

Flemming, N. C., N. M. G. Czartoryska, and P. M. Hunter. 1973. "Archeological Evidence for Vertical Earth Movements in the Region of the Aegean Island Arc," in *Science Diving International,* ed. N. C. Flemming, London, pp. 47–65.

Floyd, C. R. 1998. *The Plateia Building* (University Museum Monograph 94), Philadelphia.

Forbes, H. A. 1995. "The Identification of Pastoralist Sites within the Context of Estate-Based Agriculture in Ancient Greece," *BSA* 90, pp. 325–338.

———. 2000a. "The Agrarian Economy of the Ermionidha around 1700: An Ethnohistorical Reconstruction," in *Contingent Countryside: Settlement, Economy, and Land Use in the Southern Argolid Since 1700,* ed. S. B. Sutton, Stanford, pp. 41–70.

———. 2000b. "Landscape Exploitation via Pastoralism: Examining the 'Landscape Degradation' versus Sustainable Economy Debate in the Post-Mediaeval Southern Argolid," in Halstead and Frederick 2000, pp. 95–109.

Forbes, R. J. 1950. *Metallurgy in Antiquity,* Leiden.

———. 1972. *Studies in Ancient Technology* 9, Leiden.

Ford, C. E. 1981. "Parental Liquids of the Skaergaard Intrusion Cumulates," *Nature* 291:7, pp. 21–25.

———. 1992. "Software for Education in Geology," http://www.glg.ed.ac.uk/glgsoft.

Forenhaber, S. 1994. "Blowpipes Versus Bellows in Ancient Metallurgy," *JFA* 21, pp. 345–350.

Fortuin, A. R. 1978. "Late Cenozoic History of Eastern Crete and Implications for the Geology and Geodynamics of the Southern Aegean Area," *Geologie en Mijnbouw* 57, pp. 451–464.

Foster, B. R. 1977. "Commercial Activity in Sargonic Mesopotamia," *Iraq* 39, pp. 31–43.

———. 1987. "The Late Bronze Age Palace Economy: A View from the East," in Hägg and Marinatos 1987, pp. 11–16.

Foster, K. P. 1978. "The Mount Holyoke Collection of Minoan Pottery," *TUAS* 3, pp. 1–30.

Fotou, V. 1993. *New Light on Gournia: Unknown Documents of the Excavation at Gournia and Other Sites on the Isthmus of Ierapetra by Harriet Boyd Hawes,* Liège.

Foxhall, L. 1996. "Feeling the Earth Move: Cultivation Techniques on Steep Slopes in Classical Antiquity," in *Human Landscapes in Classical Antiquity: Environment and Culture,* ed. G. Shipley and J. Salmon, London, pp. 44–67.

Frederick, C. D., and A. Krahtopoulou. 2000. "Deconstructing Agricultural Terraces: Examining the Influence of Construction Method on Stratigraphy, Dating and Archaeological Visibility," in Halstead and Frederick 2000, pp. 79–94.

Freestone, I. C. 1988. "Melting Points and Viscosities of Ancient Slags: A Contribution to the Discussion," *Journal of the Historical Metallurgy Society* 22, pp. 49–51.

French, C. A., and T. M. Whitelaw. 1999. "Soil Erosion, Agricultural Terracing and Site Formation Processes at Markiani, Amorgos, Greece: The Micromorphological Perspective," *Geoarchaeology* 14, pp. 151–189.

Furness, A. 1953. "The Neolithic Pottery of Knossos," *BSA* 48, pp. 94–134.

Gale, D. 1991. "The Surface Artefact Assemblage for a Prehistoric Copper Mine, Austria," in Budd et al. 1991, pp. 143–150.

Gale, D., and B. Ottaway. 1991. "Geophysical Survey and Surface Artefact Assemblage of Prehistoric Copper Mining/Working Areas in Austria," in Pernicka and Wagner 1991, pp. 55–64.

Gale, N. H. 1989. "Lead Isotope Analyses Applied to Provenance Studies: A Brief Review," in *Proceedings of the 25th International Symposium on Archaeometry,* ed. Y. Maniatis, Athens, pp. 469–502.

———. 1990. "The Provenance of Metals for EBA Crete—Local or Cycladic?" in Πεπραγμένα του ΣΤ′ Διεθνούς Κρητολογικού Συνεδρίου Α′, Chania, pp. 299–316.

———. 1991. "Copper Oxhide Ingots: Their Origin and Their Place in the Bronze Age Metals Trade in the Mediterranean," in *Trade in the Bronze Age Mediterranean* (*SIMA* 90), ed. N. H. Gale, Jonsered, pp. 197–239.

———. 1996. "Comments on F. Begemann, E. Pernicka, S. Schmitt-Strecker, 'Thermi on Lesbos: A Case Study of Changing Trade Patterns.'" *OJA* 15, pp. 113–120.

———. 1999. "Lead Isotope Characterisation of the Ore Deposits of Cyprus and Sardinia and Its Application to the Discovery of the Sources of Copper for Late Bronze Age Oxhide Ingots," in *Metals in Antiquity* (*BAR-IS* 792), ed. S. M. M. Young, A. M. Pollard, P. Budd, and R. A. Ixer, Oxford, pp. 110–121.

Gale, N. H., A. Papastamataki, Z. A. Stos-Gale, and K. Leonis. 1985. "Copper Sources and Copper Metallurgy in the Aegean Bronze Age," in Craddock and Hughes 1985, pp. 81–93.

Gale, N. H., and Z. A. Stos-Gale. 1981a. "Cycladic Lead and Silver Metallurgy," *BSA* 76, pp. 169–224.

———. 1981b. "Lead and Silver in the Ancient Aegean," *Scientific American* 244, pp. 176–192.

———. 1982. "Bronze Age Copper Sources in the Mediterranean: A New Approach," *Science* 216, pp. 11–19.

———. 1986. "Oxhide Copper Ingots in Crete and Cyprus and the Bronze Age Metals Trade," *BSA* 81, pp. 81–100.

———. 1989. "Some Aspects of Early Cycladic Copper Metallurgy," in *Minería y Metalurgía en las Antiguas Civilizaciones Mediterráneas y Europeas* 1, ed. C. Domergue, Madrid, pp. 21–37.

Gale, N. H., Z. A. Stos-Gale, and J. L. Davis. 1984. "The Provenance of Lead Used at Ayia Irini, Keos," *Hesperia* 53, pp. 389–406.

Gale, N. H., Z. A. Stos-Gale, and G. R. Gilmore. 1985. "Alloy Types and Copper Sources of Anatolian Copper Alloy Artifacts," *AnatSt* 35, pp. 143–173.

Gale, S. J., and P. G. Hoare. 1991. *Quaternary Sediments: Petrographic Methods for the Study of Unlithified Rocks,* New York.

Gamble, C. 1982. "Animal Husbandry, Population and Urbanism," in *An Island Polity: The Archaeology of Exploitation in Melos,* ed. C. Renfrew and M. Wagstaff, Cambridge, pp. 161–171.

Garelli, P. 1963. *Les Assyriens en Cappadoce,* Paris.

Garnsey, P. 1992. "Yield of the Land," in Wells 1992, pp. 147–153.

Gat, J. R., and M. Magaritz. 1980. "Climatic Variations in the Eastern Mediterranean Sea Area," *Naturwissenschaften* 67, pp. 80–87.

Geologic Map of Greece = *Geologic Map of Greece 1:500,000,* 2nd ed., Athens 1983.

Gesell, G. C. 1990. "The Late Minoan IIIC Period at Kavousi (Ierapetra)," in Πεπραγμένα του ΣΤ΄ Διεθνούς Κρητολογικού Συνεδρίου Α΄, Chania, pp. 317–332.

Giannouli, E. 1995. "ΠΜ και ΜΜ αψιδωτά και πεταλόσχημα κτίρια και πιθανή τους σημασία," in Πεπραγμένα του Ζ΄ Διεθνούς Κρητολογικού Συνεδρίου Α΄ 1, Rethymnon, pp. 125–147.

Gimbutas, M. 1977. "Varna: A Sensationally Rich Cemetery of the Karanovo Civilization," *Expedition* 19:4, pp. 39–47.

Gladfelter, B. G. 1977. "Geoarchaeology: The Geomorphologist and Archaeology," *AmerAnt* 42, pp. 519–538.

Glotz, G. 1923. *La civilization égéenne,* Paris.

Glowacki, K. T. 2004. "Household Analysis in Dark Age Crete," in Day, Mook, and Muhly 2004, pp. 125–136.

Golden, J., T. E. Levy, and A. Hauptmann. 2001. "Recent Discoveries Concerning Chalcolithic Metallurgy at Shiqmim, Israel," *JAS* 28, pp. 951–963.

Görsdorf, J. von, and J. Bojadžiev. 1996. "Zur absoluten chronologie der bulgarischen Urgeschichte," *Eurasia Antiqua* 2, pp. 105–173.

Gowland, W. 1930. *The Metallurgy of the Non-Ferrous Metals,* London.

Grammenos, D. B. 1997. Νεολιθική Μακεδονία, Athens.

Greig, J. R. A., and P. M. Warren. 1974. "Early Bronze Age Agriculture in Western Crete," *Antiquity* 48, pp. 130–132.

Hadjianastasiou, O., and S. [J. A.] MacGillivray. 1988. "An Early Bronze Age Copper Smelting Site on the Aegean Island of Kythnos, Part Two: The Archaeological Evidence," in *Aspects of Ancient Mining and Metallurgy: Acta of a British School at Athens Centenary Conference at Bangor, 1986,* ed. J. E. Jones, Gwynedd, pp. 31–34.

Hägg, R., ed. 1997. *The Function of the "Minoan Villa,"* Stockholm.

Hägg, R., and N. Marinatos, eds. 1987. *The Function of the Minoan Palaces,* Stockholm.

Haggis, D. C. 1992. "The Kavousi-Thriphti Survey: An Analysis of Settlement Patterns in an Area of Eastern Crete in the Bronze Age and Early Iron Age" (diss. Univ. of Minnesota).

———. 1993a. "The Early Minoan Burial Cave at Ayios Antonios and Some Problems in Early Bronze Age Chronology," *SMEA* 31, pp. 7–34.

———. 1993b. "Intensive Survey, Traditional Settlement Patterns, and Dark Age Crete: The Case of Early Iron Age Kavousi," *JMA* 6, pp. 131–174.

———. 1995. "Archaeological Survey

at Kavousi, Crete: Settlement Development in Middle Minoan and Late Minoan I," in Πεπραγμένα του Z΄ Διεθνούς Κρητολογικού Συνεδρίου Α΄ 1, Rethymnon, pp. 369–381.

———. 1996a. "The Port of Tholos in Eastern Crete and the Role of a Roman Horreum along the Egyptian 'Corn Route,'" *OJA* 15, pp. 183–209.

———. 1996b. "Archaeological Survey at Kavousi, East Crete: Preliminary Report," *Hesperia* 65, pp. 373–432.

———. 1996c. "Excavations at Kalo Khorio, East Crete," *AJA* 100, pp. 645–681.

———. 1997. "The Typology of the Early Minoan I Chalice and the Cultural Implications of Form and Style in Early Bronze Age Ceramics," in Laffineur and Betancourt 1997, pp. 291–299.

———. 2000. "Coarse Ware Ceramic Distribution in the North Isthmus of Ierapetra in the Bronze Age," in Πεπραγμένα του Η΄ Διεθνούς Κρητολογικού Συνεδρίου Α΄ 1, Herakleion, pp. 535–543.

———. 2002. "Charting the Landscapes of Power in 'Minoan' Crete: A View from the Countryside in Eastern Crete," in *Labyrinth Revisited: Rethinking "Minoan" Archaeology*, ed. Y. Hamilakis, Oxford, pp. 120–142.

———. 2005. *Kavousi I: The Archaeological Survey of the Kavousi Region* (Prehistory Monographs 16), Philadelphia.

Haggis, D. C., and M. S. Mook. 1993. "The Kavousi Coarse Wares: A Bronze Age Chronology for Survey in the Mirabello Area, East Crete," *AJA* 97, pp. 265–293.

Hall, E. H. 1904–1905. "Early Painted Pottery from Gournia, Crete," *Transactions of the Department of Archaeology, Free Museum of Science and Art* 1, pp. 191–205.

———. 1908. "Early Minoan III Ware from the North Trench," in Hawes et al. 1908, p. 57.

———. 1912. *Excavations in Eastern Crete: Sphoungaras*, Philadelphia.

Halstead, P. 1981. "Counting Sheep in Neolithic and Bronze Age Greece," in *Pattern of the Past: Studies in Memory of David Clarke*, ed. I. Hodder, G. Isaac, and N. Hammond, Cambridge, pp. 307–339.

———. 1987. "Traditional and Ancient Rural Economy in Mediterranean Europe: plus ça change?" *JHS* 107, pp. 77–87.

———. 1992. "Agriculture in the Bronze Age Aegean," in Wells 1992, pp. 105–117.

———. 2000. "Land Use in Postglacial Greece: Cultural Causes and Environmental Effects," in Halstead and Frederick 2000, pp. 110–128.

Halstead, P., and C. Frederick, eds. 2000. *Landscape and Land Use in Postglacial Greece*, Sheffield.

Hanke, H. 1994. "Der Bergbau und die Mineralien von Lavrion, Griechenland," *Emser Hefte* 15:2, pp. 1–80.

Hansen, J. M. 1988. "Agriculture in the Prehistoric Aegean: Data versus Speculation," *AJA* 92, pp. 39–52.

Hanson, V. D. 1992. "Practical Aspects of Grape-Growing and the Ideology of Greek Viticulture," in Wells 1992, pp. 161–166.

Harborne, J. B., H. Baxter, and G. P. Moss, eds. 1999. *Phytochemical Dictionary*, 2nd ed., London.

Harrison, G. W. M. 1993. *The Romans and Crete*, Amsterdam.

Hassan, F. A. 1978. "Sediments in Archaeology: Methods and Implications for Paleoenvironmental and Cultural Analysis," *JFA* 5, pp. 197–213.

Hauptmann, A. 1989. "The Earliest Periods of Copper Metallurgy in Feinan, Jordan," in Hauptmann, Pernicka, and Wagner 1989, pp. 119–135.

———. 2000. *Zur frühen Metallurgie des Kupfers in Fenan/Jordanien*, Bochum.

———. 2001. "The Archaeometallurgy of Copper at Feinan, Jordan: Field Research and Analytical Work of an Ancient Ore District," in *III Congreso Nacional de Arqueometría*, ed. B. M. Gómez Tubío, M. A. Respaldiza, and M. L. Rodríguez, Seville, pp. 419–436.

Hauptmann, A., H.-G. Bachmann, and R. Maddin. 1994. "Chalcolithic Copper Smelting: New Evidence from Excavations at Feinan, Jordan," in *Archaeometry 94*, ed. S. Demirci, A. M. Özer, and G. D. Summers, Ankara, pp. 3–10.

Hauptmann, A., F. Begemann, and S. Schmitt-Strecker. 1999. "Copper Objects from Arad—Their Composition and Provenance," *BASOR* 314, pp. 1–17.

Hauptmann, A., R. Maddin, and M. Prange. 2002. "On the Structure and Composition of Copper and Tin Ingots Excavated from the Shipwreck of Uluburun," *BASOR* 328, pp. 1–30.

Hauptmann, A., E. Pernicka, T. Rehren, and Ü. Yalçin, eds. 1999. *The Beginnings of Metallurgy* (*Der Anschnitt* Beiheft 9), Bochum.

Hauptmann, A., E. Pernicka, and G. A. Wagner, eds. 1989. *Old World Archaeometallurgy* (*Der Anschnitt* Beiheft 7), Bochum.

Hauptmann, A., T. Rehren, and S. Schmitt-Strecker. 2003. "Early Bronze Age Copper Metallurgy at Shar-I-Sokhta (Iran) Reconsidered," in *Man and Mining*, ed. T. Stöllner, G. Körlin, G. Steffens, and J. Cierny, Bochum, pp. 197–213.

Hauptmann, A., S. Schmitt-Strecker, F. Begemann, and A. Palmieri. 2002. "Chemical Composition and Lead Isotopy of Metal Objects from the 'Royal' Tomb and Other Related Finds at Arslantepe, Eastern Anatolia," *Paléorient* 28, pp. 43–69.

Hauptmann, A., G. Weisgerber, and H.-G. Bachmann. 1989. "Early Copper Metallurgy in Oman," in *The Beginning of the Use of Metals and Alloys*, ed. R. Maddin, Cambridge, Mass., pp. 34–51.

Hawes, C. H., and H. Boyd Hawes. 1909. *Crete: The Forerunner of Greece*, London.

Hawes, H. Boyd, B. E. Williams, R. B. Seager, and E. H. Hall. 1908. *Gournia, Vasiliki, and Other Prehistoric Sites on the Isthmus of Hierapetra, Crete*, Philadelphia.

Hayden, B. J. 1995. "Rural Settlement of the Orientalizing through Early Classical Period: The Meseleroi

Valley, Eastern Crete," *Aegean Archaeology* 2, pp. 93–142.

———. 1997. "Evidence for 'Megalithic Farmsteads' of Late Minoan III through Early Iron Age Date," in *La Crète mycénienne*, ed. J. Driessen and A. Farnoux, Athens, pp. 195–204.

———. 2003a. "Final Neolithic–Early Minoan I–IIA Settlement in the Vrokastro Area, Eastern Crete," *AJA* 107, pp. 363–412.

———. 2003b. "The Final Neolithic–Early Minoan I/IIA Settlement History of the Vrokastro Area, Mirabello, Eastern Crete," *Mediterranean Archaeology and Archaeometry* 3, pp. 31–44.

———. 2004. *Reports on the Vrokastro Area, Eastern Crete 2: The Settlement History of the Vrokastro Area and Related Studies,* Philadelphia.

Hayden, B. J., J. A. Moody, and O. Rackham. 1992. "The Vrokastro Survey Project, 1986–1989: Research Design and Preliminary Results," *Hesperia* 61, pp. 293–353.

Healy, J. F. 1978. *Mining and Metallurgy in the Greek and Roman World,* London.

Hedges, R. E. M., R. A. Housley, C. R. Bronk, and G. J. van Klinken. 1990. "Radiocarbon Dates from the Oxford AMS System: Archaeometry Datelist 11," *Archaeometry* 32, pp. 211–237.

Hegde, K. T. M., and J. E. Ericson. 1985. "Ancient Indian Copper Furnaces," in Craddock and Hughes 1985, pp. 59–67.

Herdits, H. 2003. "Bronze Age Smelting Site in the Mitterberg Mining Area in Austria," in *Mining and Metal Production through the Ages,* ed. P. T. Craddock and J. Lang, London, pp. 69–75.

Hiller, S., and V. Nikolov. 1988. *Tell Karanovo 1988,* Salzburg.

Hood, S. 1971. *The Minoans,* London.

———. 1983. "The 'Country House' and Minoan Society," in Krzyszkowska and Nixon 1983, pp. 129–135.

———. 1986. "Evidence for Invasions in the Aegean Area at the End of the Early Bronze Age," in *The End of the Early Bronze Age in the Aegean,* ed. G. Cadogan, Leiden, pp. 31–68.

———. 1990a. "Autochthons or Settlers? Evidence for Immigration at the Beginning of the Early Bronze Age in Crete," in Πεπραγμένα του ΣΤ΄ Διεθνούς Κρητολογικού Συνεδρίου Α΄, Chania, pp. 367–375.

———. 1990b. "Settlers in Crete c. 3000 B.C.," *Cretan Studies* 2, pp. 151–158.

Hood, S., and D. Smyth. 1981. *Archaeological Survey of the Knossos Area,* London.

Hope Simpson, R. 1983. "The Limitations of Surface Surveys," in Keller and Rupp 1983, pp. 45–47.

Hutchinson, R. W. 1962. *Prehistoric Crete,* Baltimore.

Jameson, M. H. 1992. "Agricultural Labor in Ancient Greece," in Wells 1992, pp. 135–146.

Jameson, M. H., C. N. Runnels, and T. van Andel. 1994. *A Greek Countryside: The Southern Argolid from Prehistory to the Present Day,* Stanford.

Jashemski, W. F. 1999. *A Pompeian Herbal,* Austin.

Johnson, M. 1996. "Water, Animals, and Agricultural Technology: A Study of Settlement Patterns and Economic Change in Neolithic Southern Greece," *OJA* 15, pp. 267–295.

———. 1999. "Chronology of Greece and South-East Europe in the Final Neolithic and Early Bronze Age," *PPS* 65, pp. 319–336.

Jones, G. 1987. "Agricultural Practice in Greek Prehistory," *BSA* 82, pp. 115–123.

———. 1992. "Weed Phytosociology and Crop Husbandry: Identifying a Contrast between Ancient and Modern Practice," in *Festschrift for Professor van Zeist* (*Review of Palaeobotany and Palynology* 73), ed. J. P. Pals, J. Buurman, and M. van der Veen, Amsterdam, pp. 133–143.

Jones, G., A. Bogaard, M. Charles, and J. G. Hodgson. 2000. "Distinguishing the Effects of Agricultural Practices Relating to Fertility and Disturbance: A Functional Ecological Approach in

Archaeobotany," *JAS* 27, pp. 1073–1084.

Jones, G., A. Bogaard, P. Halstead, M. Charles, and H. Smith. 1999. "Identifying the Intensity of Crop Husbandry on the Basis of Weed Floras, *BSA* 94, pp. 167–189.

Jones, G., and I. Smith. Forthcoming. "The Plant Remains from Buildings AF South and AF North," in *Pseira X*.

Jovanović, B. 1978. "The Oldest Copper Metallurgy in the Balkans: A Study of the Diffusion of Copper from Asia Minor to Southeastern Europe," *Expedition* 21:1, pp. 9–17.

———. 1995. "Continuity of the Prehistoric Mining in the Central Balkans," in *Ancient Mining and Metallurgy in Southeast Europe*, ed. B. Jovanović, Belgrade, pp. 29–35.

Juleff, G. 1998. *Early Iron and Steel in Sri Lanka*, Mainz.

Kanta, A. 1980. *The Late Minoan III Period in Crete: A Survey of Sites, Pottery and their Distribution* (*SIMA* 58), Göteborg.

———. 1983. "Minoan and Traditional Crete: Some Parallels between two Cultures in the Same Environment," in Krzyszkowska and Nixon 1983, pp. 155–162.

Karageorghis, V. 1992. "The Crisis Years: Cyprus," in *The Crisis Years: The 12th Century B.C.*, ed. W. A. Ward and M. Joukowsky, Dubuque, pp. 79–86.

Karantzali, E. 1996. *Le Bronze Ancien dans les Cyclades et en Crète* (*BAR-IS* 631), Oxford.

Karrer, W. 1958–1981. *Konstitution und Vorkommen der organischen Pflanzenstoffe*, Basel.

Katsa, L. 1997. "Copper Alloys and Cycladic Metallurgy during the Bronze Age," *Ancient Greek Technology: Proceedings, 1st International Conference, Thessaloniki 1997*, Thessaloniki, pp. 73–83.

Katsarou, S., and D. U. Schilardi. 2004 "Emerging Neolithic and Early Cycladic Settlements in Paros: Koukounaries and Sklavouna," *BSA* 99, pp. 23–48.

Keller, D. R., and D. W. Rupp, eds. 1983. *Archaeological Survey in the Mediterranean Area* (*BAR-IS* 155), Oxford.

Keos I = J. E. Coleman, *Kephala: A Late Neolithic Settlement and Cemetery*, Princeton 1977.

Klein, N. L. 2004. "The Architecture of the Late Minoan IIIC Shrine (Building G) at Vronda, Kavousi," in Day, Mook, and Muhly 2004, pp. 91–101.

Knappett, C. 1999. "Assessing a Polity in Protopalatial Crete: The Malia-Lasithi State," *AJA* 103, pp. 615–639.

Kollmorgen Instruments Corporation. 1992. *Munsell Soil Color Charts*, Newburgh, N.Y.

Kommos = *Kommos: An Excavation on the South Coast of Crete*.

    I:1 = J. W. Shaw and M. C. Shaw, eds. *The Kommos Region, Ecology, and Minoan Industries*, Princeton 1995.

    II = P. P. Betancourt, *The Final Neolithic through Middle Minoan III Pottery*, Princeton 1990.

    III = L. V. Watrous, *The Late Bronze Age Pottery*, Princeton 1992.

Kongoli, F., and A. Yazawa. 2001. "Liquidus Surface of FeO-Fe_2O_3-SiO_2-CaO Slag Containing Al_2O_3, MgO and Cu_2O at Intermediate Oxygen Partial Pressures," *Metallurgical and Materials Transactions* 32B, pp. 583–592.

Korfmann, M. 2001. "Neue Aspekte zum 'Schatz des Priamos': Der Schatz A von Troia, sein Auffindungsort und seine Datierung," in *Troia-Traum und Wirklichkeit*, ed. M. Korfmann, Stuttgart, pp. 373–383.

Koucky, F. L., and A. Steinberg. 1982a. "Ancient Mining and Mineral Dressing on Cyprus," in *Early Pyrotechnology: The Evolution of the First Fire-Using Industries*, ed. T. A. Wertime and S. F. Wertime, Washington, D.C., pp. 149–180.

———. 1982b. "The Ancient Slags of Cyprus," in *Early Metallurgy in Cyprus, 4000–500 B.C.*, ed. J. D. Muhly, R. Maddin, and V. Karageorghis, Nicosia, pp. 117–141.

Kovács, T., ed. 1994. *Treasures of the Hungarian Bronze Age*, Budapest.

Krahtopoulou, A. 1997. "Agricultural Terraces in Livadi, Thessaly, Greece," in *Archaeological Sciences* (1995), ed. A. Sinclair, E. Slater, and J. Gowlett, Oxford, pp. 249–259.

Kraus, F. R. 1982. "'Karum,' ein Organ städtischer Selbstverwaltung der altbabylonischen Zeit," in *Les pouvoirs locaux en Mésopotamie et dans les régions adjacentes*, ed. A. Finet, Brussels, pp. 29–42.

Kresten, P. 1986. "Melting Points and Viscosities of Ancient Slags: A Discussion," *Journal of Historical Metallurgy Society* 20:1, pp. 43–45.

Krzyszkowska, O., and L. Nixon, eds. 1983. *Minoan Society: Proceedings of the Cambridge Colloquium 1981*, Bristol.

Laffineur, R., and P. P. Betancourt, eds. 1997. *TEXNH: Craftsmen, Craftswomen and Craftsmanship in the Aegean Bronze Age* (*Aegaeum* 16), 2 vols., Liège.

Lamb, W. 1929. *Greek and Roman Bronzes*, London.

———. 1936. *Excavations at Thermi in Lesbos*, Cambridge.

Lang, G. 1978. "Density of Liquid Elements," in *CRC Handbook of Chemistry and Physics*, 58th ed., ed. R. C. Weast, Cleveland, section B257.

Larsen, M. T. 1967. *Old Assyrian Caravan Procedures*, Istanbul.

———. 1976. *The Old Assyrian City State and its Colonies*, Copenhagen.

Leach, H. M. 1997. "The Terminology of Agricultural Origins and Food Production Systems—A Horticultural Perspective," *Antiquity* 71, pp. 135–148.

Lechtman, H. 1991. "The Production of Copper-Arsenic Alloys in the Central Andes: Highland Ores and Coastal Smelters?" *JFA* 18, pp. 43–76.

———. 1996. "Arsenic Bronze: Dirty Copper or Chosen Alloy? A View from the Americas," *JFA* 23, pp. 477–514.

Lechtman, H., and S. Klein. 1999. "The Production of Copper-Arsenic Alloys (Arsenical Bronze) by Cosmelting: Modern Experiment, Ancient Practice," *JAS* 26, pp. 497–526.

Lee, W. E. 2001. "Pylos Regional Archaeological Project, Part IV, Change and the Human Landscape

in a Modern Greek Village in Messenia," *Hesperia* 70, pp. 49–98.
Levi, D. 1976. *Festòs e la civilta minoica*, 2 vols., Rome.
Lloyd, J. 1991. "Forms of Rural Settlement in the Early Roman Empire," in Barker and Lloyd 1991, pp. 233–240.
Lohman, H. 1992. "Agriculture and Country Life in Classical Athens," in Wells 1992, pp. 29–56.
Lyon, T. L., and H. O. Buckman. 1929. *The Nature and Properties of Soils*, New York.
Mabberley, D. J. 1997. *The Plant-Book: A Portable Dictionary of the Vascular Plants*, 2nd ed., Cambridge.
MacGillivray, J. A. 1997. "The Cretan Countryside in the Old Palace Period," in Hägg 1997, pp. 21–25.
MacGillivray, J. A., L. H. Sackett, J. M. Driessen, and S. Hemingway. 1992. "Excavations at Palaikastro 1991," *BSA* 87, pp. 121–152.
Mackenzie, W. F., C. H. Donaldson, and C. Guilford. 1982. *Atlas of Igneous Rocks and Their Textures*, New York.
MacKinnon, M. 1998. "Archaeological Field Survey in the Vicinity of Monte Irsi, 1996–1997," *AJA* 102, pp. 416–417.
Maddin, R. 1996. "Some Metallurgical Considerations in the Reconstruction of Early Metallurgy," in Bagolini and Lo Schiavo 1996, pp. 9–15.
Maddin, R., J. D. Muhly, and T. S. Stech. 1999. "Early Metalworking at Çayönü," in Hauptmann et al. 1999, pp. 37–44.
Mangou, H., and P. V. Ioannou. 1998. "On the Chemical Composition of Prehistoric Greek Copper-Based Artefacts from Crete," *BSA* 93, pp. 91–102.
———. 1999. "On the Chemical Composition of Prehistoric Greek Copper-Based Artefacts from Mainland Greece," *BSA* 94, pp. 81–100.
Manning, S. W. 1994. "The Emergence of Divergence: Development and Decline on Bronze Age Crete and the Cyclades," in *Development and Decline in the Mediterranean Bronze Age* (Sheffield Archaeological Monographs 8), ed. C. Mathers and S. Stoddart, Sheffield, pp. 221–270.

Mansfeld, G. 2001. "Die 'Königsgräber' von Alaca Höyük und ihre Beziehunger nach Kaukasien," *Archäologische Mitteilungen aus Iran und Turan* 33, pp. 19–61.
Manteli, K. 1992. "The Neolithic Well at Kastelli Phournis in Eastern Crete," *BSA* 87, pp. 103–120.
Maran, J. 2000. "Das ägäische Chalkolithikum und das erste Silber in Europa," in *Studien zur Religion und Kultur Kleinasiens und des ägäischen Bereiches: Festschrift für Baki Öğün* (Asia Minor Studien 39), ed. C. Isik, Bonn, pp. 179–193.
Maréchal, J.-R. 1958. "Étude sur les propriétés mécaniques des cuivres à l'arsenic," *Métaux Corrosion Industries* 33, pp. 377–383.
Marinatos, S. 1929. "Πρωτομινωϊκός θολωτός τάφος παρά το χωρίον Κράσι Πεδιάδος," *ArchDelt* 12, pp. 102–141.
Marinos, G. 1982. "Greece," in *Mineral Deposits of Europe 2: Southeast Europe*, ed. F. W. Dunning, W. Mykura, and D. Slater, London, p. 237.
Marinos, G. P., and W. E. Petrascheck. 1956. *Lavrion* (Geological and Geophysical Research 4:1), Athens.
Mariolopoulos, E. G. 1961. *An Outline of the Climate of Greece*, Athens.
Matthiae, P., F. Pinnock, and G. Scandone Matthiae, eds. 1995. *Ebla: Alle origini della civiltà urbana: Trent'anni di scavi in Siria dell'Università di Roma "La Sapienza,"* Milan.
Maxwell, V. 2002. "Metalworking at Ftelia," in Sampson 2002b, pp. 147–149.
McGeehan-Liritzis, V. 1996. *The Role and Development of Metallurgy in the Late Neolithic and Early Bronze Age of Greece* (SIMA-PB 122), Jonsered.
McGeehan-Liritzis, V., and N. H. Gale. 1988. "Chemical and Lead Isotope Analyses of Greek Late Neolithic and Early Bronze Age Metals," *Archaeometry* 30, pp. 199–225.
Meiggs, R. 1982. *Trees and Timber in the Ancient Mediterranean World*, Oxford.
Melas, M. 1999. "The Ethnography of Minoan and Mycenaean Beekeeping," in *MELETEMATA*, vol. 2, pp. 486–491.

MELETEMATA = P. P. Betancourt, V. Karageorghis, R. Laffineur, and W.-D. Niemeier., eds. *MELETEMATA. Studies in Aegean Archaeology Presented to Malcolm H. Wiener as He Enters His 65th Year* (*Aegaeum* 20), 3 vols., Liège 1999.

Mellaart, J. 1967. *Çatal Höyük,* London.

Mellink, M. J. 1956. "The Royal Tombs at Alaca Hüyük and the Aegean World," in *The Aegean and the Near East: Festschrift Hetty Goldman,* ed. S. S. Weinberg, Locust Valley, N.Y., pp. 463–479.

———. 1986. "The Early Bronze Age in West Anatolia: Aegean and Asiatic Correlations," in *The End of the Early Bronze Age in the Aegean,* ed. G. Cadogan, Leiden, pp. 139–152.

Merkel, J. F. 1983. "Summary of Experimental Results for Late Bronze Age Copper Smelting and Refining," *MASCAJ* 2:6, pp. 173–178.

———. 1990. "Experimental Reconstruction of Bronze Age Copper Smelting Based on Archaeological Evidence from Timna," in *The Ancient Metallurgy of Copper,* ed. B. Rothenberg, London, pp. 78–122.

Merkel, J., and B. Rothenberg. 1999. "The Earliest Steps to Copper Metallurgy in the Western Arabah," in *The Beginnings of Metallurgy,* ed. A. Hauptmann, E. Pernicka, T. Rehern, and Ü. Yalçin, Bochum, pp. 149–165.

Miller, A., and R. A. Anthes. 1980. *Meteorology,* 4th ed., Columbus.

Milton, C., E. J. Diwornic, R. B. Finkelman, and P. Toulmin. 1976. "Slag from an Ancient Smelter at Timna, Israel," *Journal of the Historical Metallurgy Society* 10, pp. 24–33.

Mirtsou, E., U. Zwicker, K. Nigge, and A. A. Katsanos. 1997. "Analysis and Metallographic Examination of Dimitra's Metallic Samples," in Grammenos 1997, pp. 91–94.

Mochlos = *Excavations at Mochlos, Crete.*
  IA = J. S. Soles and C. Davaras, eds. *Period III. Neopalatial Settlement on the Coast: The Artisans' Quarter and the Farmhouse at Chalinomouri. The Sites* (Prehistory Monographs 7), Philadelphia 2003.
  IB = K. A. Barnard and T. M. Brogan, *Period III. Neopalatial Settlement on the Coast: The Artisans' Quarter and the Farmhouse at Chalinomouri. The Neopalatial Pottery* (Prehistory Monographs 8), Philadelphia 2003.
  IC = J. S. Soles and C. Davaras, eds. *Period III. Neopalatial Settlement on the Coast: The Artisans' Quarter and the Farmhouse at Chalinomouri. The Small Finds* (Prehistory Monographs 9), Philadelphia 2004.

Mohen, J.-P., and C. Éluère, eds. 1991. *Découverte du métal,* Paris.

Mohen, J.-P., and P. Walter. 1994. "Le four-creuset, une invention inédite de l'Age du Bronze européen," *Techne* 1, pp. 103–110.

Momigliano, N. 1990. "The Development of the Footed Goblet ('Egg-Cup') from EM II to MM III at Knossos," in *Πεπραγμένα του ΣΤ′ Διεθνούς Κρητολογικού Συνεδρίου* Α′, Chania, pp. 477–478.

———. 1991. "MM IA Pottery from Evans' Excavations at Knossos: A Reassessment," *BSA* 86, pp. 149–271.

———. 2000. "Knossos 1902, 1905: The Prepalatial and Protopalatial Deposits from the Room of the Jars in the Royal Pottery Stores," *BSA* 95, pp. 65–105.

Money-Coutts, M. B. 1935–1936. "Metal," in Pendlebury, Pendlebury, and Money-Coutts 1935–1936, pp. 102–109.

Moody, J. A. 1987a. "The Environmental and Cultural Prehistory of the Khania Region of West Crete: Neolithic through Late Minoan III" (diss. Univ. of Minnesota).

———.1987b. "The Minoan Palace as a Prestige Artifact," in Hägg and Marinatos 1987, pp. 235–240.

———. 1997. "The Cretan Environment: Abused or Just Misunderstood?" in *Aegean Strategies: Studies of Culture and Environment on the European Fringe,* ed. P. N. Kardulias and M. T. Shutes, Lanham, Md., pp. 61–79.

———. 2000. "Holocene Climate Change in Crete," in Halstead and Frederick 2000, pp. 52–61.

Moody, J. A., and A. T. Grove. 1990. "Terraces and Enclosure Walls in the Cretan Landscape," in *Man's Role in the Shaping of the Eastern Mediterranean Landscape,* ed. S. Bottema, G. Entjes-Nieborg, and W. van Zeist, Rotterdam, pp. 183–191.

Moody, J., O. Rackham, and G. Rapp Jr. 1996. "Environmental Archaeology of Prehistoric NW Crete," *JFA* 23, pp. 273–297.

Mook, M. S., and D. C. Haggis. 1990. "The Kavousi-Thriphti Survey, 1988–1989," *AJA* 94, p. 323.

Moorey, P. R. S. 1988. "Early Metallurgy in Mesopotamia," in *The Beginnings of the Use of Metals and Alloys,* ed. R. Maddin, Cambridge, Mass., pp. 28–33.

Morris, M. W. 1994. "A Pedological Investigation of Catchment Basins below Late Minoan Period Archaeological Sites of Eastern Crete, Greece" (diss. Univ. of Tennessee, Knoxville).

———. 2002. *Soil Science and Archaeology: Three Test Cases from Minoan Crete,* Philadelphia.

Mosso, A. 1908. "Le armi più antiche di ramo e di bronzo," *MemLinc,* 5th ser., 12:6, pp. 479–582.

———. 1910. *The Dawn of Mediterranean Civilisation,* London.

Moulherat, C., M. Tengberg, J.-F. Haquet, and B. Mille. 2002. "First Evidence of Cotton at Neolithic Mehrgarh, Pakistan: Analysis of Mineralized Fibres from a Copper Bead," *JAS* 29, pp. 1393–1401.

Mozsolics, A. 1967. *Bronzefunde des Karpatenbeckens Depotfundhorizonte von Hajdúsámson und Kosziderpadlás,* Budapest.

Muhly, J. D. 1973. *Copper and Tin: The Distribution of Mineral Resources and the Nature of the Metals Trade in the Bronze Age,* Hamden, Conn.

———. 1978. "New Evidence for Sources of and Trade in Bronze Age Tin," in *The Search for Ancient Tin,* ed. A. D. Franklin, J. S. Olin, and T. A. Wertime, Washington, D.C., pp. 43–48.

———. 1985a. "Beyond Typology: Aegean Metallurgy in its Historical Context," in *Contributions to Aegean Archeology: Studies in Honor of William A. McDonald* (Publications in Ancient Studies 1), ed. N. Wilkie and W. D. E. Coulson, Minneapolis, pp. 109–141.

———. 1985b. "Sources of Tin and the Beginnings of Bronze Metallurgy," *AJA* 89, pp. 275–291.

———. 1989. "Çayönü Tepesi and the Beginnings of Metallurgy in the Ancient World," in Hauptmann, Pernicka, and Wagner 1989, pp. 1–11.

———. 1991. "Copper in Cyprus: the Earliest Phase," in Mohen and Éluère 1991, pp. 357–374.

———. 1993. "Early Bronze Age Tin and the Taurus," *AJA* 97, pp. 239–253.

———. 1996. "The First Use of Metals in the Aegean," in Bagolini and Lo Schiavo 1996, pp. 75–84.

———. 1997. "Recent Works in Archaeometallurgy," *AJA* 101, pp. 771–773.

———. 1999. "Copper and Bronze in Cyprus and the Eastern Mediterranean," in *The Archaeometallurgy of the Asian Old World* (*MASCAP* 16), ed. V. C. Pigott, Philadelphia, pp. 15–25.

———. 2002. "Early Metallurgy in Greece and Cyprus," in *Anatolian Metal 2* (*Der Anschnitt* Beiheft 15), ed. Ü. Yalçin, Bochum, pp. 77–82.

———. 2004a. Review of Elster and Renfrew 2003, *BMCR*, June 21, 2004, pp. 1–6.

———. 2004b. "Chrysokamino and the Beginnings of Metal Technology on Crete and in the Aegean," in Day, Mook, and Muhly 2004, pp. 283–289.

Muhly, J. D., and E. Pernicka. 1992. "Early Trojan Metallurgy and Metals Trade," in *Heinrich Schliemann: Grundlagen und Ergebnisse moderner Archäologie 100 Jahre nach Schliemanns Tod*, ed. J. Herrmann, Berlin, pp. 309–318.

Muhly, J. D., and T. Stech. 2003. "The Metallurgy of Ninevite 5," in *The Origins of North Mesopotamian Civilization: Ninevite 5 Chronology, Economy, Society* (Subartu 9), ed. E. Rova and H. Weiss, Turnhout, Belgium, pp. 417–428.

Mullen, G. E. 1977. "Magnetic Susceptibility of the Soil and Its Significance in Soil Science, a Review," *Journal of Soil Science* 28, pp. 223–246.

Müller-Karpe, A. 1994. *Altanatolisches Metallhandwerk* (Offa-Bücher 75), Neumünster.

Myer, G. H. 1979. "X-Ray Powder Diffraction and Petrographic Thin Sections," in Betancourt 1979, p. 6.

Myer, G. H., and P. P. Betancourt. 1981. "The Composition of Vasilike Ware and the Production of the Mottled Colours of the Slip," in *Scientific Studies in Ancient Ceramics*, ed. M. J. Hughes, London, pp. 51–55.

Myer, G. H., K. G. McIntosh, and P. P. Betancourt. 1995. "Appendix A," in *Pseira* I, pp. 143–153.

Nakou, G. 1995. "The Cutting Edge: A New Look at Aegean Metallurgy," *JMA* 8, pp. 1–32.

———. 1997. "The Role of Poliochni and the North Aegean in the Development of Aegean Metallurgy," in *Poliochni e l'antica età del bronzo nell'Egeo settentrionale*, ed. C. G. Doumas and V. La Rosa, Athens, pp. 634–648.

Nathan, H. D., and C. K. Van Kirk. 1978. "A Model of Magmatic Crystallization," *Journal of Petrology* 19, pp. 66–94.

Nevros, K. I., and I. A. Zvorykin. 1936. "Investigations of Red Soils of Attica, Greece," *Soil Science* 41, pp. 397–421.

Niemeier, W.-D. 1997. "The Origins of the Minoan 'Villa' System," in Hägg 1997, pp. 15–19.

Nixon, L. 1987. "Neo-Palatial Outlying Settlements and the Function of the Minoan Palaces," in Hägg and Marinatos 1987, pp. 95–98.

Noble, J. V. 1988. *The Techniques of Painted Attic Pottery*, rev. ed., London.

Nodarou, E. 1998. "Soils and Sediments from the Habitation Site at Chrysokamino, East Crete: A Geoarchaeological Study" (master's thesis, Univ. of Sheffield).

Noll, W. 1982. "Mineralogie und Technik der keramiken Altkretas," *Neues Jahrbuch für Mineralogie* 143, pp. 150–199.

Noll, W., R. Holm, and L. Born. 1971. "Chemie und Techniken altkretischer Vasenmalerei vom Kamares-

Typ," *Die Naturwissenschaften* 58, pp. 615–618.

Northover, J. P. 1989. "Properties and Use of Arsenic-Copper Alloys," in Hauptmann, Pernicka, and Wagner 1989, pp. 111–118.

Nowicki, K. 1996a. "Report on Investigations in Greece X: Studies in 1993 and 1994," *Archeologia* 46 (1995), pp. 63–70.

———. 1996b. "Lasithi (Crete): One Hundred Years of Archaeological Research," *Aegean Archaeology* 3, pp. 27–47.

———. 2002. "A Middle Minoan II Deposit at the Refuge Site of Monastiraki Katalimata," *Aegean Archaeology* 5 (2001), pp. 27–45.

———. 2004. "South of Kavousi, East of Mochlos: The West Siteia Mountains at the End of the Bronze Age," in Day, Mook, and Muhly 2004, pp. 265–280.

O'Brien, W. 1996. *Bronze Age Copper Mining in Britain and Ireland,* Buckinghamshire.

Okafor, E. E. 1993. "New Evidence for Early Iron Smelting in South-East Nigeria," in *The Archaeology of Africa,* ed. T. Shaw, P. Sinclair, B. Aandah, and A. Okpoko, London, pp. 432–448.

Olsen, F. L. 1973. *The Kiln Book,* Bassett, Calif.

Oppenheimer, S. 1999. *Eden in the East: The Drowned Continent of Southeast Asia,* London.

Ottaway, B. S. 2001. "Innovation, Production, and Specialization in Early Prehistoric Copper Metallurgy," *EJA* 4:1, pp. 87–112.

Özdoğan, M., and A. Özdoğan. 1999. "Archaeological Evidence on the Early Metallurgy at Çayönü Tepesi," in Hauptmann et al. 1999, pp. 13–22.

Palmer, R. 1999. "Perishable Goods in Mycenaean Texts," in *Floreant Studia Mycenaea. Akten des 10. Internationalen Mykenologischen Kolloquiums in Salzburg vom 1.–5. Mai 1995,* ed. S. Deger-Jalkotzy, S. Hiller, O. Panagl, and G. Nightingale, Vienna, pp. 463–485.

———. 2001. "Bridging the Gap: The Continuity of Greek Agriculture from the Mycenaean to the Historical Period," in *Prehistory and History: Ethnicity, Class and Political Economy,* ed. D. W. Tandy, Montreal, pp. 41–84.

Palmieri, A., and A. Hauptmann. 2000. "Metals from Ebla: Chemical Analyses of Metal Artefacts from the Bronze and Iron Ages," in *Proceedings of the First International Congress on the Archaeology of the Ancient Near East,* ed. P. Matthiae, A. Enea, L. Peyronel, and F. Pinnock, Rome, pp. 1259–1272.

Palmieri, A. M., K. Sertok, and E. Chernykh. 1993. "From Arslantepe Metalwork to Arsenical Copper Technology in Eastern Anatolia," in *Between the Rivers and Over the Mountains: Archaeologica Anatolica et Mesopotamica Alba Palmieri Dedicata,* ed. M. Frangipane, H. Hauptmann, M. Liverani, P. Matthiae, and M. Mellink, Rome, pp. 573–599.

Papadimitriou, G. 2001. "Simulation Study of Ancient Bronzes: Their Mechanical and Metalworking Properties," in *Archaeometry Issues in Greek Prehistory and Antiquity,* ed. Y. Bassiakos, E. Aloupi, and Y. Facorellis, Athens, pp. 713–733.

Papanthimou, A., and A. Papasteriou. 1993. "Ο προϊστορικός οικισμός στο Μάνδαλο: Νέα στοιχεία στην προϊστορία της Μακεδονίας," in *Αρχαία Μακεδονία* 2, pp. 1207–1216.

Papastamataki, A. 1986. *A Study of Ancient Slags and their Contribution to the Revealing of Ancient Metallurgical Technology,* Athens.

Papastamatiou, J., et al. 1959. *Geological Map of Greece: Ierapetra,* Athens.

Papathanassopoulos, G. 1981. *Neolithic and Cycladic Civilization,* trans. A. Doumas, Athens.

Papathanassopoulos, G., ed. 1996. *Neolithic Culture in Greece,* Athens.

Paramonov, E. A., A. Z. Khalilova, V. N. Odinokov, and L. M. Khalilov. 2001. "Identification and Biological Activity of Volatile Organic Compounds Isolated from Plants and Insects 3: Chromatography-Mass Spectrometry of Volatile Compounds of *Aegopodium podagraria,*" *Chemistry of Natural Compounds* 26:6, pp. 584–586.

Pare, C. 2000. "Bronze and the Bronze Age," in *Metals Make the World Go Round: The Supply and Circulation of Metals in Bronze Age Europe. Proceedings of a Conference Held at the University of Birmingham in June 1997,* ed. C. F. E. Pare, Oxford, pp. 1–38.

Pendlebury, J. D. S. 1939. *The Archaeology of Crete,* London.

Pendlebury, J. D. S., H. W. Pendlebury, and M. B. Money-Coutts. 1935–1936. "Excavations in the Plain of Lasithi I: The Cave of Trapeza," *BSA* 36, pp. 5–131.

Perlès, C., and K. D. Vitelli. 1999. "Craft Specialization in the Neolithic of Greece," in *Neolithic Society in Greece,* ed. P. Halstead, Sheffield, pp. 108–120.

Perlin, J. 1989. *A Forest Journey: the Role of Wood in the Development of Civilization,* New York.

Pernicka, E. 1987. "Erzlagerstätten in der Ägäis und ihre Ausbeutung in Altertum. Geochemische Untersuchungen zur Herkunftsbestimmung archäologischer Metallobjekte," *JRGZM* 34, pp. 607–714.

Pernicka, E., F. Begemann, S. Schmitt-Strecker, and A. P. Grimanis. 1990. "On the Composition and Provenance of Metal Artefacts from Poliochni on Lemnos," *OJA* 9, pp. 263–298.

Pernicka, E., F. Begemann, S. Schmitt-Strecker, H. Todorova, and I. Kuleff. 1997. "Prehistoric Copper in Bulgaria: Its Composition and Provenance," *Eurasia Antiqua* 3, pp. 41–180.

Pernicka, E., F. Begemann, S. Schmitt-Strecker, and G. A. Wagner. 1993. "Eneolithic and Early Bronze Age Copper Artefacts from the Balkans and their Relation to Serbian Ores," *PZ* 68, pp. 1–54.

Pernicka, E., T. C. Seeliger, G. A. Wagner, F. Begemann, S. Schmitt-Strecker, S. Eibner, Ö. Öztunali, and I. Baranyl. 1984. "Archäometallurgische Untersuchungen in Nordwestanatolien," *JRGZM* 31, pp. 533–599.

Pernicka, E., G. A. Wagner, K. Assimenos, C. Doumas, F. Begemann, and W. Todt. 1983. "An Analytical Study of Prehistoric Lead and Silver Objects from the Aegean," in *Proceedings of the 22nd Interna-*

tional Symposium on Archaeometry, ed. A. Aspinall and S. E. Warren, Bradford, pp. 292–302.

Pernicka, E., and G. A. Wagner, eds. 1991. *Archaeometry 90,* Basel.

Peters, J. M., S. R. Troelstra, and D. van Harten. 1985. "Late Neogene and Quaternary Vertical Movements in Eastern Crete and their Regional Significance," *Journal of the Geological Society of London* 142, pp. 501–513.

Pettegrew, D. K. 2001. "Chasing the Classical Farmstead: Assessing the Formation and Signature of Rural Settlement in Greek Landscape Archaeology," *JMA* 14, pp. 189–209.

Phelps, W. W., G. J. Varoufakis, and R. E. Jones. 1979. "Five Copper Axes from Greece," *BSA* 74, pp. 175–184.

Pinnock, F. 1988. "Observations on the Trade of Lapis Lazuli in the IIIrd Millennium B.C.," in *Wirtschaft und Gesellschaft von Ebla* (Heidelberger Studien zum Alten Orient 2), ed. H. Waetzoldt and H. Hauptmann, Heidelberg, pp. 107–110.

Pirazzoli, P. A. 1988. "Sea-Level Changes and Crustal Movements in the Hellenic Arc (Greece): The Contribution of Archaeological and Historical Data," in *Archaeology of Coastal Changes* (*BAR-IS* 404), ed. A. Raban, Oxford, pp. 157–184.

Pirazzoli, P. A., J. Thommeret, Y. Thommeret, J. Laborel, and L. F. Montaggioni. 1982. "Crustal Block Movements from Holocene Shorelines: Crete and Antikythira (Greece)," *Tectonophysics* 86, pp. 27–43.

Pitty, A. F. 1978. *Geography and Soil Properties,* London.

Platon, L. 2002. "The Political and Cultural Influence of the Zakros Palaces on Nearby Sites and in a Wider Context," in *Monuments of Minos: Rethinking the Minoan Palaces* (*Aegaeum* 23), ed. J. Driessen, I. Schoep, and R. Laffineur, Liège, pp. 145–155.

Plog, S. 1974. *The Study of Prehistoric Change,* New York.

Podzuweit, C. 1979. *Trojanische Gefässformen der Frühbronzezeit in Anatolien, der Ägäis und anzugrenden Gebieten: Ein Beitrag zur vergleichended Stratigraphie,* Mainz.

Pollard, A. M., R. G. Thomas, D. P. Ware, and P. A. Williams. 1991. "Experimental Smelting of Secondary Copper Minerals: Implications for Early Bronze Age Metallurgy in Britain," in Pernicka and Wagner 1991, pp. 127–136.

Pollard, A. M., R. G. Thomas, and P. A. Williams. 1991. "Some Experiments Concerning the Smelting of Arsenical Copper," in Budd et al. 1991, pp. 169–174.

Popham, M. R. 1967. "Late Minoan Pottery, a Summary," *BSA* 62, pp. 337–351.

———. 1969. "The Late Minoan Goblet and Kylix," *BSA* 64, pp. 299–304.

Poursat, J.-C. 1985. "Ateliers et artisans minoens," in Πεπραγμένα του Ε΄ Διεθνούς Κρητολογικού Συνεδρίου A΄, Herakleion, pp. 297–300.

———. 1987. "Town and Palace at Malia in the Protopalatial Period," in Hägg and Marinatos 1987, pp. 75–76.

———. 1990. "Hieroglyphic Documents and Sealings from Malia Quartier Mu: A Functional Analysis," in *Aegean Seals, Sealings and Administration* (*Aegaeum* 5), ed. T. G. Palaima, Liège, pp. 25–32.

———. 1996. *Fouilles exécutées à Mallia: Le Quartier Mu III. Artisans minoens: Les maisons-Ateliers du Quartier Mu* (*Étcrét* 32), Paris.

Proudfoot, B. 1976. "The Analysis and Interpretation of Soil Phosphorus in Archaeological Contexts," in *Geoarchaeology, Earth Science, and the Past,* ed. D. A. Davidson and M. L. Shackley, Boulder, Colo., pp. 93–113.

Pseira = *Excavations at Pseira, Crete,* Philadelphia

    I = P. P. Betancourt and C. Davaras, eds. *The Minoan Buildings on the West Side of Area A* (University Museum Monograph 90), 1995.

    II = P. P. Betancourt and C. Davaras, eds. *Building AC (the "Shrine") and Other Buildings in Area A* (University Museum Monograph 94), 1998.

    IV = P. P. Betancourt and C. Davaras, eds. *Buildings in Areas B, C, D, and F* (University Museum Monograph 105), 1999.

V = J. C. McEnroe, *The Architecture of Pseira* (University Museum Monograph 109), 2001.

VI = P. P. Betancourt and C. Davaras, eds. *The Pseira Cemetery* 1: *The Surface Survey* (Prehistory Monographs 5), 2002.

VII = P. P. Betancourt and C. Davaras, eds. *The Pseira Cemetery* 2: *Excavation of the Tombs* (Prehistory Monographs 6), 2003.

VIII = P. P. Betancourt, C. Davaras, and R. Hope Simpson, eds., *The Archaeological Survey of Pseira Island,* Part 1 (Prehistory Monographs 11), 2004.

IX = P. P. Betancourt, C. Davaras, and R. Hope Simpson, eds., *The Archaeological Survey of Pseira Island,* Part 2: *The Intensive Surface Survey* (Prehistory Monographs 12), 2005.

X = P. P. Betancourt and C. Davaras, eds. *Pseira X, Block AF,* forthcoming.

Pye, K. 1992. "Aeolian Dust Transport and Deposition over Crete and Adjacent Parts of the Mediterranean Sea," *Earth Surface Processes and Landforms* 17, pp. 271–288.

Rackham, O. 2001. *Trees, Wood, and Timber in Greek History: The Twentieth J. L. Myres Memorial Lecture,* Oxford.

Rackham, O., and J. A. Moody. 1992. "Terraces," in Wells 1992, pp. 123–130.

———. 1996. *The Making of the Cretan Landscape,* Manchester.

Rapoport, A. 1969. *House Form and Culture,* Englewood Cliffs, N.J.

Ravich, I. G., and N. V. Ryndina. 1995. "Early Copper-Arsenic Alloys and the Problems of their Use in the Bronze Age of the North Caucasus," *Bulletin of the Metals Museum, The Japan Institute of Metals* 23, pp. 1–18.

Read, D. W. 1986. "Sampling Procedures for Regional Surveys: A Problem of Representativeness and Effectiveness," *JFA* 13, pp. 477–491.

Renfrew, A. C. 1967. "Cycladic Metallurgy and the Aegean Early Bronze Age," *AJA* 71, pp. 1–20.

———. 1969. "The Autonomy of the South-East European Copper Age," *PPS* 35, pp. 12–47.

———. 1971. "Carbon 14 and the Prehistory of Europe," *Scientific American* 225:10, pp. 63–72.

———. 1972. *The Emergence of Civilisation: The Cyclades and the Aegean in the Third Millennium B.C.,* London.

———. 1973. *Before Civilization: The Radiocarbon Revolution and Prehistoric Europe,* London.

———. 1986a. "The Sitagroi Sequence: The Excavated Areas," in *Excavations at Sitagroi: A Prehistoric Village in Northeast Greece,* ed. C. Renfrew, M. Gimbutas, and E. Elster, Los Angeles, pp. 147–222.

———. 1986b. "Varna and the Emergence of Wealth in Prehistoric Europe," in *The Social Life of Things: Commodities in Cultural Perspective,* ed. A. Appadurai, Cambridge, pp. 141–168.

Renfrew, C., and E. A. Slater. 2003. "Metal Artifacts and Metallurgy," in Elster and Renfrew 2003, pp. 301–318.

Renfrew, J. 1972. "Appendix V: The Plant Remains," in P. M. Warren, *Myrtos: An Early Bronze Age Settlement in Crete,* London, pp. 315–317.

Riley, F. R. 2002. "Olive Oil Production on Bronze Age Crete: Nutritional Properties, Processing Methods and Storage Life of Minoan Olive Oil," *OJA* 21:1, pp. 63–75.

Roeder, J. F., J. F. Sculac, and M. R. Notis. 1984. "The Precipitation of Iron in Early Smelted Copper from Timna," in *Microbeam Analysis 1984,* ed. A. D. Romig Jr. and J. I. Goldstein, San Francisco, pp. 243–246.

Rose, M. 1996. "Fishing at Minoan Pseira: Formation of a Bronze Age Fish Assemblage from Crete," *Archaeofauna* 5, pp. 135–140.

Rosenberg, M. 1994. "Hallan Çemi Tepesi: Some Further Observations Concerning Stratigraphy and Material Culture," *Anatolica* 20, pp. 121–140.

———. 1998. "Cheating at Musical Chairs: Territoriality and Sedentism in an Evolutionary Context," *CurrAnthr* 39, pp. 653–681.

Rosenberg, M., R. Mark Nesbitt, R. W. Redding, and T. F. Strasser. 1995. "Hallan Çemi Tepesi: Some Preliminary Observations Concerning Early Neolithic Subsistence Behaviors in Eastern Anatolia," *Anatolica* 21, pp. 1–12.

Rostoker, W., and J. R. Dvorak. 1991. "Some Experiments with Co-Smelting to Copper Alloys," *Archaeomaterials* 5, pp. 5–20.

Rostoker, W., V. C. Pigott, and J. R. Dvorak. 1989. "Direct Reduction to Copper Metal by Oxide-Sulfide Mineral Interaction," *Archaeomaterials* 3, pp. 69–87.

Rothenberg, B. 1972. *Timna: Valley of the Biblical Copper Mines,* London.

———. 1985. "Copper Smelting Furnaces in the Arabah, Israel: The Archaeological Evidence," in Craddock and Hughes 1985, pp. 123–150.

Runnels, C. N., and J. Hansen. 1986. "The Olive in the Prehistoric Aegean: The Evidence for Domestication in the Early Bronze Age," *OJA* 5:3, pp. 299–308.

Russell, J. C. 1985. *Control of Late Ancient and Medieval Population* (*TAPS* 48:3), Philadelphia.

Ryndina, N., G. Indenbaum, and V. Kolossova. 1999. "Copper Production from Polymetallic Sulfide Ores in the Northeastern Balkan Eneolithic Culture," *JAS* 26, pp. 1059–1068.

Sampson, A. 1996. "Euboea," in Papathanassopoulos 1996, pp. 73–74.

———. 2002a. "The Neolithic Period in the Aegean," in Sampson 2002b, pp. 161–167.

Sampson, A., ed. 2002b. *The Neolithic Settlement at Ftelia, Mykonos,* Rhodes.

Sanders, I. F. 1982. *Roman Crete,* Warminster.

Sarpaki, A. 1992. "The Palaeoethnobotanical Approach: The Mediterranean Triad or Is It a Quartet?" in Wells 1992, pp. 61–76.

Sarpaki, A., and J. Bending. 2004. "Archaeobotanical Assemblages," in *Mochlos* IC, pp. 126–131.

Schachermeyr, F. 1938. "Vorbericht über eine Expedition nach Ostkreta," *AA* 1938, cols. 465–480.

Schickler, H. 1981. "'Neolithische' Zinnbronzen," in *Studien zur Bronzezeit: Festschrift für Wilhelm Albert v. Brunn,* ed. H. Lorenz, Mainz, pp. 419–445.

Schlager, N. 1999. "'A Town of Castles': An MM/LM Fortified Site at Aspro Nero in the Far East of Crete," in *POLEMOS: Le contexte guerrier en Égée à l'âge du Bronze* (*Aegaeum* 19), vol. 1, ed. R. Laffineur, Liège, pp. 171–177.

———. 2002. "Pleistozäne, neolithische, bronzezeitliche und rezente Befunde und Ruinen im fernen osten Kretas," *ÖJh* 70 (2001), pp. 157–220.

Schlager, N., and L. Dollhofer. 1998. "Die minoische Grabhöhle Chosto im Kastri Goudourou Sitias," *ÖJh* 67, pp. 4–35.

Schmitt-Strecker, S., F. Begemann, and E. Pernicka. 1992. "Chemische Zusammensetzung und Bleiisotopenverhältnisse der Metallfunde vom Hassek Höyük," in *Hassek Höyük: Naturwissenschaftliche Untersuchungen und lithische Industrie* (Istanbuler Forschungen 38), ed. M. R. Behm-Blancke, Tübingen, pp. 108–123.

Seager, R. B. 1904–1905. "Excavations at Vasiliki, 1904," *Transactions of the Department of Archaeology, Free Museum of Science and Art* 1, pp. 207–220.

———. 1909. "Excavations on the Island of Mochlos, Crete, in 1908," *AJA* 13, pp. 273–303.

———. 1910. *Excavations on the Island of Pseira*, Philadelphia.

———. 1912. *Explorations in the Island of Mochlos*, Boston.

———. 1916. *The Cemetery of Pachyammos, Crete*, Philadelphia.

Seidel, E., M. Okrusch, H. Kreuzer, H. Raschka, and W. Haare. 1981. "Eo-Alpine Metamorphism in the Uppermost Unit of the Cretan Nappe System—Petrology and Geochronology, Part 2: Synopsis of High-Temperature Metamorphics and Associated Ophiolites," *Contributions to Mineralogy and Petrology* 76, pp. 351–361.

Sfikas, G. 1987. *Wild Flowers of Crete*, Athens.

Shaw, H. R. 1972. "Viscosities of Magmatic Silicate Liquids: An Empirical Method of Prediction," *American Journal of Science* 272, pp. 870–893.

Sherratt, S. 2000. *Catalogue of Cycladic Antiquities in the Ashmolean Museum: The Captive Spirit*, 2 vols., Oxford.

Shimada, I., and J. F. Merkel. 1991. "Copper-Alloy Metallurgy in Ancient Peru," *Scientific American* 252:2, pp. 39–47.

Shugar, A. N. 2003. "Reconstructing the Chalcolithic Metallurgical Process at Abu Matar, Israel" (paper, Milan 2003).

Sikka, I. 2002. "Νεολιθικό χρυσάφι στη Μύκονο," *Καθημερινή*, December 24, 2002, p. 15.

Sjoberg, A. 1976. "Phosphate Analysis of Anthropic Soils," *JFA* 3, pp. 447–454.

Slater, E. 2003. "Sherds with Copper Deposits," in Elster and Renfrew 2003, p. 303.

Smith, P. E. L. 1972. "Changes in Population Pressure in Archaeological Explanation," *WorldArch* 4, pp. 5–18.

SMPSAS = *Standard Method for Particle-Size Analysis of Soils* (ASTM D 422-63), Las Vegas 1972.

Soles, J. S. 1978. "Mochlos, a New Look at Old Excavations: The University Museum's Work on Crete," *Expedition* 20, pp. 5–15.

———. 1991. "The Gournia Palace," *AJA* 95, pp. 17–78.

———. 1992. *The Prepalatial Cemeteries at Mochlos and Gournia*, Princeton.

———. 1997. "A Community of Craft Specialists at Mochlos," in Laffineur and Betancourt 1997, vol. 2, pp. 425–431.

———. 2003. "The Artisan's Quarter: Building A," in *Mochlos* IA, pp. 7–40.

Soles, J. S., and C. Davaras. 1994. "Excavations at Mochlos, 1990–1991," *Hesperia* 63, pp. 391–436.

———. 1996. "Excavations at Mochlos, 1992–1993," *Hesperia* 65, pp. 175–230.

Sotirakopoulou, P. 1993. "The Chronology of the 'Kastri Group' Reconsidered," *BSA* 88, pp. 5–20.

Spanakis, S. G. 1991. Πόλεις και χωριά της Κρήτης στο πέρασμα των αιώνων 1, Herakleion.

———. 1993. Πόλεις και χωριά της Κρήτης στο πέρασμα των αιώνων 2, Herakleion.

Sperl, G. 1990. "Urgeschichte des Bleis," *Zeitschrift für Metallkunde* 81, pp. 799–801.

Spitaels, P. 1984. "The Early Helladic Period Mine 3 (Theatre Sector)," *Thorikos* 8, pp. 151–174.

Spratt, T. A. B. 1865. *Travels in Crete*, London.

Sridhar, R., J. M. Togurl, and S. Simeonov. 1997. "Copper Losses and Thermodynamic Considerations in Copper Smelting," *Metallurgical and Materials Transactions* 28B, pp. 191–200.

Stampert, D. H., and O. H. Tuorinen. 1998. "Biodegradation of the Acetanilide Herbicides Alachlor, Metolachlor, and Propachlor," *Critical Reviews in Microbiology* 24:1, pp. 1–22.

Stech, T. 1999. "Aspects of Early Metallurgy in Mesopotamia and Anatolia," in *The Archaeometallurgy of the Asian Old World* (*MASCAP* 16), ed. V. C. Pigott, Philadelphia, pp. 59–71.

Stech, T., and V. C. Pigott. 1986. "The Metals Trade in Southwest Asia in the Third Millennium B.C.," *Iraq* 48, pp. 39–64.

Stefanovich, M., and H. A. Bankoff. 1998. "Kamenska Cuka 1993–1995," in *In the Steps of James Harvey Gaul* 1: *James Harvey Gaul, in Memoriam*, ed. M. Stefanovich, H. Todorova, and H. Hauptmann, Sofia, pp. 255–338.

Stein, J. K. 1992. "Organic Matter in Archaeological Contexts," in *Soils in Archaeology*, ed. V. T. Holliday, Washington, D.C., pp. 193–205.

Stos-Gale, Z. A. 1989. "Cycladic Copper Metallurgy," in Hauptmann, Pernicka, and Wagner 1989, pp. 279–292.

———. 1992. "The Origin of Metal Objects from the Early Bronze Age Site of Thermi on the Island of Lesbos," *OJA* 11, pp. 155–177.

———. 1993. "The Origin of Metal Used for Making Weapons in Early and Middle Minoan Crete," in *Trade and Exchange in Prehistoric Europe* (Oxbow Monograph 33), ed. C. Scarre and F. Healy, Oxford, pp. 115–129.

———. 1998. "The Role of Kythnos and other Cycladic Islands in the Origins of Early Minoan Metallurgy," in *Kea-Kythnos: History and Archaeology*, ed. L. G. Mendoni and A. Mazarakis-Ainian, Athens, pp. 717–735.

———. 2003. "Origin of Metals from Sitagroi as Determined by Lead Isotope Analysis. Appendix 8.3," in Elster and Renfrew 2003, pp. 325–329.

Stos-Gale, Z. A., and N. H. Gale. 1984. "The Minoan Thalassocracy and the Aegean Metal Trade," in *The Minoan Thalassocracy: Myth and Reality*, ed. R. Hägg and N. Marinatos, Stockholm, pp. 59–63.

———. 2003. "Lead Isotopic and Other Isotopic Research in the Aegean," in *METRON: Measuring the Aegean Bronze Age* (*Aegaeum* 24), ed. K. P. Foster and R. Laffineur, Liège, pp. 83–101.

Stos-Gale, Z. A., N. H. Gale, and N. Annetts. 1996. "Lead Isotope Analyses of Ores from the Aegean," *Archaeometry* 38:2, pp. 381–390.

Stos-Gale, Z. A., and C. F. Macdonald. 1991. "Sources of Metals and Trade in the Bronze Age Aegean," in *Trade in the Bronze Age Mediterranean* (*SIMA* 90), ed. N. H. Gale, Jonsered, pp. 248–280.

Stos-Gale, Z. A., A. Sampson, and E. Mangou. 1996. "Analyses of Metal Artifacts from the Early Helladic Cemetery of Manika on Euboea," *Aegean Archaeology* 3, pp. 49–62.

Strasser, T. F. 2002. "The Polished Stone Axes from Knossos and their Implications for the Aegean Neolithic," *AJA* 106, p. 251.

Sutton, S. B. 1994. "Settlement Patterns, Settlement Perceptions: Rethinking the Greek Village," in *Beyond the Site: Regional Studies in the Aegean Area*, ed. P. N. Kardulias, Lanham, Md., pp. 313–335.

Swann, C. P., S. Ferrence, and P. P. Betancourt. 2000. "Analysis of Minoan White Pigments Used on Pottery from Palaikastro," *Nuclear Instruments and Methods in Physics Research B* 164–165, pp. 714–717.

Swann, C. P., and S. J. Fleming. 1988. "Proton-Induced X-Ray Emission Spectrometry in Archaeology," *Scanning Microscopy* 2, p. 197.

———. 1990. "Selective Filtering in PIXE Spectrometry," *Nuclear Instruments and Methods in Physics Research B* 49, p. 65.

Tamhane, R. V., D. P. Motiramani, Y. P. Bali, and R. L. Donahue. 1964. *Soils: Their Chemistry and Fertility in Tropical Asia*, New Delhi.

Terrell, J. E. 2000. "Lost Continent: An Alternative View of the Rise of Civilizations," *Archaeology* 53:2, pp. 70–73.

Theocharis, D. R. 1955. "Ανασκαφή εν Αραφήνι," *Prakt* 1952, pp. 129–151.

Thirgood, J. V. 1981. *Man and the Mediterranean Forest: A History of Resource Depletion*, New York.

Thompson, F. C. 1958. "The Early Metallurgy of Copper and Bronze," *Man* 58, pp. 1–7.

Thornton, C. P., C. C. Lamberg-Karlovsky, M. Liezers, and S. M. N. Young. 2002. "On Pins and Needles: Tracing the Evolution of Copper-Base Alloying at Tepe Yahya, Iran, via ICP-MS Analysis of Common-Place Items," *JAS* 29, pp. 1451–1460.

Tilly, B., ed. 1973. *Varro: The Farmer*, London.

Timpson, M. L. 1992. "An Investigation of the Pedogenesis of Soils Developed in Quaternary Alluvial Deposits of Eastern Crete" (diss. Univ. of Tennessee, Knoxville).

Timucin, M., and A. E. Morris. 1970. "Phase Equilibria and Thermodynamic Studies in the System CaO-FeO-Fe_2O_3-SiO_2," *Metallurgical Transactions* 1, pp. 3193–3201.

Todorova, H. 1978. *The Eneolithic in Bulgaria* (*BAR-IS* 49), Oxford.

Tsipopoulou, M., and A. Papacostopoulou. 1997. "'Villas' and Villages in the Hinterland of Petras, Siteia," in Hägg 1997, pp. 203–214.

Turkdogan, E. T. 1983. *Physicochemical Properties of Molten Slags and Glasses*, London.

Turner, J. 1978. "The Vegetation of Greece during Prehistoric Times: The Palynological Evidence," in *Thera and the Aegean World: Papers Presented at the Second International Scientific Congress, Santorini, Greece,*

August 1978, ed. C. Doumas, London, pp. 760–773.

Turner, J., and J. R. A. Grieg. 1975. "Some Holocene Pollen Diagrams from Greece," *Review of Palaeobotany and Palynology* 20, pp. 171–204.

Tylecote, R. F. 1976. *A History of Metallurgy,* London.

———. 1981. "From Pot Bellows to Tuyeres," *Levant* 13, pp. 107–118.

———. 1991. "Early Copper Base Alloys: Natural or Man-Made?" in Mohen and Éluère 1991, pp. 213–221.

Tylecote, R. F., and J. F. Merkel. 1985. "Experimental Smelting Techniques: Achievements and Future," in Craddock and Hughes 1985, pp. 3–20.

Tzedakis, Y., S. Chryssoulaki, Y. Venieri, and M. Avgouli. 1990. "Les routes minoennes: Le poste de Χοιρομάνδρες et le contrôle des communications," *BCH* 114, pp. 43–62.

Tzedakis, Y., S. Chryssoulaki, S. Voutsaki, and Y. Venieri. 1989. "Les routes minoennes: Rapport préliminaire: défense de la circulation ou circulation de la défense?" *BCH* 113, pp. 43–75.

Vagnetti, L. 1973. "Tracce di due insediamenti neolitici nel territorio dell'antica Gortina," in *Antichità Cretesi: Studi in Onore di Doro Levi,* ed. G. P. Carratelli and G. Rizza, Catania, pp. 1–9.

———. 1975. "L'insediamento neolitico di Festòs," *ASAtene,* n.s. 34–35 (1972–1973), pp. 7–138.

———. 1996. "The Final Neolithic: Crete Enters the Wider World," *Cretan Studies* 5, pp. 29–39.

Vagnetti, L., and P. Belli. 1978. "Characters and Problems of the Final Neolithic in Crete," *SMEA* 19, pp. 125–163.

Vagnetti, L., A. Christopoulou, and Y. Tzedakis. 1989. "Saggi negli strati neolitici," in *Scavi a Nerokourou, Kydonias,* Rome, pp. 9–97.

Vallat, J.-P. 1991. "Survey Archaeology and Rural History—A Difficult but Productive Relationship," in Barker and Lloyd 1991, pp. 10–17.

van Andel, T., C. Runnels, and K. Pope. 1986. "Five Thousand Years of Land Use and Abuse in the Southern Argolid, Greece," *Hesperia* 55, pp. 103–128.

van Andel, T., E. Zangger, and A. Demitrack. 1990. "Land Use and Soil Erosion in Prehistoric and Historical Greece," *JFA* 17, pp. 379–396.

Veenhof, K. R. 1972. *Aspects of Old Assyrian Trade and its Terminology,* Leiden.

Vickery, K. F. 1936. *Food in Early Greece,* Urbana.

Wachendorf, H., A. Baumann, W. Gwosdz, and W. Schneider. 1974. "Die 'Phyllite-Serie' Ostkretas—Eine Mélange," *Zeitschrift der Deutscher Geologischer Gesellschaft* 25, pp. 237–251.

Wachendorf, H., G. Best, and W. Gwosdz. 1975. "Geodynamische Interpretation Ostkretas," *Geologische Rundschau* 64, pp. 728–750.

Waetzoldt, H., and H.-G. Bachmann. 1984. "Zinn und Arsenbronzen in den Texten aus Ebla und aus dem Mesopotamien des 3. Jahrtausends," *OA* 23, pp. 1–18.

Wagner, G. A., and G. Weisgerber. 1985. *Silber, Blei, und Gold auf Sifnos,* Bochum.

Wagstaff, M., and S. Augustson. 1982. "Traditional Land Use," in *An Island Polity: The Archaeology of Exploitation in Melos,* ed. C. Renfrew and M. Wagstaff, Cambridge, pp. 106–133.

Walberg, G. 1976. *Kamares: A Study of the Character of Palatial Middle Minoan Pottery,* Uppsala.

———. 1983. *Provincial Middle Minoan Pottery,* Mainz.

Walls, J. 1982. *Combating Desertification in China,* Nairobi.

Warren, P. M. 1965. "The First Minoan Stone Vases and Early Minoan Chronology," *CretChron* 19, pp. 7–43.

———. 1972. *Myrtos: An Early Bronze Age Settlement in Crete,* London.

———. 1980. "Problems of Chronology in Crete and the Aegean in the Third and Earlier Second Millennium B.C.," *AJA* 84, pp. 487–499.

———. 1987. "The Genesis of the Minoan Palace," in Hägg and Marinatos 1987, pp. 47–56.

Watrous, L. V. 1977. "Aegean Settlements and Transhumance," *TUAS* 2, pp. 2–6.

———. 1982. *Lasithi: A History of Settlement on a Highland Plain in Crete,* Princeton.

———. 2001. "Addendum: 1994–1999," in *Aegean Prehistory: A Review,* ed. T. Cullen, Boston, pp. 216–223.

Watrous, L. V., and H. Blitzer. 1999. "The Region of Gournia in the Neopalatial Period," in *MELETEMATA,* vol. 3, pp. 905–911.

Weeks, L. R. 2003. *Early Metallurgy of the Persian Gulf: Technology, Trade, and the Bronze Age World,* Leiden.

Weisgerber, G. 2003. "Spatial Organization of Mining and Smelting at Feinan, Jordan: Mining Archaeology Beyond the History of Technology," in *Mining and Metal Production through the Ages,* ed. P. T. Craddock and J. Lang, London, pp. 76–89.

Weiss, H. 1997. "Late Third Millennium Abrupt Climate Change and Social Collapse in West Asia and Egypt," in *Third Millennium B.C. Climate Change and Old World Collapse,* ed. N. Dalfes, G. Kukla, and H. Weiss, Berlin, pp. 711–723.

Weiss, H., and M.-A. Courty. 1993. "The Genesis and Collapse of the Akkadian Empire," in *Akkad, the First World Empire: Structure, Ideology, Traditions,* ed. M. Liverani, Padua, pp. 131–155.

Wells, B. 1996. "Introduction," in *The Berbati-Limnes Archaeological Survey,* ed. B. Wells, Stockholm, pp. 9–22.

Wells, B., ed. 1992. *Agriculture in Ancient Greece,* Stockholm.

Wells, B., C. Runnels, and E. Zangger. 1990. "The Berbati-Limnes Archaeological Survey: The 1988 Season," *OpAth* 18:15, pp. 207–238.

Werner, K. 1993. *The Megaron during the Aegean and Anatolian Bronze Age* (*SIMA* 108), Jonsered.

Wertime, T. A. 1983. "The Furnace versus the Goat: The Pyrotechnologic Industries and Mediterranean Deforestation in Antiquity," *JFA* 10, 445–452.

Whallon, R. 1983. "Methods of Controlled Surface Collection in Archaeological Survey," in Keller and Rupp 1983, pp. 73–83.

Wheeler, T. S., R. Maddin, and J. D. Muhly. 1975. "Ingots and the Bronze Age Copper Trade in the Mediterranean," *Expedition* 17:4, pp. 31–39.

Whitelaw, T., P. M. Day, E. Kiriatzi, and V. Kilikoglou. 1997. "Ceramic Traditions in EM IIB Myrtos, Fournou Korifi," in Laffineur and Betancourt 1997, vol. 2, pp. 265–274.

Whittaker, C. R., ed. 1988. *Pastoral Economies in Classical Antiquity,* Cambridge.

Wilkinson, P. 1981. "Population, Resources, and Explanation in Prehistory," in *Pattern of the Past: Studies in Memory of David Clarke,* ed. I. Hodder, G. Isaac, and N. Hammond, Cambridge, pp. 251–259.

Wilkinson, T. J. 1982. "The Definition of Ancient Manured Zones by Means of Extensive Sherd-Sampling Techniques," *JFA* 9, pp. 323–333.

———. 1989. "Extensive Sherd Scatters and Land-Use Intensity: Some Recent Results," *JFA* 16, pp. 31–46.

Willetts, R. F. 1955. *Aristocratic Society in Ancient Crete,* London.

———. 1990. "Aspects of Land Tenure in Dorian Crete," *Cretan Studies* 2, pp. 221–230.

Wilson, D. E. 1985. "The Pottery and Architecture of the EM IIA West Court House at Knossos," *BSA* 80, pp. 281–364.

Wilson, D. E., and P. M. Day. 1994. "Ceramic Regionalism in Prepalatial Central Crete: The Mesara Imports from EM I to EM IIA Knossos," *BSA* 89, pp. 1–87.

———. 2000. "EM I Chronology and Social Practice: Pottery from the Early Palace Tests at Knossos," *BSA* 95, pp. 21–63.

Wines, R. A. 1985. *Fertilizer in America,* Philadelphia.

Xanthoudides, S. 1918. "Μέγας πρωτομινωικός τάφος Πύργου," *ArchDelt* 4, pp. 136–170.

———. 1924. *The Vaulted Tombs of Mesará,* London.

Yalçin, Ü. 1998. "Der Keulenkopf von Can Hasan (TR). Naturwissenschaftliche Untersuchung und neue Interpretation," in *Metallurgica Antiqua, in Honour of Hans-Gert Bachmann and Robert Maddin* (*Der Anschnitt* Beiheft 8), ed. T. Rehren, A. Hauptmann, and J. D. Muhly, Bochum, pp. 279–289.

———. 2000. "Anfänge der Metalverwendung in Anatolien," in *Anatolian Metal* 1 (*Der Anschnitt* Beiheft 13), ed. Ü. Yalçin, Bochum, pp. 17–30.

Yalçin, Ü., and E. Pernicka. 1999. "Frühneolithische Metallurgie von Aşıklı Höyük," in Hauptmann et al. 1999, pp. 45–54.

Yazawa, A. 1980. "Distribution of Various Elements between Copper, Matte and Slag," *Erzmetall* 33:7/8, pp. 377–382.

Yener, K. A. 2000. *The Domestication of Metals: The Rise of Complex Metal Industries in Anatolia,* Leiden.

Zohary, M., and G. Orshan. 1965. "An Outline of the Geobotany of Crete," *Israel Journal of Botany* 14, pp. 1–49 and suppl.

Zois, A. A. 1968. "Υπάρχει ΠΜ ΙΙΙ εποχή;," in *Πεπραγμένα του Β′ Διεθνούς Κρητολογικού Συνεδρίου Α′,* Athens, pp. 141–156.

———. 1993. "Ανασκαφή Βασιλικής Ιεράπετρας," *Prakt* (1990), pp. 315–341.

Zwicker, U. 1991. "Natural Copper-Arsenic Alloys and Smelted Arsenic Bronzes in Early Metal Production," in Mohen and Éluère 1991, pp. 331–340.

INDEX

Abietadiene, 422, 425–426
abietic acid, 420, 422, 425
acetanilide, 419, 421
a-cetone. *See* isomethyl-a-ionone
acid esters, 416, 418, 423
Aegopodium podagraria. See ground elder
Afghanistan, 169–170
agrimi or deer, 212
Agriomandra, 8–9, 21, 30–31, 33, 37, 41, 100, 142, 181, 185, 198, 200–201, 203–204, 221–223, 226, 235, 237, 239–241, 243, 249, 251, 253–255, 301
Agriospelio, 13–14
akermanite, 292
Akkadian empire, 162
Aksaray, 161
Alaca Höyük, 170, 175
aldehydes, 415, 418, 419, 421, 422, 425–426
Alepotrypa cave, 157–158, 165
Aleppo pine, 420, 423
alkanes, 413, 416–419, 421–424, 426
allspice, 418–419
almonds, 258
anthropogenic features
  AF 1, 199, 355, 357–358
  AF 2, 199, 276, 355, 357–358
  AF 3, 199, 355, 358–359
  AF 4, 199, 275, 355, 359–360
  AF 5, 199, 275, 355, 359–361
  AF 6, 199, 275, 355, 360–361
  AF 7, 199, 275, 355, 361, 362
  AF 8, 199, 355, 361–362, 378
  AF 9, 39, 199, 274, 276, 355, 361, 363, 368, 399–402
  AF 10, 199, 275, 355, 363, 394, 397
  AF 11, 199, 275–276, 355, 363
  AF 12, 199, 275, 355, 363–364
  AF 13, 199, 275, 355, 363–364
  AF 14, 199, 355, 364
  AF 15, 199, 355, 364
  AF 16, 199, 275, 355, 364–366, 397
  AF 17, 199, 275–276, 355, 365–366, 377–378, 394, 397
  AF 18, 199, 275, 356, 366
  AF 19, 199, 356, 366
  AF 20, 199, 275–276, 356, 366
  AF 21, 199, 356, 366–368, 370, 378
  AF 22, 39, 199, 251, 273–274, 276, 356, 363, 368, 372, 374, 394, 399–402, 406–408, 411–412
  AF 23, 199, 259, 356, 368
  AF 24, 199, 356, 369
  AF 25, 199, 356, 369
  AF 26, 199, 356, 369
  AF 27, 199, 356, 369–370
  AF 28, 199, 276, 356, 370
  AF 29, 199, 259, 356, 361–363, 370–372, 378
  AF 30, 199, 251, 253, 356, 372, 374
  AF 31, 199, 251, 253, 356, 372
  AF 32, 199, 201, 243–246, 249, 268, 276, 356, 361–362, 372, 378–389, 394
  AF 33, 199, 356, 373, 386, 389
  AF 34, 199, 259, 356, 373
  AF 35, 199, 356, 373
  AF 36, 199, 356, 372, 373
  AF 37, 199, 356, 373–374
  AF 38, 199, 356, 374
  AF 39, 199, 356, 372, 374
  AF 40, 199, 356, 374–375
  AF 41, 199, 356, 365, 375
aluminum, 328, 332, 335
Alvania, 149, 151
Alykomouri, 217, 225, 236–237
Amnissos cave, 157
Amorgos, 319
amphoras, 73, 393–396
Anatolia, 61
Anavisos, 157

Andiskari, 303
Andros, 157
animal fats, 420–422, 424, 427
anisaldehyde, 423–425
anise, 423–425, 427
anorthite, 305, 307, 309
Ano Varsamonero, 310–311
anteisopentadecanoic acid, 420
anteisopentanoic acid, 422
antimony, 322, 327–328
apricot, 426
apsidal structure, main discussion, 55–66
arsenic, 65, 145–147, 157, 163–169, 171–177, 185–186, 188, 300, 304, 321–324, 327–328, 332, 338, 340–344, 346–347, 349–352, 413, 428
arseniosiderite, 145
arsenopyrite, 145
Arslantepe, 166–167, 171
Aşıklı Höyük, 161–162
Asimina. See pawpaw
Aspros Pyrgos. *See* Siphnos
Atomic Emission Spectrometer, 38
augite, 289, 292, 305, 307
Avgo, 227–229, 231
Avyssalos. *See* Seriphos
awls, 156, 159, 161, 189, 318
axe-adzes, 159
axes, 101, 155, 159, 164, 170–171, 174–175, 301, 319
azurite, 142, 144, 183–186, 302–303, 330, 332

Balkans, 155–160, 164
Balomir, 159
barium, 325, 327, 332
barley, 110, 153–154, 184, 223, 233, 241, 247, 250, 260, 262, 275
basil, 417
basins, 79, 93–94, 391, 415, 422
basket impression, 71
beads, 158, 161–162
bears, 161
Bebonas, 227
Beer Ora. *See* Timna
bees, 391–392
beeswax, 422
bell cups, 208
bellows, 61, 68, 83–84, 86, 94, 96, 112, 125–132, 177, 183–184, 186–187, 351, 413, 415, 422
benzoic acid, 420
Beycesultan, 166–167
biotite, 430–432
Bir Safadi, 142
birds, 212, 254
bismuth, 172
blackberry, 421

bladelet, 108
blowpipe, 112
blueberry, 425, 426
boldo tree, 426
borneol, 418
boundary walls for agricultural fields, 4, 197, 275–276, 278, 355–359, 366–367, 369–370, 375
bowls, 68, 70, 72–73, 75–76, 78–79, 84, 87, 91, 93, 97, 209, 212, 216–217, 393, 413, 415, 419, 425
bracelet, 174
brick making, 38
bridge-spouted jar, 78, 83, 87, 89, 94, 97, 381–382, 415, 426
Britain, 142, 163
bronze jug at farmstead, 213
bronze use, main discussion of, 169–176
bucket jar, 79, 81, 87, 93, 97, 415, 416, 418, 424
buckle, 212
Bulgaria, 157, 159
burial cave, 3, 15. *See also* Theriospelio
Byblos, 170

Camphor, 65, 416–419, 421, 423, 425–427
campo. *See* Kambos
capric acid, 418, 420, 422
caproic acid, 419, 422
caprylic acid, 418, 419, 421, 422
Capsicum frutescens, 424
carbonate inclusions in slags, 285–286, 290–292, 331
carinated cup, 208, 267
carinated open vessel, 75–76
carob, 221
cassiterite, 176
Çatal Höyük, 160, 162
catmint, 425, 427
cattle, 212, 234, 248, 271–272
Caucasus, 146, 175
Çayönü Tepesi, 160–161
cedar, 221
cerith, 149
Cerithium, 149, 151
chaff, 110, 113, 133–135, 153–154, 180–182, 184, 233, 241, 250, 262, 290–291
Chalandriani, 165, 319
chalcocite, 141, 332
chalcopyrite, 141, 300, 332
Chalepa, 8–9, 197–198, 200–201, 203–204, 221–223, 226, 237, 239, 243, 248–249, 251, 253, 255, 272
chalice, 70, 216–220
Chalinomouri, 229, 269–270
Chamaizi, 4, 269–270

champaign cups, 212
Chania, 303–304, 319, 419
Charonia, 213
cheese, 221, 426
"cheese pot", 79
Cheiromandres, 269–270
chert, 284–286, 291–292, 430–432
chimney fragments, main discussion, 109–123
chisels, 165, 172, 174, 189, 212, 318–319
chloride, 146, 327
chlorine, 146
chlorite, 144, 290–292
Chomatas, 3, 5, 9–10, 14, 47, 198, 200, 202–203, 205–213, 221–233, 235, 237–239, 243, 251, 253, 255, 261, 278, 358, 365, 374
Chondrovolakes, 226, 229
Chordakia, 9–10, 12, 14, 16, 203–204, 221–232, 235, 238–239, 278, 365
chrysocolla, 142, 184–186, 330
Chrysokamino territory, area of, 204
Chrysostomos, 303–304, 310–311
church in ravine, 204, 254
Chylopittes. *See* travertine formation
Cinnamomum aromaticum, 425
cinnamon, 419
citrus, 418, 419, 421, 425
clinoclase, 185
clinopyroxene, 282, 289
coal deposit, 40
Columbella, 149, 151–152
combed decoration, 71
Congolese furnaces, 110, 349
conical cup, 72, 88, 97, 213, 385
cooking dishes, 65, 68, 72–73, 79, 84, 87, 91, 93, 97, 183, 209, 212, 377, 380, 387
cooking pots, 65–67, 73, 183, 209, 377–378, 381, 385, 387–389, 419
coriander, 418, 421, 425, 427
Coriander sativum. *See* coriander
Cornwall, 143
corrosion studies, 321–324
covellite, 332–333
crab, 149, 152
cranberry, 425
cristobalite, 309
crucibles, 14, 16, 63–64, 113, 137, 158, 163, 165–166, 173, 177, 182, 184, 263, 300, 302
cumin, 423–425
Cuminum cyminum. *See* cumin
cuprite, 141, 144, 146, 282–283, 287–289, 292, 338, 344, 348
cups, 65, 70, 72–73, 77–78, 87–89, 97, 183, 208–209, 212, 380, 385, 413, 415, 425

cuttings in bedrock, 99, 101, 104–105, 350
Cyclades, 70–71, 79, 99, 142, 145–146, 163, 166–168, 171, 177, 179, 181, 247, 255, 299, 313–316, 319, 330
Cyprus, 16, 175–176, 188, 273, 311

DAGGERS, 159, 163–166, 170, 172, 189, 208–209, 301, 318
Debla, 154
decanal, 418, 419, 421, 425, 427
decanoic acid, 424–426
deep water fish, 255
dehydroabietic acid, 420
delafossite, 146, 292, 344, 348
diacetylbenzene, 420, 421
Diaskari, 4
diethyltoluamide (DEET), 423–425
dihydroisophorone, 427
Dikili Tash, 156
dill, 421
Dimini, 164
Dimitra, 156
diopside, 292
Dioscorides, 423
ditolyl ether, 419, 423
Diyarbakır, 161
dodecanal, 418, 425
dodecane, 418, 419, 421, 424–426
Dorian farming methods, 21
dove shell, 149, 151–152
drill, 174
Durankulak, 164

EAST CRETAN WHITE-ON-DARK WARE. *See* White-on-Dark Ware
Ebla, 169–170
Egypt, 125
eicosane, 423
elaidic acid (C18:1 trans), 419
elemental iron, 288–289, 292
Emporio, 157
enanthic acid, 422
entrapped gas bubbles, 140
epicuticular waxes, 421, 423, 426
epidote, 291–292
Ermioni, 316
Eskiyapar, 170
ethyl salicylate, 424
Euboia, 157, 164, 319
Euripides, cave of, 157

FAIENCE, 137
farmstead at Chrysokamino-Chomatas, 3, 5, 7, 12–13, 16–17, 33, 181–184, 193, 196–198, 200, 201–202, 205–213, 215, 221–256, 259, 262, 267–274, 330, 356, 369–372, 375, 391–393, 397, 408, 410–411

Fasciolaria, 149, 152
fatty acids, 413, 416, 418, 420–426, 428
faunal remains, main discussion, 149–152
fayalite, 140–141, 188, 282–284, 286–289, 292, 296–297, 305, 309
Feinan, 142–143, 177, 180, 188
feldspar, 291–292, 431–432
fennel, 423–425, 427
ferruginous siltstone, 291–292
field houses, 197, 252, 274–276, 278, 355–356, 360–365, 369, 373
fields used for agriculture, main discussion, 250–253
field walls. *See* boundary walls for agricultural fields
figs, 278
Filipendula ulmaria. *See* meadowsweet
Final Neolithic period. *See* Neolithic period
Fine Gray Ware, 217
fish and fish bones, 149–151, 255, 426
fishhooks, 156, 161, 165, 318
fishing for sport, 276
fish tanks, 28
flux, 143, 145, 183–186, 330, 346
Foeniculum vulgare. *See* fennel
foraminifera, 291
Fournou Korifi. *See* Myrtos Fournou Korifi
freeze/thaw cycles, 138
fruit trees, 247, 258
Ftelia, 156
furnace lining, 133–134, 136

GALENA BEADS, 162
gamma-decanolactone, 423
gamma-hexalactone, 426
gamma-lactones, 421, 426
gamma-nonalactone, 420, 423, 426
gamma-octalactone, 420, 423, 426
gardens, main discussion, 243–248
gas chromatography, 413–432
gastropod, 149, 151
Gavdos, 301
gehlenite, 305
geology, main discussion, 22–36
Geoscan Electron Microprobe, 16
germander, 420
Giali, 158
ginger, 421
Glazed Painted Ware, 393
Glazed Ware, 397
Glycyrrhiza glabra. *See* licorice root
Göbekli Tepe Ziyaret, 161
goblet, 71, 76–77
goethite. *See* iron oxides/hydroxides
gold, 296, 310, 344
Gournes, 157

Gournia, 4–5, 18, 71, 73, 76, 100, 127, 181–182, 215, 218, 228, 231, 256, 259, 264–267, 270, 273
grapes, 247–248, 252, 258, 262, 266, 271, 278
graters, 391
grazing lands, main discussion, 248–250
Greenland, 162
ground elder, 424, 426–427
Gumilnitsa culture, 157
gypsum, 36–37, 303

HABITATION LOCATION. *See* farmstead at Chrysokamino-Chomatas
Hagia Eireini. *See* Kea
Hagia Photia, 165–168, 171, 217–218, 262, 269, 318
Hagia Triada, 70, 166, 168, 304, 311, 318–319
Hagios Antonios, 70, 226–227, 229, 231, 236
Hagios Ioannis, 8
Hagios Onouphrios, 168
Hallan Çemi Tepesi, 161
hares. *See Lepus*
Harrapan sites, 170
Hassek Höyük, 167
hearth, 210–211
hedenbergite, 141, 286, 292, 338
Hellenic arc, 23
heptadecane, 417, 421, 423, 424, 426
herbal medicines, 65–66
hercynite, 307, 309
hexanoic acid, 419, 426
hides. *See* leather
hieroglyphic tablets, 267
hook, 212
Hordeum. *See* barley
horned jar, 75–76
horned vessel, 70, 75–76
horn shell, 149, 151
horses, 161
House of the Tiles, 171
house walls near modern road, 374–375
human bones, 215
Hungary, 160
hyacinth, 424
Hyacinthus orientalis. *See* hyacinth
hydrated copper chloride, 286
hydroxy-derivative, 423
hyssop, 417, 420

IERAPETRA, 4, 22–23, 26, 28, 223, 226, 255
Ilıpınar, 171
Indus River, 163, 170
ingots, 187–188, 349, 352
Ionian Sea, 19

Iran, 167
Ireland, 142
iron oxides/hydroxides, 143, 328, 330–333, 341, 344–346, 350–352
irrigation pipe, 51
isomethyl a-ionone, 424, 425, 427
isomethyl b-ionone, 424, 425, 427
isophorone, 420, 421, 426–427
Isthmus of Ierapetra, 4–5

JADEITE, 170
jars, 61, 65, 70, 75–79, 81, 83–84, 86–89, 93–94, 96–97, 128, 209, 212, 382, 388, 413, 415, 419, 421–424, 426
Jordan, 142–143, 177, 180, 188
jugs, 65, 81, 87, 89, 97, 208, 212, 220

KALAMI, 127
Kalathiana, 168, 318
kalathoi, 209
Kalo Chorio, 24, 70, 220, 228
Kamares Ware, 266
Kambanos, 304, 310–311
Kambos, 9–13, 30, 221–224, 226, 229–231, 234, 248, 250, 254–256, 266, 277–278, 365
Kamenska Cuka, 75
Kanli Kastelli, 218
kaolinite, 184, 245
Karanovo, 157
Karoumes, 269–270
Karpathos, 391
Karum Kanesh, 182
Kastelli Phourni, 70
Kastri. See Kythera
Katalimata. See Monastiraki Katalimata
Katanga. See Congolese furnaces
Katharades, 70
Katharo, 301–302
Kato Syme, 391
Kato Zakros, 4, 125, 304, 311
Katsoprinos, 9–10, 12, 16, 33, 198, 200, 203, 222–223, 229, 238–239, 243, 249, 251, 253, 268
Kavousi, 4–7, 12, 14, 17–18, 22, 27, 33–34, 40, 47, 52, 70, 179–182, 197, 205, 217, 236, 249, 254, 256, 258, 262, 264, 267, 273, 276, 375
Kavousianos River, 12
Kavousi regional survey, main discussion, 221–232
Kea, 165, 299, 314, 319
Kefala. See Seriphos
Kephala. See Kea
Kephalolimnos, 9–10, 198, 221–224, 226, 238–240, 243, 249, 276
ketones, 415, 418, 419, 425–427
knives, 165, 172, 212

Knossos, 70, 76, 155–156, 159, 166, 220, 252, 259, 264, 266, 268, 270, 316, 347
Kokkino Phroudi, 269–270
Kolonospilio, 14
Kommos, 69, 125, 165, 391
Kondouro. See Seriphos
Konya basin, 162
Koumasa, 76, 160, 165, 168, 318–319
Krasi, 170–171, 318
Kritsa, 301–302
kylix, 212, 378
Kythera, 171–172, 314
Kythnos, 17, 20, 41, 112, 142–143, 145, 167–168, 177, 179–181, 185, 255, 263, 299–301, 305, 313–317, 319, 350

LACONIA, 314
lactones, 420, 423, 426–427
ladle, 213
Lakkos Ambelliou, 9, 12–13, 17, 34, 36, 38, 42, 110, 113, 184, 197–198, 201, 203–204, 221–223, 226, 229, 235, 237–239, 243–244, 246, 248–249, 251, 253–256, 260, 264, 267, 378, 408, 410–411
lapis lazuli, 169–170
larnax, 93
Lasaia, 302, 304, 310–311, 313
laurel, 221
lauric acid, 420, 422
lavender, 417
Lavrion, 17, 41, 142, 145, 175, 181, 185, 255, 263, 313, 315–317
lead, 145, 168, 175–176, 186, 304
lead bead, 166
lead isotope analysis, 168, 175–176, 310–319
leather, 272
Lebena, 304, 310–311
Lebena Ware, 70
Lefkandi, 171–172
lekane, 394, 397
Lemnos, 319
lemon, 425
Lenda. See Lebena
Lepus, 42, 149–150, 212, 254
Lerna, 171
Lesbos, 166, 171, 173–174, 176
Levant, 167
licorice root, 417–420, 423, 424–427
lime, 137, 139
limekiln, 204
Limogardion, 299
limonite, 303, 310
limpet, 149–152, 213
Linear B, 10, 14, 247, 263
lions, 161

löllingite, 347
loomweights, 65, 209, 213, 245, 271
MACEDONIA, 158, 164
Magnesia, 15
magnetite, 139, 282–287, 289, 292, 296–297, 305, 338, 340, 345, 347–349
malachite, 142, 144, 183–187, 254, 301–303, 311, 330–332, 344
Malia, 264–268, 270, 304
Mandalo, 158, 164
Mandra. See sheepfold
manganese, 145
Mani, 157
Manika, 171–172, 319
map of East Crete, 4
Marathokephalo, 168, 318
marble vases, 157
Mari, 169–170
marine resources, 43, 52, 65, 213, 235, 245, 271
Marshall Plan, 278, 365
mass balance, 345
meadowsweet, 424
mean temperatures in Crete, 19–20
Mehrgarh, 161
mellite, 309
Melos, 100
Mesara, 155
Mesopotamia, 24–25, 161–162, 167, 169
metacarbonate, 24–25
methyl benzoate, 420, 423
methyl decanoate, 422, 426
methyl dehydroabietate, 418, 421, 423
methyl esters, 415, 418, 419–426, 428
methyl ketone, 413
methyl nonanoate, 423
methyl 2-ethylhexaoate, 426
Miamou, 304, 311
mica, 430–432
Micromeria myrtifolia, 425
milk, 419, 422
minimum number of bellows, 128
mint, 417, 420
Mirabello Fabric, 70–71, 73, 78–79, 81, 83–84, 86–87, 91, 93–94, 96–97, 110, 127–128, 130, 132, 134, 136, 181, 377, 380
Mn oxides, 290, 292
Mochlos, 4, 26, 28, 70, 170, 186, 217–218, 229, 231, 258–259, 264, 273, 318–319
modern toponyms, main discussion, 8–14
Molai, 314
molds, 63–64, 113, 172
mollusks, 43
Monastiraki Katalimata, 268
Monodonta, 149–152, 213

INDEX

montmorillonite, 245
mortars and pestles, 186
mottled ware. *See* Vasiliki Ware
Mount Holyoke College, 17
mud daub structures, 62
murex shells, 314
Mushiston, 176
Mykonos, 156
myristic acid, 418, 420, 426
Myrtos Fournou Korifi, 88, 181, 220, 252, 302
Myrtos Pyrgos, 318

NAHAL MISHMAR, 171
Naxos, 157, 168, 319
needles, 157, 164
Neolithic period, 3, 8, 21, 42, 62, 67–70, 72, 79, 87, 109, 142, 149, 155–166, 179, 183, 193, 205, 209, 217, 225, 233–234, 236–238, 244, 248, 250, 257–262, 273, 321, 323, 325, 380, 429–432
Nepeta racemosa. See catmint
Nevali Çori, 161
nickel, 186, 332, 338, 340–343, 346–347, 349–352
Nigerian furnaces, 111, 338
Nisyros, 158
N-methylphthalidimide, 425, 427
nonanal, 418, 419, 421, 424–425, 427
nonanoic acid, 419, 423, 424, 426

OATS, 154
obsidian, 99–101, 105, 107–108, 209, 245
octadecane, 417, 419, 421, 423–426
octanoate, 422
octanoic acid, 419, 421, 424, 425, 426
oleic acid, 418, 420, 423, 426
olives (olive oil), 110, 143, 153–154, 186, 197, 204, 221, 223, 229, 234–235, 241, 246–248, 250, 252, 255, 258, 260, 262, 276, 278, 365, 414, 418–421, 423, 424, 426
olivine, 141, 282, 284–286, 289, 307, 335, 338, 344
Oman, 176
1-decanol, 421
1-dodecanol, 421
1,4-diacetylbenzene, 421
1-nonanol, 421, 423
1-octadecanol, 421
1-pentadecanol, 421
1-tetradecanol, 421
1,2-benzisothiazole, 425
oolitic structure, 291
oregano, 418
Origanum sipyleum, 425
Orkos. *See* Kea

Othrys Mountains, 299
oval structure. *See* **AF 32** *under* anthropogenic features

PACHEIA AMMOS, 4–5, 7–8, 17–18, 179, 181–182, 233, 254, 258
Pakistan, 161
Palaikastro, 4, 125, 155, 267, 318
Palestine, 142, 171
palmitic acid, 418, 420–422, 426
palmitoleic acid, 418
Panagia Skali, 229
panthers, 161
papaya, 417, 426
Papoura, 228
Paracentrotus, 149, 152
Patella, 149–152, 213
Patras, 157
pawpaw, 426
peach, 421, 426
pelargonic acid, 419–422
Peloponnese, 23
pentadecane, 419, 422
pentadecanoic acid, 419, 421–422
pepper, 420
peppermint, 426
Perganum harmala, 419
Peru, 163
petitgrain, 425
Petras, 4
petrography of slags, main discussion, 281–292
Petrokorio, 70
Petromagoula, 166
Peumus boldus, 424, 426
phacoids, 26
Phaistos, 75, 155, 259, 266–267, 304
pharmacosiderite, 185
phosphorus, 246
Photolivos, 158
Phtiotis, 299
pigs, 212, 234, 271–272
pile-based construction, 62
Pimpinella anisum. See anise
pineapple, 426
pine resin, 418, 420–423, 425, 426
pine tar, 422, 426
pine trees, 391
pins, 156–158, 172–174, 318
Pinus halepensis. See Aleppo pine
Pisania, 149, 152
piscinae. See fish tanks
pithos, 378
PIXE, 321–324
Platanos, 165, 168, 318
Platomagoulia, 157
Platys River, 12, 37, 229–230, 255
Poliochni, 170–174, 176, 319
Poros Katsambas, 166, 347

portable hearth, 65
Porti, 168, 318
pot bellows. *See* bellows
pouring vessels. *See* jugs
Precambrian ore deposits, 177
precipitation in Crete, 19–20
Priniatikos Pyrgos, 4, 73, 265
Pseira, 4, 8, 18, 21–22, 26, 28–29, 32, 40, 43, 70, 79, 127, 154, 181, 215–217, 231, 245, 251–252, 258, 264–265, 268, 270, 273, 391
Pseiran terrace G, 2, 245
pulses, 246–247, 252, 271
punch, 174, 318–319
purpose of the apsidal structure, 63–66
Pyrgos cave, 165, 170, 218, 318
Pyrgos Chrysokaminou, 9, 14–15, 197–198, 237–238, 270
Pyrgos Myrtos, 267
Pyrgos Ware, 70, 170, 217–219
pyrite, 141, 301, 332–333
pyroxene, 140–141, 188, 282–283, 286–289, 305, 335, 338, 344
pyxis, 67, 70–73, 76, 208–209

QUARTIER MU, 268
quartz, 183, 284, 286, 290–292, 331–332, 344, 351, 430–432
querns. *See* saddle-shaped querns
quicklime, 42, 112

RACHIS, 153
Rafina, 171–172
raspberry, 425
Redox Conditions, 348
Rekh-mi-re, 125–126
resinated wine, 419, 421, 423, 426, 428
Rethymno, 304
Rhodes, 316
ring-idols, 157, 159
ritual vessel, 73
Romania, 159
rosemary, 417, 420
rounded cup, 77, 87–89, 97, 413, 425
Royal Cemetery of Ur, 169
rue, 65, 418, 425, 427
Ruta graveolens. See rue
rutile, 284

SADDLE-SHAPED QUERNS, 212–213, 241, 271
safflower, 422
saffron, 65, 420, 421, 426, 427
sage, 221, 417, 420
Salamis, 157, 170
Sargon I, 162
Satureja cuneifolia. See Turkish savory
saws, 165, 318
sea level, 21, 199

sealstone, 212–213
sea shells. *See* marine resources
sea urchins, 43, 149, 152, 254, 419
Selakano, 302
Selinou. *See* Sklavopoula
Semitic loan word, 10, 14
Seriphos, 20, 142–143, 179–181, 299, 314, 316–317
Sesklo, 164, 166
Sfaki, 299
Shaba. *See* Congolese furnaces
Shahr-I-Sokhta, 335
shale, 291–292
shallow bowls, 61, 72, 78–79, 84, 87, 91, 97, 413, 415, 419, 425
sheep and goats, 212, 234, 248–249, 253–254, 271–272, 276, 420, 423
sheepfold, 200–201, 211, 356, 370
Shiqmim, 142, 177
shrew, 212
silver, 157, 159, 166–167, 169, 172, 175, 304, 310, 314, 316
silver zoomorphic pendant, 166
Siphnos, 179, 314–317
Sitagroi, 156–159, 164, 171, 404
Siteia, 24
Skhabar, 418
Sklavopoula (Sklavopoulou), 302–304, 310–311
Skouries. *See* Kythnos
soap, 221
socketed spearhead, 166
soils and sediments, main discussion, 403–412
spatula, 319
spearhead, 318
Sphoungaras, 70, 215–218, 258
spindle whorls, 65
Sri Lanka furnaces, 177
stannite, 176
stearic acid, 418, 420, 421, 422
stirrup jars, 212
stone bowl, 209
stone tools, main discussion, 99–108
stone vases, 65
storage vessels, 67–68, 72, 183, 212–213
strawberry, 425
Strophilas, 157
sunflower, 417
swine. *See* pigs
swords, 167
Syros, 162, 165, 167, 169, 171, 319

Tajikstan, 176
tankard, 171
Tarsus, 171
Taurus Mountains, 23, 161, 167, 169–170
teapots, 65, 73, 89, 175
Tell Brak, 170
tenorite, 289
terraces, 3–4, 49, 197, 200, 202–204, 224, 246, 251, 274–278, 355, 357–358, 360–361, 363–365, 369, 372–374, 399–402, 406–408, 411–412
tetradecane, 417, 419, 421, 424
Tharrounia cave, 157, 164, 166
Thebes, Egypt, 125
Theopetra cave, 157
Therio, 9, 12, 198, 200, 203–204, 215–220
Theriospelio cave, 3, 7–9, 11–12, 15–17, 34, 193, 197–198, 203, 215–220, 222, 225–227, 233, 243, 249, 251, 253, 257, 260–262, 276, 330, 356, 373
Thermi, 166, 171, 173–174, 176
thermoluminescence dating, 17
Thessaly, 62, 164, 166
Tholos beach, 7, 11–12, 22, 27, 33, 181, 229, 231, 233, 254–255
Thrace, 62
3-bromothieno[3,2-c]pyridine, 425
threshing floors, 4, 197, 252, 274–276, 278, 355–356, 359–360, 365–366
Thriphti, 4, 70
Thriphti Argira, 228
thyme, 41, 391, 417
Timna, 142, 335
tin, 143, 163, 167, 169–176, 316
top shell, 149, 213
Trapeza cave, 165
travertine formation, 9, 11, 34–37, 40, 203
triacetin, 416–417, 427
trickle decoration, 88
tridecanoic acid, 418
tridymite, 309
tripod cooking pots. *See* cooking pots
triton shells, 213
Troy, 160, 169–176
tulip shell, 149–152
tumbler, 208, 385
Turkish savory, 417, 425, 427
tuyeres, 127, 133–134, 136, 172, 183–184, 187
tweezers
2-decanone, 425–427
2-ethylhexanoic acid, 420, 422, 424
2-ketone, 418
2-Phenoxyethanol, 419, 423
2-undecanone, 418, 419, 425–427
Tylissos, 270

Unaltered minerals, 140–141
undecanal, 418, 425
undecane, 418–419
undecanoic acid, 418, 420, 422
Ur, 169–170
Urfa, 161
Usatovo, 171

Varna, 157
Vasiliki, 4, 70, 181, 217, 258, 265, 268, 270
Vasiliki Ware, 15–16, 71, 76–77, 127, 208–209, 217–218, 220, 264–265, 300, 304
vegetable oils, 422, 427
Venetian period, 14
verbena, 420, 422, 425–426
verbenone, 420, 421, 425, 426
vervain, 420
villa system, 269
Vouliagmeni, 157
Vrokastro, 70, 258
Vronda, 226
vultures, 161–162

Wavelength Dispersive Spectrometry, 325–328
wax esters, 422
weasels, 212
weaving, 234, 272
wedges for drilling, 269
wells, 4, 197, 256, 274–276, 278, 355–358, 363, 374
wheat, 153, 234, 241, 247, 250, 260, 262
White-on-Dark Ware, 65, 71–72, 77–78, 88–89, 126, 181, 264–266
wild boars, 161
wine, 414, 418–423, 425–427
wollastonite, 305, 307, 309
wustite, 141, 309, 338, 345

Xerambela, 227, 229
Xeropolis. *See* Lefkandi

Yannitsa, 157
Younger Dryas, 162

Zaire. *See* Congolese furnaces
Zakros. *See* Kato Zakros
Zas cave, 157, 164, 168
Zoghaki. *See* Kythnos
zoomorphic pendant. *See* silver zoomorphic pendant